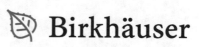

Ludmila Bourchtein • Andrei Bourchtein

Theory of Infinite Sequences and Series

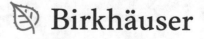 Birkhäuser

Ludmila Bourchtein
Institute of Physics and Mathematics
Federal University of Pelotas
Pelotas, Brazil

Andrei Bourchtein
Institute of Physics and Mathematics
Federal University of Pelotas
Pelotas, Brazil

ISBN 978-3-030-79430-9   ISBN 978-3-030-79431-6   (eBook)
https://doi.org/10.1007/978-3-030-79431-6

Mathematics Subject Classification: 40-01, 40A05, 26-01

This book is published under the imprint Birkhäuser, www.birkhauser-science.com, by the registered company Springer Nature Switzerland AG.
The registered company address is: Gewerbestrasse 11, 6330 Cham, Switzerland

To Victoria for stimulating mathematical discussions
To Valentina for excellence in mathematical tests
To Maxim and Natalia for very deep learning and dedication
To Haim and Maria for everything

Preface

This book covers a major part of the theory of real infinite sequences and series at the level corresponding to advanced undergraduate/initial graduate courses of mathematical analysis.

This theory is a natural part of real analysis, but its place in the major subject is ambiguous. On the one hand, the entire course of the real analysis can be based on the infinite sequences and series, but on the other hand, there are expositions of the real analysis topics with no explicit reference to the sequences and series. Some textbooks consider sequences and series as the first topic to be studied, while others leave it for more advanced parts of the text. The level of exposition also varies significantly. The calculus textbooks usually contain the basic exposition of the theory of sequences and series, but they are more concerned with the well-known tests for series of numbers and representations of functions in power series, which provide useful tools for the solution of different problems. However, the logical development of the material and rigorous proofs of the presented results are left aside due to obvious limits of time and the reader's expertise. Many real analysis texts contain a logical and deep exposition of principal parts of the theory of sequences and series, but they naturally adjust the representation of these topics for the main purpose of the book, which usually leads to the separation of the continuous theory into the parts related to sequences of numbers, series of numbers, then to sequences and series of functions, and finally to the power series. Thus, due to different reasons, calculus and analysis books usually fail to provide the presentation of the theory of sequences and series with the logical continuity and integrity inherent in this theory. At the same time, there are a few books dedicated especially to the sequences and series, and each one of them covers a part of the subject due to a wide range of its topics.

Our intention is to provide a text that covers the majority of traditional topics of infinite sequences and series and represents the theory of this subject in its integrity, logical sequence, and sufficient depth. It starts from the very beginning—the definition and elementary properties of sequences of numbers, and ends with advanced results of uniform convergence and power series. The entire text is developed at two levels: the basic level covers the undergraduate topics, while the additional material (marked as Complement) addresses more advanced subjects. The reader can choose what level of the exposition to follow, since the explanation of more complex problems does not interfere with the development and comprehension of the topics of the basic level.

This book is aimed at university students specializing in mathematics and natural sciences, and at readers interested in infinite sequences and series. It is designed for the reader who has a good working knowledge of calculus. No additional prior knowledge is required. All the initial concepts and results related to sequences and series and required for the development of the theory are covered in the initial part of the text and, when necessary, at the starting point of each topic. This makes the book self-sufficient and the reader independent of any other text on this subject. As a consequence, this book can be used both as a textbook for advanced undergraduate/early graduate courses and as a source for self-study of selected (or all) topics presented in the text.

The text is divided into five chapters which can be grouped into two parts: the first two chapters are concerned with the sequences and series of numbers, while the remaining three chapters are devoted to the sequences and series of functions, including the power series.

Each chapter is comprised of a number of sections and subsections, which are numbered separately within each chapter, bearing the sequential number of section inside a chapter and that of subsection inside a section. For instance, Sect. 3 can be found in each of the five chapters, while Sect. 3.2 means the second subsection inside the third section. When referring to a section or subsection inside the current chapter, we do not use the chapter number, otherwise the chapter number is provided. The formulas and figures are numbered sequentially within each chapter regardless of the section where they are found. In this way, formula (2.5) is the reference to the fifth formula in the second chapter, while Fig. 3.4 indicates the fourth figure in the third chapter. The theorems, propositions, lemmas, definitions, and examples are numbered independently within each section.

Throughout the text, the most important, fundamental, and useful definitions and results are highlighted, but this reflects only the personal opinion of the authors and can be disputed. Brief historical comments on some significant results and interesting facts on the development of the theory of sequences and series are provided at the end of individual sections.

Within each major topic, the exposition is, as a rule, inductive and starts with rather simple definitions and/or examples, becoming more compressed and sophisticated as the course progresses. Each important notion and result is illustrated with examples explained in detail. Some more complicated topics and results are marked as complements and can be omitted on a first reading without loss of subject continuity. The mathematical level of the exposition corresponds to advanced undergraduate courses of mathematical analysis and/or early graduate introduction to the discipline.

The first chapter introduces the initial concepts of sequences of numbers and their convergence. It studies different properties of convergent sequences, first similar to the limit properties of general functions and then specific properties of sequences. At the end of this chapter, we explain the methods of solution of indeterminate forms, frequently used in subsequent chapters. Based on the developed material of the sequences, the second chapter deals with the series of numbers. We present a standard set of the convergence tests that can be found in calculus and analysis textbooks (integral, comparison, ratio, and root tests for positive series, and Leibniz's test for alternating series) in primitive and refined forms. We also investigate the two principal chains of more elaborated tests for positive series—the Kummer and Cauchy chains, and more strong tests for arbitrary series—Dirichlet's and Abel's tests. At the end of this chapter, we analyze the associative and commutative properties of series, including the famous Dirichlet and Riemann theorems.

The part of the text devoted to the study of sequences and series of functions starts with the third chapter. The principal theme of this chapter is the uniform convergence of sequences, the methods of its investigation, and the conditions which guarantee "nice" properties of limit functions, such as boundedness, continuity, integration, and differentiation by parameter. This sets the stage for developing the material on series of functions covered in the fourth chapter. The uniform convergence continues to be the focus of the study, and the exposition revolves on the one hand around the conditions (criteria and tests) that ensure the uniform convergence, and on the other hand around hypotheses based on the uniform convergence that provide the desired properties of the sums of series.

The fifth chapter deals with the theory of power series, which, as a particular case of the series of functions, inherits many properties of the latter, but also has its specific characteristics. Different techniques of finding power series expansions are considered and employed to derive the representation of many elementary functions. At the end of the chapter, the standard applications of power series are considered, including the problems of approximation, calculation of limits, integration, and solving ordinary differential equations.

The text contains a large number of problems and exercises, which should make it suitable for both classroom use and self-study. Many standard exercises are included in each section to develop basic techniques and test understanding of concepts. Other problems are more theoretically oriented and illustrate more intricate points of the theory or provide counterexamples to false propositions which seem to be natural at first glance. Some harder exercises of theoretical interest are also included as examples or applications of theoretical results, but they may be omitted in courses addressed to less advanced students. Many additional problems are proposed as homework tasks at the end of each chapter. Their level ranges from straightforward, but not overly simple, exercises to problems of considerable difficulty, but of comparable interest. These more involved and challenging problems are marked with an asterisk.

The presented text has the following features:

1. Completeness: the text covers a major part of the traditional topics of real sequences and series at the advanced undergraduate/initial graduate level.
2. Self-sufficiency: all the background topics related to sequences and series are covered in the text, and, consequently, this work can be used as both a textbook and a source for self-study.

3. Generality: we have endeavored to present all the results in a more general form while avoiding major complications of their proofs.
4. Accessibility: all the topics are covered in a rigorous mathematical manner while keeping the exposition at a level acceptable for advanced undergraduate courses.
5. Two-level approach: the text is systematically developed at two levels—the basic and more advanced, with the possibility to choose what level of exposition to follow.
6. Exercises: there are a large number of problems and exercises, solved and proposed, which should make the book suitable for both classroom use and self-study.

Pelotas, Brazil   Ludmila Bourchtein
                                                                                   Andrei Bourchtein

Contents

1 Sequences of Numbers 1
  1 Convergence and Introductory Examples 1
    1.1 Definition of a Sequence and Trivial (Pre-limit) Properties 1
    1.2 Convergence of a Sequence 3
  2 Common Properties of Convergent Sequences 8
    2.1 Uniqueness of the Limit 8
    2.2 Comparison Properties 8
    2.3 Arithmetic and Analytic Properties 11
  3 Special Properties of Convergent Sequences 13
    3.1 Convergence of Function and Corresponding Sequence 13
    3.2 Relationship Between Convergence and Boundedness 13
    3.3 Subsequences and Their Convergence. Bolzano-Weierstrass
        Theorem 14
    3.4 Cauchy Criterion for Convergence 18
    3.5 Sequences of the Arithmetic and Geometric Means 19
  4 Indeterminate Forms and Techniques of Their Solution 22
    4.1 Definition of Indeterminate Forms 22
    4.2 Techniques of Solution of Indeterminate Forms 24
    4.3 Various Indeterminate Forms and Examples 36
  Exercises 38

2 Series of Numbers 43
  1 Convergence and Introductory Examples 43
    1.1 Definition of a Series. Partial Sums and Convergence 43
    1.2 Elementary Examples of Series of Numbers 46
  2 Elementary Properties of Convergent Series 52
    2.1 Arithmetic Properties 53
    2.2 Cauchy Criterion for Convergence 54
    2.3 Necessary Condition of Convergence (Divergence Test) 55
    2.4 Series and Its Remainder 55

3   Convergence of Positive Series 57
     3.1   General Criterion for Convergence 58
     3.2   Integral Test (Cauchy-Maclauren Test) 58
     3.3   The Comparison Tests 63
     3.4   The Cauchy Condensation Test 69
     3.5   D'Alembert's Tests (The Ratio Tests) 72
     3.6   Cauchy's Tests (The Root Tests) 76
     3.7   Comparison Between D'Alembert's and Cauchy's Tests 79
     3.8   Complement: Finer Forms of D'Alembert's and Cauchy's Tests 82
     3.9   Complement: The Kummer Chain of Tests 85
     3.10   Complement: The Cauchy Chain of Tests 97
4   Series of Different Types 104
     4.1   Alternating Series 104
     4.2   Dirichlet's and Abel's Tests 108
     4.3   Absolute and Conditional Convergence 114
     4.4   Product of Two Series 115
5   Associative and Commutative Properties of Series 118
     5.1   Positive and Negative Parts of Series 118
     5.2   Associative Property of Convergent Series 120
     5.3   Commutative Property of Absolutely Convergent Series 122
     5.4   Commutative Property of Conditionally Convergent Series 124
6   Complement: Double and Repeated Series 131
Exercises 135

3   Sequences of Functions 141
1   Pointwise Convergence and Introductory Examples 141
2   Uniform and Non-uniform Convergence 149
     2.1   Concept of the Uniform and Non-uniform Convergence 149
     2.2   Arithmetic Properties of Uniform Convergence 155
     2.3   Cauchy Criterion for Uniform Convergence 158
3   Dini's Theorem 160
4   Properties of Limit Functions Under Uniform Convergence 161
     4.1   Boundedness of Limit Function 161
     4.2   Limit of the Limit Function 163
     4.3   Continuity of the Limit Function 166
     4.4   Integrability of the Limit Function (Integration by Parameter) 169
     4.5   Differentiability of the Limit Function (Differentiation
            by Parameter) 176
5   Complement: The Weierstrass Approximation Theorem 182
Exercises 185

4 Series of Functions 191
  1 Pointwise Convergence and Introductory Examples 191
  2 Uniform and Non-uniform Convergence....................................... 197
    2.1 Concept of Uniform and Non-uniform Convergence 197
    2.2 Arithmetic Properties of Uniform Convergence 202
    2.3 The Cauchy Criterion for Uniform Convergence 202
    2.4 Uniform and Absolute Convergence.................................... 203
  3 Sufficient Conditions for Uniform Convergence of Series 205
    3.1 Comparison Tests.. 205
    3.2 Dirihlet's and Abel's Tests 207
    3.3 Dini's Theorem 214
  4 Properties of the Sum of Uniformly Convergent Series 215
    4.1 Boundedness of a Sum 215
    4.2 Limit of a Sum... 218
    4.3 Continuity of a Sum 220
    4.4 Integrability of a Sum (Integration Term by Term) 221
    4.5 Differentiability of a Sum (Differentiation Term by Term)............. 223
  5 Complement: The Weierstrass Function—Everywhere Continuous and
    Nowhere Differentiable Function 227
  Exercises... 231

5 Power Series 239
  1 Introduction.. 239
  2 Set of Convergence of a Power Series 242
    2.1 Convergence of a Power Series 242
    2.2 Determining the Radius of Convergence 245
    2.3 Convergence of the Series of Derivatives 248
    2.4 Behavior at the Endpoints of the Interval of Convergence 250
  3 Properties of Power Series and Their Sums................................... 251
    3.1 Arithmetic Properties.. 251
    3.2 Functional Properties.. 254
    3.3 Analytic Properties 260
    3.4 Uniqueness of Power Series Expansion, Analytic Functions.......... 266
  4 Taylor Series... 269
    4.1 Taylor Coefficients and Taylor Series 269
    4.2 Relation Between the Taylor Series and Formula...................... 271
    4.3 Conditions of Expansion in the Taylor Series 274
  5 Power Series Expansion of Elementary Functions 280
    5.1 Using Analytic Properties of Power Series............................. 281
    5.2 Finding the Sum of Power Series via Differential Relations 284
    5.3 Method of the Taylor Coefficients 287

5.4   Taylor Series for Various Functions 296
5.5   The List of the Derived Formulas of Taylor Series 304
6   Applications of Taylor Series... 307
6.1   Approximation of Functions 308
6.2   Numerical Approximations 321
6.3   Finding Sums of Series of Functions 326
6.4   Sums of Series of Numbers 328
6.5   Calculation of Limits 329
6.6   Calculation of Integrals 336
6.7   Solution of Ordinary Differential Equations 339
6.8   Complement: The Number e Is Irrational 349
6.9   Complement: The Number π Is Irrational 350
7   Complement: Borel's Theorem.. 352
7.1   Smooth Non-analytic Function 353
7.2   Transition Function.. 355
7.3   Borel's Theorem 358
Exercises... 361

Bibliography 369

Index.. 373

Sequences of Numbers

Sequences are fundamental to many areas of mathematics. If you can understand them and how they are distributed, it leads to the solution of many other questions.
Manjul Bhargava, 2014

1 Convergence and Introductory Examples

1.1 Definition of a Sequence and Trivial (Pre-limit) Properties

> ▶ **Definition** (**Sequence of Numbers**) A *sequence of numbers* is a relation between the set of natural numbers $\mathbb{N}$ and the set of real numbers $\mathbb{R}$ such that with each element (index) of the domain $n \in \mathbb{N}$ there is associated a unique number $a_n \in \mathbb{R}$. Hence, a sequence of numbers is a function whose domain is $\mathbb{N}$ and the image is a subset of $\mathbb{R}$.

This definition is usually extended to the case of a more general domain $\mathbb{N}_0$, which represents a subset of the integers $\mathbb{Z}$ such that $\mathbb{N}_0 = \{k_1, k_2 = k_1 + 1, k_3 = k_2 + 1, \ldots, k_{i+1} = k_i + 1, \ldots\}$, where $k_1 \in \mathbb{Z}$. The point of this extension is to make possible the counting of the sequence elements starting from an arbitrary initial index (k_1 instead of 1) still keeping the structure and order of natural numbers: each next index (the element

Electronic Supplementary Material The online version of this article (https://doi.org/10.1007/978-3-030-79431-6_1) contains supplementary material, which is available to authorized users.

of the domain) is obtained adding 1 to the current index. (Evidently, in this extension, the number of negative indices is finite or empty.) If the domain $\mathbb{N}_0$ of a sequence is not defined explicitly, we consider (by convention) that the domain is the set of natural numbers $\mathbb{N}$ or the largest of its subsets in which all the values of a sequence are determined. When there is no necessity, we will simply refer to the indices as elements of the domain without specification of $\mathbb{N}_0$.

The usual notations for a sequence are $a_1, a_2, \ldots, a_n, \ldots$ or $\{a_n\}_{n=1}^{\infty}$, or $\{a_n\}$, or simply a_n. In the most cases, when it will be clear from a context, we will use the last notation both for a sequence and its elements. Also we will usually call a sequence of numbers by a sequence if it will not cause any ambiguity.

▶ **Definition** (**Bounded Sequence**) A sequence a_n is *bounded above* if there exists a number M such that $a_n \leq M$ for all indices n. The number M is called an *upper bound* of a_n. In the same way, a sequence a_n is *bounded below* if there exists a number m such that $a_n \geq m$ for all indices n. The number m is called a *lower bound* of a_n. A sequence is *bounded* if it is bounded above and below. Otherwise, a sequence is *unbounded*.

Evidently, a bounded sequence can be also defined as a sequence, which admits a number C such that $|a_n| \leq C$ for all indices n. (If one has the upper and lower bounds M and m, then C can be determined as $C = \max\{|m|, |M|\}$; if C is known, one can choose $m = -C$ and $M = C$.)

▶ **Definition** (**Monotone Sequence**) A sequence is *increasing (strictly increasing)* if for any index n the inequality $a_n \leq a_{n+1}$ $(a_n < a_{n+1})$ is true. In a similar manner, a sequence is *decreasing (strictly decreasing)* if $a_n \geq a_{n+1}$ $(a_n > a_{n+1})$ for any index n. A (strictly) increasing or decreasing sequence is called *(strictly) monotone*.

▶ **Definition** (**Positive/Negative Sequence**) A sequence is *positive (non-negative)* if for any index n the inequality $a_n > 0$ $(a_n \geq 0)$ is true. In a similar manner, a sequence is *negative (non-positive)* if $a_n < 0$ $(a_n \leq 0)$ for any index n.

▶ **Definition** (**Alternating Sequence**) A sequence is *alternating* if it can be expressed in the form $a_n = (-1)^n b_n$ or $a_n = (-1)^{n+1} b_n$, where $b_n > 0$ for all n.

Like any function, a sequence can be represented geometrically, but the graph is usually rather unrevealing, since most of a sequence cannot be fit on the page. A more convenient representation of a sequence is frequently obtained by simply labeling the points a_n on a line, like it is made for the sequence $a_n = \frac{1}{n}$ in Fig. 1.1. In some cases, this sort of picture can show where the sequence "is going", which is the main question to be studied in this chapter.

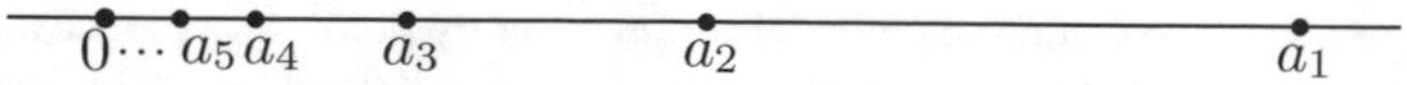

Fig. 1.1 The sequence $a_n = \frac{1}{n}$ "going to 0"

1.2   Convergence of a Sequence

The purpose now is to introduce the concept of the limit of a sequence. Recall that the indispensable condition for the definition of the limit of any function at a point x_0 is that this point should be a limit point (accumulation point) of the domain of a given function, that is, in any neighborhood of x_0 should exist points of the domain different from x_0. (Recall that a neighborhood of a point x_0 is an open interval centered at x_0, that is, $(x_0 - \delta, x_0 + \delta)$, where $\delta > 0$ is called a radius of the neighborhood.) For instance, $x = 1$ and $x = 0$ are the limit points of the domain of the function $f(x) = \frac{1}{\sqrt{x}}$, but the point $x = -1$ is not. This condition allows us to approach x_0 using points of the domain of a function. Otherwise (if x_0 is not a limit point), we cannot find the points of the domain as close as desired to x_0, and then there is no sense in talking about a tendency of the values of a function.

Taking into account this requirement to the domain, let us see what are the limit points of $\mathbb{N}$ (or $\mathbb{N}_0$). First, any integer does not have another integer in a neighborhood of the radius $\delta = \frac{1}{2}$. Therefore, none of these points (including the points of the domain) is a limit point of $\mathbb{N}$ (or $\mathbb{N}_0$). If we take now any non-integer, then we can measure the distance d between the chosen number and the closest integer and taking the radius of neighborhood $\delta = \frac{d}{2}$ we guarantee that in this neighborhood there is no element of the domain. Thus, no real point is admissible as a limit point of the domain of a sequence. Therefore, the last remaining option is to consider infinity. Evidently, there is no point of $\mathbb{N}$ (or $\mathbb{N}_0$) near $-\infty$: in a formal mode, taking $D = k_1 - 1$, where k_1 is the first (i.e., the smallest) element of $\mathbb{N}_0$, we see that there is no point k of the domain $\mathbb{N}_0$, which belongs to D-neighborhood of $-\infty$, that is, satisfies the condition $k < D$. However, we can approach $+\infty$ using the points of the domain: whatever large a number $D > 0$ happens to be, we can always find elements of $\mathbb{N}_0$ which are even larger—$n > D$, that is, in formal terms, in any neighborhood of $+\infty$ there exist elements of $\mathbb{N}_0$. This means that $+\infty$ is the only limit point (in the sense we have specified above) of the domain of a sequence, and, therefore, the unique type of limit that can be considered for sequences is the limit at infinity, more specifically, at $+\infty$. (Since $-\infty$ cannot be approached by the points of domain, in what follows we will simplify notation from $+\infty$ to ∞.)

Let us recall the definition of the limit at infinity for general functions and then "translate" it into more natural language for sequences.

▶ **Definition** (**Limit of a Function at Infinity**) A function $f(x)$ has a limit equal to a, as x approaches $+\infty$, if for $\forall \varepsilon > 0$ there exists $D = D(\varepsilon) > 0$ such that for all $x > D$ it follows that $|f(x) - a| < \varepsilon$.

For a sequence a_n we can repeat this definition point by point: a sequence a_n has a limit equal to a if for $\forall \varepsilon > 0$ there exists $D = D(\varepsilon) > 0$ such that $|a_n - a| < \varepsilon$ whenever $n > D$. (Here, the mention that n approaches $+\infty$ is unnecessary and usually is omitted, because this is the only possible tendency of the variable of a domain of a sequence). However, since the values of the variable of a domain of a sequence are natural numbers (or mostly natural numbers), it is a tradition to use the natural parameter $N \in \mathbb{N}$ instead of parameter D. With this cosmetic (but common) correction, we arrive at the following definition of the limit of a sequence and its convergence.

> ▶ **Definition** (**Limit of a Sequence and Its Convergence**) A sequence a_n has a
> *(finite) limit* equal to a, if for $\forall \varepsilon > 0$ there exists $N = N(\varepsilon) > 0$ such that for all
> indices $n > N$ it follows that $|a_n - a| < \varepsilon$. A general notation is $\lim\limits_{n\to\infty} a_n = a$ or
> $a_n \underset{n\to\infty}{\to} a$.
>
> A sequence a_n is called *convergent* if it has a (finite) limit. If the limit of a
> sequence is equal to a then we say that the *sequence converges to a*. Otherwise (if
> there is no limit or the limit is infinite) a sequence is called *divergent*. In the most
> cases, we will call the finite limit simply a limit, when the context makes clear what
> is meant.

Remark Unlike the limit of a general function, the convergence of a sequence means that every ε-neighborhood of the limit a contains all but a finite number of the terms a_n.

The illustration of the concept of the limit is given in Fig. 1.2, where all the elements a_n starting from $n = 4$ are located in ε-neighborhood of the limit a.

An elementary example of a convergent sequence is $a_n = \frac{1}{n}$ (see Fig. 1.1). To prove the convergence by definition, it is necessary first to make a (intuitive) guess about possible value of the limit. Observing a general behavior of this sequence or calculating some values ($a_1 = 1, a_2 = \frac{1}{2}, a_3 = \frac{1}{3}$, etc.) we can suppose that its limit is equal to 0. To corroborate this, we employ the provided definition. At this stage, the main problem is to show the estimate $|a_n - a| < \varepsilon$ of the definition for sufficiently large values of n. To do this for the sequence $a_n = \frac{1}{n}$, notice that $a_n > 0$, that the guessed value of the limit is $a = 0$ and write the main inequality in the form $|a_n - a| = a_n = \frac{1}{n} < \varepsilon$. The last inequality is equivalent to $n > \frac{1}{\varepsilon}$ (when solved with respect to n). In other words, the main inequality is valid for

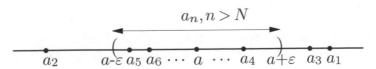

Fig. 1.2 The limit of a sequence a_n

any $n > \frac{1}{\varepsilon}$. This means that choosing $N = \left[\frac{1}{\varepsilon}\right]$ for any given $\varepsilon > 0$, we have the main estimate of the definition satisfied for all $n > N$, and consequently the definition holds. (Recall that the symbol $[x]$ means the integer part of the number x.)

For later use, let us formulate in exact $\varepsilon - N$ terms the definition of the situation when a sequence does not have a finite limit (that is, diverges).

▶ **Definition** (**Divergent Sequence**) A sequence a_n *diverges* (or, equivalently, it does not have a finite limit) if whatever a is chosen, there exists $\varepsilon_a > 0$ such that for any N always can be found an index $n > N$ such that $|a_n - a| \geq \varepsilon_a$.

An elementary example of this kind is a sequence $a_n = (-1)^n$. To prove that this sequence does not have a limit, we notice that the choice of $\varepsilon_0 = 1$ guarantees that all the elements with even indices $a_{2k} = 1$, $k \in \mathbb{N}$ or those with odd indices $a_{2k-1} = -1$, $k \in \mathbb{N}$ stay out of ε_0-neighborhood of an arbitrary real number. Indeed, if the distance from $a_{2k} = 1$ to the alleged limit a (whatever a is) is larger than or equal to ε_0, then $|a_{2k} - a| \geq \varepsilon_0 = 1$ for infinite number of indices. Otherwise (when $|a_{2k} - a| < 1$), since the distance between $a_{2k} = 1$ and $a_{2k-1} = -1$ is equal to 2, all the elements with odd indices a_{2k-1} stay out of ε_0-neighborhood of a, or analytically: $|a_{2k-1} - a| \geq |a_{2k-1} - a_{2k}| - |a_{2k} - a| > 2 - 1 = 1$.

Let us also define the *infinite limit* following again a general formulation for usual functions, but making some natural adjustments for sequences.

▶ **Definition** (**An Infinite Limit of a Sequence**) A sequence a_n has the *limit* $+\infty$ if for $\forall E > 0$ there exists $N = N(E) > 0$ such that for all indices $n > N$ it follows that $a_n > E$. In a similar mode, a sequence a_n has the *limit* $-\infty$ if for $\forall E < 0$ there exists $N = N(E) > 0$ such that $a_n < E$ whenever $n > N$. The notation in these cases is the same as for general functions: $\lim_{n \to \infty} a_n = +\infty$ or $a_n \underset{n \to \infty}{\to} +\infty$ in the first case, and $\lim_{n \to \infty} a_n = -\infty$ or $a_n \underset{n \to \infty}{\to} -\infty$ in the second.

Sometimes the concept of a *general infinite limit* is used, which happens when the absolute value of a function approaches $+\infty$: a sequence a_n has general infinite limit if for $\forall E > 0$ there exists $N = N(E) > 0$ such that $|a_n| > E$ for all indices $n > N$.

A trivial example of an infinite limit is the limit of the sequence $a_n = n$. In this case, the main inequality of the definition takes the form $a_n = n > E$ and, consequently, for all $n > N = [E]$ this inequality is satisfied. An elementary example of a general infinite limit is provided by the sequence $a_n = (-1)^n n$.

Recall that the existence of an infinite limit means that the sequence diverges.

Let us look at some examples, which illustrate the introduced concepts.

Example 1a $a_n = 1$
For any natural n the value of the sequence elements is 1 (this is a constant sequence). The given sequence is bounded ($m = M = 1$) and (non-strictly) monotone. Its limit exists and is equal to 1—one can use any N in the definition of a limit.

Example 2a $a_n = (-1)^n$
This is an alternating sequence (its sign changes from one term to the next one). The sequence is bounded ($m = -1$, $M = 1$) and non-monotone. The sequence has no limit, because choosing $\varepsilon_0 = 1$ we leave outside of the ε_0-neighborhood of any candidate for supposed limit an infinite number of elements of the sequence (additional details of the proof of divergence are provided in the illustration to the Definition of divergent sequence).

Example 3a $a_n = \frac{1}{n}$
This sequence is bounded ($m = 0$, $M = 1$) and strictly decreasing. Its limit is 0, that was already shown using the definition.

Example 4a $a_n = \frac{(-1)^n}{n}$
This sequence is bounded ($m = -1$, $M = 1$) and non-monotone (it is alternating). Its convergence to 0 is proved in the same way as in Example 3a, because the main inequality in the definition with $a = 0$ leads to the same condition: $|a_n - a| = |a_n| = \frac{1}{n} < \varepsilon$. Therefore, the rule for the choice of N is the same: for $N = \left[\frac{1}{\varepsilon}\right]$ the definition of the limit holds.

Example 5a $a_n = \frac{1}{n^2}$
This sequence is bounded ($m = 0$, $M = 1$) and strictly decreasing ($a_n = \frac{1}{n^2} > a_{n+1} = \frac{1}{(n+1)^2}$). It has the limit equal to 0 according to the definition. To show this, notice that $a_n > 0$, that the value of the supposed limit is $a = 0$ and simplify the left-hand side of the main inequality in the definition to the form $|a_n - a| = a_n = \frac{1}{n^2}$. Then, the main inequality $\frac{1}{n^2} < \varepsilon$ is valid for any $n > \frac{1}{\sqrt{\varepsilon}}$, that is, choosing $N = \left[\frac{1}{\sqrt{\varepsilon}}\right]$, we have the definition satisfied.

Example 6a $a_n = \sin \frac{\pi n}{6}$
The given sequence is bounded ($m = -1$, $M = 1$) and non-monotone: for $n = 1 + 12k, k \in \mathbb{N}$ one has $a_n = \sin(\frac{\pi}{6} + 2k\pi) = \frac{1}{2}$, while for $n = 7 + 12k, k \in \mathbb{N}$ one gets $a_n = \sin(\frac{7\pi}{6} + 2k\pi) = -\frac{1}{2}$. The sequence has no limit, since its values are infinitely oscillating between -1 and 1. In fact, let us assume, for contradiction, that a is a limit of the sequence. Choose $\varepsilon_0 = \frac{1}{2}$ and calculate the distance from a to 1. If this distance is greater than or equal to $\frac{1}{2}$, then at all the points with indices $n = 3 + 12k, k \in \mathbb{N}$ we have $|a_n - a| = |\sin(\frac{\pi}{2} + 2k\pi) - a| = |1 - a| \geq \varepsilon_0 = \frac{1}{2}$, that is, whatever N is chosen, there always exist indices in the form $n = 3 + 12k$ (with k sufficiently large) such that $n > N$, but $|a_n - a| \geq \varepsilon_0$. On the other hand, if the distance from a to 1 is smaller than

$\frac{1}{2}$, then the distance from a to -1 is greater than $\frac{1}{2}$, and, consequently, at all the points $n = 9 + 12k, k \in \mathbb{N}$ we get $|a_n - a| = |\sin(\frac{3\pi}{2} + 2k\pi) - a| = |-1 - a| > \varepsilon_0 = \frac{1}{2}$, which means that for any N there exist indices $n = 9 + 12k$ (with k large enough) such that $n > N$, but $|a_n - a| \geq \varepsilon_0$.

Example 7a $a_n = q^n, q \in \mathbb{R}$

This is a *geometric sequence*—geometric progression with the ratio q. The behavior of the sequence depends on the value of the ratio q. Let us analyze separately different situations.

(1) If $q > 1$, then the sequence is strictly increasing, unbounded above, while bounded below ($m = 1$). The sequence is divergent, but it has the infinite limit: $\lim_{n \to \infty} q^n = +\infty$. The proof of the last assertion is easy. According to the definition, we should find the indices n such that the inequality $q^n > E$ is true, or solving for n: $n > \frac{\ln E}{\ln q}$ (we can always choose $E > 1$, and, since $q > 1$ implies that $\ln q > 0$, we can divide keeping the sign). Then, choosing $N = \left[\frac{\ln E}{\ln q}\right]$, we ensure that $q^n > E$ for $\forall n > N$.

(2) If $q = 1$, then we have the constant sequence considered in Example 1: it is bounded, (non-strictly) monotone and convergent to 1.

(3) If $0 < q < 1$, then the sequence is strictly decreasing, bounded ($m = 0, M = 1$) and convergent to 0. The last property can be easily proved by definition. Solving the inequality $|a_n - a| = q^n < \varepsilon$ for n, we find $n > \frac{\ln \varepsilon}{\ln q}$ (notice that $0 < q < 1$ implies that $\ln q < 0$ and dividing we change the sign; besides, we can always suppose that $\varepsilon < 1$ to guarantee the negative sign of $\ln \varepsilon$ and positive sign of the right-hand side of the inequality for n). Then, choosing $N = \left[\frac{\ln \varepsilon}{\ln q}\right]$, we ensure that $q^n < \varepsilon$ for $\forall n > N$.

(4) If $q = 0$, then the sequence is a constant and, consequently, it is bounded, (non-strictly) monotone and has the limit equal to 0.

(5) If $-1 < q < 0$, then the sequence is non-monotone (it is alternating), bounded ($m = -1, M = 1$) and converges to 0. The last property can be demonstrated in the same way as in the case $0 < q < 1$.

(6) If $q = -1$, then we have the sequence of Example 2a, which is non-monotone (alternating), bounded ($m = -1, M = 1$) and divergent.

(7) If $q < -1$, then the sequence is non-monotone (alternating), unbounded (both above and below) and divergent. It can be shown that the sequence has two different tendencies to infinity: it approaches $+\infty$ by even indices and $-\infty$ by odd indices.

Historical Remarks Intuitive notion of a convergent infinite sequence can be traced back to the Hellenic mathematicians and philosophers. Democritus, Eudoxus, Euclid and Archimedes developed the method of exhaustion, which uses an infinite sequence of approximations to determine the area or volume of a figure. One of the first use of the mathematical concept of a convergent sequence can be found in the 1677 works of Gregory and a few years later in the works of Newton, albeit without rigor required in modern mathematics. In 1821 "Cours d'analyse" Cauchy stated the definition of convergence in terms of limits, but the proper concept of limit (not only in the case of a sequence) still had not a due elaboration. Only Weierstrass in his university lectures and research works during the 1870s formulated a completely satisfactory formal definition of the limit of a function, including that of a sequence.

2   Common Properties of Convergent Sequences

Since sequences of numbers are a special type of functions, whose domain is the set of natural numbers $\mathbb{N}$ or its extension $\mathbb{N}_0$, many properties of convergent sequences are analogous to the properties of the limits of general functions.

The majority of proofs for this group of properties follows the standards of limits of general functions, but, for completeness we reproduce them here.

2.1   Uniqueness of the Limit

Theorem About the Uniqueness of the Limit *If the limit exists, then it is unique.*

Proof Assume, for contradiction, that a sequence a_n has two different limits $a < b$. Then calculate the distance $d = b - a$ between a and b, and consider $\varepsilon = \frac{d}{2}$-neighborhood of a and b in the definition of both limits. On the one hand, for this $\varepsilon = \frac{d}{2}$ there exists N_a such that for all the indices $n > N_a$ it follows that $|a_n - a| < \frac{d}{2}$. On the other hand, for the same $\varepsilon = \frac{d}{2}$ there exists N_b such that for all $n > N_b$ one gets $|a_n - b| < \frac{d}{2}$. If now we take $N = \max\{N_a, N_b\}$, then for the same indices $n > N$ we obtain $|a_n - a| < \frac{d}{2}$ and also $|a_n - b| < \frac{d}{2}$. However, the first inequality implies that $a_n < a + \frac{d}{2} = a + \frac{b-a}{2} = \frac{a+b}{2}$, while the second—that $a_n > b - \frac{d}{2} = b - \frac{b-a}{2} = \frac{a+b}{2}$. These two statements cannot both be true. Hence, we arrive at the contradiction, and, therefore, our initial assumption about existence of two different limits is false.   □

2.2   Comparison Properties

Property 1c *If* $\lim\limits_{n \to \infty} a_n = a$, $\lim\limits_{n \to \infty} b_n = b$ *and* $a < b$, *then there exists* N *such that* $a_n < b_n$ *for* $\forall n > N$.

Proof We can employ a similar idea of the Uniqueness Theorem, while arguing directly (not appealing to contradiction). Calculate the distance $d = b - a$ between a and b, and consider $\varepsilon = \frac{d}{2}$-neighborhood of a and b in the definition of the two limits. Then, by the first definition, for $\varepsilon = \frac{d}{2}$ there exists N_a such that $|a_n - a| < \frac{d}{2}$ for all $n > N_a$. At the same time, by the second definition, for the same $\varepsilon = \frac{d}{2}$ there exists N_b such that $|b_n - b| < \frac{d}{2}$ whenever $n > N_b$ (see Fig. 1.3). If we take now $N = \max\{N_a, N_b\}$, then for the indices $n > N$ we have $|a_n - a| < \frac{d}{2}$ and $|b_n - b| < \frac{d}{2}$ at the same time. The first inequality implies

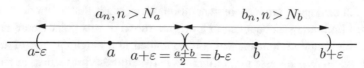

Fig. 1.3 The comparison Property 1c

that $a_n < a + \frac{d}{2} = a + \frac{b-a}{2} = \frac{a+b}{2}$, while the second—that $b_n > b - \frac{d}{2} = b - \frac{b-a}{2} = \frac{a+b}{2}$. Joining the last two inequalities, we get $a_n < \frac{a+b}{2} < b_n$ for all $n > N$. $\qquad\square$

The illustration of this comparison property is given in Fig. 1.3.

Property 2c *If* $\lim\limits_{n\to\infty} a_n = a$, $\lim\limits_{n\to\infty} b_n = b$ *and* $a_n \leq b_n$ *for* $\forall n > N$, *then* $a \leq b$.

Proof Suppose, for contradiction, that $a > b$. Then, according to Property 1c, there exists N such that $a_n > b_n$ for $\forall n > N$, but this leads to the contradiction with hypothesis that $a_n \leq b_n$. $\qquad\square$

Remark In a certain sense, these two properties are converse to each other (the first one transfer the inequality from the limit to the elements of the sequences, while the second acts in the opposite direction). However, an attempt to make these two properties even closer by using the same kind of the sign in both inequalities is inadmissible. This means that in Property 1c the inequality $a \leq b$ does not imply $a_n \leq b_n$ (consider, for instance, $a_n = \frac{1}{n}$ and $b_n = 0$, both with the limit equal to 0); and, in Property 2c, the inequality $a_n < b_n$ does not imply $a < b$ (consider, for instance, $a_n = 0$ and $b_n = \frac{1}{n}$, both convergent to 0).

Property 3c: Boundedness of a Convergent Sequence *If* a_n *converges, then it is bounded.*

Proof This result follows directly from Property 1c of comparison. We only show that a convergent sequence is bounded above, since the boundedness below is demonstrated using the same arguments. Choose any constant $b > a = \lim\limits_{n\to\infty} a_n$. A constant sequence $b_n = b$ is convergent with the limit equal to b and, consequently, applying Property 1c, we conclude that $a_n < b$ for all $n > N$. Among the remaining N elements of the sequence a_n, which do not enter in this evaluation, we can always choose the maximum number— $a_M = \max\{a_1, \ldots, a_N\}$. Finally, taking $M = \max\{a_M, b\}$, we find an upper bound for the sequence a_n: $a_n \leq M$ for all n. $\qquad\square$

Remark 1 The converse is not true: for instance, consider the sequence $a_n = (-1)^n$ of Example 2a.

Remark 2 An analogous property of general functions states that a function $f(x)$, which has a (finite) limit at x_0, is bounded in a neighborhood of this point (more precisely: in that part of the function domain, which is contained in a neighborhood of x_0). Despite this similarity, there is an essential difference between these properties: in the case of a sequence, not only its part, but the entire sequence is bounded.

Property 4c: The Squeeze Theorem *If* $\lim\limits_{n \to \infty} a_n = \lim\limits_{n \to \infty} b_n = a$ *and* $a_n \le c_n \le b_n$ *for* $\forall n > N$, *then the sequence* c_n *converges and* $\lim\limits_{n \to \infty} c_n = a$.

Proof Take an arbitrary $\varepsilon > 0$ and consider the definitions of the two given limits. For a_n we have: for chosen ε there exists N_a such that if $n > N_a$ then $|a_n - a| < \varepsilon$; and for b_n: for the same ε there exists N_b such that if $n > N_b$ then $|b_n - b| < \varepsilon$. If we take now $N = \max\{N_a, N_b\}$, then for all the indices $n > N$ it follows both $|a_n - a| < \varepsilon$ and $|b_n - b| < \varepsilon$. The inequality $|a_n - a| < \varepsilon$ implies that $a_n > a - \varepsilon$, and the inequality $|b_n - b| < \varepsilon$ implies that $b_n < a + \varepsilon$. Using the last two inequalities together with the theorem condition that $a_n \le c_n \le b_n$, we conclude that $a - \varepsilon < a_n \le c_n \le b_n < a + \varepsilon$, which is equivalent to the inequality in the definition of the limit: $|c_n - a| < \varepsilon$. Since $\varepsilon > 0$ is an arbitrary positive number, the definition of the limit for c_n is satisfied.   □

The illustration of the squeeze theorem is shown in Fig. 1.4.

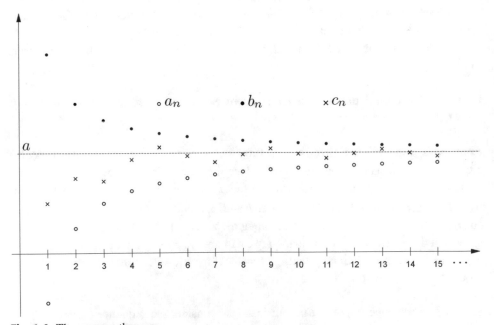

Fig. 1.4 The squeeze theorem

2.3 Arithmetic and Analytic Properties

Property 1a: Four Arithmetic Properties *If* $\lim\limits_{n\to\infty} a_n = a$ *and* $\lim\limits_{n\to\infty} b_n = b$, *then the sequences* $a_n + b_n$, $a_n - b_n$, $a_n \cdot b_n$ *and* $\frac{a_n}{b_n}$ *are convergent (the last under the additional condition that* $b \neq 0$), *and* $\lim\limits_{n\to\infty} (a_n + b_n) = a + b$, $\lim\limits_{n\to\infty} (a_n - b_n) = a - b$, $\lim\limits_{n\to\infty} (a_n \cdot b_n) = a \cdot b$ *and* $\lim\limits_{n\to\infty} \frac{a_n}{b_n} = \frac{a}{b}$.

Proof The proofs of the first two properties are trivial and those of the last two are simple and follow the patterns of general functions. Since the proofs of the last two properties are similar, we show only the third property of this group.

Let us start by writing below the main inequality in the definition of the limit that we should prove: $|a_n b_n - ab| < \varepsilon$. To demonstrate this evaluation, we have no option but to use the available information about a_n and b_n: $|a_n - a| < \varepsilon_a$ and $|b_n - b| < \varepsilon_b$. This means that the left-hand side of the desired inequality should be transformed into the form containing the factors $|a_n - a|$ and $|b_n - b|$. To achieve this, we make the following manipulations:

$$|a_n b_n - ab| = |a_n b_n - a_n b + a_n b - ab| \le |a_n b_n - a_n b| + |a_n b - ab| = |a_n||b_n - b| + |b||a_n - a|.$$

In the second summand, the constant $|b|$ multiplied by the small quantity $|a_n - a|$ result in a small value of this product. To evaluate the first summand in a similar way, we need the estimate for the coefficient $|a_n|$, which follows from the convergence of a_n: according to Property 3c of comparison, the convergence of a_n ensures its boundedness, that is, there exists a constant $C > 0$ such that $|a_n| < C$. Now, to arrive at the required definition, we choose $\varepsilon_a = \frac{\varepsilon}{2(|b|+1)}$, $\varepsilon_b = \frac{\varepsilon}{2(C+1)}$, next we find the respective indices N_a and N_b in the definitions of the limits for a_n and b_n, and calculate $N = \max\{N_a, N_b\}$. Then, for all $n > N$ both inequalities—$|a_n - a| < \varepsilon_a$ and $|b_n - b| < \varepsilon_b$—hold simultaneously, and advancing further on the evaluation of $|a_n b_n - ab|$, we obtain

$$|a_n b_n - ab| \le |a_n||b_n - b| + |b||a_n - a| < C\frac{\varepsilon}{2(C+1)} + |b|\frac{\varepsilon}{2(|b|+1)} < \frac{\varepsilon}{2} + \frac{\varepsilon}{2} = \varepsilon.$$

The last inequality is valid for all $n > N$, which means, according to the definition, that the limit of $a_n b_n$ exists and equals ab. $\qquad\square$

Property 2a: Theorem of Composite Function *If* $\lim\limits_{n\to\infty} a_n = a$, *a function* $f(x)$ *is continuous at* a *and* a_n *belong to the domain* X *of* $f(x)$, *then* $\lim\limits_{n\to\infty} f(a_n) = f\left(\lim\limits_{n\to\infty} a_n\right) = f(a).$

Proof Since the function $f(x)$ is continuous, for any $\varepsilon_f > 0$ there exists $\delta > 0$ such that for all x in X which satisfy $|x - a| < \delta$ it follows that $|f(x) - f(a)| < \varepsilon_f$. On the other hand, the convergence of a_n means that for an arbitrary $\varepsilon_a > 0$ there exists N such that $|a_n - a| < \varepsilon_a$ for all $n > N$.

Now consider an arbitrary $\varepsilon > 0$ and choose $\varepsilon_f = \varepsilon$. Then, we find the corresponding $\delta > 0$ from the definition of continuity and take $\varepsilon_a = \delta$. Next, we find N from the definition of convergence of a_n such that $|a_n - a| < \delta$ for all $n > N$. Therefore, for these indices n we obtain $|f(a_n) - f(a)| < \varepsilon$, because $a_n \in X$. Summarizing, for any $\varepsilon > 0$ we are able to determine N such that the inequality $|f(a_n) - f(a)| < \varepsilon$ holds for all $n > N$, that is, the sequence $f(a_n)$ converges to $f(a)$. $\square$

Properties 3a: Additional Analytic Properties

Different important and useful results follow immediately from Property 2a about a composite function. Some of them are indicated below.

Property of Absolute Value *First, using the continuous in* $\mathbb{R}$ *function* $f(x) = |x|$, *we deduce that if* $\lim\limits_{n\to\infty} a_n = a$, *then* $\lim\limits_{n\to\infty} |a_n| = |a|$. *This property is also easy to prove by definition.*

Notice that the converse statement is not true: for instance, consider the sequence $a_n = (-1)^n$ from Example 2a, which has no limit, but whose absolute value is a constant sequence $|a_n| = 1$ with the limit equal to 1. Nevertheless, in the case $a = 0$, it is easy to show that the converse is also true.

Property of Power *Second, recall that* $f(x) = x^c$ *is a continuous function in the three important cases: when* $x \in \mathbb{R}$ *and* $c \in \mathbb{N}$; *when* $x \neq 0$ *and* $c \in \mathbb{Z}$; *and also when* $x > 0$ *and* $c \in \mathbb{R}$. *Therefore, we can conclude that in these three cases the existence of the limit* $\lim\limits_{n\to\infty} a_n = a$ *ensures that* $\lim\limits_{n\to\infty} a_n^c = a^c$.

Property of Exponent *Third, recalling that* $f(x) = c^x, c > 0$ *is a continuous in* $\mathbb{R}$ *function, we can conclude that if* $\lim\limits_{n\to\infty} a_n = a$, *then* $\lim\limits_{n\to\infty} c^{a_n} = c^a$.

3   Special Properties of Convergent Sequences

3.1   Convergence of Function and Corresponding Sequence

Property 1s: Property of Generating Function *If* $\lim\limits_{x \to +\infty} f(x) = a$ *and* $a_n = f(n), \forall n$, *then* $\lim\limits_{n \to \infty} a_n = a$.

Proof Under the property conditions, the sequence n just represents one of possible ways (paths) for x approaching $+\infty$. Since any form of the tendency of x to $+\infty$ guarantees the same tendency of $f(x)$ to a, a specific path $x_n = n$, which gives the function values $f(n) = a_n$, results in the same tendency of $f(n)$ to a, which means convergence of a_n to a. In other words, this trivial property is nothing more than a specification of the following property of limits: if a general limit exists, then any partial limit also exists and is equal to the general limit. $\square$

3.2   Relationship Between Convergence and Boundedness

In one of the previous properties, it was shown that the convergence of a_n implies its boundedness and that the converse is not valid. It happens that adding to the boundedness the property of monotonicity we can ensure the convergence. More precisely, we get the following result.

> **Property 2s: Theorem of Monotone Sequence (Theorem on Boundedness and Convergence)** *If a_n is increasing and bounded above, then a_n is convergent. In a similar way, if a_n is decreasing and bounded below, then a_n is convergent. In general, if a_n is monotone and bounded, then it is convergent.*

Proof We focus on the proof of the first statement, since the second follows the same arguments. Since a_n is a bounded above set, this set has supremum (the least upper bound) which we denote by α. Let us show that α is the limit of a_n. In fact, since α is an upper bound, it follows that $a_n \leq \alpha$ for all n. On the other hand, since α is the supremum, for any $\varepsilon > 0$ there exists at least one element a_N such that $a_N > \alpha - \varepsilon$. However, a_n is an increasing sequence, and therefore, $a_n \geq a_N > \alpha - \varepsilon$ for all $n > N$ (see Fig. 1.5). Thus, for any $\varepsilon > 0$ we can find N (in the definition of the supremum) such that $\alpha - \varepsilon < a_N \leq a_n \leq \alpha < \alpha + \varepsilon$ for all $n > N$, that is, the definition of the limit $\lim\limits_{n \to \infty} a_n = \alpha$ is satisfied. $\square$

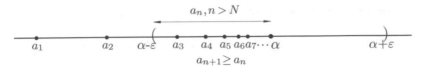

Fig. 1.5 The theorem of monotone sequence

The illustration of the Theorem of monotone sequence is shown in Fig. 1.5.

Remark 1 Both hypothesis are important for a convergence of sequence. If a_n is not monotone, then (like we have seen before) the sequence $a_n = (-1)^n$ is an example of a bounded and divergent sequence. On the other hand, if a_n is unbounded, then $a_n = n$ is an example of a monotone and divergent sequence.

Remark 2 In the opposite direction, we can notice that the convergence of a_n guarantees boundedness of the sequence (as was shown before), but a_n still may be non-monotone, like the alternating sequence $a_n = \frac{(-1)^n}{n}$ considered in Example 4a.

Remark 3 Notice also that a monotonic and unbounded sequence tends to infinity. More specifically: if a_n is increasing and unbounded (above), then $\lim_{n \to \infty} a_n = +\infty$; if a_n is decreasing and unbounded (below), then $\lim_{n \to \infty} a_n = -\infty$. Let us prove the first statement. Take an arbitrary $E > 0$. Since a_n is not bounded above, there exists N such that $a_N > E$. Then, due to the increasing of a_n, the same inequality is true for all $n > N$: $a_n \geq a_N > E$, $\forall n \geq N$. Since $E > 0$ is arbitrary, the definition of the infinite limit is satisfied.

3.3   Subsequences and Their Convergence. Bolzano-Weierstrass Theorem

A relevant concept for investigation of the properties of a sequence is a subsequence.

> ▶ **Definition (Subsequence)** A *subsequence* of a given sequence a_n is a new sequence composed of an infinite number of elements of the original sequence a_n, keeping their order. More precisely: a subsequence of a given sequence a_n is a sequence a_{n_k} in the form $a_{n_1}, a_{n_2}, \ldots, a_{n_k}, \ldots$, where $n_k, k \in \mathbb{N}$ are the indices of the domain of the original sequence ordered increasingly $n_1 < n_2 < \ldots < n_k < \ldots$. A trivial case of a subsequence is the sequence a_n itself.

The *domain of a subsequence* (different from the original one) represents a specific path of the tendency of indices n to infinity. Although this does not supply complete information about the behavior of the original sequence, a knowledge about the convergence of subsequences can be very useful. Subsequences have the same role for an original sequence as lateral and other partial limits for a general limit of function, when the limit point is approached from the left or from the right (or using any other particular path, for instance, by irrational numbers or rational numbers). An information about different partial limits can reveal what is happening with a general limit. For example, knowing that the two lateral limits are different, or one of them does not exist, one can conclude that the general limit does not exist. In the same way, the subsequences provide information about partial limits of the original sequence and this can frequently be used to draw a conclusion about the behavior of the original sequence.

> **Definition** (**Partial Limit**) If a subsequence converges, its limit is called a *partial limit* of the original sequence.

If we work simultaneously with partial limits and the limit of the original sequence, the latter is also called a *general limit* (to specify what kind of the limit we consider).

Relation Between the Limit of a Sequence and Its Subsequence (General and Partial Limits)

Property 3s: Theorem 1 of Subsequences *If a sequence a_n converges, then any of its subsequences converges to the same value.*

Proof The existence of the limit a of an original sequence a_n means that for any specific path of tendency of index n to $+\infty$ the values of a_n approach a. In particular, it occurs if we choose the indices n_k of the domain of a subsequence and move k toward $+\infty$.

More specifically: by definition, $\lim_{n\to\infty} a_n = a$ if for any $\varepsilon > 0$ there exists N such that $|a_n - a| < \varepsilon$ whenever $n > N$. For all the indices n_k of a subsequence we have $n_k \geq k$ and $n_{k+1} > n_k$, $\forall k$. Therefore, if $|a_k - a| < \varepsilon$ for all the indices $k > N$ of the original sequence, then, in particular, $|a_{n_k} - a| < \varepsilon$ for the same indices $k > N$, which means, by definition, that the subsequence a_{n_k} converges to a.   □

Remark 1 The converse is not true: for the divergent sequence $a_n = (-1)^n$, its subsequence containing all the even indices has all the elements equal to 1, and consequently, it converges: $\lim_{k\to\infty} a_{2k} = \lim_{k\to\infty} 1 = 1$.

Remark 2 Although the result of the Theorem is quite trivial, it allows us to use a convenient practical criterion for divergence of a given sequence: if two subsequences of a_n have different limits or (at least) one of them is divergent, then a_n does not have a limit. In other words: if two partial limits are different or (at least) one of them does not exist, then the general limit does not exist. Using again the sequence $a_n = (-1)^n$, we can see that $\lim\limits_{k\to\infty} a_{2k} = \lim\limits_{k\to\infty} 1 = 1$, but $\lim\limits_{k\to\infty} a_{2k-1} = \lim\limits_{k\to\infty} (-1) = -1$ that shows, one more time, that this sequence has no limit.

Now let us formulate the statement about the convergence of an original sequence under some hypotheses on its subsequences.

Property 4s: Theorem 2 of Subsequences *If a finite number of subsequences cover all the elements of an original sequence a_n and all these subsequences converge to the same limit, then a_n also converges to this limit.*

Proof Let us denote the involved subsequences by $a_{n_k}^{(1)}$, $a_{n_k}^{(2)}, \ldots, a_{n_k}^{(M)}$. Since $\lim\limits_{k\to\infty} a_{n_k}^{(m)} = a$ for all $m = 1, 2 \ldots, M$, and the number of these subsequences is finite, then in M definitions of their convergence we find M numbers N_m, $m = 1, 2 \ldots, M$ which ensure the inequalities $|a_{n_k}^{(m)} - a| < \varepsilon$ for all $n_k > N_m$. In a finite set of numbers, we always can find the maximum value, and choosing $N = \max\{N_1, N_2, \ldots, N_M\}$ we guarantee that $|a_{n_k}^{(m)} - a| < \varepsilon$ for all the subsequences whenever $n_k > N$. Since the elements of all these subsequences combined include all the elements of the original sequence, we conclude that $|a_n - a| < \varepsilon$ whenever $n > N$, that is, the definition of the limit of a_n is satisfied.   □

Remark In the formulation of the last Theorem, a finite set of subsequences cannot be substituted by an infinite set. Indeed, as was seen, the sequence $a_n = (-1)^n$ is divergent. However, we can define its subsequences in the following way:

$$a_{n_1}^{(1)} = a_1 = -1, a_{n_k}^{(1)} = a_{2(k-1)} = 1, \forall k > 1;$$

$$a_{n_1}^{(2)} = a_3 = -1, a_{n_k}^{(2)} = a_{2k} = 1, \forall k > 1;$$

$$a_{n_1}^{(3)} = a_5 = -1, a_{n_k}^{(3)} = a_{2(k+1)} = 1, \forall k > 1; \ldots ;$$

$$a_{n_1}^{(m)} = a_{2m-1} = -1, a_{n_k}^{(m)} = a_{2(k+m-2)} = 1, \forall k > 1; \ldots .$$

The elements of these subsequences combined cover all the elements of the original sequence a_n, and each of these subsequences has the same form $\{-1, 1, 1, \ldots\}$ or $\{b_n, n = 1, 2, \ldots\}$ with $b_1 = -1$ and $b_n = 1, \forall n > 1$. Since b_n is a constant sequence starting from the second term, its limit exists and is equal to 1. Hence, all the created subsequences have the same limit and the set of their elements contains all the elements of the original sequence. Nevertheless, this does not guarantee that the original sequence converges, because the number of the constructed subsequences is infinite.

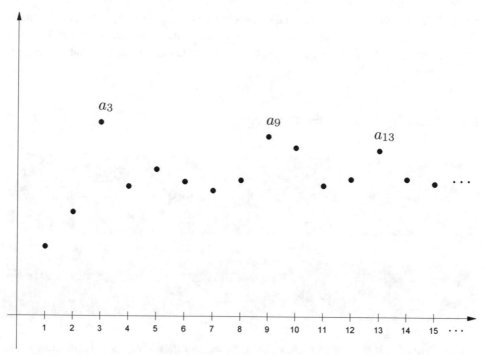

Fig. 1.6 The peak points a_3, a_9, a_{13}

Property 5s: Theorem on the Existence of a Monotone Subsequence *Any sequence has a monotone subsequence.*

Proof Let us call an element a_k a peak point of the sequence a_n if $a_m < a_k$ for all $m > k$ (see Fig. 1.6). The two cases may occur. First, if a_n has an infinite number of peak points $a_{k_1}, a_{k_2} \ldots, a_{k_p}, \ldots$, where $k_1 < k_2 < \ldots < k_p \ldots$, then, by definition, $a_{k_1} > a_{k_2} > \ldots > a_{k_p} > \ldots$, and consequently, the subsequence of the peak points a_{k_p} is the desired monotone (decreasing) subsequence. In the second case, the sequence a_n has a finite number of peak points $a_{k_1}, a_{k_2} \ldots, a_{k_p}$ ordered according to the increasing index $k_1 < k_2 < \ldots < k_p$. Then, for any $n_1 > k_p$ the element a_{n_1} is not a peak point, and consequently, there exists $n_2 > n_1$ such that $a_{n_2} \geq a_{n_1}$. Again, a_{n_2} is not a peak point, which means that there exists $n_3 > n_2$ such that $a_{n_3} \geq a_{n_2}$, etc. Continuing in this way, we obtain the desired monotone (increasing) subsequence.   □

Property 6s: Bolzano-Weierstrass Theorem *Every bounded sequence has a convergent subsequence.*

Proof The preceding Theorem guarantees the existence of a monotone subsequence. Since the original sequence is bounded, any its subsequence is also bounded. Then, the Theorem on relationship between boundedness and convergence guarantees that this subsequence is convergent.   □

3.4   Cauchy Criterion for Convergence

▶ **Definition** (**Cauchy Sequence**) A sequence a_n is called a *Cauchy sequence* if for every $\varepsilon > 0$ there exists $N = N(\varepsilon)$ such that for all $n, m > N$ it follows that $|a_n - a_m| < \varepsilon$.

Property 7s: Cauchy Criterion for Convergence *A sequence a_n converges if and only if a_n is a Cauchy sequence.*

Proof *Necessity.* If a_n converges to a, then, by definition, for $\forall \varepsilon_a > 0$ (choose $\varepsilon_a = \frac{\varepsilon}{2}$) there exists N such that $|a_n - a| < \frac{\varepsilon}{2}$ whenever $n > N$. Therefore, using the same N and considering $n, m > N$, we get:

$$|a_n - a_m| = |a_n - a + a - a_m| \le |a_n - a| + |a_m - a| < \frac{\varepsilon}{2} + \frac{\varepsilon}{2} = \varepsilon,$$

that is, the definition of the Cauchy sequence holds.

Sufficiency. First, notice that any Cauchy sequence is bounded. Indeed, taking $\varepsilon = 1$ in the definition of a Cauchy sequence, we find the corresponding N such that $|a_n - a_m| < 1$ for all $n, m > N$. In particular, $|a_n - a_{N+1}| < 1$, and consequently $|a_n| < |a_{N+1}| + 1$ for all indices $n > N$. This evaluation does not involve only a finite number of the elements of the sequence and we always can find the term with the greatest absolute value among them: $C = \max\{|a_1|, \ldots, |a_N|\}$. Therefore, for any n we have $|a_n| < M = \max\{|a_{N+1}| + 1, C\}$, that is, the sequence a_n is bounded. Second, according to the Bolzano-Weierstrass Theorem, the sequence a_n has a convergent subsequence. Finally, let us show that a Cauchy sequence that has a convergent subsequence is also convergent (to the same limit). In fact, taking in the definition of a Cauchy sequence $\varepsilon_a = \frac{\varepsilon}{2}$, we find the corresponding N_a such that $|a_n - a_m| < \frac{\varepsilon}{2}$ whenever $m, n > N_a$. At the same time, in the definition of the convergent subsequence a_{n_k} we choose the same $\varepsilon_b = \frac{\varepsilon}{2}$ and determine the corresponding N_b such that $|a_{n_k} - a| < \frac{\varepsilon}{2}$ for all $n_k > N_b$ (here a is the limit of the subsequence). Then, for $n, n_k > N = \max\{N_a, N_b\}$ both inequalities $|a_n - a_{n_k}| < \frac{\varepsilon}{2}$ and $|a_{n_k} - a| < \frac{\varepsilon}{2}$ hold, and consequently

$$|a_n - a| = |a_n - a_{n_k} + a_{n_k} - a| \le |a_n - a_{n_k}| + |a_{n_k} - a| < \frac{\varepsilon}{2} + \frac{\varepsilon}{2} = \varepsilon$$

for the same $n > N$. This means that a is the limit of the sequence a_n. $\qquad\square$

Remark The advantage of the Cauchy criterion (comparing to the definition of the convergence) is that the condition of a Cauchy sequence does not require any knowledge about the value of the limit of a sequence.

3.5 Sequences of the Arithmetic and Geometric Means

▶ **Definition** Recall that the *arithmetic mean* of n real numbers $a_1, a_2, \ldots, a_n$ is defined by the formula $\frac{a_1 + a_2 + \ldots + a_n}{n}$, and the *geometric mean* of n positive numbers $a_1, a_2, \ldots, a_n$ is defined by the formula $\sqrt[n]{a_1 a_2 \ldots a_n}$.

▶ **Definition** (**The Sequence of the Arithmetic Means**) For a sequence of numbers a_n, we define the *sequence b_n of its arithmetic means* by the formulas $b_1 = a_1$, $b_2 = \frac{a_1 + a_2}{2}, \ldots, b_n = \frac{a_1 + a_2 + \ldots + a_n}{n}, \ldots$.

▶ **Definition** (**The Sequence of the Geometric Means**) For a sequence of positive terms a_n, we define the *sequence c_n of its geometric means* by the formulas $c_1 = a_1$, $c_2 = \sqrt[2]{a_1 a_2}, \ldots, c_n = \sqrt[n]{a_1 a_2 \ldots a_n}, \ldots$.

Property 8s: The Limit of the Arithmetic Means *If a sequence a_n converges to a, then the sequence of its arithmetic means also converges to a.*

Proof The convergence of a_n implies its boundedness: there exists C such that $|a_n| \le C$ for all n. Consequently, $|a| \le C$ and $|a_n - a| \le 2C$, $\forall n$. According to the definition of the convergence, for $\forall \varepsilon_a = \frac{\varepsilon}{2}$ there exists N_a such that $|a_n - a| < \frac{\varepsilon}{2}$ whenever $n > N_a$. Let us evaluate the difference $|b_n - a|$ for the arithmetic means b_n:

$$|b_n - a| = \left| \frac{a_1 + a_2 + \ldots + a_n}{n} - a \right| = \left| \frac{a_1 - a + a_2 - a + \ldots + a_n - a}{n} \right|$$

$$\le \frac{1}{n} \left(|a_1 - a| + |a_2 - a| + \ldots + |a_n - a| \right)$$

$$= \frac{1}{n} \left(|a_1 - a| + |a_2 - a| + \ldots + |a_{N_a} - a| \right)$$

$$+ \frac{1}{n} \left(|a_{N_a+1} - a| + \ldots + |a_n - a| \right) < \frac{1}{n} 2CN_a + \frac{n - N_a}{n} \frac{\varepsilon}{2} .$$

Considering N_a fixed, we calculate a new index $N = \left[2(2C + 1) \frac{N_a}{\varepsilon} \right]$ and for any $n > N$ obtain the following evaluation:

$$|b_n - a| < \frac{1}{n} 2CN_a + \frac{n - N_a}{n} \frac{\varepsilon}{2} < \frac{\varepsilon}{2(2C + 1)N_a} 2CN_a + \frac{n - N_a}{n} \frac{\varepsilon}{2} < \frac{\varepsilon}{2} + \frac{\varepsilon}{2} = \varepsilon .$$

Hence, the definition of the limit $\lim\limits_{n \to \infty} b_n = a$ is satisfied. $\square$

Property 9s: The Limit of the Geometric Means *If a sequence of positive terms a_n converges to $a > 0$, then the sequence of its geometric means also converges to a.*

Proof Together with the sequence of the geometric means c_n, consider the sequence of its logarithms $\ln c_n = \ln \sqrt[n]{a_1 a_2 \ldots a_n} = \frac{1}{n} (\ln a_1 + \ln a_2 + \ldots + \ln a_n)$. Since $\ln x$ is a continuous function on its domain, by the Theorem of a composite function, the sequence $\ln a_n$ converges to $\ln a$: $\lim\limits_{n \to \infty} \ln a_n = \ln a$. In turn, the convergence of $\ln a_n$ implies (by the preceding Theorem) the convergence of its arithmetic means $\ln c_n$ to the same number $\ln a$: $\lim\limits_{n \to \infty} \ln c_n = \ln a$. To return to c_n, we again use the Theorem of a composite function, this time with the continuous function e^x, which guarantees that the sequence $e^{\ln c_n} = c_n$ converges to $e^{\ln a} = a$: $\lim\limits_{n \to \infty} c_n = a$. $\square$

Remark 1 In the case of a sequence with non-negative terms the second Theorem is not valid—consider, for example, the sequence $a_n = 1 + \frac{(-1)^n}{n}$ that converges to 1, but the limit of its geometric means converges to 0 due to vanishing of a_1. However, this Theorem can be extended to sequences with non-negative terms whose limit is 0. Indeed, if one of the elements of the sequence a_n is zero, then the result is trivial. If it is not so, then we can apply the arguments of the first Theorem in the following manner. Take some upper bound C of the original sequence: $|a_n| \leq C$ for all n. By the definition of the convergence of a_n to 0, for $\forall \varepsilon_a = \frac{\varepsilon^2}{C+1}$ ($\varepsilon_a < 1$) there exists N_a such that $|a_n| < \frac{\varepsilon}{2}$ whenever $n > N_a$. Let us evaluate now $|c_n|$ for $n > N_a$:

$$|c_n| = \sqrt[n]{a_1 a_2 \cdot \ldots \cdot a_n} = \sqrt[n]{a_1 \cdot \ldots \cdot a_{N_a}} \sqrt[n]{a_{N_a+1} \cdot \ldots \cdot a_n} < \sqrt[n]{C^{N_a}} \sqrt[n]{\varepsilon_a^{n - N_a}} < (C + 1)^{\frac{N_a}{n}} \varepsilon_a^{\frac{n - N_a}{n}} .$$

Considering N_a fixed, we find another index $N = 2N_a$ and notice that for all $n > N$ it follows $\frac{N_a}{n} < \frac{1}{2}$ and $\frac{n - N_a}{n} > \frac{1}{2}$, and consequently

$$|c_n| < (C+1)^{\frac{N_a}{n}} \varepsilon_a^{\frac{n - N_a}{n}} < \sqrt{C+1}\sqrt{\varepsilon_a} = \sqrt{C+1}\sqrt{\frac{\varepsilon^2}{C+1}} = \varepsilon\,.$$

Hence, the definition of the limit $\lim\limits_{n \to \infty} c_n = 0$ is satisfied.

Remark 2 The converse is not true either for the one of the two Theorems, or in the case when the arithmetic and geometric means converge at the same time. To exemplify this, consider the positive sequence $a_n = 2 + (-1)^n$. Evidently, this sequence diverges since it has two different partial limits: $\lim\limits_{k \to \infty} a_{2k} = \lim\limits_{k \to \infty} 3 = 3$, while $\lim\limits_{k \to \infty} a_{2k-1} = \lim\limits_{k \to \infty} 1 = 1$. However, the limits of the arithmetic and geometric means do exist. For the former we have

$$b_{2k} = \frac{a_1 + a_2 + \ldots a_{2k-1} + a_{2k}}{2k} = \frac{4k}{2k} = 2$$

and

$$b_{2k-1} = \frac{a_1 + a_2 + \ldots + a_{2k-1}}{2k-1} = \frac{4(k-1) + 1}{2k-1} = \frac{2(2k-1) - 1}{2k-1} = 2 - \frac{1}{2k-1}\,.$$

Since $\lim\limits_{k \to \infty} b_{2k} = \lim\limits_{k \to \infty} b_{2k-1} = 2$, the general limit exists and is equal to 2: $\lim\limits_{n \to \infty} b_n = 2$. For the latter we get

$$c_{2k} = \sqrt[2k]{a_1 a_2 \ldots a_{2k-1} a_{2k}} = \sqrt[2k]{3^k} = \sqrt{3}$$

and

$$c_{2k-1} = \sqrt[2k-1]{a_1 a_2 \ldots a_{2k-1}} = \sqrt[2k-1]{3^{k-1}} = \sqrt[2k-1]{3^{k-1/2} \cdot 3^{-1/2}} = \sqrt{3} \cdot 3^{\frac{-1}{4k-2}}\,.$$

Since $\lim\limits_{k \to \infty} c_{2k} = \lim\limits_{k \to \infty} c_{2k-1} = \sqrt{3}$, the general limit exists and is equal to $\sqrt{3}$: $\lim\limits_{n \to \infty} c_n = \sqrt{3}$.

It may even happen that the limits of the arithmetic and geometric means exist and coincide, but the original sequence diverges. For instance, the sequence $a_n = \begin{cases} 1, & n \neq 10^k \\ 2, & n = 10^k \end{cases}$ has no limit since the two partial limits are different: $\lim\limits_{k \to \infty} a_{10^k} = \lim\limits_{k \to \infty} 2 = 2$, while

$\lim\limits_{k\to\infty} a_{2k-1} = \lim\limits_{k\to\infty} 1 = 1$. But $\lim\limits_{n\to\infty} b_n = 1$ and $\lim\limits_{n\to\infty} c_n = 1$. Both limits follow from the squeeze Theorem applied to the inequalities $1 \leq b_n \leq 1 + k \cdot 10^{-k}$ and $1 \leq c_n \leq \sqrt[10^k]{2^k}$, respectively, where the inequalities hold for the indices $10^k \leq n \leq 10^{k+1} - 1$.

Historical Remarks 1 The Bolzano-Weierstrass theorem was first proved by Bolzano as an auxiliary lemma in the proof of the intermediate value theorem. However, this work was published in an obscure pamphlet in Prague in 1817 and one year later in a little-known Bohemian journal, out of reach for leading mathematical circles. More than a half century later, in 1871, Hankel made references to the Bolzano work in the article appeared in the authoritative German Encyclopedia. In 1881, Stolz republished the Bolzano results in an influential German edition, establishing his priority in the proof of the Bolzano-Weierstrass theorem, and since then this important contribution by Bolzano became widely known and acknowledged.

The first well-known proof of the Bolzano-Weierstrass theorem was presented by Weierstrass in 1860.

Historical Remarks 2 In 1821 "Cours d'analyse" Cauchy formulated his criterion and proved that a convergent sequence is a Cauchy sequence, but was not able to prove the converse. The latter proof requires the use of the completeness of the real numbers, which should be either postulated as an axiom or obtained from a construction of the real numbers, but a rigorous theory of the real numbers was not developed yet.

Four years before publication of "Cours d'analyse", Bolzano formulated the same criterion and tried to prove it. Unfortunately, his results were published in an obscure pamphlet in Prague in 1817 (in the same work where he presented his version of the Bolzano-Weierstrass theorem) and became known for mathematical community only more than half a century later.

4   Indeterminate Forms and Techniques of Their Solution

4.1   Definition of Indeterminate Forms

Recall that there are seven different types of *indeterminate forms* denoted by the symbols $\frac{0}{0}$, $\frac{\infty}{\infty}$, $\infty - \infty$, $0 \cdot \infty$, 0^0, ∞^0, 1^∞. Unlike the arithmetic rules of limits, the result of calculation of an indeterminate form cannot generally be predicted, but it depends essentially on a specification of functions or sequences involved. The first two forms in the above list are of the primarily importance for both the subjects of this text and methods of finding the limits of indeterminate forms. For this reason, we focus below on detailed explanation of their definition and techniques of solution. The remaining forms can be defined in a similar way and their solution is usually based on reducing to the first two indeterminate forms.

Indeterminate Form $\frac{0}{0}$

This form is defined as the ratio of two sequences, each of which converges to 0. More precisely, given two sequences a_n and b_n such that $\lim\limits_{n\to\infty} a_n = \lim\limits_{n\to\infty} b_n = 0$, our goal is to determine a behavior (convergence or divergence) of the sequence $c_n = \frac{a_n}{b_n}$. Let us show through examples that there is no general answer in this case, because the result crucially depends on a particular choice of the sequences a_n and b_n. That is why the limits of this type bear the name of an indeterminate form.

Example 1i

1. If $a_n = \frac{1}{n}$ and $b_n = \frac{1}{n}$, then $\lim\limits_{n\to\infty} a_n = \lim\limits_{n\to\infty} b_n = 0$ and $\lim\limits_{n\to\infty} \frac{a_n}{b_n} = \lim\limits_{n\to\infty} 1 = 1$;
2. If $a_n = \frac{C}{n}$ and $b_n = \frac{1}{n}$, where C is an arbitrary constant, then $\lim\limits_{n\to\infty} a_n = \lim\limits_{n\to\infty} b_n = 0$ and $\lim\limits_{n\to\infty} \frac{a_n}{b_n} = \lim\limits_{n\to\infty} C = C$;
3. If $a_n = \frac{1}{n}$ and $b_n = \frac{1}{n^2}$, then $\lim\limits_{n\to\infty} a_n = \lim\limits_{n\to\infty} b_n = 0$ and $\lim\limits_{n\to\infty} \frac{a_n}{b_n} = \lim\limits_{n\to\infty} n = +\infty$;
4. If $a_n = \frac{1}{n}$ and $b_n = -\frac{1}{n^2}$, then $\lim\limits_{n\to\infty} a_n = \lim\limits_{n\to\infty} b_n = 0$ and $\lim\limits_{n\to\infty} \frac{a_n}{b_n} = \lim\limits_{n\to\infty} (-n) = -\infty$;
5. If $a_n = \frac{(-1)^n}{n}$ and $b_n = \frac{1}{n}$, then $\lim\limits_{n\to\infty} a_n = \lim\limits_{n\to\infty} b_n = 0$ and $\lim\limits_{n\to\infty} \frac{a_n}{b_n} = \lim\limits_{n\to\infty} (-1)^n = \nexists$.

Thus, depending on a specific form of a_n and b_n, the sequence $c_n = \frac{a_n}{b_n}$ may converge to any constant (including 0), may diverge having the limit $\pm\infty$ or may diverge without any general tendency (although a_n and b_n converge to 0 in all these examples).

Indeterminate Form $\frac{\infty}{\infty}$

The description here is similar to the first form. There are two sequences a_n and b_n such that $\lim\limits_{n\to\infty} a_n = \lim\limits_{n\to\infty} b_n = \infty$. The problem is to determine a behavior (convergence or divergence) of the sequence $c_n = \frac{a_n}{b_n}$. Through examples we show that this case is also an indeterminate form, because there is no general answer to this question: the result depends on chosen specific sequences a_n and b_n.

Example 2i

1. If $a_n = n$ and $b_n = n^2$, then $\lim\limits_{n\to\infty} a_n = \lim\limits_{n\to\infty} b_n = \infty$ and $\lim\limits_{n\to\infty} \frac{a_n}{b_n} = \lim\limits_{n\to\infty} \frac{1}{n} = 0$;
2. If $a_n = Cn$ and $b_n = n$, where $C \neq 0$ is an arbitrary constant, then $\lim\limits_{n\to\infty} a_n = \lim\limits_{n\to\infty} b_n = \infty$ and $\lim\limits_{n\to\infty} \frac{a_n}{b_n} = \lim\limits_{n\to\infty} C = C$;
3. If $a_n = n^2$ and $b_n = n$, then $\lim\limits_{n\to\infty} a_n = \lim\limits_{n\to\infty} b_n = \infty$ and $\lim\limits_{n\to\infty} \frac{a_n}{b_n} = \lim\limits_{n\to\infty} n = +\infty$;
4. If $a_n = n^2$ and $b_n = -n$, then $\lim\limits_{n\to\infty} a_n = \lim\limits_{n\to\infty} b_n = \infty$ and $\lim\limits_{n\to\infty} \frac{a_n}{b_n} = \lim\limits_{n\to\infty} (-n) = -\infty$;
5. If $a_n = (-1)^n n$ and $b_n = n$, then $\lim\limits_{n\to\infty} a_n = \lim\limits_{n\to\infty} b_n = \infty$ and $\lim\limits_{n\to\infty} \frac{a_n}{b_n} = \lim\limits_{n\to\infty} (-1)^n = \nexists$.

Like in the case of the first indeterminate form, we obtain all possible situations related to the convergence/divergence of the sequence $c_n = \frac{a_n}{b_n}$: the sequence c_n may converge to an arbitrary constant (including 0), may have an infinite limit $\pm\infty$ or may diverge without any general tendency (despite the fact that a_n and b_n converge to ∞ in all the above examples).

To diversify exposition and show that other indeterminate forms can be analyzed similarly, let us consider one more form.

Indeterminate Form $\infty - \infty$

We follow the same line of reasoning. First, the definition of this indeterminate form is as follows: given two sequences a_n and b_n such that $\lim\limits_{n\to\infty} a_n = \lim\limits_{n\to\infty} b_n = \infty$, we would like to know whether there is a definitive result about convergence/divergence of the sequence $c_n = a_n - b_n$. The answer is "no" and we show below different cases that can occur.

Example 3i

1. If $a_n = n$ and $b_n = n$, then $\lim\limits_{n\to\infty} a_n = \lim\limits_{n\to\infty} b_n = \infty$ and $\lim\limits_{n\to\infty} (a_n - b_n) = \lim\limits_{n\to\infty} 0 = 0$;
2. If $a_n = n + C$ and $b_n = n$, where C is an arbitrary constant, then $\lim\limits_{n\to\infty} a_n = \lim\limits_{n\to\infty} b_n = \infty$ and $\lim\limits_{n\to\infty} (a_n - b_n) = \lim\limits_{n\to\infty} C = C$;
3. If $a_n = 2n$ and $b_n = n$, then $\lim\limits_{n\to\infty} a_n = \lim\limits_{n\to\infty} b_n = \infty$ and $\lim\limits_{n\to\infty} (a_n - b_n) = \lim\limits_{n\to\infty} n = +\infty$;
4. If $a_n = n$ and $b_n = 2n$, then $\lim\limits_{n\to\infty} a_n = \lim\limits_{n\to\infty} b_n = \infty$ and $\lim\limits_{n\to\infty} (a_n - b_n) = \lim\limits_{n\to\infty} (-n) = -\infty$;
5. If $a_n = n + (-1)^n$ and $b_n = n$, then $\lim\limits_{n\to\infty} a_n = \lim\limits_{n\to\infty} b_n = \infty$ and $\lim\limits_{n\to\infty} (a_n - b_n) = \lim\limits_{n\to\infty} (-1)^n = \nexists$.

Again, we arrived at the same conclusion as in the two preceding cases. Under the same condition that a_n and b_n have an infinite limit, we have found a variety of situations which can occur with convergence/divergence of the sequence $c_n = a_n - b_n$: the sequence c_n may converge to any constant, may have an infinite limit, or may diverge without any general tendency. This behavior justifies the name of an indeterminate form.

4.2   Techniques of Solution of Indeterminate Forms

To solve the first two indeterminate forms $\frac{0}{0}$ and $\frac{\infty}{\infty}$ there are two principal methods: *division by the highest power* in the case of rational or irrational functions and *L'Hospital's rules* for general functions. Since the latter requires the calculation of derivatives, it is not applied directly to the sequences, but usually there is no problem with transformation of sequences into the corresponding general functions and subsequent application of

L'Hospital's rules. In the cases when the generation of differentiable functions for L'Hospital's rules is problematic, one can apply the *Stolz-Cesàro method* for evaluation of the indeterminate forms $\frac{0}{0}$ and $\frac{\infty}{\infty}$, which is a discrete version of L'Hospital's rules. All the remaining indeterminate forms are usually reduced to one of the first two forms in the process of their solution.

Division by the Highest Power (Method of Rational Functions)

Consider a sequence represented as a ratio of polynomial terms:

$$c_n = \frac{a_n}{b_n} = \frac{n^k + \alpha_{k-1} n^{k-1} + \ldots + \alpha_0}{n^m + \beta_{m-1} n^{m-1} + \ldots + \beta_0},$$

where α_i and β_i are constant coefficients. If n is substituted by a continuous variable x, we have a definition of a rational function (a ratio of two polynomials). It is natural to presume that the limit of c_n depends only on the principal terms—n^k in the numerator and n^m in the denominator. If it is so, then it is sufficient to calculate the limit $\lim\limits_{n \to \infty} \frac{n^k}{n^m} = \lim\limits_{n \to \infty} n^{k-m}$. The last limit does not represent an indeterminate form and its value depends on the relation between the powers k and m: if $k = m$, then $\lim\limits_{n \to \infty} n^{k-m} = 1$; if $k < m$, then $\lim\limits_{n \to \infty} n^{k-m} = 0$; if $k > m$, then $\lim\limits_{n \to \infty} n^{k-m} = +\infty$. Since this case is intimately related to the indeterminate form of rational functions $R(x) = \frac{x^k + \alpha_{k-1} x^{k-1} + \ldots + \alpha_0}{x^m + \beta_{m-1} x^{m-1} + \ldots + \beta_0}$, the method of its solution is frequently called the technique of rational functions.

Our intuitive supposition can be corroborated by strict arguments. Dividing the numerator and denominator by the highest power of the denominator, we transform c_n to the following equivalent form:

$$\frac{a_n}{b_n} = \frac{n^k + \alpha_{k-1} n^{k-1} + \ldots + \alpha_0}{n^m + \beta_{m-1} n^{m-1} + \ldots + \beta_0} = \frac{n^{k-m} + \alpha_{k-1} n^{k-m-1} + \ldots + \alpha_0 n^{-m}}{1 + \beta_{m-1} n^{-1} + \ldots + \beta_0 n^{-m}}.$$

In accordance with the intuitive considerations, we split exact procedure into three cases.

Case 1 If $k = m$, then $k - m = 0$ and

$$\lim\limits_{n \to \infty} \frac{a_n}{b_n} = \lim\limits_{n \to \infty} \frac{1 + \alpha_{k-1} n^{-1} + \ldots + \alpha_0 n^{-m}}{1 + \beta_{m-1} n^{-1} + \ldots + \beta_0 n^{-m}}.$$

The last expression does not contain an indeterminate form—all the limits are finite and that of the denominator is different from zero. Therefore, we can apply arithmetic rules of limits to arrive at the following result:

$$\lim\limits_{n \to \infty} \frac{a_n}{b_n} = \lim\limits_{n \to \infty} \frac{1 + \alpha_{k-1} n^{-1} + \ldots + \alpha_0 n^{-m}}{1 + \beta_{m-1} n^{-1} + \ldots + \beta_0 n^{-m}} = \frac{1 + 0 + \ldots + 0}{1 + 0 + \ldots + 0} = 1.$$

Case 2 If $k < m$, then $k - m = -p < 0$ and

$$\lim_{n\to\infty} \frac{a_n}{b_n} = \lim_{n\to\infty} \frac{1}{n^p} \cdot \frac{1 + \alpha_{k-1} n^{-1} + \ldots + \alpha_0 n^{-k}}{1 + \beta_{m-1} n^{-1} + \ldots + \beta_0 n^{-m}}.$$

In the last expression the problem of an indeterminate form is eliminated—all the limits are finite and that of the denominator is different from zero. Therefore, it is possible to employ the arithmetic rules, that gives the following result:

$$\lim_{n\to\infty} \frac{a_n}{b_n} = \lim_{n\to\infty} \frac{1}{n^p} \cdot \frac{1 + \alpha_{k-1} n^{-1} + \ldots + \alpha_0 n^{-k}}{1 + \beta_{m-1} n^{-1} + \ldots + \beta_0 n^{-m}} = 0 \cdot \frac{1 + 0 + \ldots + 0}{1 + 0 + \ldots + 0} = 0.$$

Case 3 If $k > m$, then $k - m = p > 0$ and

$$\lim_{n\to\infty} \frac{a_n}{b_n} = \lim_{n\to\infty} n^p \cdot \frac{1 + \alpha_{k-1} n^{-1} + \ldots + \alpha_0 n^{-k}}{1 + \beta_{m-1} n^{-1} + \ldots + \beta_0 n^{-m}}.$$

The first factor n^p approaches $+\infty$, and the second is well defined to use the arithmetic rules—all its limits are finite and that of the denominator is different from zero. Therefore, we obtain the following result:

$$\lim_{n\to\infty} \frac{a_n}{b_n} = \lim_{n\to\infty} n^p \cdot \frac{1 + \alpha_{k-1} n^{-1} + \ldots + \alpha_0 n^{-k}}{1 + \beta_{m-1} n^{-1} + \ldots + \beta_0 n^{-m}} = +\infty.$$

Thus, our intuitive considerations were confirmed by the exact calculations of the limits.

Notice that the cases when the leading coefficients are different from 1 are easily reduced to the considered form by normalizing the expressions in the numerator and denominator.

The same technique can also be applied to various irrational expressions.

Example 1p $\lim_{n\to\infty} \frac{n^2-5n+4}{3n-2+4n^2}$

In this example we have the indeterminate form $\frac{\infty}{\infty}$. Dividing the numerator and denominator by n^2, we eliminate an indeterminate form, and using subsequently the arithmetic rules, we calculate the limit:

$$\lim_{n\to\infty} \frac{n^2 - 5n + 4}{3n - 2 + 4n^2} = \lim_{n\to\infty} \frac{1 - \frac{5}{n} + \frac{4}{n^2}}{\frac{3}{n} - \frac{2}{n^2} + 4} = \frac{1 - 0 + 0}{0 - 0 + 4} = \frac{1}{4}.$$

Example 2p $\lim_{n\to\infty} \frac{(n^2-2n+5)^{100}-5n^{20}+4}{\sqrt{1+n^{400}+4n^{100}}}$

In this example we have the indeterminate form $\frac{\infty}{\infty}$. Dividing the numerator and denominator by n^{200}, we reduce the original quotient to the equivalent form, which allows us to apply the arithmetic rules of limits:

$$\lim_{n\to\infty} \frac{(n^2 - 2n + 5)^{100} - 5n^{20} + 4}{\sqrt{1 + n^{400}} + 4n^{100}} = \lim_{n\to\infty} \frac{(1 - \frac{2}{n} + \frac{5}{n^2})^{100} - \frac{5}{n^{180}} + \frac{4}{n^{200}}}{\sqrt{\frac{1}{n^{400}} + 1} + \frac{4}{n^{100}}}$$

$$= \frac{(1 - 0 + 0)^{100} - 0 + 0}{\sqrt{0 + 1} + 0} = 1.$$

Example 3p $\displaystyle \lim_{n\to\infty} \frac{\sqrt{2n^3+3n}}{\sqrt[3]{2-n^5}}$

The indeterminate form $\frac{\infty}{\infty}$ of this example is solved by dividing the numerator and denominator by $n^{\frac{5}{3}}$ and applying subsequently the arithmetic rules of the limits:

$$\lim_{n\to\infty} \frac{\sqrt{2n^3 + 3n}}{\sqrt[3]{2 - n^5}} = \lim_{n\to\infty} \frac{\sqrt{\frac{2n^3+3n}{n^{10/3}}}}{\sqrt[3]{\frac{2-n^5}{n^5}}} = \lim_{n\to\infty} \frac{\sqrt{\frac{2}{n^{1/3}} + \frac{3}{n^{7/3}}}}{\sqrt[3]{\frac{2}{n^5} - 1}} = \frac{\sqrt{0 + 0}}{\sqrt[3]{0 - 1}} = 0.$$

L'Hospital's Rules

L'Hospital's rules are used to solve the indeterminate forms $\frac{0}{0}$ and $\frac{\infty}{\infty}$. Since these rules require calculation of the derivatives, their application to the sequences is usually made by transformation of the original sequences into corresponding general functions of a continuous variable. Below we recall the formulations of two L'Hospital's rules for the limit point $+\infty$: the first is concerned with the indeterminate form $\frac{0}{0}$ and the second with $\frac{\infty}{\infty}$.

Theorem 1 (First L'Hospital's Rule, the Case $\frac{0}{0}$)

Assume $f(x)$ and $g(x)$ are differentiable functions on $(a, +\infty)$ such that

(1) $\displaystyle \lim_{x\to+\infty} g(x) = 0$ *and $g'(x) \neq 0$ on $(a, +\infty)$,*

(2) $\displaystyle \lim_{x\to+\infty} f(x) = 0$,

(3) $\displaystyle \lim_{x\to+\infty} \frac{f'(x)}{g'(x)} = K$ *(where K may be either a finite number or $\pm\infty$),*

then $\displaystyle \lim_{x\to+\infty} \frac{f(x)}{g(x)} = K$.

Theorem 2 (Second L'Hospital's Rule, the Case $\frac{\infty}{\infty}$) *Assume $f(x)$ and $g(x)$ are differentiable functions on $(a, +\infty)$ such that*

(1) $\displaystyle \lim_{x\to+\infty} g(x) = \infty$ *and $g'(x) \neq 0$ on $(a, +\infty)$,*

(2) $\displaystyle \lim_{x\to+\infty} f(x) = \infty$,

(3) $\displaystyle \lim_{x\to+\infty} \frac{f'(x)}{g'(x)} = K$ *(where K may be either a finite number or $\pm\infty$),*

then $\displaystyle \lim_{x\to+\infty} \frac{f(x)}{g(x)} = K$.

Remark In what follows we refer to the first two assumptions in L'Hospital's rules as the preliminary conditions.

A traditional application of these rules to sequences consists of the following steps. First, substitute the sequences a_n and b_n, which give rise to the indeterminate form $\frac{0}{0}$ or $\frac{\infty}{\infty}$ in the sequence $c_n = \frac{a_n}{b_n}$, by the corresponding functions $f(x)$ and $g(x)$ such that $f(n) = a_n$ and $g(n) = b_n$, $\forall n$ (these relations between the sequences and functions can hold starting from some index N and the functions $f(x)$ and $g(x)$ can be defined in an interval $[a, +\infty)$, $a \geq 1$). This substitution should preserve the essential condition of the tendency of the functions: if $\lim\limits_{n\to\infty} a_n = \lim\limits_{n\to\infty} b_n = 0$ then $\lim\limits_{x\to+\infty} f(x) = \lim\limits_{x\to+\infty} g(x) = 0$; and if $\lim\limits_{n\to\infty} a_n = \lim\limits_{n\to\infty} b_n = \infty$ then $\lim\limits_{x\to+\infty} f(x) = \lim\limits_{x\to+\infty} g(x) = \infty$. Next, try to apply one of L'Hospital's rules, and if a specific result $\lim\limits_{x\to+\infty} \frac{f(x)}{g(x)} = K$ is found, then (by the property of the relationship between the limits of sequences and generating functions, Sect. 3.1) this implies that $\lim\limits_{n\to\infty} \frac{a_n}{b_n} = K$.

Let us consider some examples.

Example 1h $\lim\limits_{n\to\infty} \frac{n^2-5n+4}{3n-2+4n^2}$

This limit can be calculated using the second L'Hospital's rule. First, we introduce the functions $f(x) = x^2 - 5x + 4$ and $g(x) = 3x - 2 + 4x^2$ such that $f(n) = a_n$ and $g(n) = b_n$, and verify that the preliminary conditions of the second rule hold: the functions $f(x)$ and $g(x)$ are differentiable and $g'(x) = 3 + 8x \neq 0$ in $[1, +\infty)$; besides, $\lim\limits_{x\to+\infty} f(x) = \lim\limits_{x\to+\infty} g(x) = \infty$. Now we evaluate the limit in the principal condition of the rule: $\lim\limits_{x\to+\infty} \frac{f'(x)}{g'(x)} = \lim\limits_{x\to+\infty} \frac{2x-5}{3+8x}$. This limit is still the indeterminate form of the same type. However, we can apply to the last limit the second rule one more time: the functions $f'(x)$ and $g'(x)$ are differentiable and $g''(x) = 8 \neq 0$ in $[1, +\infty)$; besides, $\lim\limits_{x\to+\infty} f'(x) = \lim\limits_{x\to+\infty} g'(x) = \infty$. Calculating now the limit $\lim\limits_{x\to+\infty} \frac{f''(x)}{g''(x)} = \lim\limits_{x\to+\infty} \frac{2}{8} = \frac{1}{4}$ we find the definitive value. Therefore, by the second rule, $\lim\limits_{x\to+\infty} \frac{f'(x)}{g'(x)} = \lim\limits_{x\to+\infty} \frac{f''(x)}{g''(x)} = \frac{1}{4}$, and using the same rule one more time, we conclude that $\lim\limits_{x\to+\infty} \frac{f(x)}{g(x)} = \lim\limits_{x\to+\infty} \frac{f'(x)}{g'(x)} = \frac{1}{4}$. Then, it follows that $\lim\limits_{n\to\infty} \frac{n^2-5n+4}{3n-2+4n^2} = \frac{1}{4}$.

Recall that, in practice, we start with formal attempts to calculate the limit of the ratio of derivatives. If we arrive at a definite result, we return to justification of the use of L'Hospital's rules. Otherwise, we try to find the limit of the ratio of the second derivatives, and so on, expecting to obtain a definite result on one of the steps of these attempts, or conclude that L'Hospital's rules are not applicable.

Example 2h $\lim\limits_{n\to\infty} \dfrac{\sqrt{2n^3+3n}}{\sqrt[3]{2-n^5}}$

This limit was solved in Example 3p using polynomial method, but L'Hospital's rule is not applicable in this case. Indeed, defining the corresponding functions $f(x) = \sqrt{2x^3 + 3x}$ and $g(x) = \sqrt[3]{2 - x^5}$, and trying to calculate the limit of derivatives

$$\lim_{x\to+\infty} \frac{f'(x)}{g'(x)} = \lim_{x\to+\infty} \frac{\frac{1}{2}(2x^3 + 3x)^{-1/2}(6x^2 + 3)}{\frac{1}{3}(2 - x^5)^{-2/3}(-5x^4)} = \lim_{x\to+\infty} \frac{3(6x^2 + 3)\sqrt[3]{(2 - x^5)^2}}{(-10x^4)\sqrt{2x^3 + 3x}},$$

we notice that the limit expression became more complex rather than simpler, still keeping the same indeterminate form. If we proceed, the situation will not be better: the expressions for derivatives are more complicated than for functions themselves.

This Example shows that L'Hospital's rules are not universal, despite the fact that they are a very powerful (and very popular) tools which allow us to solve many limits when other techniques are useless.

Example 3h $\lim\limits_{n\to\infty} \dfrac{n^2}{2^n}$

In this Example we have the indeterminate form $\frac{\infty}{\infty}$ and the method of rational functions is not applicable. However, L'Hospital's rule works well. Introduce the functions $f(x) = x^2$ and $g(x) = 2^x$, and notice that the preliminary conditions of the second L'Hospital's rule are satisfied. Then, calculate the limit of the derivatives twice (reducing the potential function to a constant)

$$\lim_{x\to+\infty} \frac{f(x)}{g(x)} = \lim_{x\to+\infty} \frac{x^2}{2^x} = \lim_{x\to+\infty} \frac{2x}{2^x \ln 2} = \lim_{x\to+\infty} \frac{2}{2^x \ln^2 2} = 0$$

and notice that in the last limit an indeterminate form is already eliminated. This justifies the successive (twice) application of L'Hospital's rule and allow us to conclude that $\lim\limits_{n\to\infty} \frac{n^2}{2^n} = 0$. Of course, this result was expected, because this limit represents the comparison of the growth rates of a polynomial (in the numerator) and exponential (in the denominator) functions.

Example 4h $\lim\limits_{n\to\infty} \dfrac{\ln^3 n}{\sqrt{n}}$

This limit is the indeterminate form $\frac{\infty}{\infty}$. Again the method of rational functions does not work, but the second L'Hospital's rule solves the problem. Define the functions $f(x) = \ln^3 x$ and $g(x) = \sqrt{x}$ and notice that the preliminary conditions of the second L'Hospital's rule hold. Then, calculate the limit of the derivatives three times in row (to eliminate the logarithmic function), simplifying expressions inside the limits after each differentiation:

$$\lim_{x \to +\infty} \frac{f(x)}{g(x)} = \lim_{x \to +\infty} \frac{\ln^3 x}{\sqrt{x}} = \lim_{x \to +\infty} \frac{3 \ln^2 x \cdot \frac{1}{x}}{\frac{1}{2} x^{-1/2}} = \lim_{x \to +\infty} \frac{6 \ln^2 x}{\sqrt{x}}$$

$$= \lim_{x \to +\infty} \frac{12 \ln x \cdot \frac{1}{x}}{\frac{1}{2} x^{-1/2}} = \lim_{x \to +\infty} \frac{24 \ln x}{\sqrt{x}} = \lim_{x \to +\infty} \frac{24 \cdot \frac{1}{x}}{\frac{1}{2} x^{-1/2}} = \lim_{x \to +\infty} \frac{48}{\sqrt{x}} = 0.$$

In the last limit an indeterminate form is already eliminated and the arithmetic rules are applied. This procedure justifies successive (thrice) application of L'Hospital's rule and leads to the conclusion that $\lim_{n \to \infty} \frac{\ln^3 n}{\sqrt{n}} = 0$. This result was expected, because this limit compares the growth rates of a logarithmic (in the numerator) and polynomial (in the denominator) functions.

Example 5h $\lim_{n \to \infty} \frac{\sin \frac{1}{n}}{\ln(1 - \frac{1}{n})}$

This limit represents the indeterminate form $\frac{0}{0}$. The technique of rational functions is not applicable and we appeal to the first L'Hospital's rule. Define the functions $f(x) = \sin \frac{1}{x}$ and $g(x) = \ln(1 - \frac{1}{x})$, and verify that the preliminary conditions of the first L'Hospital's rule are satisfied in $[2, +\infty)$. Then, calculate the limit of the derivative:

$$\lim_{x \to +\infty} \frac{f(x)}{g(x)} = \lim_{x \to +\infty} \frac{\sin \frac{1}{x}}{\ln(1 - \frac{1}{x})} = \lim_{x \to +\infty} \frac{\cos \frac{1}{x} \cdot \frac{-1}{x^2}}{\frac{1}{1 - 1/x} \cdot \frac{1}{x^2}} = - \lim_{x \to +\infty} \cos \frac{1}{x} \cdot (1 - \frac{1}{x}) = -1.$$

Since the last limit does not contain an indeterminate form, it can be solved by the arithmetic rules. This procedure justifies the application of L'Hospital's rule and shows that $\lim_{n \to \infty} \frac{\sin \frac{1}{n}}{\ln(1 - \frac{1}{n})} = -1$.

The Stolz-Cesàro Theorems

One more useful tool for evaluating indeterminate forms $\frac{0}{0}$ and $\frac{\infty}{\infty}$ is the Stolz-Cesàro theorem, which is a discrete version of L'Hospital's rule.

Theorem 1 (The First Stolz-Cesàro Theorem, the Case $\frac{0}{0}$) *If the sequences a_n and b_n are such that*

(1) $\lim_{n \to +\infty} b_n = 0$ *and b_n is strictly monotone,*

(2) $\lim_{n \to +\infty} a_n = 0$,

(3) $\lim_{n \to +\infty} \frac{a_{n+1} - a_n}{b_{n+1} - b_n} = L$ *(where L may be either a finite number or $\pm\infty$),*

then $\lim_{n \to +\infty} \frac{a_n}{b_n} = L$.

Proof It is sufficient to consider the case of a strictly increasing sequence b_n, because the decreasing case is treated in the same way.

Assume first that L is a finite number. Since $\lim\limits_{n\to+\infty}\frac{a_{n+1}-a_n}{b_{n+1}-b_n}=L$, for $\forall\varepsilon_0>0$ (choose $\varepsilon_0=\frac{\varepsilon}{2}$ for convenience) there exists N such that $\forall n>N$ it follows that $\left|\frac{a_{n+1}-a_n}{b_{n+1}-b_n}-L\right|<\varepsilon_0$, or equivalently, $L-\varepsilon_0<\frac{a_{n+1}-a_n}{b_{n+1}-b_n}<L+\varepsilon_0$. Taking into account that b_n is a strictly increasing sequence, the last inequality can be written as follows:

$$(L-\varepsilon_0)(b_{n+1}-b_n)<a_{n+1}-a_n<(L+\varepsilon_0)(b_{n+1}-b_n),\ \forall n>N.$$

Now we take any such $n>N$ and fix it. Then we sum up all these inequalities for the indices $n,n+1,\dots,m-1$, where $m>n+1$, and obtain

$$(L-\varepsilon_0)\sum_{k=n}^{m-1}(b_{k+1}-b_k)=(L-\varepsilon_0)(b_m-b_n)<\sum_{k=n}^{m-1}(a_{n+1}-a_n)=a_m-a_n$$

$$<(L+\varepsilon_0)(b_m-b_n)=(L+\varepsilon_0)\sum_{k=n}^{m-1}(b_{k+1}-b_k),\ \forall m>n+1.$$

Since the sequences a_m and b_m converge to 0, taking the limit as $m\to+\infty$ in the last inequality, we get $(L-\varepsilon_0)(-b_n)\le-a_n\le(L+\varepsilon_0)(-b_n)$ or $L-\varepsilon<L-\varepsilon_0\le\frac{a_n}{b_n}\le L+\varepsilon_0<L+\varepsilon$ (notice that $b_n<0$ because b_n is increasing and convergent to 0). Since the last inequality holds for any $n>N$, we arrive at the definition of the limit $\lim\limits_{n\to+\infty}\frac{a_n}{b_n}=L$.

Consider now the case $L=+\infty$. Write down the definition of $\lim\limits_{n\to+\infty}\frac{a_{n+1}-a_n}{b_{n+1}-b_n}=+\infty$: for $\forall E>0$ there exists N such that $\forall n>N$ it follows that $\frac{a_{n+1}-a_n}{b_{n+1}-b_n}>E$, or equivalently, $a_{n+1}-a_n>E(b_{n+1}-b_n)$ (recall that $b_{n+1}-b_n>0$). Fix any such $n>N$ and sum up all these inequalities for $k=n,n+1,\dots,m-1$, where $m>n+1$, to get $a_m-a_n>E(b_m-b_n)$. Apply now the limit as $m\to+\infty$ to the last inequality to obtain $-a_n\ge-Eb_n$ (recall that $a_m\xrightarrow[m\to+\infty]{}0$ and $b_m\xrightarrow[m\to+\infty]{}0$). Since $b_n<0$, the last inequality is equivalent to $\frac{a_n}{b_n}\ge E$, which means that $\lim\limits_{n\to+\infty}\frac{a_n}{b_n}=+\infty$.

The case $L=-\infty$ is similar to $L=+\infty$. $\square$

Theorem 2 (The Second Stolz-Cesàro Theorem, the Case $\frac{\cdot}{\infty}$) *If the sequences a_n and b_n are such that*

(1) $\lim\limits_{n\to+\infty}b_n=+\infty$ *and b_n is strictly increasing,*

(2) $\lim\limits_{n\to+\infty}\frac{a_{n+1}-a_n}{b_{n+1}-b_n}=L$ *(where L may be either a finite number or $\pm\infty$),*

then $\lim\limits_{n\to+\infty}\frac{a_n}{b_n}=L$.

Proof Let us consider first the case when L is a finite number. Since $\lim\limits_{n\to+\infty}\frac{a_{n+1}-a_n}{b_{n+1}-b_n}=L$, for $\forall\varepsilon_0>0$ (choose $\varepsilon_0=\frac{\varepsilon}{2}$ for convenience) there exists N_0 such that for $\forall n>N_0$ it follows that $\left|\frac{a_{n+1}-a_n}{b_{n+1}-b_n}-L\right|<\varepsilon_0$, or equivalently, $L-\varepsilon_0<\frac{a_{n+1}-a_n}{b_{n+1}-b_n}<L+\varepsilon_0$. Taking into account that b_n is strictly increasing sequence, the last inequality can be written as follows:

$$(L-\varepsilon_0)(b_{n+1}-b_n)<a_{n+1}-a_n<(L+\varepsilon_0)(b_{n+1}-b_n),\ \forall n>N_0.$$

Summing these inequalities for $n=N_0+1,\ldots,m-1$, where $m>N_0+1$ (from now on we consider that N_0 is fixed), we obtain

$$(L-\varepsilon_0)\sum_{n=N_0+1}^{m-1}(b_{n+1}-b_n)=(L-\varepsilon_0)(b_m-b_{N_0+1})<\sum_{n=N_0+1}^{m-1}(a_{n+1}-a_n)=a_m-a_{N_0+1}$$

$$<(L+\varepsilon_0)(b_m-b_{N_0+1})=(L+\varepsilon_0)\sum_{n=N_0+1}^{m-1}(b_{n+1}-b_n),\ \forall m>N_0+1.$$

Dividing by $b_m>0$ and regrouping the terms, we can also rewrite this result in the form

$$-\varepsilon_0\left(1-\frac{b_{N_0+1}}{b_m}\right)+\frac{a_{N_0+1}-Lb_{N_0+1}}{b_m}<\frac{a_m}{b_m}-L<\varepsilon_0\left(1-\frac{b_{N_0+1}}{b_m}\right)+\frac{a_{N_0+1}-Lb_{N_0+1}}{b_m}.$$

Since a_{N_0+1}, L and b_{N_0+1} are fixed numbers, and $\lim\limits_{m\to+\infty}b_m=+\infty$, for $\forall\varepsilon_1>0$ (choose $\varepsilon_1=\frac{\varepsilon}{2}$ for convenience) there exists $N>N_0+1$ such that $\left|\frac{a_{N_0+1}-Lb_{N_0+1}}{b_m}\right|<\varepsilon_1$ whenever $m>N$. Taking additionally into account that $0<1-\frac{b_{N_0+1}}{b_m}<1$ (since $b_m>b_{N_0+1}$), the double inequality for $\frac{a_m}{b_m}-L$ can be written in the form

$$-\frac{\varepsilon}{2}-\frac{\varepsilon}{2}<-\varepsilon_0\left(1-\frac{b_{N_0+1}}{b_m}\right)+\frac{a_{N_0+1}-Lb_{N_0+1}}{b_m}$$

$$<\frac{a_m}{b_m}-L<\varepsilon_0\left(1-\frac{b_{N_0+1}}{b_m}\right)+\frac{a_{N_0+1}-Lb_{N_0+1}}{b_m}<\frac{\varepsilon}{2}+\frac{\varepsilon}{2},$$

that is, $\left|\frac{a_m}{b_m}-L\right|<\varepsilon$, for $\forall m>N$. This means that $\lim\limits_{m\to+\infty}\frac{a_m}{b_m}=L$.

The case $L=+\infty$ can be reduced to the analyzed above case of a finite limit. Indeed, if $\lim\limits_{n\to+\infty}\frac{a_{n+1}-a_n}{b_{n+1}-b_n}=+\infty$, it follows that $a_{n+1}-a_n>b_{n+1}-b_n$ (for sufficiently large n), and consequently, a_n is strictly increasing sequence and $\lim\limits_{n\to+\infty}a_n=+\infty$. Then, we can apply the proved statement for a finite limit to the inverse quotient $\frac{b_n}{a_n}$: $\lim\limits_{n \to +\infty} \frac{b_n}{a_n} =$

$\lim\limits_{n \to +\infty} \frac{b_{n+1}-b_n}{a_{n+1}-a_n} = 0$. Therefore, $\lim\limits_{n \to +\infty} \frac{a_n}{b_n} = +\infty$.

The case $L = -\infty$ is similar to $L = +\infty$. $\qquad\qquad\qquad\qquad\qquad\qquad\quad \square$

Remark The converse to the Stolz-Cesàro theorem is false (for both cases $\frac{0}{0}$ and $\frac{\cdot}{\infty}$). For instance, in the case $\frac{\infty}{\infty}$, consider the sequences $a_n = n + (-1)^n$ and $b_n = n$ such that b_n is strictly increasing, $\lim\limits_{n \to +\infty} b_n = +\infty$ and $\lim\limits_{n \to +\infty} \frac{a_n}{b_n} = \lim\limits_{n \to +\infty} (1 + \frac{(-1)^n}{n}) = 1$. Nevertheless, the limit of $\frac{a_{n+1}-a_n}{b_{n+1}-b_n} = 1 - 2 \cdot (-1)^n$ does not exist.

In the case $\frac{0}{0}$, we can construct a similar example: the sequences $a_n = \frac{1}{n+(-1)^n}$ and $b_n = \frac{1}{n}$ converge to 0, the latter is strictly decreasing and

$$\lim\limits_{n \to +\infty} \frac{a_n}{b_n} = \lim\limits_{n \to +\infty} \frac{n}{n + (-1)^n} = \lim\limits_{n \to +\infty} \frac{1}{1 + \frac{(-1)^n}{n}} = 1.$$

At the same time,

$$\frac{a_{n+1} - a_n}{b_{n+1} - b_n} = \frac{\frac{1}{n+1+(-1)^{n+1}} - \frac{1}{n+(-1)^n}}{\frac{1}{n+1} - \frac{1}{n}}$$

$$= \frac{(n+1)n}{(n+1+(-1)^{n+1})(n+(-1)^n)} \cdot (1 - 2 \cdot (-1)^n) = \frac{1 + \frac{1}{n}}{\left(1 + \frac{1}{n} + \frac{(-1)^{n+1}}{n}\right)\left(1 + \frac{(-1)^n}{n}\right)} \cdot (1 - 2 \cdot (-1)^n).$$

Since the limit of the first factor (the ratio) exists and equals 1, but the limit of the second $(1 - 2 \cdot (-1)^n)$ does not exist, the product has no limit.

Some examples of application of the Stolz-Cesàro theorems are presented below.

Example 1s Using the second Stolz-Cesàro theorem we can easily show an extended version of Properties 8s and 9s about arithmetic and geometric means. The former can be formulated as follows: if $\lim\limits_{n \to +\infty} a_n = a$ (where a may be either a finite number or $\pm\infty$), then the sequence of arithmetic means $b_n = \frac{a_1 + \ldots + a_n}{n}$ has the same limit $\lim\limits_{n \to +\infty} b_n = a$. Indeed, setting the sequences $p_n = a_1 + \ldots + a_n$ and $q_n = n$ for the second Stolz-Cesàro theorem, we have $\lim\limits_{n \to +\infty} \frac{p_{n+1}-p_n}{q_{n+1}-q_n} = \lim\limits_{n \to +\infty} \frac{a_{n+1}}{1} = a$, and consequently, $\lim\limits_{n \to +\infty} b_n = \lim\limits_{n \to +\infty} \frac{p_n}{q_n} = a$.

For the geometric means we have: if $a_n > 0$ and $\lim\limits_{n \to +\infty} a_n = a > 0$ (where a may be either a finite number or $+\infty$), then the sequence of geometric means $c_n = \sqrt[n]{a_1 \ldots a_n}$ has the same limit $\lim\limits_{n \to +\infty} c_n = a$. To prove this property, we reduce the geometric means to arithmetic ones: the sequence $\ln a_n$ has the limit $\ln a$ ($\ln a = +\infty$ if $a =$

$+\infty$), and applying the property of arithmetic means, we conclude that $\lim\limits_{n\to+\infty} \ln c_n = \lim\limits_{n\to+\infty} \frac{1}{n}(\ln a_1 + \ldots + \ln a_n) = \ln a$. Therefore, $\lim\limits_{n\to+\infty} c_n = a$.

Example 2s $\lim\limits_{n\to+\infty} \frac{\ln n}{n}$

To solve the indeterminate form $\frac{\infty}{\infty}$, we can employ the second Stolz-Cesàro theorem with $a_n = \ln n$ and $b_n = n$. Since the latter sequence is strictly increasing and divergent, and $\lim\limits_{n\to+\infty} \frac{a_{n+1}-a_n}{b_{n+1}-b_n} = \lim\limits_{n\to+\infty} \frac{\ln\frac{n+1}{n}}{1} = 0$, it follows from the Stolz-Cesàro theorem that $\lim\limits_{n\to+\infty} \frac{a_n}{b_n} = 0$. In practice, assuming that the ratio $\frac{a_{n+1}-a_n}{b_{n+1}-b_n}$ has some limit value, we write directly:

$$\lim_{n\to+\infty} \frac{\ln n}{n} = \lim_{n\to+\infty} \frac{\ln(n+1) - \ln n}{n+1-n} = \lim_{n\to+\infty} \frac{\ln\frac{n+1}{n}}{1} = 0.$$

Of course, this limit can also be easily solved by L'Hospital's rule.

Example 3s $\lim\limits_{n\to+\infty} \frac{n^2}{2^n}$

This limit, representing the indeterminate form $\frac{\infty}{\infty}$, was solved in Example 3h by L'Hospital's rule. We will see that the application of the second Stolz-Cesàro theorem leads to very similar calculations. Setting $a_n = n^2$ and $b_n = 2^n$ (the latter sequence is strictly increasing and divergent), instead of $\lim\limits_{n\to+\infty} \frac{a_n}{b_n}$ we consider $\lim\limits_{n\to+\infty} \frac{a_{n+1}-a_n}{b_{n+1}-b_n} = \lim\limits_{n\to+\infty} \frac{2n+1}{2^n}$. Since the indeterminate form was not eliminated yet, we apply the Stolz-Cesàro theorem once more with the sequences $\tilde{a}_n = 2n+1$ and $\tilde{b}_n = 2^n$, which yields $\lim\limits_{n\to+\infty} \frac{\tilde{a}_{n+1}-\tilde{a}_n}{\tilde{b}_{n+1}-\tilde{b}_n} = \lim\limits_{n\to+\infty} \frac{2}{2^n} = 0$. Therefore,

$$\lim_{n\to+\infty} \frac{a_n}{b_n} = \lim_{n\to+\infty} \frac{a_{n+1}-a_n}{b_{n+1}-b_n} = \lim_{n\to+\infty} \frac{\tilde{a}_n}{\tilde{b}_n} = \lim_{n\to+\infty} \frac{\tilde{a}_{n+1}-\tilde{a}_n}{\tilde{b}_{n+1}-\tilde{b}_n} = \lim_{n\to+\infty} \frac{2}{2^n} = 0.$$

In practice, expecting that the final limit will give a conclusive result, we apply the Stolz-Cesàro theorem repeatedly without special notations:

$$\lim_{n\to+\infty} \frac{n^2}{2^n} = \lim_{n\to+\infty} \frac{(n+1)^2 - n^2}{2^{n+1} - 2^n} = \lim_{n\to+\infty} \frac{2n+1}{2^n} = \lim_{n\to+\infty} \frac{2(n+1)+1-(2n+1)}{2^{n+1}-2^n}$$
$$= \lim_{n\to+\infty} \frac{2}{2^n} = 0.$$

Example 4s $\lim\limits_{n\to+\infty} \frac{1^k + 2^k + \ldots + n^k}{n^{k+1}}, k \in \mathbb{N}$

Considering $a_n = 1^k + 2^k + \ldots + n^k$ and $b_n = n^{k+1}$ (the latter sequence is strictly increasing and divergent) in the Stolz-Cesàro theorem, we have to calculate the limit

$\lim\limits_{n\to+\infty} \frac{a_{n+1}-a_n}{b_{n+1}-b_n} = \lim\limits_{n\to+\infty} \frac{(n+1)^k}{(n+1)^{k+1}-n^{k+1}}$. Using the binomial theorem in the denominator, we continue as follows:

$$\lim_{n\to+\infty} \frac{(n+1)^k}{(n+1)^{k+1}-n^{k+1}} = \lim_{n\to+\infty} \frac{(n+1)^k}{\left(1+\binom{k+1}{1}n+\ldots+\binom{k+1}{k}n^k+n^{k+1}\right)-n^{k+1}}$$

$$= \lim_{n\to+\infty} \frac{(1+\frac{1}{n})^k}{\frac{1}{n^k}+\binom{k+1}{1}\frac{1}{n^{k-1}}+\ldots+\binom{k+1}{k}} = \frac{1}{k+1}.$$

Therefore, $\lim\limits_{n\to+\infty} \frac{a_n}{b_n} = \frac{1}{k+1}$.

Example 5s $\lim\limits_{n\to+\infty} n \sum_{k=n}^{2n} \frac{1}{k^2}$

Notice first that $a_n = \sum_{k=n}^{2n} \frac{1}{k^2} < (n+1)\frac{1}{n^2} \underset{n\to+\infty}{\to} 0$. Therefore, the limit is the

indeterminate form $\infty \cdot 0$. Writing the original limit in the form $\lim\limits_{n\to+\infty} \frac{\sum_{k=n}^{2n}\frac{1}{k^2}}{\frac{1}{n}}$ we get

the indeterminate form $\frac{0}{0}$ and to solve the last limit we can apply the first Stolz-Cesàro theorem with a_n defined above and $b_n = \frac{1}{n}$:

$$\lim_{n\to+\infty} \frac{a_n}{b_n} = \lim_{n\to+\infty} \frac{a_{n+1}-a_n}{b_{n+1}-b_n} = \lim_{n\to+\infty} \frac{\frac{1}{(2n+1)^2}+\frac{1}{(2n+2)^2}-\frac{1}{n^2}}{-\frac{1}{n(n+1)}}$$

$$= \lim_{n\to+\infty} n(n+1)\left(\frac{1}{n^2}-\frac{1}{(2n+1)^2}-\frac{1}{(2n+2)^2}\right) = 1-\frac{1}{4}-\frac{1}{4} = \frac{1}{2}.$$

Historical Remarks 1 L'Hospital's rule was first formulated for the case $\frac{0}{0}$ in "Analyse des infiniment petits", the first textbook on the differential calculus, written by L'Hospital in 1696. For a long time it was considered that this is the original work by L'Hospital, including the rule for the indeterminate form $\frac{0}{0}$, despite the claims of Johann Bernoulli that the book was essentially a publication of his course notes. Although Bernoulli had an outstanding reputation as a mathematician and it was well-known that he was tutoring L'Hospital in analysis, his objections were not taken too seriously because of many other disputes he had with colleagues. Only in 1920 a manuscript copy of the course on differential calculus prepared by Johann Bernoulli for his lectures in Basel university (certainly before 1694) was found in the Basel university library and it was seen how closely the book followed the course notes. Nevertheless, the name "L'Hospital rules" was already coined in mathematical world as well as the authorship of the first textbook on the differential calculus.

Historical Remarks 2 The case $\frac{\infty}{\infty}$ of the Stolz-Cesàro theorem is stated and proved in Stolz's 1885 book and also in Cesàro's 1888 article. For the special choice of $b_n = n$ it was presented by Cauchy in 1821 "Cours d'analyse".

4.3   Various Indeterminate Forms and Examples

In practical applications of L'Hospital's rules, the substitution of a sequence by the corresponding function is frequently does not performed explicitly, noting that the function formula usually just repeats the form of the sequence with x substituted for n. For this reason, to shorten the formula writings, L'Hospital's rule is applied directly to the sequence, considering implicitly that for the purpose of differentiation n is a continuous variable, and in other steps of the problem solution n is treated as a discrete variable (index) of the sequence domain. In the following examples we use this artifice in the cases when L'Hospital's rule is applied.

Example 1v $\lim\limits_{n \to \infty} \left(\sqrt{n^2 + 2n} - \sqrt{n^2 + 1} \right)$

In this example we have the indeterminate form $\infty - \infty$. To solve the problem, we reduce this form to the indeterminate form $\frac{\infty}{\infty}$, divide by the highest power to eliminate the indeterminate form and finally apply the arithmetic rules to obtain the result:

$$\lim_{n \to \infty} \left(\sqrt{n^2 + 2n} - \sqrt{n^2 + 1} \right) = \lim_{n \to \infty} \frac{(\sqrt{n^2 + 2n} - \sqrt{n^2 + 1})(\sqrt{n^2 + 2n} + \sqrt{n^2 + 1})}{\sqrt{n^2 + 2n} + \sqrt{n^2 + 1}}$$

$$= \lim_{n \to \infty} \frac{2n - 1}{\sqrt{n^2 + 2n} + \sqrt{n^2 + 1}} = \lim_{n \to \infty} \frac{2 - \frac{1}{n}}{\sqrt{1 + \frac{2}{n}} + \sqrt{1 + \frac{1}{n^2}}} = \frac{2 - 0}{\sqrt{1 + 0} + \sqrt{1 + 0}} = 1.$$

Example 2v $\lim\limits_{n \to \infty} n \sin \frac{1}{n}$

The indeterminate form $0 \cdot \infty$ of this example can be solved by reducing it to the form $\frac{0}{0}$ and then applying the first remarkable limit with the variable $x = \frac{1}{n} \underset{n \to \infty}{\to} 0$ (we make this without explicit use of the continuous variable x):

$$\lim_{n \to \infty} n \sin \frac{1}{n} = \lim_{n \to \infty} \frac{\sin \frac{1}{n}}{\frac{1}{n}} = 1.$$

Example 3v $\lim\limits_{n \to \infty} \frac{1}{n} \sin n$

There is no indeterminate form in this example, since $\frac{1}{n} \underset{n \to \infty}{\to} 0$ and $\sin n$ has no limit but is a bounded sequence. Taking into account the boundedness of $\sin n$, we evaluate $|\frac{\sin n}{n}| \leq \frac{1}{n}$, and using the fact that the right-hand side of this inequality converges to 0, we apply the squeeze Theorem to conclude that $|\frac{\sin n}{n}| \underset{n\to\infty}{\to} 0$. Finally, the property of absolute value implies that $\frac{\sin n}{n} \underset{n\to\infty}{\to} 0$.

Example 4v $\lim\limits_{n\to\infty} \frac{n!}{n^n}$

This limit is the indeterminate form $\frac{\infty}{\infty}$, but none of the considered techniques of solution is applicable. Nevertheless, a simple specific procedure can be carried out to solve this case. For any n we have the following inequality:

$$0 < \frac{n!}{n^n} = \frac{1\cdot 2\cdot\ldots\cdot n}{n\cdot n\cdot\ldots\cdot n} < \frac{1}{n}.$$

The sequence on the right-hand side converges to 0, and consequently, by the squeeze Theorem, $\lim\limits_{n\to\infty} \frac{n!}{n^n} = 0$.

Example 5v $\lim\limits_{n\to\infty} \sqrt[n]{n}$

This limit is the indeterminate form ∞^0. It can be calculated by reducing to the indeterminate form $\frac{\infty}{\infty}$ through application of the logarithmic function: $\ln \sqrt[n]{n} = \frac{\ln n}{n}$. The limit of the last quotient is easy to find using the second L'Hospital's rule (continuous variable is not introduced here explicitly):

$$\lim_{n\to\infty} \ln \sqrt[n]{n} = \lim_{n\to\infty} \frac{\ln n}{n} = \lim_{n\to\infty} \frac{\frac{1}{n}}{1} = \lim_{n\to\infty} \frac{1}{n} = 0.$$

To return to the original limit, we use the property of a composed function with the exponential function e^x:

$$\lim_{n\to\infty} \sqrt[n]{n} = \lim_{n\to\infty} e^{\ln \sqrt[n]{n}} = e^0 = 1.$$

Example 6v $\lim\limits_{n\to\infty} \frac{2^n+(-1)^n}{2^{n+1}+(-1)^{n+1}}$

In this example we have the indeterminate form $\frac{\infty}{\infty}$. Although the corresponding continuous function is neither rational nor irrational, we still can use the method of division by the principal term:

$$\lim_{n\to\infty} \frac{2^n + (-1)^n}{2^{n+1} + (-1)^{n+1}} = \lim_{n\to\infty} \frac{1 + \frac{(-1)^n}{2^n}}{2 + \frac{(-1)^{n+1}}{2^n}} = \frac{1+0}{2+0} = \frac{1}{2}.$$

Example 7v $\lim\limits_{n\to\infty} \left(1 + \sin\frac{2}{n}\right)^{3n}$

This limit is the indeterminate form 1^∞. A general line of solution consists of reducing the original limit to the second remarkable limit and use the analytic properties of limits. First we rewrite the given limit in the form

$$\lim_{n\to\infty}\left(1+\sin\frac{2}{n}\right)^{3n}=\lim_{n\to\infty}\left(\left(1+\sin\frac{2}{n}\right)^{\frac{1}{\sin\frac{2}{n}}}\right)^{\sin\frac{2}{n}\cdot3n}.$$

Now we calculate separately the two auxiliary limits. The first one we reduce to the second remarkable limit with the variable $x=\sin\frac{2}{n}\underset{n\to\infty}{\to}0$:

$$\lim_{n\to\infty}\left(1+\sin\frac{2}{n}\right)^{\frac{1}{\sin\frac{2}{n}}}=\lim_{x\to0}(1+x)^{\frac{1}{x}}=e.$$

The second limit we solve by applying the first remarkable limit with $t=\frac{2}{n}\underset{n\to\infty}{\to}0$:

$$\lim_{n\to\infty}\sin\frac{2}{n}\cdot3n=\lim_{t\to0}6\frac{\sin t}{t}=6.$$

Since both auxiliary limits are finite and the first one is positive, we can apply the analytic properties of limits to obtain

$$\lim_{n\to\infty}\left(1+\sin\frac{2}{n}\right)^{3n}=\lim_{n\to\infty}\left(\left(1+\sin\frac{2}{n}\right)^{\frac{1}{\sin\frac{2}{n}}}\right)^{\sin\frac{2}{n}\cdot3n}=e^{6}.$$

Exercises

1. Verify if the sequence is bounded, monotone and convergent using the definition:
   (1) $a_n=(-2)^n$;
   (2) $a_n=\frac{(-2)^n}{n}$;
   (3) $a_n=\left(-\frac{1}{2}\right)^n$;
   (4) $a_n=\frac{4}{n+6}$;
   (5) $a_n=\frac{n}{n+1}$;
   (6) $a_n=\frac{4n-3}{3n+2}$;
   (7) $a_n=\frac{3n^2+1}{6n^2+5}$;
   (8) $a_n=\frac{n^2-3}{n+2}$;
   (9) $a_n=\cos\frac{\pi n}{4}$;
   (10) $a_n=\tan\frac{\pi n}{3}$.

2. Let a_n be a convergent sequence and b_n divergent. What can be said about the sequence:
   (1) $a_n + b_n$;
   (2) $a_n \cdot b_n$.

3. Let a_n and b_n be divergent sequences. What can be said about the sequence:
   (1) $a_n + b_n$;
   (2) $a_n \cdot b_n$.

4. Let $\lim_{n \to \infty} a_n b_n = 0$. Does it imply that at least one of the sequences a_n or b_n converges to 0?

5. Show that the limit exists or not and find its value if it exists:
   (1) $a_n = \frac{2-n^2}{2n^2+1}$;
   (2) $a_n = \frac{2n^3+5n-7}{3n^3+6n^2+n}$;
   (3) $a_n = \frac{\sqrt{n^6+2n^2+3}+\sqrt[3]{n^9+7n^5+n}}{2n^3-5n^2+3n+1}$;
   (4) $a_n = \sqrt{n+2} - \sqrt{n+1}$;
   (5) $a_n = \sqrt{n^2+n-2} - n$;
   (6) $a_n = \sqrt{n^4+7n^2-3n-2} - \sqrt{n^4-3n^2+2}$;
   (7) $a_n = \frac{\sqrt{n^4+3n^3+2n}-\sqrt{n^4-5n^2+4}}{7n+2}$;
   (8) $a_n = \frac{\sqrt{3n^8+7n^7-5n^3-2}-\sqrt{3n^8+6n^6-2n^2+1}}{5n^3-3n^2+2n-7}$;
   (9) $a_n = \frac{\sqrt{2n^8+5n^6-3n^3+1}-\sqrt{2n^8-3n^2-n}}{3n^3+2n^2-5n+1}$;
   (10) $a_n = \sqrt[3]{n^6+4n^5-3n^4-1} - \sqrt[3]{n^6+4n^5-2n^2+1}$;
   (11) $a_n = \sqrt{n+3}\left(\sqrt{n+4} - 2\sqrt{n+1} + \sqrt{n}\right)$;
   (12) $a_n = (n+2)\sqrt{n(n+1)} - 2(n+1)\sqrt{n(n+2)} + n\sqrt{(n+1)(n+2)}$;
   (13) $a_n = \frac{n}{n+1} - \frac{n+1}{n}$;
   (14) $a_n = \frac{2^n+(-1)^n}{2^{n+1}+(-1)^{n+1}}$;
   (15) $a_n = \frac{\sqrt{n^2+4n+3}}{n!}$;
   (16) $a_n = \frac{(-5)^n}{n!}$;
   (17) $a_n = \frac{\ln(2+e^n)}{3n}$;
   (18) $a_n = \frac{\ln^4 n}{\sqrt[3]{n}}$;
   (19) $a_n = \frac{1-\cos\frac{4}{n}}{\tan^2\frac{3}{n}}$;
   (20) $a_n = \left(\cos\frac{1}{n}\right)^{\cot^2\frac{3}{n}}$;
   (21) $a_n = \frac{1}{\sqrt{n}}\sin 2n$;
   (22) $a_n = \frac{1}{\sqrt{n}}\sin\frac{1}{2n}$;
   (23) $a_n = \sqrt{n}\sin\frac{1}{2n}$;
   (24) $a_n = n\sin\frac{1}{\sqrt{2n}}$;
   (25) $a_n = \ln(n+1) - \ln(n+2)$;
   (26) $a_n = n^2\left(\ln(5+3n^2) - \ln 3n^2\right)$;

(27) $a_n = \sqrt[n]{a}, \ a > 0$;

(28) $a_n = \sqrt[n]{n+1}$;

(29) $a_n = \sqrt[n]{n^2 + 3}$;

(30) $a_n = \frac{1 + 2 + 2^2 + \ldots + 2^n}{1 + 3 + 3^2 + \ldots + 3^n}$;

(31) $a_n = \frac{1 + \frac{1}{2} + \frac{1}{2^2} + \ldots + \frac{1}{2^n}}{1 + \frac{1}{3} + \frac{1}{3^2} + \ldots + \frac{1}{3^n}}$;

(32) $a_n = \frac{1}{n^2} + \frac{2}{n^2} + \ldots + \frac{n-1}{n^2}$;

(33) $a_n = 1 + \frac{1}{1 \cdot 2} + \frac{1}{2 \cdot 3} + \ldots + \frac{1}{n \cdot (n+1)}$;

(34) $a_n = \frac{1}{1 \cdot 3} + \frac{1}{3 \cdot 5} + \frac{1}{5 \cdot 7} \ldots + \frac{1}{(2n-1) \cdot (2n+1)}$;

(35) $a_n = \frac{1}{n} - \frac{2}{n} + \frac{3}{n} - \ldots + \frac{(-1)^{n-1} n}{n}$;

(36) $a_n = \sqrt{5} \cdot \sqrt[4]{5} \cdot \sqrt[8]{5} \cdot \ldots \sqrt[2^n]{5}$;

(37) $a_n = \sqrt[3]{3} \cdot \sqrt[9]{3} \cdot \sqrt[27]{3} \cdot \ldots \sqrt[3^n]{3}$;

(38) $a_1 = \sqrt{2}, a_2 = \sqrt{2\sqrt{2}}, a_3 = \sqrt{2\sqrt{2\sqrt{2}}}, \ldots, a_n = \underbrace{\sqrt{2\sqrt{2 \ldots \sqrt{2}}}}_{n}$;

(39) $a_n = \frac{(2n)!}{n^{2n}}$.

6. Show that the following sequence converges or diverges applying the Cauchy criterion:

(1) $a_n = \frac{\sin 1}{2} + \frac{\sin 2}{2^2} + \ldots + \frac{\sin n}{2^n}$;

(2) $a_n = 1 + \frac{1}{2^2} + \ldots + \frac{1}{n^2}$;

(3) $a_n = 1 + \frac{1}{2} + \ldots + \frac{1}{n}$;

(4) $a_n = 1 + \frac{1}{\sqrt{2}} + \ldots + \frac{1}{\sqrt{n}}$.

7. Demonstrate that the following sequence converges using the theorem of monotone sequence and find the value of the limit:

(1) $a_n = \frac{n}{3^n}$;

(2) $a_n = \frac{3^n}{n!}$;

(3*) $a_1 = \sqrt{2}, a_2 = \sqrt{2 + \sqrt{2}}, a_3 = \sqrt{2 + \sqrt{2 + \sqrt{2}}}, \ldots, a_n = \underbrace{\sqrt{2 + \sqrt{2 + \ldots + \sqrt{2}}}}_{n}$;

(4*) $a_1 = \sqrt{6}, a_2 = \sqrt{6 + \sqrt{6}}, a_3 = \sqrt{6 + \sqrt{6 + \sqrt{6}}}, \ldots, a_n = \underbrace{\sqrt{6 + \sqrt{6 + \ldots + \sqrt{6}}}}_{n}$;

(5*) $a_1 = \sqrt[3]{6}, a_2 = \sqrt[3]{6 + \sqrt[3]{6}}, a_3 = \sqrt[3]{6 + \sqrt[3]{6 + \sqrt[3]{6}}}, \ldots, a_n = \underbrace{\sqrt[3]{6 + \sqrt[3]{6 + \ldots + \sqrt[3]{6}}}}_{n}$;

(6*) $a_1 = \sqrt{a}, a_2 = \sqrt{a + \sqrt{a}}, a_3 = \sqrt{a + \sqrt{a + \sqrt{a}}}, \ldots, a_n = \underbrace{\sqrt{a + \sqrt{a + \ldots + \sqrt{a}}}}_{n}, \ a > 0$;

(7*) $a_1 = \sqrt{2}, a_2 = \sqrt{2 - \sqrt{2}}, a_3 = \sqrt{2 - \sqrt{2 - \sqrt{2}}}, \ldots, a_n = \underbrace{\sqrt{2 - \sqrt{2 - \ldots - \sqrt{2}}}}_{n}$.

8*. Prove that the sequence $a_n = \left(1 + \frac{1}{n}\right)^n$ is strictly increasing and the sequence $b_n = \left(1 + \frac{1}{n}\right)^{n+1}$ is strictly decreasing and that both are bounded. Deduce that these sequences have the same limit (denoted by e).

9. Using the result of Exercise 8, show that $\frac{1}{n+1} < \ln\left(1 + \frac{1}{n}\right) < \frac{1}{n}, \forall n \in \mathbb{N}$.

10*. Show that the sequence $a_n = 1 + \frac{1}{2} + \frac{1}{3} + \ldots + \frac{1}{n} - \ln n$ converges.

11*. Show that the sequence $a_n = \frac{1}{n+1} + \frac{1}{n+2} + \ldots + \frac{1}{2n}$ converges and find its limit.

12. Find the limit by applying the Stolz-Cesàro theorem:

(1) $\displaystyle\lim_{n \to +\infty} \frac{p^n}{n}, p \in \mathbb{R}$;

(2) $\displaystyle\lim_{n \to +\infty} \frac{1}{n} \sum_{k=1}^{n} \frac{(k-1)^2}{k^2+1}$;

(3) $\displaystyle\lim_{n \to +\infty} \frac{1}{(n+1)!} \sum_{k=1}^{n} 2k \cdot k!$;

(4) $\displaystyle\lim_{n \to +\infty} \frac{\ln(n!) - n \ln n}{n}$;

(5) $\displaystyle\lim_{n \to +\infty} \frac{1}{n} \left(1 + \frac{2}{1+\sqrt{2}} + \frac{3}{1+\sqrt{2}+\sqrt{3}} + \cdots + \frac{n}{1+\sqrt{2}+\sqrt{3}+\ldots+\sqrt{n}}\right)$;

(6) $\displaystyle\lim_{n \to +\infty} \frac{1^k + 3^k + \ldots + (2n+1)^k}{n^{k+1}}, k \in \mathbb{N}$.

13. Prove the following "converse" to Stolz-Cesàro theorem in the case $\frac{\infty}{\infty}$: if the sequences a_n and b_n are such that

(1) $\displaystyle\lim_{n \to +\infty} b_n = +\infty$, b_n is strictly increasing and $\displaystyle\lim_{n \to +\infty} \frac{b_n}{b_{n+1}} = B \neq 1$;

(2) $\displaystyle\lim_{n \to +\infty} \frac{a_n}{b_n} = L$ (L is a finite number),

then $\displaystyle\lim_{n \to +\infty} \frac{a_{n+1} - a_n}{b_{n+1} - b_n} = L$.

14. A very particular, but elegant technique of calculating the limit of a sequence is based on its representation (when it is possible) as a Riemann sum of a Riemann-integrable function. Find the following limits by reducing to the Riemann integral:

(1) $\displaystyle\lim_{n \to +\infty} \frac{1^k + 2^k + \ldots + n^k}{n^{k+1}}, k \geq 0$;

(2) $\displaystyle\lim_{n \to +\infty} \left(\frac{1}{3n} + \frac{1}{3n+1} + \cdots + \frac{1}{4n}\right)$;

(3) $\displaystyle\lim_{n \to +\infty} n \left(\frac{1}{n^2+1^2} + \frac{1}{n^2+2^2} + \cdots + \frac{1}{2n^2}\right)$;

(4) $\displaystyle\lim_{n \to +\infty} \frac{1^p + 3^p + \ldots + (2n-1)^p}{n^{p+1}}, p \geq 0$;

(5) $\displaystyle\lim_{n \to +\infty} \sum_{k=1}^{n} \frac{k}{n^2+k^2}$;

(6) $\displaystyle\lim_{n \to +\infty} \frac{1}{n} \sum_{k=1}^{2n} \frac{k}{\sqrt{n^2+k^2}}$.

15. Find all the limits of subsequences of a given sequence:

(1) $3 \cdot \left(1 - \frac{1}{n}\right) + 2 \cdot (-1)^{n+1}$;

(2) $1, -\frac{1}{2}, 1, -\frac{1}{3}, 1, -\frac{1}{4}, \ldots$;

(3) $1, 2, 1, 2, 3, 1, 2, 3, 4, 1, 2, 3, 4, 5, \ldots$;

(4) $\frac{1}{2}, \frac{1}{3}, \frac{2}{3}, \frac{1}{4}, \frac{2}{4}, \frac{3}{4}, \frac{1}{5}, \frac{2}{5}, \frac{3}{5}, \frac{4}{5}, \ldots$.

16. Let a_n be a convergent sequence with the limit a. What can be said about the limit of
    the following sequence:
    (1) $a_n a_{n+1}$;
    (2) $\max\{a_n, a_{n+1}\}$;
    (3) $\operatorname{sgn} a_n$;
    (4) $(a_{n+1} - a_n)^n$.

17. What can be said about the sequence a_n if it satisfies the following property:
    (1) $a_n^2 \underset{n \to \infty}{\to} 2$;
    (2) $a_n + \frac{1}{a_n} \underset{n \to \infty}{\to} 2$;
    (3*) $a_n^2 - a_n \underset{n \to \infty}{\to} 2$;
    (4*) $\frac{a_n}{\sin a_n} \underset{n \to \infty}{\to} 2$.
    Consider the case of a positive sequence ($a_n > 0$) and also a general case.

18. What can be said about the sequence a_n if it satisfies the following property
    (recurrence relation):
    (1) $a_{n+1} = \frac{1}{2} a_n$;
    (2) $a_{n+1} = -a_n$;
    (3) $a_{n+1} = 2a_{n-1}$;
    (4) $a_{n+1} = a_n^2$.
    Determine the initial values of the sequence for which the given relation converges.

19*. Verify whether the following statement is true or false:
    (1) if a sequence a_n is positive and unbounded, then $\lim\limits_{n \to \infty} a_n = +\infty$;
    (2) if a sequence a_n is positive, bounded and $\lim\limits_{n \to \infty} (a_{n+1} - a_n) = 0$, then a_n
        converges;
    (3) if a sequence a_n is positive, bounded and $\lim\limits_{n \to \infty} \frac{a_{n+1}}{a_n} = 1$, then a_n converges;
    (4) if a sequence a_n has the property that $\lim\limits_{n \to \infty} (a_{n+p} - a_n) = 0$ for any fixed $p \in \mathbb{N}$,
        then a_n converges.

Series of Numbers

2

> *It is strange that the immense variety in nature can be resolved into a series of numbers.*
> William Henry Bragg, 1923

1   Convergence and Introductory Examples

> *I have been forced to admit some propositions which will seem, perhaps, hard to accept. For instance, that a divergent series has no sum.*
> Augustin Louis Cauchy, 1821

1.1   Definition of a Series. Partial Sums and Convergence

The theory of series of numbers is based on the theory of sequences of numbers. This dependence is the result of both the definition of a series as a sum of all the elements of a sequence and the definition of its convergence as the convergence of a special sequence called the sequence of partial sums. The last point is of extreme importance in the construction of the theory of series.

Electronic Supplementary Material The online version of this article (https://doi.org/10.1007/978-3-030-79431-6_2) contains supplementary material, which is available to authorized users.

▶ **Definition** (**Series of Numbers**) Let a_n, $n \in \mathbb{N}$ be a sequence of numbers. The corresponding *series of numbers* is an infinite sum of all the elements of the sequence a_n in the order prescribed by the sequence:

$$a_1 + a_2 + \ldots + a_n + \ldots.$$

To shorten the writing the following notation is used

$$a_1 + a_2 + \ldots + a_n + \ldots \equiv \sum_{n=1}^{\infty} a_n.$$

The element a_n is called a *general term* of the series of numbers. In many cases, when there is no ambiguity, we call a series of numbers simply series.

If a sequence a_n is defined on an extended set $\mathbb{N}_0 = \{k_1, k_2 = k_1 + 1, \ldots, k_{i+1} = k_i + 1, \ldots\}$, $k_1 \in \mathbb{Z}$, the corresponding series follows the same variation of indices:

$$a_{k_1} + a_{k_1+1} + \ldots + a_n + \ldots \equiv \sum_{n=k_1}^{\infty} a_n.$$

In the following exposition, for simplicity of notation, we will mostly use the series with natural indices, but the same definitions can be extended to the series with indices in $\mathbb{N}_0$. In some cases when the variation of the index has no importance, we will use the short symbol $\sum a_n$.

▶ **Definition** (**Partial Sum and Remainder**) The sum of the first n term of a series is called *n-th partial sum* (or simply a *partial sum*):

$$s_n = \sum_{k=1}^{n} a_k,$$

and the sum of all the remaining terms is called *n-th remainder* (or simply a *remainder*) of a series:

$$r_n = \sum_{k=n+1}^{\infty} a_k.$$

> ▶ **Definition** (**Convergence of a Series**) If the sequence of the partial sums s_n converges, then we say that the *series* $\sum_{n=1}^{\infty} a_n$ *converges*, and the limit s of the sequence s_n is called the *sum of this series*
>
> $$\sum_{n=1}^{\infty} a_n = \lim_{n \to \infty} s_n = s.$$
>
> Otherwise, the *series diverges*.

Remark The used definition of a series of numbers, although not being completely rigorous, is comprehensible and serves well the objectives of this text. The weak point of this definition is the reference to the infinite sum whose interpretation is left to intuitive understanding. A strict and more formal definition can be formulated as follows. A series $\sum_{n=1}^{\infty} a_n$ is a pair of two sequences of numbers a_n and s_n, where the former is an original sequence and the latter (called the sequence of partial sums) is defined by the formula $s_n = \sum_{k=1}^{n} a_k$. If the sequence s_n converges to s, then we say that the series $\sum_{n=1}^{\infty} a_n$ converges to the sum s and we write $\sum_{n=1}^{\infty} a_n = s$. In this manner, the conceptual reference to the infinite sum is eliminated, but the meaning continue to be the same.

Many results studied in this part of the text are valid for special types of series, whose definitions are provided below.

▶ **Definition** (**Positive/Negative Series**) If the elements of a series $\sum a_n$ satisfy the condition $a_n > 0$ ($a_n \geq 0$), $\forall n$, then this is the *series of positive (non-negative) terms*, also called *positive (non-negative) series*. In the same way, if the elements of a series $\sum a_n$ satisfy the inequality $a_n < 0$ ($a_n \leq 0$), $\forall n$, then the *series is called negative (non-positive)*.

▶ **Definition** (**Alternating Series**) If a series can be represented in the form $\sum a_n = \sum (-1)^n b_n$ or $\sum a_n = \sum (-1)^{n+1} b_n$, where $b_n > 0$, $\forall n$, then this *series is called alternating*. Notice that in an alternating series the change of sign occurs from the current element a_n to the next one a_{n+1}. Other forms of the sign changing are not considered as alternating series.

1.2   Elementary Examples of Series of Numbers

Let us look at some examples of series.

Example 1a $\sum_{n=1}^{\infty} 0$

The original series is constant $a_n = 0$, $\forall n$ and the partial sums are $s_1 = 0$, $s_2 = 0 + 0 = 0, \ldots, s_n = \sum_{k=1}^{n} 0 = 0, \ldots$. Then, the sequence of the partial sums is also a constant (null) sequence and its limit is 0, which means that the series converges to the sum 0.

Example 2a $\sum_{n=1}^{\infty} 1$

The original sequence is constant $a_n = 1$, $\forall n$ and the partial sums are $s_1 = 1$, $s_2 = 1 + 1 = 2, \ldots, s_n = \sum_{k=1}^{n} 1 = n, \ldots$. Evidently, this sequence diverges (it tends to $+\infty$) and, respectively, the series is divergent.

Example 3a $\sum_{n=1}^{\infty} (-1)^n$

This series is alternating (originated by the alternating sequence $a_n = (-1)^n$), since its sign changes from the current term to the nearest one. Its partial sums are $s_1 = -1$, $s_2 = -1 + 1 = 0, \ldots, s_n = \sum_{k=1}^{n} (-1)^n = \begin{cases} -1, & n \text{ impar} \\ 0, & n \text{ par} \end{cases} = \frac{(-1)^n - 1}{2}, \ldots$. This sequence diverges (it has two different partial limits -1 and 0) and, consequently, the series diverges.

Example 4a $\sum_{n=0}^{\infty} q^n$, $q \in \mathbb{R}$

This important *series is called geometric* (it is the sum of the elements of the geometric sequence). For $q > 0$ this series is positive and for $q < 0$—alternating. For $q = 1$ we have the divergent series of constants 1 of Example 2a, and for $q \neq 1$ the convergence of the geometric series can be studied through direct calculation of its partial sums $s_n = \sum_{k=0}^{n} q^k = \frac{1-q^{n+1}}{1-q}$. (The last formula is well-known and can be easily deduced multiplying s_n by q—$q s_n = \sum_{k=0}^{n} q^{k+1}$—and noting that $s_n - q s_n = 1 - q^{n+1}$.) In the expression of s_n the only term that depends on n is q^{n+1}, which represents a geometric sequence, whose properties were already studied. Therefore,

(1) if $|q| < 1$, then $\lim_{n \to \infty} s_n = \lim_{n \to \infty} \frac{1-q^{n+1}}{1-q} = \frac{1}{1-q}$ and the series converges to the sum $s = \frac{1}{1-q}$;

(2) if $|q| \geq 1$, then a finite limit $\lim_{n \to \infty} s_n$ does not exists and the geometric series diverges (the case $q = 1$ can be included in this option noting that the series $\sum_{n=0}^{\infty} 1$ diverges).

The behavior of the geometric series with different values of q are shown in Fig. 2.1.

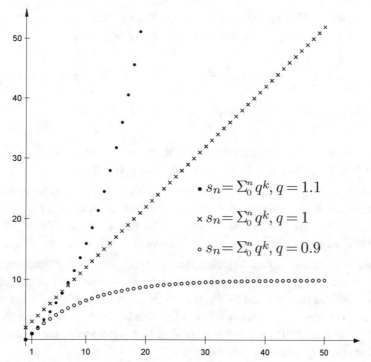

Fig. 2.1 Example 4a: convergent and divergent geometric series

Example 5a $\sum_{n=1}^{\infty} \frac{1}{n}$

This positive *series is called harmonic* and plays important role in the theory of series. Its general term converges to zero: $a_n = \frac{1}{n} \underset{n \to \infty}{\to} 0$ that may give an impression that this series has a finite sum. However, actually this series diverges and it can be proved by definition using the partial sums of selected indices. Indeed, consider initially the following four sums:

$$s_1 = 1, s_2 = 1 + \frac{1}{2}, s_4 = 1 + \frac{1}{2} + \frac{1}{3} + \frac{1}{4} = s_2 + \frac{1}{3} + \frac{1}{4} > 1 + \frac{1}{2} + 2 \cdot \frac{1}{4} = 1 + \frac{2}{2},$$

$$s_8 = 1 + \frac{1}{2} + \frac{1}{3} + \frac{1}{4} + \frac{1}{5} + \frac{1}{6} + \frac{1}{7} + \frac{1}{8} = s_4 + \frac{1}{5} + \frac{1}{6} + \frac{1}{7} + \frac{1}{8} > 1 + \frac{2}{2} + 4 \cdot \frac{1}{8} = 1 + \frac{3}{2}.$$

At this point, it may arise a supposition that $s_{2^k} \geq 1 + \frac{k}{2}, \forall k \in \mathbb{N}$. Let us check this inequality by induction. Since the result is already shown for the first indices, let us assume that it holds for $n = 2^{k-1}$ and prove that this implies its validity for $n = 2^k$. To do this, represent the partial sum s_{2^k} in the same form as for the first sums:

$$s_{2^k} = 1 + \frac{1}{2} + \ldots + \frac{1}{2^{k-1}} + \frac{1}{2^{k-1}+1} + \frac{1}{2^{k-1}+2} + \ldots + \frac{1}{2^k} = s_{2^{k-1}} + \frac{1}{2^{k-1}+1}$$

$$+ \frac{1}{2^{k-1}+2} + \ldots + \frac{1}{2^k} > 1 + \frac{k-1}{2} + 2^{k-1} \cdot \frac{1}{2^k} = 1 + \frac{k}{2}.$$

Hence, the supposition is proved. It implies that the chosen subsequence of the partial sums s_{2^k} approaches $+\infty$, since $1 + \frac{k}{2} \underset{k \to \infty}{\to} +\infty$. Therefore, the sequence of the partial sums s_n does not have a finite limit, that is, the series diverges. (More precisely, the sequence s_n tends to $+\infty$, since for any $n \in \mathbb{N}$ we can find the index k such that $n > 2^k$, and consequently, the positiveness of the series implies that $s_n > s_{2^k}, \forall n > 2^k$, and then from the obtained result $\lim_{k \to \infty} s_{2^k} = +\infty$ it follows that $\lim_{n \to \infty} s_n = +\infty$.) This is a very interesting feature of this important series: it may appear that the divergence of the series contradicts the convergence of the general term to zero, since the last fact means that for advanced values of n we add to the previously obtained partial sum the terms with arbitrary small values. However, we will see that this behavior of the harmonic series is an illustration of a general result that reveals a relationship between the convergence of a series and the convergence of its general term.

Example 6a $\sum_{n=1}^{\infty} \frac{1}{n^p}, p \in \mathbb{R}$

This is one more important *positive series, called the p-series*, which represents a generalization of the harmonic series. Let us investigate its convergence/divergence. For $p = 1$ the series is harmonic and divergent. If $p < 1$, then $s_n = \sum_{k=1}^{n} \frac{1}{n^p} \geq \sum_{k=1}^{n} \frac{1}{n}$. Since the partial sums of the harmonic series diverge to $+\infty$, it follows that $\lim_{n \to \infty} s_n = +\infty$, that is, the series diverges.

If $p > 1$, we can use the approach similar to the harmonic series, collecting the terms of the partial sums in groups of $2^k - 1$ summands. Considering the following evaluations

$$s_1 = 1, s_3 = 1 + \frac{1}{2^p} + \frac{1}{3^p} < 1 + 2 \cdot \frac{1}{2^p} = 1 + \frac{1}{2^{p-1}},$$

$$s_7 = 1 + \frac{1}{2^p} + \ldots + \frac{1}{7^p} = s_3 + \frac{1}{4^p} + \frac{1}{5^p} + \frac{1}{6^p} + \frac{1}{7^p} < 1 + \frac{1}{2^{p-1}} + 4 \cdot \frac{1}{4^p} = 1 + \frac{1}{2^{p-1}} + \frac{1}{(2^2)^{p-1}},$$

we can suppose that the inequality

$$s_{2^k-1} < 1 + \frac{1}{2^{p-1}} + \frac{1}{4^{p-1}} + \ldots + \frac{1}{(2^{k-1})^{p-1}} = \sum_{j=0}^{k-1} \frac{1}{(2^j)^{p-1}}$$

is true for any $k \in \mathbb{N}$. Indeed, we can prove this result by induction. Since the result is already shown for the first indices, let us assume that it holds for $n = 2^{k-1} - 1$ and prove that it is valid for $n = 2^k - 1$:

$$s_{2^k-1} = s_{2^{k-1}-1} + \frac{1}{(2^{k-1})^p} + \frac{1}{(2^{k-1}+1)^p} + \ldots + \frac{1}{(2^k-1)^p}$$

$$< \sum_{j=0}^{k-2} \frac{1}{(2^j)^{p-1}} + 2^{k-1} \cdot \frac{1}{(2^{k-1})^p} = \sum_{j=0}^{k-2} \frac{1}{(2^j)^{p-1}} + \frac{1}{(2^{k-1})^{p-1}} = \sum_{j=0}^{k-1} \frac{1}{(2^j)^{p-1}}.$$

On the right-hand side we have the sum of the finite geometric sequence, which is easily calculated: $g_k = \sum_{j=0}^{k-1} \frac{1}{(2^{p-1})^j} = \frac{1 - \frac{1}{(2^{p-1})^k}}{1 - \frac{1}{2^{p-1}}}$. Since $0 < \frac{1}{2^{p-1}} < 1$, the sequence g_k is increasing and convergent to $g = \frac{1}{1 - \frac{1}{2^{p-1}}}$. Notice also that any sequence of partial sums s_n (including subsequence s_{2^k-1}) is increasing because the series is positive. Hence, the sequence of the partial sums s_{2^k-1} is increasing and bounded $s_{2^k-1} < g_k \leq g, \forall k \in \mathbb{N}$, and consequently, by the Theorem of monotone sequence (Property 2s in Sect. 3.2 of Chap. 1), the sequence s_{2^k-1} converges: $\lim_{k \to \infty} s_{2^k-1} = s$. To extend this result to any sequence of the partial sums s_n, notice that for $\forall n$ there exists k such that $2^{k-1} - 1 \leq n \leq 2^k - 1$ and the corresponding partial sums satisfy the inequality $s_{2^{k-1}-1} \leq s_n \leq s_{2^k-1}$. Taking the limit as n tends to $+\infty$ in this double inequality and applying the squeeze Theorem, we conclude that $\lim_{n \to \infty} s_n = s$.

The behavior of p-series with different values of p are shown in Fig. 2.2.

Example 7a $\sum_{n=1}^{\infty} \frac{(-1)^{n-1}}{n}$

This *series is called alternating harmonic*. Comparing to the harmonic series, its terms alternate their signs, and consequently, the next term compensate at some degree the

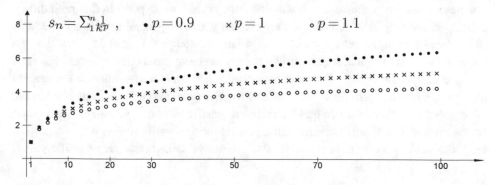

Fig. 2.2 Example 6a: convergent and divergent p-series

contribution of the current term. It happens that this is sufficient to convert the harmonic series into a convergent one. Let us prove the convergence of this series employing a simplified prototype of a general method of investigation of alternating series (which will be studied later). Consider, first, the even partial sums s_{2n}. Using a suitable grouping of their terms, we can show that these sums are bounded:

$$s_{2n} = \sum_{k=1}^{2n} \frac{(-1)^{k-1}}{k} = 1 - \frac{1}{2} + \frac{1}{3} - \frac{1}{4} + \ldots + \frac{1}{2n-1} - \frac{1}{2n}$$

$$= 1 - \left(\frac{1}{2} - \frac{1}{3}\right) - \ldots - \left(\frac{1}{2n-2} - \frac{1}{2n-1}\right) - \frac{1}{2n} < 1.$$

On the other hand, these sums form an increasing sequence:

$$s_{2n} = s_{2n-2} + \frac{1}{2n-1} - \frac{1}{2n} > s_{2n-2}.$$

Since s_{2n} is an increasing and bounded above sequence, it is convergent (according to Property 2s of Sect. 3.2 in Chap. 1), and we denote its limit by s: $\lim_{n\to\infty} s_{2n} = s$. Notice now that the odd partial sums $s_{2n+1} = s_{2n} + \frac{1}{2n+1}$ have the same limit, because each of two term on the right-hand side has a limit and the second limit is 0:

$$\lim_{n\to\infty} s_{2n+1} = \lim_{n\to\infty} s_{2n} + \lim_{n\to\infty} \frac{1}{2n+1} = s + 0 = s.$$

Since the subsequences of the even and odd partial sums have the same limit, it follows that the general limit exists and has the same value: $\lim_{n\to\infty} s_n = s$.

Remark 1 The examples presented in this section were selected using the criteria of their clarity, importance and simplicity. The last means that these series can be analyzed by employing rudimentary techniques, before developing different tests of convergence. Nevertheless, some suppositions arisen and primitive procedures applied in this preliminary exposition are very helpful, because they contain germs of general results and methods to be developed in the next sections. For instance, Examples 1a and 2a show that the convergence of the original sequence does not say anything about convergence/divergence of the corresponding series; Example 5a indicates that even convergence of the general term to 0 is not sufficient for convergence of a series (later this will be formalized in the Divergence test); Examples 5a and 6a exhibit a specific form of general approach based on the comparison of positive series (in a general form this will give us the Comparison test); the same two examples represent also an idea of using a few very specific partial sums and group of terms to study the convergence of the series (later this approach will be formulated as the Cauchy condensation test); finally, the alternating harmonic series of Example 7a carries the exposition of both the basic method of investigation of general alternating series, and the technique of its demonstration, which will be later presented as Leibniz's test.

Remark 2 One can clearly see the difference between convergent and divergent geometric series in Fig. 2.1, while it is very hard to make a decision if a p-series converges or diverges looking at Fig. 2.2. It is worth noting that the same visual proximity between convergent and divergent p-series is also observed for elevated values of indices. In such cases, the visual representation can be very deceptive and is useless to clarify the behavior of series.

There are many series like harmonic, which diverge very slow and cannot be distinguished visually from the series with a slow rate of convergence (like p-series with p slightly larger than 1). These series are a perfect counterexample to the misconception that convergence/divergence of a series can be checked by using a calculator. For instance, a modern supercomputer with high velocity of 100 teraflops (10^{14} floating-point operations per second) would take about 500 years to calculate the partial sum of harmonic series with the value around 55. On the other hand, to obtain the sum of the convergent p-series $\sum_{n=1}^{\infty} \frac{1}{n^{5/4}}$ with the accuracy of 10^{-6} the same supercomputer would take about 1000 years.

This occurs not only with a visual representation or numerical evaluation, but also with some basic methods of investigation of series. We will see that such popular tests as the ratio and root tests are unable to distinguish between convergent and divergent p-series, and in such cases we need to appeal to more subtle techniques.

Historical Remarks 1 A first notion of infinite series may be traced back to the ancient Greek mathematicians and philosophers: the famous paradox of Achilles and the tortoise by Zeno leads us to the concept of a convergent geometric series; Aristotle considered (implicitly) that the sum of a series of infinitely many summands can be a finite quantity; the "father of geometry" Euclid in one of the books of "Elements" expressed the partial sum of a geometric series in terms of members of the series, which is equivalent to the modern formula; Archimedes performed the first known summation of an infinite (geometric) series employing the method of exhaustion to calculate the area under the arc of a parabola.

The Indian mathematicians of Kerala school have obtained important results on infinite series in the fourteenth–sixteenth centuries, including calculation of the sum of a convergent geometric series and the use of the power series for representation of some elementary functions. However, their work was not known beyond India until the nineteenth century.

The Middle Ages philosopher Oresme worked on infinite sequences and series, and c.1350 he was first to prove that the harmonic series is divergent. However, his results were lost for several centuries and this fact was rediscovered later, by the seventeenth century mathematicians. Apparently the first of them was Mengoli: in the treatise published in 1650, he showed that the harmonic series diverges and found the sum of the convergent geometric series. Moreover, he probably was the first to consider infinite sums as the values that are approached by the sequence of partial sums.

In 1665–1668, Newton and Mercator first found the expansion of logarithmic function in an infinite series and since then the era of intensive use of power series and more general series of functions began.

Historical Remarks 2 Until the beginning of the nineteenth century the problems of convergence/divergence of series of numbers were considered only in studies of specific series and there was no systematic theory. Mathematicians had only very informal ideas about convergence and divergence of series. Probably, the first attempt to propose a general definition of the convergence was made by Leibniz in 1713 during discussion on the convergence/divergence of Grandi's series $\sum_{n=0}^{\infty}(-1)^n$. The absence of a rigorous definition led the great mathematicians to different conclusions about this series during about 100 years. Leibniz, Euler and Lagrange argued that the series converges to $\frac{1}{2}$, while Varignon, Riccati and D'Alembert thought that it diverges.

The situation has changed when in 1821 "Cours d'analyse" Cauchy gave the first modern definition of convergence of a series through convergence of its partial sums, and called the limit of the sequence of the partial sums the sum of the series. The formulated general definition of convergence/divergence and largely accepted ban imposed by Cauchy and Abel on the use of divergent series, established a solid basis for the subsequent development of the theory of convergence of series (and closed the problem of Grandi's series in the classical theory of convergent/divergent series).

Historical Remarks 3 The name "harmonic series" apparently was introduced by Brouncker, the first president of the Royal Society, around 1660 as the reference to the property of the three subsequent terms $a = \frac{1}{n-1}$, $b = \frac{1}{n}$, $c = \frac{1}{n+1}$ of this series, which form so-called harmonic proportion: $\frac{a}{c} = \frac{a-b}{b-c}$.

The divergence of the harmonic series was first demonstrated in the fourteenth century, c.1350, by Oresme, applying the method similar to that used in Example 5a. However, his achievement fell into obscurity, and in the seventeenth century new proofs were given by Mengoli and Johann Bernoulli.

---

2   Elementary Properties of Convergent Series

> *With the exception of the geometric series, there does not exist in all of mathematics a single infinite series the sum of which has been rigorously determined. In other words, the things which are the most important in mathematics are also those which have the least foundation.*
> *Niels Henrik Abel, 1826*

According to the definition, the convergence of a series is understood as the convergence of the sequence of its partial sums. Therefore, different properties of convergent sequences can be transferred to the case of series. Some of these results are presented below.

2.1   Arithmetic Properties

Linear Combination

The *property of a linear combination* is frequently split into two parts—the *property of sum* and that of *multiplication by a constant*. Both properties have simple formulation and can be easily proved.

> **Property of linear combination** *If $\sum_{n=1}^{\infty} a_n = a$ and $\sum_{n=1}^{\infty} b_n = b$ are two convergent series, then*
>
> *(1) $\sum_{n=1}^{\infty}(a_n + b_n) = a + b$*
> *(2) $\sum_{n=1}^{\infty}(ca_n) = ca, \forall c \in \mathbb{R}$.*

Proof

(1) The convergence of the series $\sum_{n=1}^{\infty} a_n = a$ and $\sum_{n=1}^{\infty} b_n = b$ means that $\lim_{n\to\infty} s_n^{(a)} = a$ and $\lim_{n\to\infty} s_n^{(b)} = b$, where $s_n^{(a)}$ and $s_n^{(b)}$ are the sequences of the partial sums of the first and second series, respectively. Then, by the properties of sequences, there exists the limit of the sequence

$$s_n = s_n^{(a)} + s_n^{(b)} = \sum_{k=1}^{n} a_k + \sum_{k=1}^{n} b_k = \sum_{k=1}^{n}(a_k + b_k)$$

equal to $a + b$. But s_n is exactly the sequence of the partial sums of the series $\sum_{n=1}^{\infty}(a_n + b_n)$. Therefore, the last series converges to the sum $a + b$.

(2) The convergence of the series $\sum_{n=1}^{\infty} a_n = a$ means that $\lim_{n\to\infty} s_n^{(a)} = a$ where $s_n^{(a)}$ is the sequence of its partial sums. Then, by the sequence properties, for any constant $c \in \mathbb{R}$ the sequence $s_n = cs_n^{(a)} = \sum_{k=1}^{n}(ca_k)$ converges to ca. But s_n is the sequence of the partial sums of the series $\sum_{n=1}^{\infty}(ca_n)$. Hence, the last series converges to the sum ca.

$\square$

Product of Two Series

The situation is more complex with the *product of two series*. First, there are different ways to define the product. If we desire to obtain such product of two convergent series $\sum_{n=0}^{\infty} a_n = a$ and $\sum_{n=0}^{\infty} b_n = b$ which is a convergent series $\sum_{n=0}^{\infty} c_n$ with the sum ab, then a straightforward definition of the general term of the product series by $c_n = a_n b_n$ will not assure this result. Indeed, consider two geometric (convergent) series with $a_n = \frac{1}{2^n}$ and

$b_n = \frac{1}{3^n}$: $\sum_{n=0}^{\infty} \frac{1}{2^n} = \frac{1}{1-\frac{1}{2}} = 2$, $\sum_{n=0}^{\infty} \frac{1}{3^n} = \frac{1}{1-\frac{1}{3}} = \frac{3}{2}$. Then, the series with the general term $c_n = a_n b_n = \frac{1}{2^n} \frac{1}{3^n} = \frac{1}{6^n}$ is a convergent series, but its sum is different from $ab = 3$: $\sum_{n=0}^{\infty} \frac{1}{6^n} = \frac{1}{1-\frac{1}{6}} = \frac{6}{5} \neq 3$.

One of the ways to define the product of two series, which satisfies the property $c = ab$ for convergent series, is to use the definition of the Cauchy product:

▶ **Definition** (**Cauchy Product**) The *Cauchy product of two series* $\sum_{n=0}^{\infty} a_n$ and $\sum_{n=0}^{\infty} b_n$ is the series $\sum_{n=0}^{\infty} c_n$ with the general term defined by the formula

$$c_n = \sum_{k=0}^{n} a_k b_{n-k} = a_0 b_n + a_1 b_{n-1} + \ldots + a_{n-1} b_1 + a_n b_0.$$

For such defined product of two series, the convergence of each of two original series still does not guarantee the convergence of the product. However a slightly stronger requirement to the original series—the condition of the convergence of the series $\sum_{n=0}^{\infty} |a_n|$ or $\sum_{n=0}^{\infty} |b_n|$ (the absolute convergence of one of the series)—is sufficient to ensure the property $c = ab$. A formulation of the respective property is simple, but its proof is not accessible at the moment, because it is based on some results that were not presented yet. For this reason, we give here only the formulation of the property, but postpone its proof until Sect. 4 of this chapter.

Property of the product Let $\sum_{n=0}^{\infty} a_n = a$ and $\sum_{n=0}^{\infty} b_n = b$ be two convergent series. If $\sum_{n=0}^{\infty} |a_n|$ converges, then the Cauchy product of these series $\sum_{n=0}^{\infty} c_n$, $c_n = \sum_{k=0}^{n} a_k b_{n-k}$ also converges and $\sum_{n=0}^{\infty} c_n = ab$.

2.2   Cauchy Criterion for Convergence

The *Cauchy criterion* for sequences leads directly to the corresponding criterion for series.

Theorem (Cauchy Criterion) *A series* $\sum_{n=1}^{\infty} a_n$ *converges if and only if the sequence* s_n *of its partial sums is a Cauchy sequence, that is,*

$$\forall \varepsilon > 0 \, \exists N_\varepsilon \text{ such that for } \forall m > n > N_\varepsilon \text{ it follows that } |s_m - s_n| = \left| \sum_{k=n+1}^{m} a_k \right| < \varepsilon.$$

$$(2.1)$$

2.3   Necessary Condition of Convergence (Divergence Test)

Theorem *If a series $\sum_{n=1}^{\infty} a_n$ converges, then its general term a_n tends to 0:*
$\lim_{n \to \infty} a_n = 0$.
   This result is called the Divergence Test because its equivalent contrapositive is as follows: If $\lim_{n \to \infty} a_n \neq 0$, then $\sum_{n=1}^{\infty} a_n$ diverges.

Proof From the definition of the partial sums we have $a_n = s_n - s_{n-1}$. The convergence of the series means the convergence of its partial sums: $\lim_{n \to \infty} s_n = s$. Taking into account that s_{n-1} is the same sequence of the partial sums, we obtain

$$\lim_{n \to \infty} a_n = \lim_{n \to \infty} (s_n - s_{n-1}) = \lim_{n \to \infty} s_n - \lim_{n \to \infty} s_{n-1} = s - s = 0.$$

□

Remark This result provides a very simple technique for verification of divergence of a series: we can draw a conclusion based only on the limit of the general term without need to evaluate the partial sums. However, it is important to notice that the convergence of the general term to 0 does not guarantee the convergence of a series. Indeed, the classic example of this situation is the harmonic series $\sum_{n=1}^{\infty} \frac{1}{n}$ already analyzed by definition: its general term $\frac{1}{n}$ approaches zero, but the series itself diverges.

2.4   Series and Its Remainder

Convergence of the Original and Modified Series

Theorem *A modification, elimination or addition of a finite number of terms of a series does not change its convergence/divergence (but can change its sum).*
   In short, the convergence of a series is not affected by a finite number of terms.

Proof Suppose that M terms $a_{n_1}, \ldots, a_{n_M}$ of the original series $\sum_{n=1}^{\infty} a_n$ were substituted by $a'_{n_1}, \ldots, a'_{n_M}$ (it includes elimination of these terms, which corresponds to the choice $a'_{n_1} = 0, \ldots, a'_{n_M} = 0$). In this manner, a new series $\sum_{n=1}^{\infty} a'_n$ was formed with the elements $a'_n = a_n, n \neq n_1, \ldots, n_M$. Considering the original partial sums s_n for the indices $n > N = \max\{n_1, \ldots, n_M\}$ (recall that the first N elements of a sequence has no effect on its limit), notice that the difference between s_n and the partial sums of the new series s'_n is a constant: $s'_n = s_n + C$, where $C = a'_{n_1} + \ldots + a'_{n_M} - (a_{n_1} + \ldots + a_{n_M})$. Therefore, the existence of the limit of the sequence s_n is equivalent to the existence of the limit of s'_n, that is, the series $\sum_{n=1}^{\infty} a_n$ converges if and only if the new series $\sum_{n=1}^{\infty} a'_n$ converges.

Finally, an addition of a finite number of terms $a'_{n_1}, \ldots, a'_{n_M}$ to the original series can be treated as the case of elimination of these elements from the series $\sum_{n=1}^{\infty} a'_n$. Reverting the roles of the original and modified series (considering $\sum_{n=1}^{\infty} a'_n$ to be the original one), we immediately convert this case into the one already proven (the case of elimination of a finite number of terms). $\qquad\qquad\square$

Corollary *A series $\sum_{n=1}^{\infty} a_n$ converges if and only if the series $\sum_{n=N+1}^{\infty} a_n$ converges.*

Proof This is a particular case of the above Theorem (when the first N terms were eliminated from the original series). $\qquad\qquad\square$

Remark The Corollary shows the equivalence between convergence/divergence of a series and its fixed remainder (the number N is fixed), that is, a finite number of the terms of a series does not affect its convergence/divergence (but can change the sum of a series). From this statement it follows the following important note regarding formulations of the properties of convergence/divergence: although the hypotheses with respect to the terms of a series are frequently imposed in a general form (involving all the terms of a series), actually the same results continue to hold (qualitatively, for convergence/divergence) if these hypotheses are not satisfied for a finite number of the elements of a series. In what follows, we usually formulate the properties using conditions for all the terms of a series, but it is implicitly understood that the very same properties hold when the imposed conditions are satisfied starting from some index.

Criterion for Convergence Through Remainders

Theorem *A series $\sum a_n$ converges if and only if the sequence of its remainders converges to 0.*

Proof *Necessity*. For a convergent series $\sum_{n=1}^{\infty} a_n$ with a sum s, its specific remainder $\sum_{k=n+1}^{\infty} a_k$ is also a convergent series (by the above Corollary). Besides, the sum r_n of the last series satisfies the relation $r_n = s - s_n$, where s_n is the partial sum $s_n = \sum_{k=1}^{n} a_k$. Since this relation holds for any index n, we can pass to the limit and obtain $\lim\limits_{n\to\infty} r_n = \lim\limits_{n\to\infty} (s - s_n) = s - s = 0$.

Sufficiency. If $r_n = \sum_{k=n+1}^{\infty} a_k \underset{n\to\infty}{\to} 0$, then by the definition of the limit of a sequence, for any $\varepsilon > 0$ (choose, for convenience, $\frac{\varepsilon}{2}$) there exists N such that for all $n > N$ it follows that $|r_n| < \frac{\varepsilon}{2}$. Take now $\forall m > n > N$ and evaluate the sum in the Cauchy criterion:

$$\left| \sum_{k=n+1}^{m} a_k \right| = \left| \sum_{k=n+1}^{\infty} a_k - \sum_{k=m+1}^{\infty} a_k \right| = |r_n - r_m| \leq |r_n| + |r_m| < \frac{\varepsilon}{2} + \frac{\varepsilon}{2} = \varepsilon.$$

This shows that the series satisfies the Cauchy criterion, and consequently, it converges.

$\square$

Remark The given above formal proof of the sufficiency is actually redundant, because the proper assumption that the sequence of the remainders converges to 0 implies that this sequence is defined (at least starting from some index N). However, if at least one remainder, say r_N, exists, then r_N is a specific number, which means that the series $\sum_{n=N+1}^{\infty} a_n$ converges to r_N. Hence, the conditions of the preceding Corollary are satisfied, and consequently, the original series converges.

Historical Remarks 1 During the seventeenth–eighteenth centuries there was no rigorous theory of series and usually their convergence/divergence was considered intuitively, sometimes identifying the convergence with the tendency of the general term to zero. However, at the end of the eighteenth century, it was well known that a series with a general term approaching zero need not be convergent.

Historical Remarks 2 The criterion named after Cauchy appeared first in a paper of Euler in 1734 who did not return to it in his later work. In that paper, Euler did not use the terms like convergence or divergence and employed the language of infinitesimals instead of limits. Probably because of this, the paper was not widely known to mathematicians of later generations. Independently, Bolzano formulated the same criterion and tried to prove it. Unfortunately, his results were published in an obscure pamphlet in Prague in 1817 (and one year later in a little-known Bohemian journal) and became known for mathematical community only more than half a century later. In 1821 Cauchy presented his criterion in "Cours d'analyse", a seminal textbook, popular at that time. He established that a convergent series is a Cauchy series, but did not provide the proof of the converse, which is based on the completeness of the real numbers, the notion still to be developed in the second half of the nineteenth century.

3   Convergence of Positive Series

> *Cauchy is crazy, and there is no way of getting along with him, even though right now he is the only one who knows how mathematics should be done. What he is doing is excellent, but very confusing.*
> *Niels Henrik Abel, 1826*

3.1   General Criterion for Convergence

Theorem *A non-negative series converges if and only if the sequence of its partial sums is bounded.*

Proof By definition, a series converges if the sequence of its partial sums converges, and any convergent sequence is bounded.

For the converse, notice that a non-negative series ($a_n \geq 0$, $\forall n$) has an increasing sequence of the partial sums: $s_{n+1} = s_n + a_{n+1} \geq s_n$, $\forall n$. If additionally s_n is bounded, then the sequence s_n converges according to the result in Sect. 3.2 of Chap. 1, which means that the series converges.   □

Corollary *If a non-negative series diverges, the limit of its partial sums s_n is $+\infty$.*

Proof An increasing sequence of the partial sums s_n is unbounded above (if not, it would be convergent by the above Theorem). This means that for any constant $C > 0$, there exists an element s_N such that $s_N > C$. Besides, since s_n is increasing, it follows that $s_n \geq s_N > C$ for $\forall n > N$. The last relation is the definition of the limit $+\infty$.   □

The general criterion itself is rarely applied directly for analysis of the convergence of a series (the evaluation of partial sums is not a simple problem). However, this criterion and its versions are frequently used in the process of deriving the convergence tests which have immediate practical applications. One of such examples is the demonstration of the Integral test in the next subsection.

3.2   Integral Test (Cauchy-Maclauren Test)

Theorem (The Integral Test) *Let function $f(x)$ be Riemann integrable on $[1, b]$, $\forall b > 1$, and positive and decreasing on $[1, +\infty)$. If $f(n) = a_n$, $\forall n$, then the series $\sum_{n=1}^{\infty} a_n$ converges if and only if the improper integral $\int_1^{+\infty} f(x)dx$ converges.*

Proof We start with the preliminary comparison between the partial sums of the series and the Riemann integral on the corresponding interval. Consider $f(x)$ on the interval $[n, n+1]$. According to the function properties we have $a_n = f(n) \geq f(x) \geq f(n+1) = a_{n+1}$. Then, integrating this inequality on $[n, n + 1]$, we obtain (due to the comparison

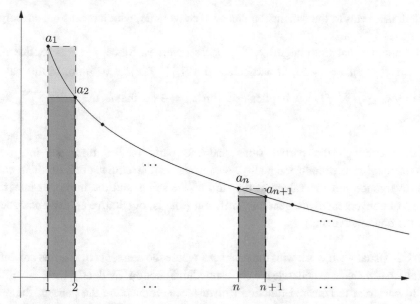

Fig. 2.3 Comparison of the areas of the rectangles and the area under $f(x)$ in the Integral test

properties of the Riemann integral):

$$a_n = \int_n^{n+1} a_n dx \geq \int_n^{n+1} f(x)dx \geq \int_n^{n+1} a_{n+1}dx = a_{n+1}$$

(see the illustration in the terms of corresponding areas in Fig. 2.3). This relation is true for any $n \in \mathbb{N}$, therefore, adding up all these inequalities from $k = 1$ to $k = n$, we get

$$s_n = \sum_{k=1}^n a_k \geq \sum_{k=1}^n \int_k^{k+1} f(x)dx = \int_1^{n+1} f(x)dx \geq \sum_{k=1}^n a_{k+1} = s_{n+1} - a_1.$$

Omitting the auxiliary calculations in the last formula, we arrive at the following main inequality used in the next two parts of the proof:

$$s_n \geq \int_1^{n+1} f(x)dx \geq s_{n+1} - a_1.$$

In the first part of the proof, we suppose that the integral $\int_1^{+\infty} f(x)dx$ converges. Then, there exists a finite limit $\lim_{n \to \infty} \int_1^{n+1} f(x)dx = I \equiv \int_1^{+\infty} f(x)dx$. Notice that the sequence $I_n = \int_1^n f(x)dx$ is increasing, and consequently, $I_n \leq I$, $\forall n \in \mathbb{N}$. Together with the inequality $s_{n+1} \leq a_1 + \int_1^{n+1} f(x)dx$ this shows the boundedness of the partial sums: $s_{n+1} \leq a_1 + I$, $\forall n \in \mathbb{N}$. Hence, the sequence of the partial sums s_n is increasing and bounded, that leads to the conclusion that s_n is convergent, which means the convergence of the series $\sum_{n=1}^{\infty} a_n$.

Now assume that the integral $\int_1^{+\infty} f(x)dx$ diverges. Since $f(x) > 0$, this means that $\lim_{b \to +\infty} \int_1^b f(x)dx = +\infty$, in particular, $\lim_{n \to \infty} \int_1^{n+1} f(x)dx = +\infty$. In this case, the inequality $s_n \geq \int_1^{n+1} f(x)dx$ implies that $\lim_{n \to \infty} s_n = +\infty$, that is, the series $\sum_{n=1}^{\infty} a_n$ also diverges. $\square$

Remark 1 Although the partial sums and the parts of the improper integral are related through the double inequality, which shows the equivalence of their convergence/divergence, but the values of the sum of the series and the improper integral (in the case of convergence) are generally different (that is, qualitative equivalence does not imply the quantitative one).

Remark 2 Usually we work with the problems where the elements of a series are defined, and from that we have to generate the corresponding function. Evidently, there are different ways to construct the desired function (for instance, connecting the pairs of the nearest values $f(n)$ and $f(n + 1)$ by line segments), but there usually exists a natural way to determine the function $f(x)$ induced by the form of a_n.

Application of the Integral Test makes the study of different series much easier than using other methods. One of such examples is the harmonic series.

Example 1i The harmonic series $\sum_{n=1}^{\infty} \frac{1}{n}$ has the positive decreasing term $\frac{1}{n}$ that can be associated with the function $f(x) = \frac{1}{x}$ that satisfies the required properties of the Integral Test: $f(x)$ is continuous, positive and decreasing on $[1, +\infty)$. Recall that the improper integral of $f(x) = \frac{1}{x}$ diverges, and it can be easily demonstrated:

$$\int_1^{+\infty} \frac{1}{x}dx = \lim_{b \to +\infty} \int_1^b \frac{1}{x}dx = \lim_{b \to +\infty} \ln x \big|_1^b = \lim_{b \to +\infty} \ln b = +\infty.$$

Consequently, the series $\sum_{n=1}^{\infty} \frac{1}{n}$ also diverges.

Notice that applying the Integral Test we obtain the result in much easier way than using the elementary technique in Example 5a of Sect. 1.2.

Example 2i In the same way as in Example 1i, the Integral Test is applicable to the study of a family of series called the *p-series*, which includes the harmonic series: $\sum_{n=1}^{\infty} \frac{1}{n^p}$, where $p \in \mathbb{R}$. These series have an important role for investigation of many other series through their comparison, that we will see later. Since the harmonic series was already analyzed, we will not consider it again, and this will simplify the integration. First, notice that for $p \leq 0$ the general term does not approach 0, and consequently, the series is divergent (by the Divergence Test). If $p > 0$, $p \neq 1$, then the general term $\frac{1}{n^p}$ can be naturally related to the function $f(x) = \frac{1}{x^p}$, which is continuous, positive and decreasing on $[1, +\infty)$. Recalling the evaluations of improper integrals, we obtain:

$$\int_1^{+\infty} \frac{1}{x^p} dx = \lim_{b \to +\infty} \int_1^b \frac{1}{x^p} dx = \lim_{b \to +\infty} \frac{-1}{p-1} \frac{1}{x^{p-1}} \Big|_1^b = \lim_{b \to +\infty} \frac{1}{p-1} \left(1 - \frac{1}{b^{p-1}} \right)$$

$$= \begin{cases} \frac{1}{p-1}, & p > 1 \\ +\infty, & p < 1 \end{cases}.$$

Hence, the p-series converges when $p > 1$ and diverges when $p \leq 1$.

Recall that we have derived the same result in Example 6a of Sect. 1.2, but the application of the Integral test is more straightforward.

Example 3i One more example of the same type is the series $\sum_{n=2}^{\infty} \frac{1}{n \ln n}$, whose general term $\frac{1}{n \ln n}$ is smaller than $\frac{1}{n}$ of the harmonic series, but still greater than $\frac{1}{n^{1+\delta}}$ for any $\delta > 0$. Therefore, this general term stay on the border between the convergent p-series (with $p > 1$) and divergent (with $p \leq 1$). Applying the Integral Test in the interval $[2, +\infty)$ with the corresponding function $f(x) = \frac{1}{x \ln x}$ (which is continuous, positive and decreasing on $[2, +\infty)$), we have

$$\int_2^{+\infty} \frac{1}{x \ln x} dx = \lim_{b \to +\infty} \int_2^b \frac{1}{x \ln x} dx = \lim_{b \to +\infty} \ln(\ln x)|_2^b = \lim_{b \to +\infty} (\ln(\ln b) - \ln(\ln 2)) = +\infty.$$

Since the improper integral diverges, the given series also diverges.

Example 4i The preceding Example can be promptly generalized to the series $\sum_{n=2}^{\infty} \frac{1}{n \ln^p n}$, $p \in \mathbb{R}$. The case $p = 1$ was already considered, and we will assume that $p \neq 1$. Introducing the corresponding function $f(x) = \frac{1}{x \ln^p x}$, we see that it is continuous and positive on $[2, +\infty)$. Since its derivative is negative for sufficiently large x:

$$f'(x) = \left(\frac{1}{x \ln^p x} \right)' = \frac{-p \ln^{-p-1} x \cdot \frac{1}{x} \cdot x - \ln^{-p} x \cdot 1}{x^2} = -\frac{p + \ln x}{x^2 \ln^{p+1} x} < 0, \ \forall x > e^{-p},$$

the function $f(x)$ is decreasing on $[e^{-p}, +\infty)$. Therefore, for any $p \neq 1$ we can apply the Integral Test on $[a, +\infty)$, $a = [e^{-p}] + 2$ with the function $f(x) = \frac{1}{x \ln^p x}$. Calculating the improper integral, we get

$$\int_a^{+\infty} \frac{1}{x \ln^p x} dx = \lim_{b \to +\infty} \int_a^b \frac{1}{x \ln^p x} dx = \lim_{b \to +\infty} \frac{1}{1-p} \ln^{1-p} x |_a^b = \begin{cases} +\infty, & p < 1 \\ \frac{1}{p-1} \ln^{1-p} a, & p > 1 \end{cases}.$$

Thus, the improper integral diverges for $p < 1$ and converges for $p > 1$. Together with the result of Example 3i, we arrive at the conclusion that the series diverges for $p \leq 1$ and converges for $p > 1$.

Example 5i The investigation of the series $\sum_{n=2}^{\infty} \frac{1}{a^{\ln n}}, a > 0$ is also simple if we apply the Integral Test. First, if $a = e$, we have the divergent harmonic series. If $a \leq 1$, then the general term does not approach 0 and the series diverges. Therefore, in the application of the Integral test we consider only $a > 1$, $a \neq e$. Introducing the corresponding function $f(x) = \frac{1}{a^{\ln x}}$, we see that it is continuous, positive and decreasing on $[2, +\infty)$, that is, satisfies the conditions of the test. To calculate the improper integral, we change the variable $t = \ln x$ and obtain:

$$\int_2^{+\infty} \frac{1}{a^{\ln x}} dx = \int_{\ln 2}^{+\infty} \left(\frac{e}{a}\right)^t dt = \lim_{b \to +\infty} \left(\left(\frac{e}{a}\right)^t \cdot \frac{1}{\ln(e/a)}\right)\Big|_{\ln 2}^{b} = \begin{cases} -\left(\frac{e}{a}\right)^{\ln 2} \cdot \frac{1}{\ln(e/a)}, & a > e \\ +\infty, & a \leq e \end{cases}.$$

This means that the integral and the original series converge when $a > e$ and diverge when $a \leq e$.

Evaluation of the Remainder in the Integral Test

Under the conditions of the Integral Test, we take advantage of the inequality $a_n \geq \int_n^{n+1} f(x)dx \geq a_{n+1}, \forall n \in \mathbb{N}$, derived in the course of the proof, to obtain the *evaluation of the remainder of a convergent series*. Summing up all these inequalities starting from $k = n$, we get

$$r_{n-1} = \sum_{k=n}^{\infty} a_k \geq \sum_{k=n}^{\infty} \int_k^{k+1} f(x)dx = \int_n^{+\infty} f(x)dx \geq \sum_{k=n}^{\infty} a_{k+1} = r_n, \forall n \in \mathbb{N}.$$

The last formula can be rewritten to evaluate the remainder r_n both below and above:

$$\int_{n+1}^{+\infty} f(x)dx \leq r_n \leq \int_n^{+\infty} f(x)dx.$$

Then, the sum s of the series can be evaluated as follows:

$$\int_{n+1}^{+\infty} f(x)dx + s_n \leq s \leq \int_n^{+\infty} f(x)dx + s_n.$$

Example 6i The estimates of this kind allow us to evaluate the degree of proximity between the sum of the series and its partial sums. Let us consider the problem of approximation of the sum of the series $\sum_{n=1}^{\infty} \frac{1}{n^5}$ with the accuracy at least 10^{-3} using the least possible number of the terms of this series. From the remainder estimate $r_n \leq \int_n^{+\infty} \frac{1}{x^5} dx$ it follows that the required accuracy is ensured if $\int_n^{+\infty} \frac{1}{x^5} dx < 10^{-3}$.

Calculating the integral, we get $\lim\limits_{b\to+\infty} -\frac{1}{4x^4}|_n^b = \frac{1}{4n^4} < 10^{-3}$, whence $n^4 > \frac{10^3}{4}$ or
$n > 3.98$. Therefore, the partial sum $s_4 = 1 + \frac{1}{2^5} + \frac{1}{3^5} + \frac{1}{4^5} \approx 1.03634$ deviates from the
exact sum of the series by less then 10^{-3}.

Example 7i Verify whether the series $\sum_{n=0}^{\infty} \frac{1}{1+n^2}$ converges and if so, find the number
of terms required to approximate its sum with the accuracy of 10^{-3}. The investigation
of the convergence of this series is simple if we apply the Integral Test. Considering the
corresponding function $f(x) = \frac{1}{1+x^2}$ on the interval $[1, +\infty)$ or $[0, +\infty)$, we see that this
function satisfies the conditions of the test: it is continuous, positive and decreasing on
$[0, +\infty)$. Then, the calculation of the improper integral gives the following result:

$$\int_0^{+\infty} \frac{1}{1+x^2}dx = \lim_{b\to+\infty} \int_0^b \frac{1}{1+x^2}dx = \lim_{b\to+\infty} \arctan x|_0^b = \lim_{b\to+\infty} \arctan b = \frac{\pi}{2}.$$

This means that the integral and the original series converge. To estimate the remainder of
this series we use the inequality

$$r_n \le \int_n^{+\infty} \frac{1}{1+x^2}dx = \lim_{b\to+\infty} \arctan x|_n^b = \frac{\pi}{2} - \arctan n.$$

The required accuracy $r_n < 10^{-3}$ is attained for n that satisfies the inequality $\frac{\pi}{2} - \arctan n < 10^{-3}$, or equivalently, $n > \tan\left(\frac{\pi}{2} - 10^{-3}\right) \approx 999,9997$. This means that
we need $n = 1000$ first terms of the original series to approximate its sum with the error
smaller than 10^{-3}. Evidently, this elevated number of terms, as compared to Example 6i,
is caused by slower convergence of this series: starting from $n = 2$, the general term $\frac{1}{n^5}$
has much smaller values than $\frac{1}{1+n^2}$.

3.3   The Comparison Tests

Theorem 1 (The Comparison Test Without Limit) *Let $\sum_{n=1}^{\infty} a_n$ and $\sum_{n=1}^{\infty} b_n$ be
two non-negative series. If $a_n \le b_n$, $\forall n$, then the convergence of the series $\sum_{n=1}^{\infty} b_n$
implies the convergence of the series $\sum_{n=1}^{\infty} a_n$ (or, equivalently, from the divergence
of $\sum_{n=1}^{\infty} a_n$ it follows the divergence of $\sum_{n=1}^{\infty} b_n$).*

Proof Denoting the partial sums by $s_n = \sum_{k=1}^n a_k$ and $p_n = \sum_{k=1}^n b_k$, we have $s_n \le p_n$,
$\forall n$. If the series $\sum_{n=1}^{\infty} b_n$ converges, then its partial sums p_n are bounded—$p_n \le M$, and
consequently $s_n \le M$. Then, by the Criterion of convergence for non-negative series, the
series $\sum_{n=1}^{\infty} a_n$ converges.   □

Remark As was shown before, the properties of convergence/divergence are not affected by a finite number of the elements of a series. Therefore, the result continue hold if the required relation between the elements of the two series is satisfied starting from some index: $a_n \leq b_n, \forall n > N$.

Theorem 2 (The Comparison Test with Limit) *Let $\sum_{n=1}^{\infty} a_n$ and $\sum_{n=1}^{\infty} b_n$ be two non-negative series and $b_n > 0$ starting from some index. The following relations between the two series hold:*

(1) if $\lim\limits_{n\to\infty} \frac{a_n}{b_n} = c > 0$, then the series $\sum_{n=1}^{\infty} a_n$ converges if and only if the series $\sum_{n=1}^{\infty} b_n$ converges (equivalently, the series $\sum_{n=1}^{\infty} a_n$ diverges if and only if the series $\sum_{n=1}^{\infty} b_n$ diverges);

(2) if $\lim\limits_{n\to\infty} \frac{a_n}{b_n} = 0$, then the convergence of $\sum_{n=1}^{\infty} b_n$ implies the convergence of $\sum_{n=1}^{\infty} a_n$ (equivalently, the divergence of $\sum_{n=1}^{\infty} a_n$ implies the divergence of $\sum_{n=1}^{\infty} b_n$), but the converse is not true;

(3) if $\lim\limits_{n\to\infty} \frac{a_n}{b_n} = +\infty$, then the convergence of $\sum_{n=1}^{\infty} a_n$ implies the convergence of $\sum_{n=1}^{\infty} b_n$ (equivalently, the divergence of $\sum_{n=1}^{\infty} b_n$ implies the divergence of $\sum_{n=1}^{\infty} a_n$), but the converse is not valid.

Proof Consider the first statement. Since $\lim\limits_{n\to\infty} \frac{a_n}{b_n} = c > 0$, there exists N such that $\left| \frac{a_n}{b_n} - c \right| < \varepsilon_c \equiv \frac{c}{2}, \forall n > N$. Equivalently, $\frac{c}{2} < \frac{a_n}{b_n} < \frac{3c}{2}$, and consequently, $a_n < \frac{3c}{2} b_n$ and $b_n < \frac{2}{c} a_n, \forall n > N$. Under the hypothesis that the series $\sum_{n=1}^{\infty} b_n$ converges, the series $\sum_{n=1}^{\infty} \frac{3c}{2} b_n$ also converges and, by the preceding Theorem 1, the series $\sum_{n=1}^{\infty} a_n$ also converges due to the evaluation $a_n < \frac{3c}{2} b_n$. In the same manner, if $\sum_{n=1}^{\infty} a_n$ converges, then $\sum_{n=1}^{\infty} \frac{2}{c} a_n$ also converges, and the incquality $b_n < \frac{2}{c} a_n$ guarantees the convergence of the series $\sum_{n=1}^{\infty} b_n$ according to Theorem 1.

Move on to the second statement. If $\lim\limits_{n\to\infty} \frac{a_n}{b_n} = 0$, then $\left| \frac{a_n}{b_n} \right| < \varepsilon_1 \equiv 1, \forall n > N$, that is, $a_n < b_n$, which guarantees the convergence of $\sum_{n=1}^{\infty} a_n$ if $\sum_{n=1}^{\infty} b_n$ is convergent. At the same time, the example of the series $\sum_{n=1}^{\infty} a_n = \sum_{n=1}^{\infty} \frac{1}{n^2}$ and $\sum_{n=1}^{\infty} b_n = \sum_{n=1}^{\infty} \frac{1}{n}$, whose terms satisfy the relation $\lim\limits_{n\to\infty} \frac{a_n}{b_n} = \lim\limits_{n\to\infty} \frac{n}{n^2} = 0$, shows that the second series can diverge while the first converges.

Finally, in the case $\lim\limits_{n\to\infty} \frac{a_n}{b_n} = +\infty$, the result follows straightforward from the second statement, noting that $\lim\limits_{n\to\infty} \frac{a_n}{b_n} = +\infty$ implies that $\lim\limits_{n\to\infty} \frac{b_n}{a_n} = 0$.   □

Remark Since $a_n \geq 0$ and $b_n \geq 0$, the formulation of Theorem 2 includes all the options that can occur with the limit $\lim\limits_{n\to\infty} \frac{a_n}{b_n}$ if it exists (in finite or infinite form). The only situation that was left out of consideration is when this limit does not exist. This case is discussed below.

Relationship Between the Comparison Tests with and Without Limit and Examples

The proper proof of Theorem 2 reveals that when Theorem 2 is applicable Theorem 1 also works. The converse is not true, because Theorem 2 requires the existence of the limit $\lim_{n \to \infty} \frac{a_n}{b_n}$, which is not used in Theorem 1. Therefore, there are cases when Theorem 1 works, but Theorem 2 does not, which is illustrated by the next examples.

Example 1c The two series $\sum_{n=1}^{\infty} a_n = \sum_{n=1}^{\infty} \frac{2+(-1)^n}{n^2}$ and $\sum_{n=1}^{\infty} b_n = \sum_{n=1}^{\infty} \frac{4}{n^2}$ converge, but the other two series $\sum_{n=1}^{\infty} a_n = \sum_{n=1}^{\infty} \frac{2+(-1)^n}{n}$ and $\sum_{n=1}^{\infty} b_n = \sum_{n=1}^{\infty} \frac{4}{n}$ diverge. However, in both cases the ratio $\frac{a_n}{b_n} = \frac{1}{4}(2 + (-1)^n)$ is the same and its limit does not exist, which means that Theorem 2 is powerless in both cases. Nevertheless, Theorem 1 is applicable directly in the two cases: for the first pair, we have $a_n \leq b_n$ which allows us to conclude that the convergence of $\sum_{n=1}^{\infty} \frac{4}{n^2}$ implies the convergence of $\sum_{n=1}^{\infty} \frac{2+(-1)^n}{n^2}$; for the second pair, the inequality is inverted—$a_n \geq \frac{b_n}{4}$, and therefore, from the divergence of $\sum_{n=1}^{\infty} \frac{1}{n}$ it follows the divergence of $\sum_{n=1}^{\infty} \frac{2+(-1)^n}{n}$.

Example 2c One more example shows that series $\sum_{n=1}^{\infty} a_n$ and $\sum_{n=1}^{\infty} b_n$ may have different behavior when the limit $\lim_{n \to \infty} \frac{a_n}{b_n}$ does not exists. If $a_n = \begin{cases} \frac{1}{n}, & n = 2k-1 \\ \frac{1}{n^2}, & n = 2k \end{cases}, k \in \mathbb{N}$ and $b_n = \frac{1}{n^2}$, then $\frac{a_{2k-1}}{b_{2k-1}} = 2k-1$ and $\frac{a_{2k}}{b_{2k}} = 1$, which indicates that the limit $\lim_{n \to \infty} \frac{a_n}{b_n}$ does not exist. In this example, the series $\sum_{n=1}^{\infty} a_n$ diverges while $\sum_{n=1}^{\infty} b_n$ converges. If we choose $\tilde{b}_n = \frac{1}{n}$, then $\frac{a_{2k-1}}{\tilde{b}_{2k-1}} = 1$ and $\frac{a_{2k}}{\tilde{b}_{2k}} = \frac{1}{2k}$, which shows that again the limit $\lim_{n \to \infty} \frac{a_n}{b_n}$ does not exist, but in this case both series diverge. Notice that for these two pairs of the series Theorem 1 has also no practical use (we cannot draw any conclusion about convergence/divergence of more complicated series $\sum_{n=1}^{\infty} a_n$ using information about the behavior of a simpler series $\sum_{n=1}^{\infty} b_n$).

Although the test without limit has a larger range of applicability, the one with limit can be of a simpler practical use for different series. The examples below illustrate these situations.

Example 3c It is easy to compare the series $\sum_{n=1}^{\infty} \frac{1}{n^2+4n+3}$ with the simpler convergent series $\sum_{n=1}^{\infty} \frac{1}{n^2}$ to conclude that the former also converges: $\frac{1}{n^2+4n+3} \leq \frac{1}{n^2}$. But choosing the very similar series $\sum_{n=1}^{\infty} \frac{1}{n^2-4n-3}$, which intuitively should have the same behavior as the first two series (since the principal term n^2 in the denominator is the same in all the series), we have no possibility to make such simple comparison as in the first case. We have to perform some preliminary evaluations, noting that $\frac{n^2}{2} \geq 4n$ and $\frac{n^2}{4} \geq 3$

for all $n \geq 8$, and only after this we can compare the general term in the following way:
$\frac{1}{n^2-4n-3} = \frac{1}{\frac{n^2}{4}+\frac{n^2}{2}-4n+\frac{n^2}{4}-3} \leq \frac{4}{n^2}$. Therefore, from the convergence of the series $\sum_{n=1}^{\infty} \frac{4}{n^2}$
it follows the convergence of the original series.

However, the intuitive impression does not fail in this case: both series $\sum_{n=1}^{\infty} \frac{1}{n^2+4n+3}$
and $\sum_{n=1}^{\infty} \frac{1}{n^2-4n-3}$ converge with the same rate as the simplest series of this type $\sum_{n=1}^{\infty} \frac{1}{n^2}$.
This can be evidenced in an elementary way by using the test with limit: choosing $a_n =$
$\frac{1}{n^2}$ and $b_n = \frac{1}{n^2+4n+3}$, we have $\frac{a_n}{b_n} = \frac{n^2+4n+3}{n^2} = 1 + \frac{4}{n} + \frac{3}{n^2} \underset{n\to\infty}{\to} 1$, and for the pair
$a_n = \frac{1}{n^2}$ and $\bar{b}_n = \frac{1}{n^2-4n-3}$ we arrive at the same conclusion with the very same easiness
$\frac{a_n}{\bar{b}_n} = \frac{n^2-4n-3}{n^2} = 1 - \frac{4}{n} - \frac{3}{n^2} \underset{n\to\infty}{\to} 1$. In both cases the two secondary terms $\frac{4}{n}$ and $\frac{3}{n^2}$ have
no effect on the limit calculation, that confirms in exact form our intuitive guess that only
the principal term n^2 (the same in all the three series) determines the qualitative behavior
of the series. Hence, a small technical complication in the application of the test without
limit was avoid by applying the test with limit. This complication can be considered as a
practical deficiency of the formulation of the former test.

Example 4c Just as in the preceding Example 3c, let us consider the following two series:
$\sum_{n=3}^{\infty} \frac{\sqrt{2n+5}}{(3n^2-4n-7)^{2/3}}$ and $\sum_{n=3}^{\infty} \frac{\sqrt{2n-5}}{(3n^2+4n+7)^{2/3}}$. A preliminary intuitive evaluation may say
that the two series have the same qualitative behavior, because the principal terms in the
numerator and denominator are the same—$\sqrt{2n}$ and $(3n^2)^{2/3}$, respectively. Therefore,
both series should have the same property of convergence/divergence as the simplest series
of this type $\sum_{n=1}^{\infty} \frac{\sqrt{2n}}{(3n^2)^{2/3}} = \sum_{n=1}^{\infty} \frac{\sqrt{2}}{3^{2/3}} \frac{\sqrt{n}}{n^{4/3}} = \frac{\sqrt{2}}{3^{2/3}} \sum_{n=1}^{\infty} \frac{1}{n^{5/6}}$, and this last series diverges
since it is a p-series with $p = \frac{5}{6} < 1$.

Utilizing the test without limit for the first series we immediately have $\frac{\sqrt{2n+5}}{(3n^2-4n-7)^{2/3}} \geq$
$\frac{\sqrt{2n}}{(3n^2)^{2/3}} = \frac{\sqrt{2}}{3^{2/3}} \frac{1}{n^{5/6}}$, and consequently, this series diverges. However, to apply the same test
to the second series, we have to perform a preliminary technical job: notice that $2n-5 \geq n$,
$\forall n \geq 5$ and $3n^2+4n+7 = 5n^2 - (n^2-4n) - (n^2-7) \leq 5n^2, \forall n \geq 4$. Then, we obtain the
following evaluation $\frac{\sqrt{2n-5}}{(3n^2+4n+7)^{2/3}} \geq \frac{\sqrt{n}}{(5n^2)^{2/3}} = \frac{1}{5^{2/3}} \frac{1}{n^{5/6}}$, which reveals the divergence of
the second series. Evidently, in the case of more complicated form of the secondary terms,
the preliminary work may be more involved.

However, the application of the test with limit is straightforward for both series and
it even shows clearly that our initial intuition does not fail. Indeed, choosing for the
comparison the series $\sum_{n=1}^{\infty} \frac{1}{n^{5/6}}$, we obtain for the first series:

$$\frac{\frac{1}{n^{5/6}}}{\frac{\sqrt{2n+5}}{(3n^2-4n-7)^{2/3}}} = \frac{(3n^2-4n-7)^{2/3}}{n^{5/6}\sqrt{2n+5}} = \frac{(3n^2-4n-7)^{2/3}}{n^{4/3}} \cdot \frac{n^{1/2}}{\sqrt{2n+5}}$$

$$= \left(3 - \frac{4}{n} - \frac{7}{n^2}\right)^{2/3} \cdot \frac{1}{\sqrt{2+\frac{5}{n}}} \underset{n\to\infty}{\to} \frac{3^{2/3}}{\sqrt{2}} > 0.$$

In the same way we get for the second series:

$$\frac{\frac{1}{n^{5/6}}}{\frac{\sqrt{2n-5}}{(3n^2+4n+7)^{2/3}}} = \frac{(3n^2+4n+7)^{2/3}}{n^{5/6}\sqrt{2n-5}} = \frac{(3n^2+4n+7)^{2/3}}{n^{4/3}} \cdot \frac{n^{1/2}}{\sqrt{2n-5}}$$

$$= \left(3 + \frac{4}{n} + \frac{7}{n^2}\right)^{2/3} \cdot \frac{1}{\sqrt{2 - \frac{5}{n}}} \xrightarrow{n\to\infty} \frac{3^{2/3}}{\sqrt{2}} > 0.$$

These calculations show that both series diverge and that the secondary terms in numerator and denominator of both series have no effect on these results.

Example 5c $\sum_{n=2}^{\infty} \frac{\sqrt[n]{n}}{n \ln^2 n}$

Notice first that the limit of the numerator is 1: by L'Hospital's rule $\lim\limits_{x\to+\infty} \frac{\ln x}{x} = \lim\limits_{x\to+\infty} \frac{\frac{1}{x}}{1} = 0$, which implies that $\lim\limits_{n\to+\infty} \frac{\ln n}{n} = \lim\limits_{n\to+\infty} \ln \sqrt[n]{n} = 0$, and consequently, $\lim\limits_{n\to+\infty} \sqrt[n]{n} = \lim\limits_{n\to+\infty} e^{\ln \sqrt[n]{n}} = e^0 = 1$. Second, recall that the series $\sum_{n=2}^{\infty} \frac{1}{n \ln^2 n}$ was already analyzed (see Example 4i) and it was shown that the series converges. Therefore, it is suitable to compare the given series with $\sum_{n=2}^{\infty} \frac{1}{n \ln^2 n}$ in the limit form: $\frac{\frac{\sqrt[n]{n}}{n \ln^2 n}}{\frac{1}{n \ln^2 n}} = \sqrt[n]{n} \xrightarrow{n\to\infty} 1$. Hence, the original series converges.

Notice that the application of the test without limit follows a similar line of reasoning: first, show that $\lim\limits_{n\to+\infty} \sqrt[n]{n} = 1$, from which it follows that $\sqrt[n]{n}$ is a bounded sequence, that is, there exists a constant M such that $\sqrt[n]{n} \leq M, \forall n$. Then, $\frac{\sqrt[n]{n}}{n \ln^2 n} \leq \frac{M}{n \ln^2 n}$ and the convergence of the series $\sum_{n=2}^{\infty} \frac{M}{n \ln^2 n}$ implies the convergence of the original series.

Complement: Nonexistence of an Universal Series for Comparison

Here we show that it does not exist a unique positive series that can be used in the Comparison test to verify convergence/divergence of any positive series.

Let us start with convergence and introduce first the notion of the relative rate of convergence of two series.

▶ **Definition** Let $\sum_{n=1}^{\infty} a_n$ and $\sum_{n=1}^{\infty} a_n'$ be two positive convergent series with the remainders r_n and r_n', respectively. We say that the second *series converges more slowly* than the first one if their remainders satisfy the relation $\lim\limits_{n\to\infty} \frac{r_n}{r_n'} = 0$.

Let us show that for any positive convergent series $\sum_{n=1}^{\infty} a_n$ it can be constructed a positive series that converges more slowly than a given series. Indeed, consider the series $\sum_{n=1}^{\infty} a_n'$, $a_n' = \sqrt{r_{n-1}} - \sqrt{r_n}$, where we define additionally that $r_0 = s = \sum_{n=1}^{\infty} a_n$. Since $a_n > 0$, it follows that $r_{n-1} > r_n$ and $a_n' > 0$. The new series is telescopic, and consequently, it is easy to calculate its partial sums:

$$s_n' = \sum_{k=1}^{n} a_k' = (\sqrt{r_0} - \sqrt{r_1}) + (\sqrt{r_1} - \sqrt{r_2}) + \ldots + (\sqrt{r_{n-2}} - \sqrt{r_{n-1}}) + (\sqrt{r_{n-1}} - \sqrt{r_n})$$

$$= \sqrt{r_0} - \sqrt{r_n} = \sqrt{s} - \sqrt{r_n}.$$

Therefore, the sum of this series is $s' = \sum_{n=1}^{\infty} a_n' = \lim_{n \to \infty} s_n' = \sqrt{s}$ and its remainders are $r_n' = s' - s_n' = \sqrt{r_n}$. Then, substituting r_n and r_n' in the limit of the above Definition, we have

$$\lim_{n \to \infty} \frac{r_n}{r_n'} = \lim_{n \to \infty} \frac{r_n}{\sqrt{r_n}} = \lim_{n \to \infty} \sqrt{r_n} = 0,$$

which means that the series $\sum_{n=1}^{\infty} a_n'$ converges more slowly than $\sum_{n=1}^{\infty} a_n$.

It follows from this result that the original series $\sum_{n=1}^{\infty} a_n$ cannot be used in the Comparison Test to reveal the behavior of the new series $\sum_{n=1}^{\infty} a_n'$, because

$$\lim_{n \to \infty} \frac{a_n}{a_n'} = \lim_{n \to \infty} \frac{r_{n-1} - r_n}{\sqrt{r_{n-1}} - \sqrt{r_n}} = \lim_{n \to \infty} (\sqrt{r_{n-1}} + \sqrt{r_n}) = 0.$$

Rewording, whatever positive convergent series is given, there always exists another positive convergent series whose behavior cannot be revealed by comparison with the first series.

In a similar mode it can be shown that it does not exist a universal positive series that can be used in the Comparison test to verify divergence of any other positive series. To this end, let us introduce the notion of the relative rate of divergence of two series.

▶ **Definition** Let $\sum_{n=1}^{\infty} a_n$ and $\sum_{n=1}^{\infty} a_n'$ be two positive divergent series with the partial sums s_n and s_n', respectively. We say that the second *series diverges more slowly* than the first one if their partial sums satisfy the relation $\lim_{n \to \infty} \frac{s_n'}{s_n} = 0$.

Now we prove that for any positive divergent series $\sum_{n=1}^{\infty} a_n$ there always exists another positive divergent series that diverges more slowly than the given series. One of such series can be defined in the form $\sum_{n=1}^{\infty} a_n'$, $a_n' = \sqrt{s_n} - \sqrt{s_{n-1}}$, where $s_0 = 0$. Then, it can be shown that $s_n' = \sqrt{s_n}$, and consequently $\lim_{n \to \infty} \frac{s_n'}{s_n} = \lim_{n \to \infty} \frac{1}{\sqrt{s_n}} = 0,$

that is, the second series diverges more slowly that the first one. Besides, $\lim\limits_{n \to \infty} \frac{a_n}{a_n'} = \lim\limits_{n \to \infty} (\sqrt{s_n} + \sqrt{s_{n-1}}) = +\infty$, which means that the comparison of the second series with the given series does not reveal any information about the former.

We leave it to the reader to complete the details of the last proof.

3.4   The Cauchy Condensation Test

Let us recall the first "naive" study of the harmonic series $\sum_{n=1}^{\infty} \frac{1}{n}$ in Example 5a of Sect. 1.2: choosing the very particular set of partial sums s_{2^k} and using for evaluation of these sums the very selective sample of the terms $a_{2^k} = \frac{1}{2^k}$ we were able to discover the divergence of this series. A similar method was applied to examine the convergence/divergence of the p-series $\sum_{n=1}^{\infty} \frac{1}{n^p}$ in Example 6a of Sect. 1.2. It happens that this approach, which take into consideration a very few partial sums and terms of the original series, can be extended to general positive series with monotonically decreasing terms. The striking feature of the corresponding result, called the *Cauchy condensation theorem*, is that a rather "thin" subsequence of the original sequence a_n determines the convergence or divergence of the series $\sum a_n$.

Theorem (The Cauchy Condensation Test) *Let a_n be a decreasing non-negative sequence. Then, the series $\sum_{n=1}^{\infty} a_n$ converges if and only if the series $\sum_{n=0}^{\infty} 2^n a_{2^n} = a_1 + 2a_2 + 4a_4 + 8a_8 + \ldots$ converges.*

Proof By the general criterion for convergence of positive series (Sect. 3.1), it suffices to show that the partial sums $s_n = a_1 + a_2 + \ldots + a_n$ and $t_k = a_1 + 2a_2 + \ldots + 2^k a_{2^k}$ are bounded (or unbounded) at the same time. Taking into account that a_n is a decreasing sequence, for $n < 2^k$ we have

$$s_n \leq a_1 + (a_2 + a_3) + \ldots + (a_{2^k} + \ldots + a_{2^{k+1}-1}) \leq a_1 + 2a_2 + \ldots + 2^k a_{2^k} = t_k.$$

On the other hand, if $n \geq 2^k$, then

$$s_n \geq a_1 + a_2 + (a_3 + a_4) + \ldots + (a_{2^{k-1}+1} + \ldots + a_{2^k}) \geq \frac{1}{2} a_1 + a_2 + 2a_4 + \ldots + 2^{k-1} a_{2^k} = \frac{1}{2} t_k.$$

According to these two inequalities, the sequences s_n and t_k are either both bounded or both unbounded. This completes the proof. $\square$

Example 1s $\sum_{n=1}^{\infty} \frac{1}{n^p}$
For $p \leq 0$ the series diverges since its general term does not approach 0. For $p > 0$, the sequence $\frac{1}{n^p}$ is decreasing (and positive), and applying the Cauchy condensation test, we arrive at the geometric series $\sum_{k=0}^{\infty} 2^k \frac{1}{2^{kp}} = \sum_{k=0}^{\infty} 2^{(1-p)k}$, whose behavior is already

well known: the series converges if and only if $1 - p < 0$. Therefore, the original series converges for $p > 1$ and diverges for $p \leq 1$.

Recall that the same result was derived in Example 6a of Sect. 1.2 using rudimentary techniques and then obtained by a straightforward application of the Integral test in Example 2i of Sect. 3.2.

Example 2s $\sum_{n=2}^{\infty} \frac{1}{n \ln^p n}$

If $p \leq 0$ the series diverges by comparison with the harmonic series. If $p > 0$, the sequence $\frac{1}{n \ln^p n}$ is decreasing and positive, and the Cauchy condensation test gives us p-series

$$\sum_{k=1}^{\infty} 2^k \frac{1}{2^k \ln^p 2^k} = \sum_{k=1}^{\infty} \frac{1}{k^p \ln^p 2} = \frac{1}{\ln^p 2} \sum_{k=1}^{\infty} \frac{1}{k^p},$$

whose behavior was studied in the previous example (and in Sect. 3.2). (Alternatively, we can apply the Cauchy condensation test one more time and reduce the problem to the geometric series.) Therefore, the original series converges if $p > 1$ and diverges if $p \leq 1$.

In Example 4i of Sect. 3.2 the same result was obtained using the Integral test.

Example 3s $\sum_{n=3}^{\infty} \frac{1}{n \ln n \ln \ln n}$

Since the sequence $\frac{1}{n \ln n \ln \ln n}$ is decreasing and positive, we can employ the Cauchy condensation test:

$$\sum_{k=2}^{\infty} 2^k \frac{1}{2^k \ln 2^k \ln \ln 2^k} = \sum_{k=2}^{\infty} \frac{1}{k \ln 2 \ln(k \ln 2)} = \frac{1}{\ln 2} \sum_{k=2}^{\infty} \frac{1}{k \ln(k \ln 2)}.$$

The following inequality holds for the general term $\frac{1}{k \ln(k \ln 2)} > \frac{1}{k \ln k}$, $\forall k \geq 2$ and the positive series $\sum_{k=2}^{\infty} \frac{1}{k \ln k}$ diverges (see the previous example). Therefore, by the Comparison test, the series $\sum_{k=2}^{\infty} \frac{1}{k \ln(k \ln 2)}$ also diverges, and consequently, the original series diverges.

The same result can be found by application of the Integral test (see Exercise 9.9)).

Complement: Schlömilch's Test

The following generalization by Schlömilch shows that we can choose other subsequences of a_n keeping the same result.

Theorem (Schlömilch's Test) *Let a_n be a decreasing non-negative sequence, and let n_k be a strictly increasing sequence of positive integers such that $\frac{n_{k+1} - n_k}{n_k - n_{k-1}}$ is bounded, that is, $\frac{n_{k+1} - n_k}{n_k - n_{k-1}} \leq C$, $\forall k$. Then, the series $\sum_{n=1}^{\infty} a_n$ converges if and only if the series $\sum_{k=0}^{\infty} (n_{k+1} - n_k) a_{n_k}$ converges.*

Proof By the general criterion for convergence of positive series (Sect. 3.1), it is sufficient to demonstrate that the partial sums

$$s_n = a_1 + a_2 + \ldots + a_n$$

and

$$t_k = (n_1 - n_0)a_{n_0} + (n_2 - n_1)a_{n_1} + \ldots + (n_k - n_{k-1})a_{n_{k-1}} + (n_{k+1} - n_k)a_{n_k}$$

are bounded (or unbounded) at the same time. Taking into account that a_n is a decreasing sequence, if $n < n_k$ we get

$$s_n \leq (a_1 + \ldots + a_{n_0-1}) + (a_{n_0} + \ldots + a_{n_1-1}) + (a_{n_1} + \ldots + a_{n_2-1}) + \ldots + (a_{n_{k-1}} + \ldots + a_{n_k-1})$$

$$\leq s_{n_0-1} + (n_1 - n_0)a_{n_0} + (n_2 - n_1)a_{n_1} + \ldots + (n_k - n_{k-1})a_{n_{k-1}} = s_{n_0-1} + t_{k-1}.$$

On the other hand, if $n \geq n_k$, then

$$s_n \geq (a_1 + \ldots + a_{n_0}) + (a_{n_0+1} + \ldots + a_{n_1}) + (a_{n_1+1} + \ldots + a_{n_2}) + \ldots + (a_{n_{k-1}+1} + \ldots + a_{n_k})$$

$$\geq s_{n_0} + (n_1 - n_0)a_{n_1} + (n_2 - n_1)a_{n_2} + \ldots + (n_k - n_{k-1})a_{n_k}$$

$$\geq s_{n_0} + \frac{1}{C}(n_2 - n_1)a_{n_1} + \frac{1}{C}(n_3 - n_2)a_{n_2} + \ldots + \frac{1}{C}(n_{k+1} - n_k)a_{n_k} = s_{n_0} + \frac{1}{C}(t_k - t_0).$$

According to the derived inequalities, the sequences s_n and t_k are either both bounded or both unbounded. This completes the proof. $\square$

Remark The Cauchy condensation test is a special case of Schlömilch's test under the choice $n_k = 2^k$.

Example 4s $\sum_{n=1}^{\infty} \frac{1}{2\sqrt{n}}$

The sequence $\frac{1}{2\sqrt{n}}$ is decreasing and positive. Choose the strictly increasing sequence of integers $n_k = k^2$ and notice that $\frac{(k+1)^2 - k^2}{k^2 - (k-1)^2} = \frac{2k+1}{2k-1} \leq 3$, $\forall k \geq 0$. Hence, the conditions of Schlömilch's test are satisfied, and the convergence of the original series is equivalent to the convergence of the series $\sum_{k=1}^{\infty} \frac{(k+1)^2 - k^2}{2\sqrt{k^2}} = \sum_{k=1}^{\infty} \frac{2k+1}{2^k}$. Since $\frac{2k+1}{2^k} < \frac{4k}{2^k} = \frac{k}{2^{k-2}} < \frac{1}{2^{k/2}} = \frac{1}{\sqrt{2}^k}$ for sufficiently large k (for instance, for $\forall k \geq 12$) and the geometric series $\sum_{k=1}^{\infty} \frac{1}{\sqrt{2}^k}$ converges, by the Comparison test, the series $\sum_{k=1}^{\infty} \frac{2k+1}{2^k}$ also converges and, by Schlömilch's test, so does the original series.

Historical Remarks 1 In his famous and popular textbook "Cours d'analyse" published in 1821, Cauchy provided a theoretical basis for a general investigation of series, posed the problem of searching for the test of convergence/divergence and gave different tests with rigorous proofs.

Historical Remarks 2 Cauchy introduced and proved the integral test in the paper of 1827. However, the use of an integral for evaluation of the sum of a series was already known to mathematicians of the 18th century and appeared explicitly in the works of Maclaurin and Euler.

The Comparison test was in common use and considered obvious by many mathematicians in the eighteenth century. For instance, in the 1768 study of the binomial series, D'Alembert used the assumption that convergence/divergence of a series can be determined by comparison with another series whose convergence/divergence is known. In "Cours d'analyse" of 1821, Cauchy did not explicitly state the Comparison test, but he used it for particular series and proved it in this case.

The condensation test was first given by Cauchy in his "Cours d'analyse" in 1821. He came to this result analyzing the proof of divergence of the harmonic series by the Oresme method (see Example 5a in Sect. 1.2) and a similar technique of the proof of convergence of p-series with $p > 1$ (see Example 6a in Sect. 1.2).

3.5   D'Alembert's Tests (The Ratio Tests)

Theorem 1D (D'Alembert's (Ratio) Test Without Limit) *Suppose that $\sum_{n=1}^{\infty} a_n$ is a positive series and denote by $D_n = \frac{a_{n+1}}{a_n}$ the D'Alembert variable. Then,*

(1) if $D_n \le q < 1$, $\forall n \ge N$, where q is a constant, $0 < q < 1$, then the series converges;
(2) if $D_n \ge 1$, $\forall n \ge N$, then the series diverges.

Proof Under the first hypothesis, we have $D_n = \frac{a_{n+1}}{a_n} \le q$, $\forall n \ge N$, that is, $a_{n+1} \le q a_n$, $\forall n \ge N$. Writing this inequality for the first terms, we obtain

$$a_{N+1} \le q a_N, \ a_{N+2} \le q a_{N+1} \le q^2 a_N, \ \dots \ , \ a_{N+k} \le q a_{N+k-1} \le \dots \le q^k a_N.$$

Since $0 < q < 1$, the geometric series $\sum_{k=1}^{\infty} a_N q^k$ converges, which implies (by the Comparison Test) the convergence of the series $\sum_{k=1}^{\infty} a_{N+k}$. Therefore, the original series converges.

Under the second hypothesis, we get $D_n = \frac{a_{n+1}}{a_n} \ge 1$, $\forall n \ge N$, that is, $a_{n+1} \ge a_n$, $\forall n \ge N$. This means that

$$a_{N+1} \ge a_N, \ a_{N+2} \ge a_{N+1} \ge a_N, \ \dots \ , \ a_{N+k} \ge a_{N+k-1} \ge \dots \ge a_N.$$

Since a_N is a positive constant and $a_{N+k} \geq a_N > 0$, $\forall k$, we conclude that the general term does not approach 0, and consequently, the series diverges. $\qquad\square$

Theorem 2D (D'Alembert's (Ratio) Test with Limit) *Suppose that $\sum_{n=1}^{\infty} a_n$ is a positive series and there exists the limit $\lim_{n\to\infty} D_n = D$ (including the case of the infinite limit $D = +\infty$), where $D_n = \frac{a_{n+1}}{a_n}$ is the D'Alembert variable. Then,*

(1) if $D < 1$, the series converges;
(2) if $D > 1$, the series diverges;
(3) if $D = 1$, the case is inconclusive.

Proof The first two cases can be demonstrated by reducing them to Theorem 1D. If $D < 1$, then there exists a number q such that $D < q < 1$. Therefore, by the comparison properties of limits, starting from some index N the elements of the sequence D_n satisfy the same inequality: $D_n < q < 1$, $\forall n \geq N$. Then, all the conditions of Theorem 1D hold, and consequently, the series converges.

If $D > 1$, then, by the comparison properties of limits, $D_n > 1$, $\forall n \geq N$, and the second result of Theorem 1D guarantees the divergence of the series.

In the last case, we provide two examples of simple series—harmonic $\sum_{n=1}^{\infty} \frac{1}{n}$ and $p = 2$-series $\sum_{n=1}^{\infty} \frac{1}{n^2}$, which have the same limit $D = 1$, but the former is divergent while the latter is convergent. (For the harmonic series we have $D = \lim_{n\to\infty} D_n = \lim_{n\to\infty} \frac{n}{n+1} = \lim_{n\to\infty} \frac{1}{1+\frac{1}{n}} = 1$, and for the $p = 2$-series we obtain $D = \lim_{n\to\infty} D_n = \lim_{n\to\infty} \frac{n^2}{(n+1)^2} = \lim_{n\to\infty} \frac{1}{\left(1+\frac{1}{n}\right)^2} = 1$.) $\qquad\square$

Remark 1 If the limit $\lim_{n\to\infty} D_n$ does not exist, then we cannot draw any conclusion. Indeed, for the series $\sum_{n=1}^{\infty} \frac{2+(-1)^n}{n^2}$ the required limit does not exist because $D_{2k-1} = \frac{\frac{3}{(2k)^2}}{\frac{1}{(2k-1)^2}} = 3\frac{(2k-1)^2}{(2k)^2} \xrightarrow{k\to\infty} 3$, while $D_{2k} = \frac{\frac{1}{(2k+1)^2}}{\frac{3}{(2k)^2}} = \frac{1}{3}\frac{(2k)^2}{(2k+1)^2} \xrightarrow{k\to\infty} \frac{1}{3}$. The same happens with the series $\sum_{n=1}^{\infty} \frac{2+(-1)^n}{n}$: $D_{2k-1} = \frac{\frac{3}{2k}}{\frac{1}{2k-1}} = 3\frac{2k-1}{2k} \xrightarrow{k\to\infty} 3$, but $D_{2k} = \frac{\frac{1}{2k+1}}{\frac{3}{2k}} = \frac{1}{3}\frac{2k}{2k+1} \xrightarrow{k\to\infty} \frac{1}{3}$. Under these circumstances, the first series converges (compare it with the convergent series $\sum_{n=1}^{\infty} \frac{3}{n^2}$), but the second diverges (compare it with the harmonic series $\sum_{n=1}^{\infty} \frac{1}{n}$). Notice that Theorem 1D is also not applicable to these two series.

Remark 2 The method of the proof of D'Alembert's test shows that both forms of the test are not very subtle. The convergence case is based on the comparison with the geometric series, whose general term has the exponential rate of decreasing. This leaves many other series with slower decreasing of the general term out of scope of D'Alembert's test. The divergence case is even more primitive by deducing the series divergence from the fact that the general term does not tend to 0. In such cases it is frequently simpler and more straightforward to apply the Divergence Test. Despite its limitations, D'Alembert's test (in both forms) is a powerful and popular tool for investigating the behavior of many important and complicated series.

Relationship Between the Tests with and Without Limit and Some Examples

The method of the proof of the test with limit by reducing it to the test without limit shows that if the former is conclusive then the latter is also applicable. There are series that cannot be treated by either of the two forms of the test. For example, the two already considered series—$\sum_{n=1}^{\infty} \frac{1}{n}$ and $\sum_{n=1}^{\infty} \frac{1}{n^2}$—for which the test with limit is inconclusive since $\lim_{n\to\infty} D_n = 1$, cannot also be solved by Theorem 1D: for both series $D_n < 1$, but it does not exist constant $q < 1$ that separates D_n from 1 for all the indices $n \geq N$. However, there are series, like $\sum_{n=1}^{\infty} n$, for which the test without limit is applicable ($D_n = \frac{n+1}{n} \geq 1$, $\forall n$), but that with limit does not work ($D = \lim_{n\to\infty} \frac{n+1}{n} = 1$).

Generalizing the discussion for the case of the p-series $\sum_{n=1}^{\infty} \frac{1}{n^p}$, we see from the used proof that D'Alembert's test is based on the comparison with the geometric series, whose general term q^n is of the exponential type. Therefore, it is hard to expect that this test will work in the cases when the general term of a given series has the rate of decreasing quite smaller, like the term $\frac{1}{n^p}$ in the p-series which is of polynomial type. In fact, calculating the limit we have $\lim_{n\to\infty} D_n = \lim_{n\to\infty} \frac{n^p}{(n+1)^p} = 1$, $\forall p$, that is, the test is inconclusive. For the test without limit the situation is slightly different. If $p > 0$, then $D_n = \frac{n^p}{(n+1)^p} < 1$, but it does not exist $q < 1$ that separates D_n from 1 for all $n \geq N$, and consequently, the test does not work. However, if $p \leq 0$, then $D_n = \frac{n^p}{(n+1)^p} \geq 1$, which means that the series diverges. Even so, the last case is a trivial one when the general term of the p-series does not approach 0, and consequently, the divergence can be detected by the simpler Divergence Test.

Another interesting example is the series $\sum_{n=1}^{\infty} a_n = \frac{1}{3} + \frac{1}{3\cdot4} + \frac{1}{3^2\cdot4} + \frac{1}{3^2\cdot4^2} + \ldots$, that is, $a_n = \begin{cases} \frac{1}{3^k\cdot4^{k-1}}, & n = 2k-1 \\ \frac{1}{3^k\cdot4^k}, & n = 2k \end{cases}$, $\forall k \in \mathbb{N}$. In this case, $D_{2k-1} = \frac{a_{2k}}{a_{2k-1}} = \frac{3^k\cdot4^{k-1}}{3^k\cdot4^k} = \frac{1}{4}$ and $D_{2k} = \frac{a_{2k+1}}{a_{2k}} = \frac{3^k\cdot4^k}{3^{k+1}\cdot4^k} = \frac{1}{3}$. Since the even and odd subsequences have different limits, the limit of D_n does not exist and the test with limit does not work. But the test without limit has an immediate application: $D_n \leq \frac{1}{3}$, $\forall n \in \mathbb{N}$, that shows the convergence of the series. Notice that this series is of the geometric type (evidently, $\frac{1}{4^n} < a_n < \frac{1}{3^n}$, which implies directly the convergence of the series), but even so, the test with limit fails because of non-existence of the required limit.

Based on the provided examples and discussion we can conclude that the test without limit has a wider scope of applicability. Nevertheless, in many cases the test with limit is easier to use. Sometimes it is worth to combine the two forms of D'Alembert's test.

Let us consider some examples.

Example 1D $\sum_{n=1}^{\infty} \frac{n^3}{3^n}$
For the test without limit we have

$$D_n = \frac{a_{n+1}}{a_n} = \frac{\frac{(n+1)^3}{3^{n+1}}}{\frac{n^3}{3^n}} = \frac{(n+1)^3}{n^3} \cdot \frac{3^n}{3^{n+1}} = \frac{1}{3}\left(1 + \frac{1}{n}\right)^3 \leq \frac{1}{2}, \; \forall n \geq 7.$$

For the test with limit we obtain

$$D_n = \frac{a_{n+1}}{a_n} = \frac{1}{3}\left(1 + \frac{1}{n}\right)^3 \xrightarrow[n \to \infty]{} \frac{1}{3} < 1.$$

Both forms indicate that the series converges.

Example 2D $\sum_{n=1}^{\infty} \frac{a^n}{n!}, a > 0$
The form without limit:

$$D_n = \frac{a_{n+1}}{a_n} = \frac{\frac{a^{n+1}}{(n+1)!}}{\frac{a^n}{n!}} = \frac{a}{n+1} < \frac{1}{2}, \; \forall n \geq 2a.$$

The form with limit:

$$D_n = \frac{a_{n+1}}{a_n} = \frac{a}{n+1} \xrightarrow[n \to \infty]{} 0.$$

Both forms reveal that the series converges for any $a > 0$.

Example 3D $\sum_{n=1}^{\infty} n! \left(\frac{a}{n}\right)^n, a > 0$
The D'Alembert variable has the form

$$D_n = \frac{a_{n+1}}{a_n} = \frac{(n+1)!\left(\frac{a}{n+1}\right)^{n+1}}{n!\left(\frac{a}{n}\right)^n} = a(n+1)\frac{n^n}{(n+1)^{n+1}} = a\frac{1}{\left(1 + \frac{1}{n}\right)^n}.$$

Applying the limit we have $D_n = a\frac{1}{\left(1 + \frac{1}{n}\right)^n} \xrightarrow[n \to \infty]{} \frac{a}{e}$. Then, for $a < e$ the series converges and for $a > e$ diverges. If $a = e$, the test with limit is inconclusive, but noting that $\left(1 + \frac{1}{n}\right)^n$ approximates e by the values smaller than e, we conclude that $D_n(e) = \frac{e}{\left(1 + \frac{1}{n}\right)^n} > 1, \forall n$, and consequently, the series diverges by the test without limit.

3.6 Cauchy's Tests (The Root Tests)

The formulation of these tests, their treatment and application is quite similar to that of D'Alembert's tests.

Theorem 1C (Cauchy's (Root) Test Without Limit) *Suppose that $\sum_{n=1}^{\infty} a_n$ is a positive series and denote by $C_n = \sqrt[n]{a_n}$ the Cauchy variable. Then,*

(1) if $C_n \leq q < 1, \forall n \geq N$, where q is a constant, $0 < q < 1$, then the series converges;
(2) if $C_n \geq 1, \forall n \geq N$, then the series diverges.

Proof In the first case we have $C_n = \sqrt[n]{a_n} \leq q < 1$, and then $a_n \leq q^n, \forall n \geq N$. Since the geometric series $\sum_{k=N+1}^{\infty} q^k$ converges, by the Comparison Test, the series $\sum_{k=N+1}^{\infty} a_k$ also converges, which implies the convergence of the original series.

In the second case, from $C_n = \sqrt[n]{a_n} \geq 1$ it follows that $a_n \geq 1, \forall n \geq N$, which means that the general term does not converge to 0, and consequently, the series diverges. □

Theorem 2C (Cauchy's (Root) Test with Limit) *Suppose that $\sum_{n=1}^{\infty} a_n$ is a positive series and there exists the limit $\lim_{n\to\infty} C_n = C$ (including the case of the infinite limit $C = +\infty$), where $C_n = \sqrt[n]{a_n}$ is the Cauchy variable. Then,*

(1) if $C < 1$, the series converges;
(2) if $C > 1$, the series diverges;
(3) if $C = 1$, the case is inconclusive.

Proof The first two cases can be demonstrated by reducing them to Theorem 1C. If $C < 1$, then there exists a constant q such that $C < q < 1$. Therefore, by the comparison properties of limits, starting from some index N the elements of the sequence C_n satisfy the same inequality: $C_n < q < 1, \forall n \geq N$. Then, all the conditions of Theorem 1C are met, and consequently, the series converges.

If $C > 1$, then, by the comparison properties of limits, $C_n > 1, \forall n \geq N$, and the second result of Theorem 1C guarantees the divergence of the series.

In the third case, we can use the two series, one convergent and another divergent, of Theorem 2D: the divergent harmonic series $\sum_{n=1}^{\infty} \frac{1}{n}$ and the convergent $p = 2$-series $\sum_{n=1}^{\infty} \frac{1}{n^2}$. For the first series we have $C = \lim_{n\to\infty} \sqrt[n]{\frac{1}{n}}$. To solve the indeterminate form 0^0, we introduce the variable $t = \frac{1}{n}$ and calculate the limit of the function $\ln t^t = t \ln t$ as t approaches 0 by applying L'Hospital's rule: $\lim_{t\to 0} t \ln t = \lim_{t\to 0} \frac{\ln t}{\frac{1}{t}} = \lim_{t\to 0} \frac{\frac{1}{t}}{-\frac{1}{t^2}} = \lim_{t\to 0} (-t) = 0$. Taking into account the relation $t = \frac{1}{n}$ and using the composite property of limits, we obtain $C = \lim\limits_{n\to\infty} \sqrt[n]{\frac{1}{n}} = e^0 = 1$. For the second series we get $C = \lim\limits_{n\to\infty} \sqrt[n]{\frac{1}{n^2}}$ and following the same reasoning, we introduce the function $\ln t^{2t} = 2t \ln t$ and reduce the indeterminate form to the limit $\lim\limits_{t\to 0} 2t \ln t = 0$. Again, applying the composite property of limits, we arrive at the same result: $C = \lim\limits_{n\to\infty} \sqrt[n]{\frac{1}{n^2}} = e^0 = 1$. $\qquad\square$

Remark 1 If the limit $\lim\limits_{n\to\infty} C_n$ does not exist, we cannot draw any conclusion. Indeed, for the series $\sum_{n=1}^{\infty} \left(\frac{1}{2} + (-1)^n \frac{1}{4}\right)^n$ the limit of $C_n = \frac{1}{2} + (-1)^n \frac{1}{4}$ does not exists, because the odd subsequence has the limit $\frac{1}{4}$, but the limit of the even subsequence is $\frac{3}{4}$. Nevertheless, a simple comparison with the convergent geometric series $\sum_{n=1}^{\infty} \left(\frac{3}{4}\right)^n$ shows that $\left(\frac{1}{2} + (-1)^n \frac{1}{4}\right)^n \leq \left(\frac{3}{4}\right)^n$, $\forall n$, and consequently, the original series converges. A similar effect we have for the series $\sum_{n=1}^{\infty} \left(2 + (-1)^n\right)^n$: the limit of $C_n = 2 + (-1)^n$ does not exist since the partial limit of the odd subsequence is 1 and that of the even is 3, but the series evidently diverges, because its general term does not tend to 0 (it is greater than or equal to 1 for any index n). Thus, these two series does not have the limit $\lim\limits_{n\to\infty} C_n$, but the first series converges while the second diverges.

Remark 2 The same observations made for D'Alembert's test in Remark 2 of the preceding section are true for Cauchy's test. This test is not very sensitive in analysis of many series. Its convergence case is based on the comparison with the geometric series, that leaves many series with slower decreasing of the general term out of scope of applicability. The divergence case is even more primitive since it is based on the fact that the general term does not approach 0. Despite these limitations, Cauchy's test (in both forms) is powerful and popular tool for investigating the behavior of many important and complicated series.

Relationship Between the Tests with and Without Limit and Some Examples

Here we follow the same reasoning and arrive at the same conclusions as for the two forms of D'Alembert's test.

First notice that the proof of Theorem 2C reveals that in all cases when the test with limit is conclusive, the one without limit is also applicable. There are series that cannot be treated by either of the two forms of Cauchy's test. For example, the two series indistinguishable by the test with limit—$\sum_{n=1}^{\infty} \frac{1}{n}$ and $\sum_{n=1}^{\infty} \frac{1}{n^2}$—are also not solvable by the test without limit: for both of them $C_n < 1$, $\forall n \geq 2$, but it does not exist a number $q < 1$ which separates C_n from 1 for all $n \geq N$. However, there are series, like $\sum_{n=1}^{\infty} n$, for which the test without limit works ($C_n = \sqrt[n]{n} \geq 1$, $\forall n$), but that with limit does not ($C = \lim\limits_{n\to\infty} \sqrt[n]{n} = 1$).

The same conclusion about the p-series $\sum_{n=1}^{\infty} \frac{1}{n^p}$ made for D'Alembert's test is also valid for Cauchy's test. Indeed, the convergence case of both forms is based on comparison with the convergent geometric series, whose general term q^n has the exponential rate of decreasing, while the general term of the p-series has the polynomial form. Therefore, it is natural to guess that Cauchy's test is not able to distinguish between different polynomial rates of variation. In fact, the test with limit is inconclusive for any p, because $\lim_{n \to \infty} C_n = \lim_{n \to \infty} \sqrt[n]{\frac{1}{n^p}} = 1$. The form without limit does not differentiate the behavior of the p-series when $p > 0$, because $C_n = \sqrt[n]{\frac{1}{n^p}} < 1, \forall n$, but there is no constant $q < 1$ that can separate C_n from 1 for all $n \geq N$. Only when $p \leq 0$ the test without limit is conclusive since $C_n = \sqrt[n]{\frac{1}{n^p}} \geq 1$. However, in the last case, it is more straightforward to apply the simpler Divergence Test.

Other examples of series for which the two forms of Cauchy's test work differently are the series $\sum_{n=1}^{\infty} \left(\frac{1}{2} + \frac{(-1)^n}{4} \right)^n$ and $\sum_{n=1}^{\infty} \left(2 + (-1)^n \right)^n$. It was shown in the above Remark 1 that the test with limit is inconclusive for both series just because the required limit does not exist. However, both series can be analyzed in a simple way by the test without limit. For the first series $C_n = \frac{1}{2} + \frac{(-1)^n}{4} \leq \frac{3}{4}$ that guarantees its convergence, and for the second—$C_n = 2 + (-1)^n \geq 1$, that show its divergence. Notice that both series are of the geometric type with the ratio oscillating between $\frac{1}{4}$ and $\frac{3}{4}$ in the first series (which shows immediately its convergence) and with the ratio oscillating between 1 and 3 in the second series (which implies directly its divergence). Even so, the test with limit fails for these two series.

Based on the provided examples and discussion we can conclude that the test without limit has wider scope of use. Nevertheless, in many cases the test with limit is easier to apply. Sometimes it is worth to combine the two forms of Cauchy's test.

Let us consider some examples. To compare the use of the Cauchy's and D'Alambert's tests, we include the previous examples.

Example 1C $\sum_{n=1}^{\infty} \frac{n^3}{3^n}$
This is Example 1D of D'Alembert's test. To solve it using Cauchy's test, calculate first $C_n = \sqrt[n]{a_n} = \sqrt[n]{\frac{n^3}{3^n}} = n^{3/n} \frac{1}{3}$. The limit of $n^{3/n}$ can be reduced to the limit of the function $\ln t^{3/t} = \frac{3}{t} \ln t$ which can be found by L'Hospital's rule: $\lim_{t \to +\infty} \frac{3 \ln t}{t} = \lim_{t \to +\infty} \frac{3}{t} = 0$. Then, $\lim_{n \to \infty} n^{3/n} = 1$, and consequently $C = \lim_{n \to \infty} n^{3/n} \frac{1}{3} = \frac{1}{3} < 1$, which means that the series converges.

We can see that both D'Alembert's and Cauchy's tests are working in this case, but Cauchy's test requires a slightly larger volume of technical manipulations.

Example 2C $\sum_{n=1}^{\infty} \frac{a^n}{n!}, a > 0$
This is Example 2D of D'Alembert's test. First, let us derive a suitable evaluation of $n!$ in the form: $n! > \left(\frac{n}{e} \right)^n$. To do this, we start from the properties of the sequence $\left(1 + \frac{1}{n} \right)^n$: this sequence is increasing and convergent to e. This implies, in particular, that $\left(1 + \frac{1}{n} \right)^n < e$

or, equivalently, $\left(\frac{n}{n+1}\right)^n > \frac{1}{e}$, $\forall n \in \mathbb{N}$. To prove the desired evaluation of $n!$, we use the method of induction. For $n = 1$ the inequality is trivial: $1! > \frac{1}{e}$. Let us suppose that the inequality is true for n and demonstrate that this implies its validity for $n + 1$. Utilizing the induction hypothesis and the inequality $\left(\frac{n}{n+1}\right)^n > \frac{1}{e}$, we obtain

$$(n+1)! = n!(n+1) \geq \left(\frac{n}{e}\right)^n (n+1) = \left(\frac{n+1}{e}\right)^n \left(\frac{n}{n+1}\right)^n (n+1) > \left(\frac{n+1}{e}\right)^n \frac{1}{e}(n+1)$$

$$= \left(\frac{n+1}{e}\right)^{n+1},$$

which confirms that the desired evaluation for $n!$ is true.

Turn now to the two forms of Cauchy's test. The Cauchy variable takes the form $C_n = \sqrt[n]{a_n} = \sqrt[n]{\frac{a^n}{n!}} = \frac{a}{\sqrt[n]{n!}}$. According to the evaluation $n! > \left(\frac{n}{e}\right)^n$, we get $C_n = \frac{a}{\sqrt[n]{n!}} < \frac{a}{\frac{n}{e}} = \frac{ae}{n} < \frac{1}{2}$, $\forall n > 2ae$. This means that the series converges for any $a > 0$. Calculating the limit $C_n = \frac{a}{\sqrt[n]{n!}} < \frac{ae}{n} \underset{n \to \infty}{\to} 0$, we arrive at the same result.

We conclude that although both D'Alembert's and Cuachy's tests are applicable to this series, Cauchy's test requires more involved technical evaluations.

Example 3C $\sum_{n=1}^{\infty} \frac{(n^3+2)3^{2n}}{(n^4+n^2)5^n}$

The variable C_n is found in the form $C_n = \sqrt[n]{a_n} = \sqrt[n]{\frac{n^3+2}{n^4+n^2} \cdot \left(\frac{3^2}{5}\right)^n} = \sqrt[n]{\frac{n^3+2}{n^4+n^2}} \cdot \frac{9}{5}$. Since $\lim_{n \to \infty} \sqrt[n]{\frac{n^3+2}{n^4+n^2}} = 1$, it follows that $C = \lim_{n \to \infty} C_n = 1 \cdot \frac{9}{5} = \frac{9}{5} > 1$, which means that the series diverges.

3.7 Comparison Between D'Alembert's and Cauchy's Tests

The examples solved so far show that when both tests are applicable, a technical work required in Cauchy's test is more involved. Generalizing, D'Alembert's test is frequently easier to employ than Cauchy's test, since it is usually easier to compute ratios than n-th roots. However, in compensation, Cauchy's test can be applied to the series which are not solvable by D'Alembert's test.

Let us analyze some examples. Consider the series

$$\sum_{n=1}^{\infty} a_n = \frac{1}{4} + \frac{1}{2} + \frac{1}{16} + \frac{1}{8} + \frac{1}{64} + \frac{1}{32} + \dots = \sum_{n=1}^{\infty} 2^{(-1)^n - n}.$$

Its general term has different expressions for even and odd indices:
$$a_n = \begin{cases} \frac{1}{2^{n+1}}, & n = 2k-1 \\ \frac{1}{2^{n-1}}, & n = 2k \end{cases}, \forall k \in \mathbb{N}.$$ Correspondingly, the D'Alembert variable has also two different forms:

$$D_{2k-1} = \frac{a_{2k}}{a_{2k-1}} = \frac{2^{1-2k}}{2^{-1-2k+1}} = 2 > 1, \quad D_{2k} = \frac{a_{2k+1}}{a_{2k}} = \frac{2^{-1-2k-1}}{2^{1-2k}} = \frac{1}{8} < 1, \forall k \in \mathbb{N}.$$

Then, both forms of D'Alembert's test (without limit or with limit) are inconclusive about the behavior of this series. At the same time, Cauchy's test can be applied in either of the two forms. Indeed, for $\forall k \in \mathbb{N}$ we have

$$C_{2k-1} = \sqrt[2k-1]{2^{-1-2k+1}} = \left(\frac{1}{2}\right)^{\frac{2k}{2k-1}} = \frac{1}{2} \cdot \left(\frac{1}{2}\right)^{\frac{1}{2k-1}} < \frac{1}{2}$$

and

$$C_{2k} = \sqrt[2k]{2^{1-2k}} = \left(\frac{1}{2}\right)^{\frac{2k-1}{2k}} = \frac{1}{2} \cdot \left(\frac{1}{2}\right)^{-\frac{1}{2k}} \leq \frac{1}{\sqrt{2}}.$$

Therefore, $C_n \leq \frac{1}{\sqrt{2}} < 1$, $\forall n$ and the series is convergent. The same conclusion is obtained using the limit form: $\lim\limits_{k \to \infty} C_{2k-1} = \lim\limits_{k \to \infty} \frac{1}{2} \cdot \left(\frac{1}{2}\right)^{\frac{1}{2k-1}} = \frac{1}{2}$ and $\lim\limits_{k \to \infty} C_{2k} = \lim\limits_{k \to \infty} \frac{1}{2} \cdot \left(\frac{1}{2}\right)^{-\frac{1}{2k}} = \frac{1}{2}$, whence $\lim\limits_{n \to \infty} C_n = \frac{1}{2} < 1$, and consequently, the series converges.

A similar situation may occur with divergent series. Consider the series

$$\sum_{n=1}^{\infty} a_n = 4 + 2 + 16 + 8 + 64 + 32 + \ldots = \sum_{n=1}^{\infty} 2^{n-(-1)^n}.$$

Again, the general term is defined in a piecewise manner: $a_n = \begin{cases} 2^{n+1}, & n = 2k-1 \\ 2^{n-1}, & n = 2k \end{cases}$, $\forall k \in \mathbb{N}$, and consequently, the D'Alembert variable has two different expressions:

$$D_{2k-1} = \frac{a_{2k}}{a_{2k-1}} = \frac{2^{2k-1}}{2^{2k-1+1}} = \frac{1}{2} < 1, \quad D_{2k} = \frac{a_{2k+1}}{a_{2k}} = \frac{2^{2k+1+1}}{2^{2k-1}} = 8 > 1, \forall k \in \mathbb{N}.$$

Then, neither of the two forms of D'Alembert's test is applicable to this series. On the other hand, both forms of Cauchy's test can be used. Indeed, for $\forall k \in \mathbb{N}$ one has

$$C_{2k-1} = \sqrt[2k-1]{2^{2k-1+1}} = 2^{\frac{2k}{2k-1}} = 2 \cdot 2^{\frac{1}{2k-1}} > 2$$

and

$$C_{2k} = \sqrt[2k]{2^{2k-1}} = 2^{\frac{2k-1}{2k}} = 2 \cdot 2^{-\frac{1}{2k}} \geq \sqrt{2}.$$

Therefore, $C_n \geq \sqrt{2} > 1$, $\forall n$ which means that the series diverges. The same result can be obtained using the limit form: $\lim_{k\to\infty} C_{2k-1} = \lim_{k\to\infty} 2 \cdot 2^{\frac{1}{2k-1}} = 2$ and $\lim_{k\to\infty} C_{2k} = \lim_{k\to\infty} 2 \cdot 2^{-\frac{1}{2k}} = 2$, whence $\lim_{n\to\infty} C_n = 2 > 1$, and consequently, the series diverges.

Naturally, the question whether there are series solvable by D'Alembert's test but not treatable by Cauchy's test arises. The answer to this question is negative and it follows from the following relation between the two limits involved in these tests: if $\lim_{n\to\infty} D_n$ exists, then $\lim_{n\to\infty} C_n$ also exists, but the converse is not true. The fact that the converse is not valid was substantiated by the two last examples. Let us prove now the general statement about the implication between the two limits.

Theorem (Relationship Between the Limits of D'Alembert and Cauchy) *Let $\sum_{n=1}^{\infty} a_n$ be a positive series, $D_n = \frac{a_{n+1}}{a_n}$ the D'Alembert variable and $C_n = \sqrt[n]{a_n}$ the Cauchy variable. If $\lim_{n\to\infty} D_n$ exists, then $\lim_{n\to\infty} C_n$ also exists and the two limits are equal.*

Proof By the condition, $\lim_{n\to\infty} D_n = D$. Then, the auxiliary positive sequence $P_n = D_{n-1}$, $n = 2, 3 \ldots$ with the first element $P_1 = a_1$ has the same limit: $\lim_{n\to\infty} P_n = D$ (the choice of the first element has no effect on the convergence of the sequence P_n and on the value of its limit). Therefore, by the Theorem of the geometric means, the sequence of its geometric means converges to the same limit. Noting that these geometric means are exactly the elements of the sequence C_n:

$$\sqrt[n]{P_1 \cdot P_2 \cdot \ldots \cdot P_n} = \sqrt[n]{a_1 \cdot \frac{a_2}{a_1} \cdot \ldots \cdot \frac{a_n}{a_{n-1}}} = \sqrt[n]{a_n},$$

we conclude that $\lim_{n\to\infty} C_n = D$. □

Concluding, we have shown the following result: whenever the ratio test reveals the convergence/divergence, the root test does too; whenever the root test is inconclusive, the ratio test is too.

Historical Remarks The first use of the ratio test (in implicit form) can be found in the Euler textbook on analysis "Introductio" in 1748. Then it was employed by D'Alembert in 1768 in the study of the convergence of the binomial series, and later, in 1821, it was extended to general positive series by Cauchy in his famous textbook "Cours d'analyse". In the same textbook, Cauchy also presented his test (the root test). Differently from modern calculus textbooks, which provide two independent and similar proofs for the ratio and root tests, Cauchy first derived the former and then established the relationship between the limits of D'Alembert and Cauchy (see the Theorem in this section), arriving in this way at the latter test.

3.8　Complement: Finer Forms of D'Alembert's and Cauchy's Tests

One of the major deficiencies of D'Alembert's and Cauchy's tests in the form with limits is that the required limits may not exist. This problem can be overcome by employing the finer test formulations with the use of the upper and lower limits. We start presentation of these finer forms of the tests by recalling the notion and basic properties of the upper and lower limits.

Upper and Lower Limits and Their Properties

▶ **Definition** Let a_n be a sequence of numbers and E the set of all the limits of its subsequences (including $+\infty$ and $-\infty$). The *upper limit* $\limsup\limits_{n\to\infty} a_n$ *of the sequence* a_n is the supremum (the least upper bound) of the set E: $\limsup\limits_{n\to\infty} a_n = \sup E$. In a similar way, the *lower limit* $\liminf\limits_{n\to\infty} a_n$ *of the sequence* a_n is the infimum (the greatest lower bound) of the set E: $\liminf\limits_{n\to\infty} a_n = \inf E$.

Property 1 *The upper and lower limits are specific partial limits of the original sequence* a_n. *In other words,* $\sup E$ *and* $\inf E$ *are the elements—the maximum and minimum, respectively, of the set* E.

Proof Let us show this proposition for $\limsup\limits_{n\to\infty} a_n$. First, consider the case when $\sup E$ is a finite number. Assume, for contradiction, that it does not exist a subsequence of a_n with the limit equal to $\sup E$. Then, there exists an ε_0-neighborhood of $\sup E$ such that it contains at most a finite number of the elements of a_n. Therefore, any point of this neighborhood, including any element c of the interval $(\sup E - \varepsilon_0, \sup E]$, cannot be a partial limit of a_n, since there exists an ε_c-neighborhood of the element c, which contains at most a finite number of elements of a_n (choose, for instance, $\varepsilon_c = c - (\sup E - \varepsilon_0)$ for any $c \in (\sup E - \varepsilon_0, \sup E]$). However, this contradicts the definition of the supremum of a set. Hence, $\sup E$ should be a partial limit of a_n.

Second, if $\sup E = +\infty$, then for any constant $C > 0$ there exists a partial limit c of a_n greater than C. Therefore, there is an infinite number of elements of a_n greater than C. Due to the arbitrariness of C, we conclude that $+\infty$ is a partial limit of the sequence a_n. Finally, if $\sup E = -\infty$, then it implies that the general limit of a_n is $-\infty$, since all the partial limits are equal to $-\infty$.

The reasoning for $\inf E$ is quite similar.　　　　　　　　　　　　　　　　□

Property 2 *Since the upper and lower limits are particular cases of partial limits, it follows immediately that if a_n converges, its general limit is equal to the upper and lower limits.*

Property 3 *The upper and lower limits always exist (considering that infinite limits are also included).*

Proof Indeed, if a sequence a_n is bounded—$|a_n| \leq C$, $\forall n$, then, by the Bolzano-Weierstrass Theorem, there exists at least one convergent subsequence (that is, the set E is not empty), and additionally, by the comparison properties of limits, any partial limit $a = \lim_{k \to \infty} a_{n_k}$ satisfies the same inequality: $|a| \leq C$. Then, the set E is bounded and it contains its supremum and infimum, which by definition are the upper and lower limits, respectively. If a_n is unbounded above, then there exists its subsequence tending to $+\infty$ and in this case the upper limit is $+\infty$. Analogously, if a_n is unbounded below, then there exists a subsequence with the limit equal to $-\infty$, and consequently, the lower limit is $-\infty$. $\square$

Property 4 *If $\overline{a} = \limsup_{n \to \infty} a_n < q$, then there exists N such that $a_n < q$, $\forall n \geq N$. Similarly, if $\underline{a} = \liminf_{n \to \infty} a_n > q$, then there exists N such that $a_n > q$, $\forall n \geq N$.*

Proof Suppose, for contradiction, that there is an infinite number of elements $a_n \geq q$, and form from these elements a subsequence with the partial limit $\tilde{a} \geq q$ (according to the comparison properties of limits). However, in this case, $\tilde{a} > \overline{a}$, that contradicts the definition of the upper limit $\overline{a}$.

The proof for the lower limit is the same. $\square$

Theorem 3D (D'Alembert's Test with Upper and Lower Limits) *Suppose that $\sum_{n=1}^{\infty} a_n$ is a positive series and denote $\overline{D} = \limsup_{n \to \infty} D_n$, $\underline{D} = \liminf_{n \to \infty} D_n$, where $D_n = \frac{a_{n+1}}{a_n}$ is the D'Alembert variable. Then,*

(1) if $\overline{D} < 1$, the series converges;
(2) if $\underline{D} > 1$ (in particular, $\underline{D} = +\infty$), the series diverges;
(3) if $\overline{D} = 1$ or $\underline{D} = 1$, no conclusion can be drawn.

Proof The proof follows the same line of reasoning as in the test with the general limit (Theorem 2D). If $\overline{D} < 1$, then there exists a number q such that $\overline{D} < q < 1$. Therefore, by Property 4 of the upper limits, a similar inequality holds for all the elements of the sequence D_n starting from some index N: $D_n < q < 1$, $\forall n \geq N$. In this way, the conditions of Theorem 1D, which guarantee the series convergence, are met.

If $\underline{D} > 1$, then, by Property 4 of the lower limits, a similar inequality is satisfied for all the elements D_n starting from some index N: $D_n > 1$, $\forall n \geq N$. Therefore, those conditions of Theorem 1D, which guarantee the series divergence, are met.

In the inconclusive case, we can use the previous two series: the divergent harmonic series $\sum_{n=1}^{\infty} \frac{1}{n}$ and the convergent $p = 2$-series $\sum_{n=1}^{\infty} \frac{1}{n^2}$. For both series we have $\limsup\limits_{n\to\infty} D_n = \liminf\limits_{n\to\infty} D_n = \lim\limits_{n\to\infty} D_n = 1$.   □

Theorem 3C (Cauchy's Test with Upper Limit) *Suppose that $\sum_{n=1}^{\infty} a_n$ is a positive series and denote $\overline{C} = \limsup\limits_{n\to\infty} C_n$, where $C_n = \sqrt[n]{a_n}$ is the Cauchy variable. Then,*

(1) if $\overline{C} < 1$, the series converges;
(2) if $\overline{C} > 1$ (in particular, $\overline{C} = +\infty$), the series diverges;
(3) if $\overline{C} = 1$, no conclusion can be drawn.

Proof The proof follows the same line of reasoning as in the test with the general limit (Theorem 2C). If $\overline{C} < 1$, then there exists a number q such that $\overline{C} < q < 1$. Therefore, by Property 4 of the upper limits, a similar inequality holds for all the elements of the sequence C_n starting from some index N: $C_n < q < 1$, $\forall n \geq N$. In this way, the conditions of Theorem 1C, which guarantee the series convergence, are met.

If $\overline{C} > 1$, then, by the comparison properties of limits, starting from some index N for all the elements of the respective subsequence C_{n_k} we have $C_{n_k} = \sqrt[n_k]{a_{n_k}} > 1$, $\forall k \geq N$, or equivalently, $a_{n_k} > 1$, $\forall k \geq N$. Therefore, the subsequence a_{n_k} does not approach 0, which means that the general limit of a_n is different from 0. Hence, the necessary condition of the convergence does not hold.

In the inconclusive case, we can use the same two series: the divergent harmonic series $\sum_{n=1}^{\infty} \frac{1}{n}$ and the convergent $p = 2$-series $\sum_{n=1}^{\infty} \frac{1}{n^2}$. For both series we have $\limsup\limits_{n\to\infty} C_n = \lim\limits_{n\to\infty} C_n = 1$.   □

Remark It is important to note that, unlike the case of the general limits (Theorems 2D and 2C), the statements of the last two tests (Theorems 3D and 3C) are not completely similar. The formulation of D'Alembert's test requires the use of both upper and lower limits. The upper limit $\overline{D}$ cannot substitute the lower limit $\underline{D}$ in the second condition, neither $\underline{D}$ can substitute $\overline{D}$ in the first condition. The illustrative examples were already considered before in the section about comparison between D'Alembert's and Cauchy's tests. Let us recall them briefly. The first series is

$$\sum_{n=1}^{\infty} a_n = \frac{1}{4} + \frac{1}{2} + \frac{1}{16} + \frac{1}{8} + \frac{1}{64} + \frac{1}{32} + \ldots = \sum_{n=1}^{\infty} 2^{(-1)^n - n},$$

for which the D'Alembert variable has two different expressions for even and odd indices:

$$D_{2k-1} = \frac{a_{2k}}{a_{2k-1}} = \frac{2^{1-2k}}{2^{-1-2k+1}} = 2, \; D_{2k} = \frac{a_{2k+1}}{a_{2k}} = \frac{2^{-1-2k-1}}{2^{1-2k}} = \frac{1}{8}, \; \forall k \in \mathbb{N}.$$

Therefore,

$$\overline{D} = \limsup_{n \to \infty} D_n = \lim_{k \to \infty} D_{2k-1} = 2 > 1, \underline{D} = \liminf_{n \to \infty} D_n = \lim_{k \to \infty} D_{2k} = \frac{1}{8} < 1.$$

Hence, the upper limit $\overline{D} > 1$, but the series converges (notice that D'Alembert's test is inconclusive in either form). The second series is

$$\sum_{n=1}^{\infty} a_n = 4 + 2 + 16 + 8 + 64 + 32 + \ldots = \sum_{n=1}^{\infty} 2^{n-(-1)^n}$$

with the D'Alembert variable

$$D_{2k-1} = \frac{a_{2k}}{a_{2k-1}} = \frac{2^{2k-1}}{2^{2k-1+1}} = \frac{1}{2}, \; D_{2k} = \frac{a_{2k+1}}{a_{2k}} = \frac{2^{2k+1+1}}{2^{2k-1}} = 8, \forall k \in \mathbb{N}.$$

Therefore,

$$\overline{D} = \limsup_{n \to \infty} D_n = \lim_{k \to \infty} D_{2k} = 8 > 1, \; \underline{D} = \liminf_{n \to \infty} D_n = \lim_{k \to \infty} D_{2k-1} = \frac{1}{2} < 1.$$

Hence, the lower limit $\underline{D} < 1$, but the series diverges (notice that D'Alembert's test is inconclusive in either form).

3.9   Complement: The Kummer Chain of Tests

The known result about non-existence of a universal test for convergence/divergence, which works for any series (see Sect. 3.3), poses the problem of construction of a hierarchy (chain) of tests with systematic refinement, in such a way that each subsequent test of the constructed family is applicable to a wider range of series than its predecessor. In this section we study such kind of the chain of the tests based on the Kummer Theorem. We start with the proof of this theorem and proceed by constructing a family of tests for positive series, called the Kummer chain, each of which is a consequence of the Kummer Theorem. The first and the simplest test in this chain is D'Alembert's test, and with each following step we obtain more sophisticated and more general tests. We give illustrations of the application of the constructed tests and also analyze the situations when the tests of the Kummer chain are inconclusive.

The Kummer Theorem (Test) *Suppose $\sum a_n$ is a positive series and $\sum d_n$ is an auxiliary series, which is positive and divergent. Introduce the Kummer characteristic K_n by the formula*

$$K_n = \frac{1}{d_n} \cdot \frac{a_n}{a_{n+1}} - \frac{1}{d_{n+1}}, \quad d_n > 0, \tag{2.2}$$

and suppose that the following limit exists: $\lim\limits_{n \to \infty} K_n = K$ *(including the infinite limits $K = \pm\infty$). Then:*

(1) if $K > 0$, the series $\sum a_n$ converges;
(2) if $K < 0$, the series $\sum a_n$ diverges.

Proof

(1) Consider first the case $\lim\limits_{n \to +\infty} K_n = K > 0$. It follows from the definition of limit that for $\forall \varepsilon > 0$ (in particular, for $\varepsilon = \frac{K}{2}$) there exists N such that for $\forall n > N$ we have:

$$\left(\frac{1}{d_n} \cdot \frac{a_n}{a_{n+1}} - \frac{1}{d_{n+1}} \right) > K - \varepsilon = \frac{K}{2},$$

implying that $a_{n+1} < \frac{2}{K} \left(\frac{a_n}{d_n} - \frac{a_{n+1}}{d_{n+1}} \right)$. Therefore, we obtain

$$\sum_{k=N+1}^{n+1} a_k < \frac{2}{K} \left(\frac{a_N}{d_N} - \frac{a_{n+1}}{d_{n+1}} \right) < \frac{2}{K} \cdot \frac{a_N}{d_N}.$$

Then, the corresponding partial sums are bounded:

$$s_{n+1} = \sum_{k=1}^{n+1} a_k = \sum_{k=1}^{N} a_k + \sum_{k=N+1}^{n+1} a_k < s_N + \frac{2}{K} \cdot \frac{a_N}{d_N} = C, \ \forall n,$$

where C is a constant. Since s_n is an increasing and bounded above sequence, it has a finite limit $\lim\limits_{n \to +\infty} s_n = s$, which implies the convergence of $\sum a_n$.

(2) Consider the case $\lim\limits_{n \to +\infty} K_n = K < 0$. It follows from the definition of limit that for $\forall \varepsilon > 0$ (in particular, for $\varepsilon = -\frac{K}{2}$) there exists N such that for $\forall n > N$, we have:

$$\left(\frac{1}{d_n} \cdot \frac{a_n}{a_{n+1}} - \frac{1}{d_{n+1}} \right) < K + \varepsilon = \frac{K}{2} < 0,$$

and consequently, $a_{n+1} > \frac{a_n}{d_n}d_{n+1}$, $\forall n > N$. Then, $a_{N+m} > \frac{a_N}{d_N}d_{N+m}$, $\forall m \in \mathbb{N}$. Since $a_n > c \cdot d_n$ and $\sum d_n$ diverges, by the comparison property, the series $\sum a_n$ also diverges.

$\square$

Remark 1 Notice that the divergence of the series $\sum d_n$ was used only in the second part of the proof. In the first part, $\sum d_n$ can be arbitrary positive series.

Remark 2 The Kummer Test, as well as any specific test following from this general theorem, can also be formulated in the form without limits. It is interesting to note that this was the form used in the original formulation of the test by Kummer in 1835 and also in the improved modern version presented by Dini in 1867. However, in this text we confine ourselves to the form with limit, which is more used in practice.

Attributing different values to d_n, it is possible to construct specific tests with different level of refinement. We start with the simpler cases and then gradually increase the level of complexity and generality of tests.

Corollary 1 *Choosing $d_n = 1$ in the Kummer test (notice that the series $\sum d_n = \sum 1$ diverges), we have*

$$K_n = \frac{a_n}{a_{n+1}} - 1 \equiv A_n - 1,$$

where

$$A_n = \frac{a_n}{a_{n+1}} = \frac{1}{D_n} \tag{2.3}$$

is the D'Alembert characteristic. Then the Kummer Theorem takes the specific form equivalent to D'Alembert's test.

D'Alembert's Test *Suppose $\sum a_n$ is a positive series and $\lim_{n\to\infty} A_n = A$ (including the infinite limit $A = +\infty$; the case $A = -\infty$ may not occur). Then:*

(1) if $K = A - 1 > 0$ (that is, $D < 1$), the series $\sum a_n$ converges;
(2) if $K = A - 1 < 0$ (that is, $D > 1$), the series $\sum a_n$ diverges.

(Recall that $D = \lim_{n\to+\infty} D_n = \lim_{n\to+\infty} \frac{a_{n+1}}{a_n}$ in the original formulation of D'Alembert's test.)

Corollary 2 *If we choose* $d_n = \frac{1}{n}$ *in the Kummer Test (notice that the series* $\sum \frac{1}{n}$ *diverges), then*

$$K_n = n \cdot \frac{a_n}{a_{n+1}} - (n+1) = n(A_n - 1) - 1 \equiv R_n - 1,$$

where

$$R_n = n(A_n - 1) = n\left(\frac{a_n}{a_{n+1}} - 1\right) \tag{2.4}$$

is the Raabe characteristic. In this case, the Kummer Theorem takes the form called the Raabe test.

Raabe Test *Suppose* $\sum a_n$ *is a positive series and* $\lim\limits_{n \to \infty} R_n = R$ *(including the infinite limits* $R = \pm\infty$*). Then:*

(1) if $R > 1$ *(that is,* $K = R - 1 > 0$*), the series* $\sum a_n$ *converges;*
(2) if $R < 1$ *(that is,* $K = R - 1 < 0$*), the series* $\sum a_n$ *diverges.*

Corollary 3 *Setting* $d_n = \frac{1}{n \ln n}$ *in the Kummer Test (it was already shown in Example 3i that the series* $\sum \frac{1}{n \ln n}$ *diverges according to the Integral test), we find*

$$K_n = n \ln n \cdot \frac{a_n}{a_{n+1}} - (n+1)\ln(n+1) = n \ln n \cdot (A_n - 1) + n \ln n - (n+1)\ln(n+1)$$

$$= \ln n \cdot (R_n - 1) - (n+1)(\ln(n+1) - \ln n) \equiv B_n - (n+1)\ln\left(1 + \frac{1}{n}\right),$$

where

$$B_n = \ln n \cdot (R_n - 1) = \ln n \cdot \left[n\left(\frac{a_n}{a_{n+1}} - 1\right) - 1\right] \tag{2.5}$$

is the Bertrand characteristic. Since $\lim\limits_{n \to \infty} (n+1)\ln\left(1 + \frac{1}{n}\right) = \lim\limits_{n \to \infty} \ln\left(1 + \frac{1}{n}\right)^{n+1} = \ln e = 1$ *(we employ here the second remarkable limit), the existence of* $\lim\limits_{n \to \infty} K_n$ *and* $\lim\limits_{n \to \infty} B_n$ *are equivalent facts. Therefore, in this case, the Kummer Test can be formulated in the form called the Bertrand Test.*

Bertrand Test *Suppose that* $\sum a_n$ *is a positive series and* $\lim\limits_{n \to \infty} B_n = B$ *(including the infinite limits* $B = \pm\infty$*). Then:*

(1) if $B > 1$ *(that is,* $K = B - 1 > 0$*), the series* $\sum a_n$ *converges;*
(2) if $B < 1$ *(that is,* $K = B - 1 < 0$*), the series* $\sum a_n$ *diverges.*

Corollary 4 *The consequence of Corollaries 1, 2 and 3 is the Gauss test.*

Gauss Test *Suppose that the elements of the series $\sum a_n$ satisfy the relation*

$$\frac{a_n}{a_{n+1}} = \lambda + \frac{\mu}{n} + \frac{\theta_n}{n^\gamma}, \tag{2.6}$$

where $\gamma > 1$ and θ_n is a bounded sequence. Then:

(1) if $\lambda > 1$, the series $\sum a_n$ converges, and if $\lambda < 1$, the series diverges;
(2) if $\lambda = 1$ and $\mu \neq 1$, the series $\sum a_n$ converges when $\mu > 1$, and diverges when $\mu < 1$;
(3) if $\lambda = 1$ and $\mu = 1$, the series $\sum a_n$ diverges.

Proof

(1) $A = \lim\limits_{n\to\infty} A_n = \lim\limits_{n\to\infty} \frac{a_n}{a_{n+1}} = \lim\limits_{n\to\infty} \left(\lambda + \frac{\mu}{n} + \frac{\theta_n}{n^\gamma}\right) = \lambda$, where $\lambda \neq 1$ and the conclusions follow directly from D'Alembert's test.

(2) $R = \lim\limits_{n\to\infty} R_n = \lim\limits_{n\to\infty} n\left(\frac{a_n}{a_{n+1}} - 1\right) = \lim\limits_{n\to\infty} \left(\mu + \frac{\theta_n}{n^{\gamma-1}}\right) = \mu$, where $\mu \neq 1$ and the Raabe test provides the results.

(3) $B = \lim\limits_{n\to\infty} B_n = \lim\limits_{n\to\infty} \ln n \left(n\left(\frac{a_n}{a_{n+1}} - 1\right) - 1\right) = \lim\limits_{n\to\infty} \ln n \left(n\left(\frac{1}{n} + \frac{\theta_n}{n^\gamma}\right) - 1\right) = \lim\limits_{n\to\infty} \frac{\ln n}{n^{\gamma-1}}\theta_n = 0 < 1$, because θ_n is bounded and $\gamma > 1$. The last evaluation implies the divergence of the series due to the Bertrand Test.

$\square$

Remark Following this line of the construction of refined tests using the Kummer Theorem, the reader may guess that the next test in this chain requires a more sophisticated characteristic

$$E_n = \ln\ln n \cdot \{B_n - 1\} = \ln\ln n \cdot \left\{\ln n \cdot \left[n\left(\frac{a_n}{a_{n+1}} - 1\right) - 1\right] - 1\right\}$$

and it can be formulated as follows. Suppose that $\sum a_n$ is a positive series and $\lim\limits_{n\to\infty} E_n = E$ (including the infinite limits $E = \pm\infty$). Then:

(1) if $E > 1$, the series $\sum a_n$ converges;
(2) if $E < 1$, the series $\sum a_n$ diverges.

The demonstration and application of this test as well as the next tests in the Kummer chain is more technically involved and would take us beyond the scope of this text.

The goal and method of construction of the Kummer chain lead to the assumption that the next test in this hierarchy has a broader range of application than its predecessor: the

Raabe test is applicable in some cases when D'Alembert's test fails, the Bertrand test may solve some cases in which the Raabe test is inconclusive, and so on. The inverse implication is not valid.

Indeed, suppose that D'Alembert's test is conclusive for some positive series $\sum a_n$. Then, in the case of the convergence, we have $\lim_{n \to +\infty} (A_n - 1) = A - 1 > 0$, and consequently, $R = \lim_{n \to +\infty} R_n = \lim_{n \to +\infty} n(A_n - 1) = +\infty$, that is, the Raabe test also reveals the convergence. For the divergence, we get $\lim_{n \to +\infty} (A_n - 1) = A - 1 < 0$, and consequently, $R = \lim_{n \to +\infty} R_n = \lim_{n \to +\infty} n(A_n - 1) = -\infty$, which means that the Raabe test also works. The same reasoning shows that the Bertrand test is conclusive in all the cases when the Raabe test is applicable; and so on.

> Thus, the Kummer chain is a sequence of the tests of convergence which starts with the simplest result in this family, D'Alembert's test, proceeds to the more complicated and general Raabe test, then to the even more finer and broader Bertrand test, and so on.

The ratio test is based on the comparison with geometric series and, as the consequence, it can not be applied to series with a slower rate of convergence, such as the p-series and similar ones. The Raabe test solves the problem of classification of the p-series, since its formulation is based on the comparison with the p-series, and consequently, it has a broader range of applications. However, the Raabe test is inconclusive for series whose convergence/divergence rate is even slower than that of the p-series. In such cases the Bertrand test, the next in the Kummer chain, may provide a solution; and so on.

Now we give some examples which illustrate how the range of applicability extends when we pass from one test to the next in the Kummer chain.

Example 1K Series $\sum_{n=1}^{+\infty} \frac{1}{n^2}$
It was already shown that D'Alembert's (and also Cauchy's) test does not work for the p-series. In this specific case we have

$$\lim_{n \to \infty} A_n = \lim_{n \to \infty} \frac{a_n}{a_{n+1}} = \lim_{n \to \infty} \frac{1}{n^2} \cdot (n + 1)^2 = \lim_{n \to \infty} \left(1 + \frac{1}{n}\right)^2 = 1,$$

that is, the test is inconclusive. However, the Raabe test shows the convergence:

$$\lim_{n \to \infty} R_n = \lim_{n \to \infty} n(A_n - 1) = \lim_{n \to \infty} n \left(\frac{(n + 1)^2}{n^2} - 1\right) = \lim_{n \to \infty} n \frac{2n + 1}{n^2} = \lim_{n \to \infty} \left(2 + \frac{1}{n}\right) = 2 > 1.$$

Example 2K Series $\sum_{n=2}^{+\infty} \frac{1}{(n-\sqrt{n})\ln^2 n}$

The Raabe test is not applicable to this series:

$$\lim_{n\to\infty} R_n = \lim_{n\to\infty} n\left(\frac{(n+1-\sqrt{n+1})\ln^2(n+1)}{(n-\sqrt{n})\ln^2 n} - 1\right)$$

$$= \lim_{n\to\infty}\left\{n\left[\left(1+\frac{\ln\left(1+\frac{1}{n}\right)}{\ln n}\right)^2 - 1\right] + n\left(\frac{(n+1-\sqrt{n+1})}{n-\sqrt{n}} - 1\right)\cdot\left(1+\frac{\ln\left(1+\frac{1}{n}\right)}{\ln n}\right)^2\right\}$$

$$= \lim_{n\to\infty}\left\{\frac{2}{\ln n}\cdot\ln\left(1+\frac{1}{n}\right)^n + \frac{1}{n\ln^2 n}\left(\ln\left(1+\frac{1}{n}\right)^n\right)^2\right\}$$

$$+ \lim_{n\to\infty}\frac{n}{n-\sqrt{n}}\cdot\lim_{n\to\infty}\left(1+\frac{\ln\left(1+\frac{1}{n}\right)}{\ln n}\right)^2\cdot\lim_{n\to\infty}\left(1-\sqrt{n+1}+\sqrt{n}\right) = 1 - \lim_{n\to\infty}\frac{1}{\sqrt{n+1}+\sqrt{n}} = 1.$$

However, using the Bertrand test we can conclude that the series converges:

$$\lim_{n\to\infty} B_n = \lim_{n\to\infty}\ln n\left(n\frac{(n+1-\sqrt{n+1})\ln^2(n+1)-(n-\sqrt{n})\ln^2 n}{(n-\sqrt{n})\ln^2 n} - 1\right)$$

$$= \lim_{n\to\infty}\frac{1}{1-\frac{1}{\sqrt{n}}}\cdot\lim_{n\to\infty}\left((n+1)\frac{\left(\left(\ln n+\ln\left(1+\frac{1}{n}\right)\right)^2-\ln^2 n\right)}{\ln n} - \frac{\sqrt{n+1}\ln^2(n+1)-\sqrt{n}\ln^2 n}{\ln n} + \frac{\ln^2 n}{\sqrt{n}\ln n}\right)$$

$$= \lim_{n\to\infty}\left(1+\frac{1}{n}\right)\left[2\ln\left(1+\frac{1}{n}\right)^n + \frac{1}{n\ln n}\left(\ln\left(1+\frac{1}{n}\right)^n\right)^2\right]$$

$$+ \lim_{n\to\infty}\left\{-\ln n\left(\sqrt{n+1}-\sqrt{n}\right) - 2\frac{\left(1+\frac{1}{n}\right)^{1/2}}{n^{1/2}}\ln\left(1+\frac{1}{n}\right)^n - \frac{\left(1+\frac{1}{n}\right)^{1/2}}{n\sqrt{n}\ln n}\left(\ln\left(1+\frac{1}{n}\right)^n\right)^2\right\}$$

$$= 2 - \lim_{n\to\infty}\frac{\ln n}{\sqrt{n+1}+\sqrt{n}} = 2 > 1.$$

Example 3K Series $\sum_{n=1}^{+\infty}\left(\frac{1\cdot3\cdot5\cdot\ldots\cdot(2n-1)}{2\cdot4\cdot6\cdot\ldots\cdot2n}\right)^P\cdot\frac{1}{n^q}$

Let us evaluate the D'Alembert characteristic A_n:

$$A_n = \frac{a_n}{a_{n+1}} = \left(\frac{1\cdot3\cdot5\cdot\ldots\cdot(2n-1)}{2\cdot4\cdot6\cdot\ldots\cdot2n}\right)^P\cdot\frac{(n+1)^q}{n^q}\cdot\left(\frac{2\cdot4\cdot6\cdot\ldots\cdot2n(2n+2)}{1\cdot3\cdot5\cdot\ldots\cdot(2n-1)(2n+1)}\right)^P$$

$$= \frac{(2n+2)^P}{(2n+1)^P}\cdot\left(1+\frac{1}{n}\right)^q = \left(1+\frac{1}{2n+1}\right)^P\cdot\left(1+\frac{1}{n}\right)^q$$

$$= \left(1 + \sum_{k=1}^{+\infty} \frac{p(p-1)\ldots(p-k+1)}{k!} \cdot \left(\frac{1}{2n+1}\right)^k\right) \cdot \left(1 + \sum_{j=1}^{+\infty} \frac{q(q-1)\ldots(q-j+1)}{j!n^j}\right)$$

$$= 1 + \frac{(2n+1)q + pn}{n(2n+1)} + \frac{q(q-1)}{2!n^2} + \frac{pq}{n(2n+1)} + \frac{p(p-1)}{2!(2n+1)^2} + \ldots = 1 + \frac{2q+p}{2n+1} + O\left(\frac{1}{n^2}\right).$$

Therefore, D'Alembert's test fails.

Using the Raabe test

$$\lim_{n\to+\infty} R_n = \lim_{n\to+\infty} n\left(1 + \frac{2q+p}{2n+1} + O\left(\frac{1}{n^2}\right) - 1\right)$$

$$= \lim_{n\to+\infty}\left((2q+p)\cdot\left(2+\frac{1}{n}\right)^{-1} + n\cdot O\left(\frac{1}{n^2}\right)\right) = \frac{2q+p}{2},$$

we obtain that the series converges when $p > 2 - 2q$ and diverges when $p < 2 - 2q$.

In the remaining case $p = 2 - 2q$, the D'Alembert characteristic takes the form

$$A_n = \frac{a_n}{a_{n+1}} = 1 + \frac{2q+p}{2n+1} + O\left(\frac{1}{n^2}\right) = 1 + \frac{2}{2n+1} + O\left(\frac{1}{n^2}\right) = 1 + \frac{1}{n} + \frac{2}{2n+1} - \frac{1}{n} + O\left(\frac{1}{n^2}\right) = 1 + \frac{1}{n} + O\left(\frac{1}{n^2}\right).$$

Therefore, by the Gauss test with $\gamma = 2 > 1$, $\lambda = 1$ and $\mu = 1$, the series diverges when $p = 2 - 2q$.

Example 4K Series $\sum_{n=2}^{+\infty} a^{-\left(1+\frac{1}{2}+\ldots+\frac{1}{n-1}\right)}$

We analyze only the case when $a > 1$, because for $0 < a \le 1$ the general term does not approach 0, and consequently, the series diverges.

Evidently, D'Alembert's test is inconclusive, and then we try the Raabe test. To calculate the limit $\lim_{n\to+\infty} R_n = \lim_{n\to+\infty} n\left(a^{\frac{1}{n}} - 1\right)$, we change the variable $a^{\frac{1}{n}} - 1 = t \Rightarrow \frac{1}{n}\ln a = \ln(1+t)$, and taking into account that $n \to \infty \Rightarrow \frac{1}{n} \to 0 \Rightarrow a^{\frac{1}{n}} - 1 = t \to 0$, we obtain

$$\lim_{t\to 0} \frac{t\cdot\ln a}{\ln(1+t)} = \lim_{t\to 0} \frac{\ln a}{\ln(1+t)^{1/t}} = \ln a.$$

Therefore, if $a > e$, the series converges, while if $a < e$, it diverges.

In the case $a = e$ the Raabe test does not work, and we turn to the Bertrand test. Using the functions of continuous variable to apply L'Hospital's rule, we have:

$$\lim_{n\to+\infty} B_n = \lim_{n\to+\infty} \ln n\left(n\left(e^{\frac{1}{n}} - 1\right) - 1\right) = \lim_{n\to+\infty} \ln n\left(ne^{\frac{1}{n}} - (n+1)\right)$$

$$= \lim_{x\to 0^+} (-\ln x)\left(\frac{1}{x}e^x - \frac{1}{x} - 1\right) = -\lim_{x\to 0^+} \frac{e^x - 1 - x}{x(\ln x)^{-1}}$$

$$= -\lim_{x \to 0^+} \frac{e^x - 1}{(\ln x)^{-1}} \cdot \lim_{x \to 0^+} \frac{1}{1 - (\ln x)^{-1}} = -\lim_{x \to 0^+} \frac{e^x - 1}{(\ln x)^{-1}}$$

$$= \lim_{x \to 0^+} e^x \cdot \lim_{x \to 0^+} \frac{(\ln x)^2}{x^{-1}} = \lim_{x \to 0^+} \frac{2(\ln x)}{-x^{-1}} = \lim_{x \to 0^+} \frac{2x}{1} = 0 < 1 .$$

This result indicates the divergence of the series when $a = e$.

The Kummer Tests with Upper and Lower Limits

Notice that the limits required in the Kummer tests are general limits and may not exist: the same problem was already analyzed for D'Alembert's and Cauchy's tests formulated with general limits. To overcome this problem, it is suitable to formulate the Kummer tests with the use of the upper and lower limits which always exist. We present these forms of the tests below.

The Kummer Theorem (Test) with Upper and Lower Limits *Let $\sum a_n$ be a positive series and $\sum d_n$ be a positive divergent series (auxiliary series) with K_n defined by the formula (2.2). Then:*

(1) if $\liminf\limits_{n \to \infty} K_n > 0$, the series $\sum a_n$ converges;
(2) if $\limsup\limits_{n \to \infty} K_n < 0$, the series $\sum a_n$ diverges.

The proof of this Theorem follows the same line of reasoning as the Theorem with general limits.

Using the last Kummer Test, we can obtain the already formulated tests of the Kummer chain in the versions with upper and lower limits. We present below the first three tests: D'Alembert's test (one more time), the Raabe and Bertrand tests.

D'Alembert's Test with Upper and Lower Limits *Let $\sum a_n$ be a positive series and $A_n = \frac{a_n}{a_{n+1}}$ the D'Alembert characteristic (see (2.3)). Then:*

(1) if $\liminf\limits_{n \to \infty} A_n > 1$, the series $\sum a_n$ converges;
(2) if $\limsup\limits_{n \to \infty} A_n < 1$, the series $\sum a_n$ diverges.

The Raabe Test with Upper and Lower Limits *Let $\sum a_n$ be a positive series and $R_n = n(A_n - 1) = n\left(\frac{a_n}{a_{n+1}} - 1\right)$ the Raabe characteristic (see (2.4)). Then:*

(1) if $\liminf\limits_{n \to \infty} R_n > 1$, the series $\sum a_n$ converges;
(2) if $\limsup\limits_{n \to \infty} R_n < 1$, the series $\sum a_n$ diverges.

The Bertrand Test with Upper and Lower Limits *Let $\sum a_n$ be a positive series and*
$B_n = \ln n \cdot (R_n - 1) = \ln n \cdot \left[n \left(\frac{a_n}{a_{n+1}} - 1 \right) - 1 \right]$ *the Bertrand characteristic (see (2.5)).*
Then:

(1) if $\liminf\limits_{n \to \infty} B_n > 1$, the series $\sum a_n$ converges;
(2) if $\limsup\limits_{n \to \infty} B_n < 1$, the series $\sum a_n$ diverges.

The following examples illustrate the cases when the tests with general limits do not work, but it is possible to reveal the behavior of the series by applying D'Alembert's test and the Raabe test in the form with the upper and lower limits.

Example 5K   Series $\sum_{n=1}^{+\infty} \frac{n^2(3+2(-1)^n)}{3^{n+(-1)^n}}$

Noting that $a_{2n} = \frac{20n^2}{3^{2n+1}}$ and $a_{2n+1} = \frac{(2n+1)^2}{3^{2n}}$, we conclude that $\liminf A_n = \lim\limits_{n \to \infty} A_{2n} = \lim\limits_{n \to \infty} \frac{a_{2n}}{a_{2n+1}} = \frac{5}{3}$ and $\limsup\limits_{n \to \infty} A_n = \lim\limits_{n \to \infty} A_{2n-1} = \lim\limits_{n \to \infty} \frac{a_{2n-1}}{a_{2n}} = \frac{27}{5}$. D'Alembert's test with general limit is not applicable since the required limit $\lim\limits_{n \to \infty} A_n$ does not exist. However, since $\liminf\limits_{n \to \infty} A_n = \frac{5}{3} > 1$, D'Alembert's test with the lower limit indicates that the series converges.

Example 6K   Series $\sum_{n=1}^{+\infty} \frac{1}{(5n+(-1)^n)^2}$

Consider the characteristics A_n and R_n. Since $a_{2n} = \frac{1}{(10n+1)^2}$ and $a_{2n+1} = \frac{1}{(10n+4)^2}$, the odd and even subsequences have the same limit, which is the general limit: $\lim\limits_{n \to \infty} A_n = \lim\limits_{n \to \infty} A_{2n} = \lim\limits_{n \to \infty} A_{2n-1} = 1$. Therefore, the upper and lower limits in D'Alembert's test are also equal 1 and this test is inconclusive. However, the upper and lower limits of the Raabe test are different from 1:

$$\liminf_{n \to \infty} R_n = \lim_{n \to \infty} R_{2n} = \lim_{n \to \infty} 2n(A_{2n}-1) = \lim_{n \to \infty} 2n \left(\frac{(10n+4)^2}{(10n+1)^2} - 1 \right) = \lim_{n \to \infty} \frac{30n(4n+1)}{(10n+1)^2} = \frac{6}{5};$$

$$\limsup_{n \to \infty} R_n = \lim_{n \to \infty} R_{2n-1} = \lim_{n \to \infty} ((2n-1) \cdot (A_{2n-1} - 1))$$

$$= \lim_{n \to \infty} (2n-1) \cdot \left(\frac{(10n+1)^2}{(10n-6)^2} - 1 \right) = \lim_{n \to \infty} \frac{35 \cdot (2n-1) \cdot (4n-1)}{(10n-6)^2} = \frac{14}{5}.$$

Since $\liminf\limits_{n \to \infty} R_n = \frac{6}{5} > 1$, we conclude that the series converges.

Example 7K Series $\sum_{n=1}^{+\infty} \frac{1}{\sqrt{7n+2\cdot(-1)^n}}$

Since $a_{2n} = \frac{1}{\sqrt{14n+2}}$ and $a_{2n+1} = \frac{1}{\sqrt{14n+5}}$, the even and odd characteristics A_n have the same limit equal to the general limit: $\lim_{n\to\infty} A_n = \lim_{n\to\infty} A_{2n} = \lim_{n\to\infty} A_{2n-1} = 1$. Therefore, D'Alembert's test is not applicable. For the Raabe characteristic we have the following upper and lower limits:

$$\liminf_{n\to\infty} R_n = \lim_{n\to\infty} R_{2n} = \lim_{n\to\infty} 2n \left(\frac{(14n+5)^{\frac{1}{2}}}{(14n+2)^{\frac{1}{2}}} - 1 \right)$$

$$= \lim_{n\to\infty} \frac{6n}{(14n+2)^{\frac{1}{2}}} \left[(14n+5)^{\frac{1}{2}} + (14n+2)^{\frac{1}{2}} \right]^{-1}$$

$$= \frac{3}{7} \cdot \lim_{n\to\infty} \left(1 + \frac{1}{7n}\right)^{-\frac{1}{2}} \cdot \left[\left(1 + \frac{5}{14n}\right)^{\frac{1}{2}} + \left(1 + \frac{1}{7n}\right)^{\frac{1}{2}} \right]^{-1} = \frac{3}{14};$$

$$\limsup_{n\to\infty} R_n = \lim_{n\to\infty} R_{2n-1} = \lim_{n\to\infty} (2n-1) \cdot \left(\frac{(14n+2)^{\frac{1}{2}}}{(14n-9)^{\frac{1}{2}}} - 1 \right)$$

$$= \lim_{n\to\infty} \frac{11 \cdot (2n-1)}{(14n-9)^{\frac{1}{2}}} \left[(14n+2)^{\frac{1}{2}} + (14n-9)^{\frac{1}{2}} \right]^{-1}$$

$$= \lim_{n\to\infty} \frac{11 \cdot (2n-1)}{14n} \cdot \left(1 - \frac{9}{14n}\right)^{-\frac{1}{2}} \cdot \left[\left(1 + \frac{1}{7n}\right)^{\frac{1}{2}} + \left(1 - \frac{9}{14n}\right)^{\frac{1}{2}} \right]^{-1} = \frac{11}{14}.$$

Since $\limsup_{n\to\infty} R_n = \frac{11}{14} < 1$, the Raabe test indicates the divergence.

Restrictions in Application of the Kummer Hierarchy

Although we can increase infinitely the level of sophistication and generality of the tests in the Kummer chain, there are cases of convergence/divergence which cannot be solved by any of the tests of this chain even in more precise form with the upper and lower limits. The only alternative in such cases is to search for the tests of other kinds.

We can indicate two possible causes of failure of the Kummer tests. First, the Kummer Theorem itself does not give an answer when $K = 0$, which makes all the tests of the family to carry this imperfection. Second, none of the tests in the Kummer chain works if some test of this chain has the upper and lower limits separated by 1. For example, if in the D'Alembert's test $\limsup_{n\to\infty} A_n > 1 > \liminf_{n\to\infty} A_n$, then the entire chain of the tests is inconclusive. The same happens if $\limsup_{n\to\infty} R_n > 1 > \liminf_{n\to\infty} R_n$, etc.. We leave these statements without proofs, but show an example of the series for which the entire Kummer chain is inconclusive.

Example 8K $\sum_{n=1}^{\infty} 2^{(-1)^n - n}$
This series was already used to show the weakness of D'Alembert's test (see Remark to Theorems 3D and 3C). Indeed, for this series we have $a_{2n} = 2^{1-2n}$ and $a_{2n+1} = 2^{-2-2n}$. Therefore,

$$\limsup_{n\to\infty} A_n = \lim_{n\to\infty} A_{2n} = 8 > 1 > \frac{1}{2} = \lim_{n\to\infty} A_{2n-1} = \liminf_{n\to\infty} A_n,$$

and consequently, D'Alembert's test is inconclusive.

The Raabe test also fails, keeping the same separation of the upper and lower limits by 1:

$$\limsup_{n\to\infty} R_n > 1 > \liminf_{n\to\infty} R_n.$$

Let us check this by the direct calculations. On the one hand,

$$\limsup_{n\to\infty} R_n = \lim_{n\to\infty} R_{2n} = \lim_{n\to\infty} (2n \cdot (A_{2n} - 1)) = +\infty,$$

because $(A_{2n} - 1) \underset{n\to\infty}{\to} 7 > 0$, but on the other hand,

$$\liminf_{n\to\infty} R_n = \lim_{n\to\infty} R_{2n-1} = \lim_{n\to\infty} ((2n-1) \cdot (A_{2n-1} - 1)) = -\infty,$$

since $(A_{2n-1} - 1) \underset{n\to\infty}{\to} -\frac{1}{2} < 0$.
In the same manner,

$$\limsup_{n\to\infty} B_n = +\infty > 1 > -\infty = \liminf_{n\to\infty} B_n.$$

And so on: all the tests of the Kummer chain keep the same separation of the upper and lower limits by 1, and consequently, all of them are inconclusive.

However, it is easy to reveal the convergence of this series by using the comparison theorem: since $a_n = \frac{1}{2^{n-(-1)^n}} \le \frac{1}{2^{n-1}}$ and $\sum_{n=1}^{+\infty} \frac{1}{2^{n-1}}$ is a convergent geometric series, the original series converges.

A similar example in the case of divergence is the series $\sum_{n=1}^{\infty} 2^{n-(-1)^n}$. Again, the Kummer chain is inconclusive, but it is evident that the series diverges, because its general term $a_n = 2^{n-(-1)^n}$ does not approach 0.

Historical Remarks The Kummer test was first formulated and proved by Kummer in 1835. However, the first formulation in the contemporary form was proposed by Dini in 1867, who simplified both the Kummer formulation (dropping a superfluous restriction) and its proof.

Raabe formulated and proved his test in a paper of 1832. Curiously, in the same paper, he provided the formulation and proof of the integral test, apparently being unaware of Cauchy's 1827 paper where the integral test was already derived. Bertrand presented his test in a paper of 1842.

In a paper of 1813 Gauss investigated the properties of the hypergeometric series, including their convergence, which leaded him to discovery of the Gauss test. This result was notably out of the mainstream line of the test inventions which started after publication of Cauchy's "Cours d'analyse" in 1821, where the solid basis for developing the convergence theory was provided and different examples of the tests were presented. Well before the mathematical community arrived at a consensus about the general theory of series, Gauss already was aware about the importance of such theory and gave an important example of its development.

3.10   Complement: The Cauchy Chain of Tests

Similarly to the structure of the Kummer hierarchy, one can construct the Cauchy chain of tests, which starts with Cauchy's test itself and proceeds increasing step by step complexity and generality of the tests.

The Cauchy chain can be developed from the principal Theorem of this family, whose role is analogous to the Kummer Theorem.

Theorem of the Cauchy Chain *Suppose that $F(x)$ is a positive and differentiable function on the interval $[a, +\infty)$, the derivative $F'(x)$ is positive and $\frac{F'(x)}{F(x)}$ is a decreasing function on $[a, +\infty)$. Suppose also that the series $\sum F'(n)$ diverges and $\sum a_n$ is a positive series. Denote $W_n = \frac{\ln \frac{F'(n)}{a_n}}{\ln F(n)}$ and suppose the following limit exists $W = \lim\limits_{n \to \infty} W_n$ (including infinite limits $W = \pm\infty$). Then:*

(1) if $W > 1$, the series $\sum a_n$ converges;
(2) if $W < 1$, the series $\sum a_n$ diverges.

Proof In the preliminary part, we demonstrate that under the general conditions of the Theorem, the function $F(x)$ tends to $+\infty$ as $x \to +\infty$. Let us suppose, for contradiction, that $F(x)$ is a bounded function on $[a, +\infty)$, that is, $F(x) < C, \forall x \in [a, +\infty)$, where C

is a positive constant. Then, $\frac{F'(x)}{F(x)} > \frac{F'(x)}{C}$ for $\forall x \in [a, +\infty)$, in particular, $\frac{F'(n)}{F(n)} > \frac{F'(n)}{C}$ for $\forall n$. Therefore, by the comparison theorem, the divergence of $\sum F'(n)$ implies the divergence of $\sum \frac{F'(n)}{F(n)}$. Notice also that $\frac{F'(x)}{F(x)}$ is integrable on any interval $[a, x]$:

$$\int_a^x \frac{F'(t)}{F(t)} dt = \int_a^x d(\ln F(t)) = \ln F(x) - \ln F(a).$$

Since the function $\frac{F'(x)}{F(x)}$ is positive, decreasing and integrable on $[a, +\infty)$, applying the Integral test we conclude that the improper integral $\int_a^{+\infty} \frac{F'(x)}{F(x)} dx$ diverges due to the divergence of the series $\sum \frac{F'(n)}{F(n)}$. This means that

$$\int_a^x \frac{F'(t)}{F(t)} dt = \ln F(x) - \ln F(a) \underset{x \to +\infty}{\to} +\infty,$$

that is, $\ln F(x) \underset{x \to +\infty}{\to} +\infty$, and consequently $F(x) \underset{x \to +\infty}{\to} +\infty$. Thus, we arrive at the contradiction with the made supposition that $F(x)$ is a bounded function on $[a, +\infty)$. Therefore, $F(x)$ is unbounded, and since it is also increasing, it follows that $F(x)$ tends to $+\infty$ as $x \to +\infty$.

Now we proceed to the analysis of the two statements in the formulation of the Theorem.

(1) If $\lim_{n \to \infty} W_n = W > 1$, there exists $p > 1$ such that for sufficiently large indices n we have $W_n > p > 1$. Since $F(x) \underset{x \to +\infty}{\to} +\infty$, for sufficiently large n we get $F(n) > 1$. Choosing now $N \in \mathbb{N}$ such that for $\forall n > N$ both conditions—$W_n > p > 1$ and $F(n) > 1$—are satisfied, we obtain

$$\ln \frac{F'(n)}{a_n} > p \cdot \ln (F(n)) = \ln (F(n))^p,$$

that is, $\frac{F'(n)}{a_n} > (F(n))^p$, whence $a_n < \frac{F'(n)}{(F(n))^p}$.

Now let us show that the series $\sum \frac{F'(n)}{(F(n))^p}$, $p > 1$ converges. Since the functions $\frac{F'(x)}{F(x)}$ and $\frac{1}{(F(x))^{p-1}}$ are positive and decreasing (the first by the hypotheses of the Theorem and the second—due to the positiveness and increase of $F(x)$), their product $\frac{F'(x)}{(F(x))^p}$ is also a positive and decreasing function. Then, we can apply the Integral test to investigate the behavior of the positive series $\sum \frac{F'(n)}{(F(n))^p}$. The direct calculation of the improper integral reveals its convergence:

$$\int_a^{+\infty} \frac{F'(x)}{(F(x))^p}\,dx = \lim_{x\to+\infty} \int_a^x \frac{F'(v)}{(F(v))^p}\,dv = \lim_{y\to+\infty} \int_b^y \frac{dt}{t^p}$$

$$= \lim_{y\to+\infty} \frac{t^{1-p}}{1-p}\bigg|_b^y = \lim_{y\to+\infty} \frac{y^{1-p}-b^{1-p}}{1-p} = \frac{b^{1-p}}{p-1}.$$

(We utilize here the change of variable $t = F(v)$ and the property that $t = F(v) \underset{v\to+\infty}{\to} +\infty$.) Therefore, the series $\sum \frac{F'(n)}{(F(n))^p}$ converges and, by the Comparison test, the series $\sum a_n$ converges too.

(2) If $\lim_{n\to\infty} W_n = W < 1$, for sufficiently large indices n we have $W_n < 1$. Besides, using the fact that $F(x)$ tends to $+\infty$, we can choose such large indices n that $F(n) > 1$. Then, choosing such $n > N$ that both inequalities hold, we can transform $W_n < 1$ into $\frac{F'(n)}{a_n} < F(n)$, or equivalently, into $a_n > \frac{F'(n)}{F(n)}$, $\forall n > N$. Let us show that the series $\sum \frac{F'(n)}{F(n)}$ diverges. Indeed, since the function $\frac{F'(x)}{F(x)}$ is positive and decreasing, we can use the Integral test to determine the behavior of the positive series $\sum \frac{F'(n)}{F(n)}$. The direct calculation of the improper integral shows its divergence:

$$\int_a^{+\infty} \frac{F'(x)}{F(x)}\,dx = \lim_{x\to+\infty} \int_a^x \frac{F'(v)}{F(v)}\,dv = \lim_{y\to+\infty} \int_b^y \frac{dt}{t} = \lim_{y\to+\infty} \ln t\big|_b^y$$

$$= \lim_{y\to+\infty} \ln y - \ln b = +\infty.$$

(We use here the change of variable $t = F(v)$ and the tendency of $F(v)$: $t = F(v) \underset{v\to+\infty}{\to} +\infty$.) Therefore, the series $\sum \frac{F'(n)}{F(n)}$ also diverges, and, by the Comparison test, it follows that the series $\sum a_n$ diverges.

This completes the proof of the theorem.

□

Remark Notice that the condition of decreasing of $\frac{F'(x)}{F(x)}$ can be substituted, keeping all the remaining conditions of the Theorem, by the stronger condition of decreasing of $F'(x)$: since $F'(x) > 0$, it follows that $F(x)$ is increasing, and consequently, $\frac{1}{F(x)}$ is decreasing, which means that $F'(x) \cdot \frac{1}{F(x)}$ is decreasing as the product of the two decreasing (and positive) functions. However, the condition of decreasing of $\frac{F'(x)}{F(x)}$ allows the use of a wider range of functions $F(x)$, like, for example, $F(x) = e^x$.

The construction of the tests in the Cauchy chain is more intricate than in the Kummer chain. For this reason, we present only the first two representatives of this chain: Cauchy's test itself and the next test called the Jamet test.

Corollary 1 (Cauchy's Test) *Choosing $F(x) = e^x$, we have the conditions of the Theorem of the Cauchy chain satisfied: $F(x) = F'(x) = e^x > 0$, $\frac{F'(x)}{F(x)} = 1$ is decreasing (in a non-strict sense) and the series $\sum F'(n) = \sum e^n$ diverges. Therefore, for the positive series $\sum a_n$ we have the following result:*

(1) if $\lim\limits_{n \to \infty} W_n > 1$, the series $\sum a_n$ converges;

(2) if $\lim\limits_{n \to \infty} W_n < 1$, the series $\sum a_n$ diverges.

Here,

$$W_n = \frac{\ln \frac{F'(n)}{a_n}}{\ln F(n)} = \frac{\ln \frac{e^n}{a_n}}{\ln e^n} = \frac{n - \ln a_n}{n} = 1 - \ln a_n^{1/n}$$

and the limit $\lim\limits_{n \to \infty} W_n$ can be finite or infinite.

Proof To verify that this result coincides with the original Cauchy's test, it is sufficient to rewrite the two conditions in the equivalent forms:

$$\lim_{n \to \infty} W_n > 1 \Leftrightarrow \lim_{n \to \infty} (1 - \ln a_n^{1/n}) > 1 \Leftrightarrow \lim_{n \to \infty} \ln a_n^{1/n} < 0 \Leftrightarrow \lim_{n \to \infty} a_n^{1/n} < 1,$$

$$\lim_{n \to \infty} W_n < 1 \Leftrightarrow \lim_{n \to \infty} (1 - \ln a_n^{1/n}) < 1 \Leftrightarrow \lim_{n \to \infty} \ln a_n^{1/n} > 0 \Leftrightarrow \lim_{n \to \infty} a_n^{1/n} > 1.$$

$\square$

Corollary 2 (The Jamet Test) *Denote by $J_n = \frac{n}{\ln n}\left(1 - a_n^{1/n}\right)$ the Jamet characteristic and suppose that the following limit exists $J = \lim\limits_{n \to \infty} J_n$ (including the infinite limits $J = \pm\infty$). Then, for a positive series $\sum a_n$ we have the following result:*

(1) if $J > 1$, the series $\sum a_n$ converges;

(2) if $J < 1$, the series $\sum a_n$ diverges.

Proof First, consider the two simple cases when Cauchy's test is also applicable. If $\lim\limits_{n \to \infty} a_n^{\frac{1}{n}} < 1$, then the series $\sum a_n$ converges by Cauchy's test and, at the same time, $J = \lim\limits_{n \to \infty} J_n = +\infty > 1$, that is, the first statement of the Jamet Test also shows the convergence. On the other hand, if $\lim\limits_{n \to \infty} a_n^{\frac{1}{n}} > 1$, then the series $\sum a_n$ diverges by Cauchy's test and, in this case, $J = \lim\limits_{n \to \infty} J_n = -\infty < 1$, that is, the second statement of the Jamet Test is applicable.

Now turn to the case when $\lim\limits_{n \to \infty} a_n^{\frac{1}{n}} = 1$, and consequently, Cauchy's test is inconclusive. We have to search for a finer test and to find such a test we can appeal to the Theorem of the Cauchy chain with the function $F(x) = x$. Notice that this function satisfies all the conditions of the Theorem: $F(x) = x > 0$, $F'(x) = 1 > 0$, $\frac{F'(x)}{F(x)} = \frac{1}{x}$ is decreasing for $\forall x \geq 1$ and the series $\sum F'(n) = \sum 1$ diverges. Then, we have the following two results:

(1) if $W > 1$, the series $\sum a_n$ converges;
(2) if $W < 1$, the series $\sum a_n$ diverges,

where $W = \lim\limits_{n \to \infty} W_n$, $W_n = \frac{\ln \frac{F'(n)}{a_n}}{\ln F(n)} = -\frac{\ln a_n}{\ln n}$.

To connect these results with the statements of the Jamet test, let us establish the relationship between the quantities W_n and J_n. Notice first that the following chain of inequalities holds:

$$1 + \ln x \leq x \leq \frac{1}{1 - \ln x}, \forall x \in (0, e).$$

Choosing $x = a_n^{\frac{1}{n}} \in (0, e)$, that is, $a_n \in (0, e^n)$, and using this double inequality, we obtain

$$1 + \frac{\ln a_n}{n} = 1 + \ln a_n^{\frac{1}{n}} \leq a_n^{\frac{1}{n}} \leq \frac{1}{1 - \ln a_n^{\frac{1}{n}}} = \frac{1}{1 - \frac{\ln a_n}{n}}.$$

After a little algebra, the last formula can be rewritten in the form

$$-\frac{\ln a_n}{\ln n} \cdot \frac{1}{1 - \frac{\ln a_n}{n}} \leq \frac{n}{\ln n}\left(1 - a_n^{\frac{1}{n}}\right) = J_n \leq -\frac{\ln a_n}{\ln n},$$

or, using the definition of $W_n = -\frac{\ln a_n}{\ln n}$, we arrive at

$$W_n \cdot \frac{1}{1 - \frac{\ln a_n}{n}} \leq J_n \leq W_n.$$

Since $\lim\limits_{n \to \infty} a_n^{\frac{1}{n}} = 1$, we get $\lim\limits_{n \to \infty} \frac{\ln a_n}{n} = 0$. Then, it follows from the squeeze Theorem that $\lim\limits_{n \to \infty} J_n = \lim\limits_{n \to \infty} W_n$, which finishes the proof in the case when $\lim\limits_{n \to \infty} a_n^{\frac{1}{n}} = 1$.

Notice that there is no loss of generality in the made choice of $a_n^{\frac{1}{n}} \in (0, e)$, for if $a_n^{\frac{1}{n}} \geq e$, that is, $a_n \geq e^n$, then, comparing with the series $\sum e^n$, we conclude that $\sum a_n$ diverges and, at the same time, $\lim\limits_{n \to \infty} J_n = -\infty$, which is in agreement with the Jamet test.

This completes the proof of the theorem. $\square$

Remark 1 From the first part of the proof it follows that the Jamet test is applicable whenever Cauchy's test works. Although the former requires usually more involved technical calculations, it can be applied in the cases when the latter is inconclusive.

Remark 2 Following the proposed line of the construction of the tests of the Cauchy chain, the reader may probably conjecture that the next test in this family involves even more sophisticated characteristic

$$I_n = \frac{\ln n}{\ln \ln n} \cdot \{J_n - 1\} = \frac{\ln n}{\ln \ln n} \cdot \left\{ \frac{n}{\ln n}\left(1 - a_n^{1/n}\right) - 1 \right\}$$

and the test itself can be formulated in the following way. If there exists the limit $I = \lim_{n \to \infty} I_n$ (including the infinite limits $I = \pm\infty$), then for a positive series $\sum a_n$ the two statements are true:

(1) if $I > 1$, the series $\sum a_n$ converges;
(2) if $I < 1$, the series $\sum a_n$ diverges.

The demonstration of this and the next tests of the Cauchy chain is very technically involved and is out of scope of this text.

Just as for the Kummer chain, the tests in the Cauchy hierarchy starts from the simplest test in this family (in this case Cauchy's test) and increase their range of applicability and also the level of sophistication with each next test. Unlike the Kummer chain, the level of complexity and the volume of technical work growth very fast, and in practice only the first two tests of this chain are usually employed.

For these reasons we do not proceed with construction of the following tests of the Cauchy chain.

The next four examples illustrate the situations when Cauchy's test is inconclusive, but the Jamet test works.

Example 1G $\sum_{n=1}^{+\infty} \frac{1}{n^3}$
As we have already discussed, Cauchy's test is incapable to determine the behavior of the p-series. In particular, for the given p-series with $p = 3$ we get

$$\lim_{n \to \infty} a_n^{\frac{1}{n}} = \lim_{n \to \infty} \left(\frac{1}{n^3}\right)^{1/n} = \lim_{n \to \infty} e^{-\ln n^{3/n}} = \lim_{n \to \infty} e^{-\frac{3}{n} \ln n} = 1.$$

Then, we try to apply the Jamet test:

$$\lim_{n\to\infty} J_n = \lim_{n\to\infty} \frac{n}{\ln n}\left(1 - \left(\frac{1}{n^3}\right)^{1/n}\right) = \lim_{n\to\infty} \frac{1}{\ln n^{1/n}}\cdot\left(1 - \left(\frac{1}{n^{1/n}}\right)^3\right).$$

Substituting the variable $t = n^{1/n} \underset{n\to\infty}{\to} 1$ and using L'Hospital's rule, we obtain

$$\lim_{t\to1}\frac{1}{\ln t}\cdot\left(1 - \frac{1}{t^3}\right) = \lim_{t\to1}\frac{1}{t^3}\cdot\lim_{t\to1}\frac{t^3 - 1}{\ln t} = \lim_{t\to1}\frac{3t^2}{1/t} = 3\cdot\lim_{t\to1}t^3 = 3 > 1.$$

Hence, the Jamet test reveals that the series converges.

Example 2G $\sum_{n=2}^{+\infty}\frac{1}{\ln n}$
Trying to use Cauchy's test we get

$$\lim_{n\to\infty} a_n^{\frac{1}{n}} = \lim_{n\to\infty} e^{-\ln(\ln n)\frac{1}{n}} = \lim_{n\to\infty} e^{-\frac{\ln(\ln n)}{n}} = 1,$$

since $\lim_{x\to+\infty}\frac{\ln(\ln x)}{x} = 0$. Therefore, Cauchy's test fails, and we appeal to the Jamet test. Calculating the corresponding limit, we have:

$$\lim_{n\to\infty} J_n = \lim_{n\to\infty} \frac{n}{\ln n}\left(1 - \left(\frac{1}{\ln n}\right)^{1/n}\right) = \lim_{n\to\infty} \frac{e^{-\ln\ln n/n} - 1}{-\ln\ln n/n}\cdot\lim_{n\to\infty}\frac{n}{\ln n}\cdot\frac{\ln\ln n}{n} = 1\cdot0 = 0,$$

where we have used the results that $\lim_{x\to0}\frac{e^x-1}{x} = 1$ and $\lim_{x\to+\infty}\frac{\ln\ln x}{\ln x} = 0$. Hence, the Jamet test shows the divergence of the series.

Example 3G $\sum_{n=1}^{+\infty}\left(1 - 2\frac{\ln n}{n}\right)^n$
Using Cauchy's test we are not able to draw any conclusion:

$$\lim_{n\to\infty} a_n^{\frac{1}{n}} = \lim_{n\to\infty}\left(1 - 2\frac{\ln n}{n}\right) = 1.$$

Therefore, we turn to the Jamet test:

$$\lim_{n\to\infty} J_n = \lim_{n\to\infty}\frac{n}{\ln n}\left(1 - a_n^{\frac{1}{n}}\right) = \lim_{n\to\infty}\frac{n}{\ln n}\left(1 - 1 + 2\frac{\ln n}{n}\right) = \lim_{n\to\infty}\frac{n}{\ln n}\cdot2\frac{\ln n}{n} = 2.$$

Hence, we can conclude that the series converges.

Example 4G $\sum_{n=2}^{+\infty} \frac{1}{(\ln n)^{\ln n}}$

Let us start with the auxiliary limit that we can calculate using L'Hospital's rule:

$$\lim_{x \to +\infty} \frac{\ln x \cdot \ln(\ln x)}{x} = \lim_{x \to +\infty} \frac{\frac{1}{x} \ln(\ln x) + \ln x \cdot \frac{1}{\ln x} \cdot \frac{1}{x}}{1} = \lim_{x \to +\infty} \frac{\ln(\ln x) + 1}{x} = \lim_{x \to +\infty} \frac{\frac{1}{\ln x} \cdot \frac{1}{x}}{1} = 0.$$

Now we try to apply Cauchy's test:

$$\lim_{n \to \infty} a_n^{\frac{1}{n}} = \lim_{n \to \infty} (\ln n)^{-\frac{\ln n}{n}} = \lim_{n \to \infty} e^{-\frac{\ln n \cdot \ln(\ln n)}{n}} = e^0 = 1,$$

where we use the auxiliary limit to find the limit of the exponent. Therefore, Cauchy's test is inconclusive.

Next, we try the Jamet test, again using the auxiliary limit and L'Hospital's rule:

$$\lim_{n \to \infty} J_n = \lim_{n \to \infty} \frac{n}{\ln n} \left(1 - a_n^{\frac{1}{n}} \right) = \lim_{n \to \infty} \frac{n}{\ln n} \left(1 - (\ln n)^{-\frac{\ln n}{n}} \right) = \lim_{n \to \infty} \frac{1 - e^{-\frac{\ln n \cdot \ln(\ln n)}{n}}}{\ln n \cdot n^{-1}}$$

$$= \lim_{n \to \infty} \frac{e^{-\frac{\ln n \cdot \ln(\ln n)}{n}} \cdot \frac{1}{n^2} \left[n \left(\frac{1}{n} \ln(\ln n) + \ln n \cdot \frac{1}{\ln n} \cdot \frac{1}{n} \right) - \ln n \cdot \ln(\ln n) \right]}{\frac{1 - \ln n}{n^2}}$$

$$= \lim_{n \to \infty} e^{-\frac{\ln n \cdot \ln(\ln n)}{n}} \cdot \lim_{n \to \infty} \left(\ln(\ln n) + \frac{1}{1 - \ln n} \right) = +\infty.$$

(In the final evaluation we take into account that $\lim_{n \to \infty} e^{-\frac{\ln n \cdot \ln(\ln n)}{n}} = 1$, $\lim_{n \to \infty} \ln(\ln n) = +\infty$ and $\lim_{n \to \infty} \frac{1}{1 - \ln n} = 0$.) Hence, we can conclude that the series converges.

Historical Remarks The first to propose the chain of the successively refined tests based on Cauchy's test was Bonnet in a paper of 1843.

---

4 Series of Different Types

4.1 Alternating Series

Recall the definition of an alternating series given in Sect. 1.1.

▶ **Definition** (**Alternating Series**) A *series is alternating* if it can be represented in one of the two forms $\sum a_n = \sum (-1)^n b_n$ or $\sum a_n = \sum (-1)^{n+1} b_n$, where $b_n > 0$,

(continued)

$\forall n$. Notice that the change of sign occurs from the current element a_n to the next a_{n+1}. Other forms of the change of sign between the elements of the series are not considered alternating series.

Already considered examples of alternating series are $\sum_{n=1}^{\infty}(-1)^n$ and $\sum_{n=1}^{\infty}\frac{(-1)^{n-1}}{n}$, the former is divergent (because its general term does not approach 0) and the latter—convergent (it was demonstrated in Example 7a, Sect. 1.2, using the definition). The series $\sum_{n=1}^{\infty}\sin\frac{n\pi}{4}$ and $\sum_{n=1}^{\infty}\cos n$ are not alternating, despite the fact that their terms change sign an infinite number of times.

In this section we consider the *test for convergence of alternating series* known as *Leibniz's test* and the remainder evaluation following from this test.

Test for Alternating Series (Leibniz's Test). *An alternating series* $\sum a_n = \sum(-1)^n b_n$ *or* $\sum a_n = \sum(-1)^{n+1}b_n$, $b_n > 0$, $\forall n$ *converges if* b_n *approaches 0 monotonically, that is, if the following two conditions hold:*

(1) $b_{n+1} \leq b_n$, $\forall n$;
(2) $\lim_{n\to\infty} b_n = 0$.

Proof To specify reasoning, let us choose the form $\sum_1^{+\infty}(-1)^{n+1}b_n$ (for the second form the proof is the same). Consider first the even partial sums s_{2n}. Using a suitable grouping of its terms, we can see that these sums are bounded:

$$s_{2n} = \sum_{k=1}^{2n}(-1)^{k-1}b_k = b_1 - b_2 + \ldots + b_{2n-1} - b_{2n} = b_1 - (b_2 - b_3) - \ldots - (b_{2n-2} - b_{2n-1}) - b_{2n} \leq b_1.$$

At the same time, these sums constitute an increasing sequence:

$$s_{2n} = s_{2n-2} + (b_{2n-1} - b_{2n}) \geq s_{2n-2}.$$

Since s_{2n} is an increasing and bounded above sequence, it is convergent (recall Property 2s in Sect. 3.2 of Chap. 1) and we denote its limit by s: $\lim_{n\to\infty} s_{2n} = s$ (see Fig. 2.4). Notice

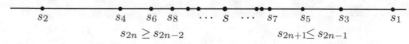

Fig. 2.4 Convergence of the even and odd partial sums of a convergent alternating series

now that the odd partial sums can be expressed through the even ones by the formula $s_{2n+1} = s_{2n} + b_{2n+1}$, and consequently, the sequence s_{2n+1} has the same limit s, because s_{2n} and b_{2n+1} are convergent sequences and the latter tends to 0: $\lim\limits_{n\to\infty} s_{2n+1} = \lim\limits_{n\to\infty} s_{2n} + \lim\limits_{n\to\infty} b_{2n+1} = s + 0 = s$. Since the even and odd subsequences have the same limit, it follows that the general limit exists and equals s: $\lim\limits_{n\to\infty} s_n = s$.   □

Remark 1 As in other tests, the posed conditions (sign alternating and monotonicity) can be satisfied starting from some index.

Remark 2 The second condition (the convergence of b_n to 0) is equivalent to the condition of the convergence of the general term $(-1)^{n-1}b_n$ to 0 and, for this reason, it is just a necessary condition of the convergence of a series: a violation of this condition leads (obligatorily) to divergence of a series.

Remark 3 Under the second condition satisfied, the first one (of monotonicity) is a sufficient condition: if it does not hold, a series still can be convergent. An example of such a series is $\sum_{n=1}^{+\infty}(-1)^{n+1}b_n$ with $b_{2n-1} = \left(\frac{1}{3}\right)^{2n-1}$ and $b_{2n} = \left(\frac{1}{2}\right)^{2n}$. However, without the first condition, an alternating series can be divergent, as it shows an example of the series $\sum_{n=1}^{+\infty}(-1)^{n+1}b_n$ with $b_{2n-1} = \frac{2}{n}$ and $b_{2n} = \frac{1}{n}$, whose even partial sums coincide with the partial sums of the divergent harmonic series:

$$s_{2n} = \sum_{k=1}^{2n}(-1)^{n+1}b_n = \left(\frac{2}{1} - \frac{1}{1}\right) + \left(\frac{2}{2} - \frac{1}{2}\right) + \ldots + \left(\frac{2}{n} - \frac{1}{n}\right) = \frac{1}{1} + \frac{1}{2} + \ldots + \frac{1}{n}.$$

Remark 4 The relevance of the first condition is reflected in the fact that the Comparison test does not work for alternating series. More precisely: the condition $0 < b_n < d_n$ and the convergence of the series $\sum_{n=1}^{+\infty}(-1)^{n+1}d_n$ do not guarantee the convergence of the series $\sum_{n=1}^{+\infty}(-1)^{n+1}b_n$. An example of this is the divergent series $\sum_{n=1}^{+\infty}(-1)^{n+1}b_n$ with $b_{2n-1} = \frac{2}{n}$ and $b_{2n} = \frac{1}{n}$, discussed in Remark 3, and the series $\sum_{n=1}^{+\infty}(-1)^{n+1}d_n$ with $d_n = \frac{5}{n}$. Although $b_n < d_n$, $\forall n$, the first series diverges, while the second—converges (by Leibniz's test).

Theorem (Estimate of the Sum and Remainder of Alternating Series) *Under the conditions of Leibniz's test, the remainder r_n of a convergent alternating series satisfies the inequality $|r_n| \leq b_{n+1}$ and the sum can be evaluated in the form $|s_n| - b_{n+1} \leq |s| \leq |s_n| + b_{n+1}$.*

Proof Again we consider an alternating series in the form $\sum_{n=1}^{+\infty}(-1)^{n+1}b_n$. It follows from the proof of Leibniz's test that $s_{2n} \leq b_1$. On the other hand,

$$s_{2n} = \sum_{k=1}^{2n}(-1)^{k-1}b_k = (b_1 - b_2) + \ldots + (b_{2n-1} - b_{2n}) \geq 0.$$

Then, $0 \leq s_{2n} \leq b_1$. For the odd partial sums we have similar evaluations:

$$s_{2n+1} = s_{2n} + b_{2n+1} > 0, \; s_{2n+1} = b_1 - (b_2 - b_3) - \ldots - (b_{2n} - b_{2n+1}) \leq b_1,$$

and consequently $0 \leq s_{2n+1} \leq b_1$. Hence, for any partial sum we have $0 \leq s_n \leq b_1$, and consequently, the sum of the series satisfies the same evaluation $0 \leq s \leq b_1$.

Now we turn to the remainder estimate:

$$r_n = s - s_n = \sum_{k=n+1}^{+\infty}(-1)^{k+1}b_k = (-1)^n \sum_{k=n+1}^{+\infty}(-1)^{k-n+1}b_k.$$

The series $\sum_{k=n+1}^{+\infty}(-1)^{k-n+1}b_k$ is alternating of the same type as the original series and its elements satisfy the same conditions. Therefore, the sum s_r is located between 0 and the first term of the series: $0 \leq s_r \leq b_{n+1}$. Hence, we obtain the desired evaluation of the reminder $|r_n| = s_r \leq b_{n+1}$. Taking into account that $s = s_n + r_n$, the corresponding estimate for the sum of the original series can be given in the form

$$|s_n| - b_{n+1} \leq |s_n| - |r_n| \leq |s| \leq |s_n| + |r_n| \leq |s_n| + b_{n+1}.$$

$\square$

Let us solve some examples using Leibniz's test.

Example 1L Alternating harmonic series $\sum_{n=1}^{\infty} \frac{(-1)^{n-1}}{n}$
It was already shown by definition that this series converges. Let us see that the application of Leibniz's test is much simpler than the use of definition: first, $b_n = \frac{1}{n} \underset{n\to\infty}{\to} 0$, and second, $b_n = \frac{1}{n} > \frac{1}{n+1} = b_{n+1}, \forall n$. Let us evaluate how many terms we have to use to guarantee an approximation of the exact sum by the partial sums with an accuracy better then 10^{-1}. According to the last Theorem we have $|r_n| \leq b_{n+1} = \frac{1}{n+1} < \frac{1}{10}$, whence $n > 9$.

Example 2L $\sum_{n=1}^{\infty} \frac{(-1)^n}{\sqrt{n+2}}$
Check the conditions of Leibniz's test: first, $b_n = \frac{1}{\sqrt{n+2}} \underset{n\to\infty}{\to} 0$, and second, $b_n = \frac{1}{\sqrt{n+2}} > \frac{1}{\sqrt{n+3}} = b_{n+1}, \forall n$. Therefore, the series converges. Let us also estimate the error in the approximation of the exact sum of the series by the sixth partial sum: $|r_6| \leq b_7 = \frac{1}{\sqrt{7+2}} = \frac{1}{3}$.

Example 3L $\sum_{n=1}^{\infty} \frac{(-1)^n}{n!}$

The conditions of Leibniz's test are satisfied: first, $b_n = \frac{1}{n!} \underset{n \to \infty}{\to} 0$, and second, $b_n = \frac{1}{n!} >$

$\frac{1}{(n+1)!} = b_{n+1}, \forall n$. Let us evaluate the number of terms needed to approximate the exact

sum by the partial sums with accuracy better than 10^{-3}: $|r_n| \le b_{n+1} = \frac{1}{(n+1)!} < \frac{1}{10^3}$,

whence $n > 5$.

Historical Remarks In the beginning of the eighteenth century, the problems of convergence of series of numbers were considered only in studies of specific series and a general theory did not exist. In this context, the result of Leibniz on convergence of a general alternating series with monotonically approaching zero general term appears to be more than surprising. He made an assumption regarding such behavior of alternating series in 1705 and provided the proof of the statement as well as the corresponding remainder estimate in 1714. To understand the deepness of this achievement, let us recall that about the same time the leading mathematicians were involved in the discussion about convergence/divergence of Grandi's series $\sum_{n=0}^{\infty} (-1)^n$, and many of them, including Leibniz himself, argued that the sum of this series is $\frac{1}{2}$. The same position was supported during the eighteenth century by many great mathematicians, including Euler and Lagrange.

4.2　Dirichlet's and Abel's Tests

In this section we prove the two strongly connected tests for series $\sum_{n=1}^{\infty} c_n$ with the general term in the form $c_n = a_n b_n$. We start with an auxiliary result called Abel's summation by parts formula which turns to be a useful tool for demonstration of both tests.

Abel's Summation by Parts Formula *Let* $\sum_{n=1}^{\infty} c_n$ *be a series with the general term in the form* $c_n = a_n b_n$ *and* $B_n = \sum_{k=1}^{n} b_k$ *be the partial sums of the series* $\sum_{n=1}^{\infty} b_n$. *Then for any* $m > n \ge 1$ *the following formula is true*

$$\sum_{k=n+1}^{m} c_k = \sum_{k=n+1}^{m} a_k b_k = \sum_{k=n+1}^{m-1} B_k (a_k - a_{k+1}) + B_m a_m - B_n a_{n+1}.$$

Proof Let us evaluate the sum $S \equiv \sum_{k=n+1}^{m} c_k = \sum_{k=n+1}^{m} a_k b_k$ by representing b_k through the partial sums:

$$S = \sum_{k=n+1}^{m} a_k b_k = \sum_{k=n+1}^{m} (B_k - B_{k-1}) \cdot a_k = \sum_{k=n+1}^{m} B_k a_k - \sum_{k=n+1}^{m} B_{k-1} a_k$$

$$= \sum_{k=n+1}^{m-1} B_k a_k + B_m a_m - \sum_{k=n+2}^{m} B_{k-1} a_k - B_n a_{n+1}.$$

Change now the summation index in the second sum from $k - 1$ to k, and then join the first and second sums:

$$S = \sum_{k=n+1}^{m-1} B_k a_k + B_m a_m - \sum_{k=n+1}^{m-1} B_k a_{k+1} - B_n a_{n+1} = \sum_{k=n+1}^{m-1} B_k (a_k - a_{k+1}) + B_m a_m - B_n a_{n+1}.$$

This gives the desired result. $\square$

Remark 1 This formula resembles the integration by parts formula if we consider a_k as the counterpart of the first function and $a_{k+1} - a_k$ as its differential, B_k as the counterpart of the second function and $b_k = B_k - B_{k-1}$ as its differential.

Remark 2 If we set $B_0 = 0$, the formula will also be true for $n = 0$:

$$\sum_{k=1}^{m} a_k b_k = \sum_{k=1}^{m-1} B_k (a_k - a_{k+1}) + B_m a_m.$$

Sometimes this particular case is called Abel's summation formula.

Dirichlet's Test *Let $\sum_{n=1}^{\infty} c_n$ be a series with the general term in the form $c_n = a_n b_n$. If*

(1) the partial sums B_n of the series $\sum_{n=1}^{\infty} b_n$ are bounded,
(2) the sequence a_n approaches monotonically 0,

then the original series $\sum_{n=1}^{\infty} c_n$ converges.

Proof Let us show that the series $\sum_{n=1}^{\infty} c_n$ satisfies the Cauchy criterion for convergence. To do this, take $m > n$ and use Abel's summation by parts formula to evaluate $S \equiv \sum_{k=n+1}^{m} c_k = \sum_{k=n+1}^{m} a_k b_k$:

$$|S| = \left| \sum_{k=n+1}^{m} a_k b_k \right| = \left| \sum_{k=n+1}^{m-1} B_k (a_k - a_{k+1}) + B_m a_m - B_n a_{n+1} \right|$$

$$\leq \sum_{k=n+1}^{m-1} |B_k| \cdot |a_k - a_{k+1}| + |B_m| \cdot |a_m| + |B_n| \cdot |a_{n+1}| \equiv S_1.$$

By the condition (1) of the Theorem, the partial sums B_n of the series $\sum_{n=1}^{\infty} b_n$ are bounded, that is, there exists $M > 0$ such that $|B_n| \leq M$ for $\forall n \in \mathbb{N}$. Then we continue the evaluation of S_1 in the following manner:

$$S_1 \leq M \left(\sum_{k=n+1}^{m-1} |a_k - a_{k+1}| + |a_m| + |a_{n+1}| \right) \equiv S_2.$$

Since the sequence a_n is monotone, all the differences $a_k - a_{k+1}$ have the same sign, and consequently, the sum of the absolute values is equal to the absolute value of the sum:

$$S_2 = M \left(\left| \sum_{k=n+1}^{m-1} (a_k - a_{k+1}) \right| + |a_m| + |a_{n+1}| \right)$$

$$= M \left(|a_{n+1} - a_{n+2} + a_{n+2} - a_{n+3} + \ldots + a_{m-1} - a_m| + |a_m| + |a_{n+1}| \right)$$

$$= M \left(|a_{n+1} - a_m| + |a_m| + |a_{n+1}| \right) \leq 2M \left(|a_{n+1}| + |a_m| \right).$$

Now apply the condition that $a_n \underset{n \to \infty}{\to} 0$. By the definition of the limit, this means that for $\forall \varepsilon > 0$ (choose $\varepsilon_1 = \frac{\varepsilon}{4M}$) there exists N_ε such that $|a_n| < \frac{\varepsilon}{4M}$ whenever $n > N_\varepsilon$. Using in the evaluation of S the indices $\forall m > n > N_\varepsilon$, we obtain

$$|S| = \left| \sum_{k=n+1}^{m} a_k b_k \right| \leq 2M \left(|a_{n+1}| + |a_m| \right) < 2M \left(\frac{\varepsilon}{4M} + \frac{\varepsilon}{4M} \right) = \varepsilon.$$

Therefore, according to the Cauchy criterion, the series $\sum_{n=1}^{\infty} a_n b_n$ converges. $\square$

Corollary *Leibniz's test is a particular case of Dirichlet's test.*

Proof Indeed, choosing in an alternating series $\sum_{n=1}^{\infty} (-1)^n c_n$, $c_n > 0$ the sequences $a_n = c_n$ and $b_n = (-1)^n$, we see that the partial sums $B_n = \sum_{k=1}^{n} (-1)^k$ are bounded and if the sequence a_n approaches zero monotonically, both conditions of Dirichlet's test are satisfied, which implies the convergence of the alternating series $\sum_{n=1}^{\infty} (-1)^n c_n$. On the other hand, since the condition of boundedness of the partial sums B_n is automatically satisfied, the only requirement to the terms of an alternating series $\sum_{n=1}^{\infty} (-1)^n c_n$ comes from the second condition of Dirihlct's test (monotonic tendency of a_n to 0), but this is also the condition of Leibniz's test. $\square$

Abel's Test *Let $\sum_{n=1}^{\infty} c_n$ be a series with the general term in the form $c_n = a_n b_n$. If*

(1) the series $\sum_{n=1}^{\infty} b_n$ converges,
(2) the sequence a_n is monotone and bounded,

then the original series $\sum_{n=1}^{\infty} c_n$ converges.

Proof Let us show that the series $\sum_{n=1}^{\infty} a_n b_n$ satisfies the Cauchy criterion for convergence. First we use Abel's summation by parts formula and the telescopic representation $a_{n+1} = \sum_{k=n+1}^{m-1} (a_k - a_{k+1}) + a_m$ in order to express the sum $S \equiv \sum_{k=n+1}^{m} c_k =$

$\sum_{k=n+1}^{m} a_k b_k$, $m > n$ in the following manner:

$$S = \sum_{k=n+1}^{m} a_k b_k = \sum_{k=n+1}^{m-1} B_k (a_k - a_{k+1}) + B_m a_m - B_n a_{n+1}$$

$$= \sum_{k=n+1}^{m-1} B_k (a_k - a_{k+1}) + B_m a_m - B_n \left(\sum_{k=n+1}^{m-1} (a_k - a_{k+1}) + a_m \right)$$

$$= \sum_{k=n+1}^{m-1} (B_k - B_n) (a_k - a_{k+1}) + (B_m - B_n) a_m.$$

The last representation of S is suitable for evaluation under the given conditions. Since the sequence a_n is bounded, there exists M such that $|a_n| \leq M$ for all $n \in \mathbb{N}$. Further, the convergence of the series $\sum_{n=1}^{\infty} b_n$ implies that it satisfies the Cauchy criterion: for $\forall \varepsilon > 0$ (choose $\varepsilon_1 = \frac{\varepsilon}{3M}$) there exists N_ε such that for all indices $\forall m > n > N_\varepsilon$ it follows

$$\left| \sum_{k=n+1}^{m} b_k \right| = |B_m - B_n| < \varepsilon_1 = \frac{\varepsilon}{3M}.$$

Let us evaluate now the sum S for the same indices $m > n > N_\varepsilon$:

$$|S| = \left| \sum_{k=n+1}^{m-1} (B_k - B_n) (a_k - a_{k+1}) + (B_m - B_n) a_m \right|$$

$$\leq \sum_{k=n+1}^{m-1} |B_k - B_n| \cdot |a_k - a_{k+1}| + |B_m - B_n| \cdot |a_m|$$

$$< \frac{\varepsilon}{3M} \left(\sum_{k=n+1}^{m-1} |a_k - a_{k+1}| + |a_m| \right) \equiv S_1 .$$

Since the sequence a_n is monotone, all the differences $a_k - a_{k+1}$ have the same sign, and consequently, the sum of the absolute values is equal to the absolute value of the sum:

$$S_1 = \frac{\varepsilon}{3M} \left(\left| \sum_{k=n+1}^{m-1} (a_k - a_{k+1}) \right| + |a_m| \right)$$

$$= \frac{\varepsilon}{3M} (|a_{n+1} - a_{n+2} + a_{n+2} - a_{n+3} + \ldots + a_{m-1} - a_m| + |a_m|)$$

$$= \frac{\varepsilon}{3M} (|a_{n+1} - a_m| + |a_m|) \leq \frac{\varepsilon}{3M} (|a_{n+1}| + 2 |a_m|) .$$

Now utilize the condition of the boundedness of a_n ($|a_n| \leq M, \forall n \in \mathbb{N}$) to arrive at the inequality

$$|S| = \left| \sum_{k=n+1}^{m} a_k b_k \right| \leq \frac{\varepsilon}{3M} \left(|a_{n+1}| + 2|a_m| \right) \leq \frac{\varepsilon}{3M} \cdot 3M = \varepsilon,$$

which shows that the series $\sum_{n=1}^{\infty} a_n b_n$ satisfies the Cauchy criterion and consequently converges. $\square$

Remark The presented proof handles Abel's test independently from Dirichlet's test. Another advantage of this proof is that it prepares the technique of verification of Abel's test for series of functions. However, if we desire to take advantage of Dirichlet's test and look for the simplest form of demonstration of Abel's test, we can reduce Abel's test to Dirichlet's test. It can be made in the following way: if the conditions of Abel's test are satisfied, then the partial sums B_n are bounded (because these are the partial sums of a convergent series), the sequence a_n converges to a (because this is a monotone and bounded sequence), and consequently, using the representation of the original series in the form $\sum_{n=1}^{\infty} a_n b_n = \sum_{n=1}^{\infty} (a_n - a) b_n + \sum_{n=1}^{\infty} a b_n$, we conclude that the first series converges by Dirichlet's test and the second converges due to the hypothesis of Abel's test.

Example 1D $\sum_{n=1}^{\infty} \frac{\ln n}{n} \sin \frac{n\pi}{4}$

Although the general term changes its sign infinite number of times, but not in the alternating form, that makes impossible the use of the simpler Leibniz's test. Let us apply Dirichlet's test with $a_n = \frac{\ln n}{n}$ and $b_n = \sin \frac{n\pi}{4}$. The sequence a_n converges to 0, which is seen by applying L'Hospital's rule to the function $f(x) = \frac{\ln x}{x}$:

$$\lim_{x \to +\infty} \frac{\ln x}{x} = \lim_{x \to +\infty} \frac{\frac{1}{x}}{1} = 0.$$

The evaluation of the derivative of $f(x)$:

$$f'(x) = \left(\frac{\ln x}{x} \right)' = \frac{1 - \ln x}{x^2} < 0, \forall x > e$$

shows that a_n is monotone for the indices $n \geq 3$. For the partial sums B_n, we notice that b_n takes three positive values followed by zero and then three negative values and one more zero, after which it returns to three positive values and continues in a periodic fashion. Therefore, the partial sums are bounded:

$$|B_n| = \left| \sum_{k=1}^{n} \sin \frac{k\pi}{4} \right| \leq \max_k \left| \sin \frac{k\pi}{4} + \sin \frac{(k+1)\pi}{4} + \sin \frac{(k+2)\pi}{4} \right| \leq 3.$$

Hence, all the conditions of Dirichlet's test hold, which implies the series convergence.

Example 2D $\sum_{n=2}^{\infty}(-1)^n \frac{1}{\ln n}\cos\frac{1}{n^2}$

Notice that the series is alternating, but direct application of Leibniz's test is somewhat involved due to the growth of the factor $\cos\frac{1}{n^2}$. The simplest way to analyze the convergence is by Abel's test, splitting the general term in the product of $a_n = \cos\frac{1}{n^2}$ and $b_n = (-1)^n\frac{1}{\ln n}$. The convergence of the series $\sum_{n=2}^{\infty}(-1)^n\frac{1}{\ln n}$ follows straightforward from Leibniz's test ($\frac{1}{\ln n}$ tends monotonically to 0) and the sequence $a_n = \cos\frac{1}{n^2}$ is bounded by 1 and monotone. Hence, all the conditions of Abel's test are met, and consequently, the series converges.

Example 3D $\sum_{n=1}^{\infty}\frac{\cos\alpha n}{n^p}$, $\alpha, p \in \mathbb{R}$

If $p \leq 0$, then the series diverges, since its general term does not approach 0. In the case $\alpha = 2m\pi$, $\forall m \in \mathbb{Z}$ we have a p-series that converges when $p > 1$ and diverges when $p \leq 1$. For $\alpha \neq 2m\pi$ and $p > 0$ we use Dirichlet's test to investigate the behavior of this series. Choose the sequence $a_n = \frac{1}{n^p}$, which tends monotonically to 0, and evaluate the partial sums B_n of the second series $\sum_{n=1}^{\infty} b_n = \sum_{n=1}^{\infty}\cos\alpha n$ in the following way:

$$B_n = \sum_{k=1}^{n}\cos\alpha k = \frac{1}{\sin\frac{\alpha}{2}}\sum_{k=1}^{n}\cos\alpha k\cdot\sin\frac{\alpha}{2} = \frac{1}{\sin\frac{\alpha}{2}}\sum_{k=1}^{n}\frac{1}{2}\left(\sin\frac{2k+1}{2}\alpha - \sin\frac{2k-1}{2}\alpha\right)$$

$$= \frac{1}{2\sin\frac{\alpha}{2}}\left(-\sin\frac{1}{2}\alpha + \sin\frac{3}{2}\alpha - \sin\frac{3}{2}\alpha + \sin\frac{5}{2}\alpha - \ldots - \sin\frac{2n-1}{2}\alpha + \sin\frac{2n+1}{2}\alpha\right)$$

$$= \frac{1}{2\sin\frac{\alpha}{2}}\left(\sin\frac{2n+1}{2}\alpha - \sin\frac{1}{2}\alpha\right).$$

(Notice that all the transformations are valid because $\sin\frac{\alpha}{2} \neq 0$ for $\alpha \neq 2m\pi$.) Then, B_n are bounded according to the following evaluation:

$$|B_n| = \left|\frac{1}{2\sin\frac{\alpha}{2}}\left(\sin\frac{2n+1}{2}\alpha - \sin\frac{1}{2}\alpha\right)\right| \leq \frac{1}{2|\sin\frac{\alpha}{2}|}\left(\left|\sin\frac{2n+1}{2}\alpha\right| + \left|\sin\frac{1}{2}\alpha\right|\right) \leq \frac{1}{|\sin\frac{\alpha}{2}|}.$$

Thus, all the conditions of Dirichlet's test are satisfied, and consequently, the original series converges when $\alpha \neq 2m\pi$, $\forall m \in \mathbb{Z}$ and $p > 0$.

Example 4D $\sum_{n=2}^{\infty}\frac{\sqrt{n^2-n+1}}{n\ln n}\cos\alpha n$, $\alpha \in \mathbb{R}$

In the case $\alpha = 2m\pi$, $\forall m \in \mathbb{Z}$, we can investigate the behavior of the series by using the Comparison test in the form with limit. Comparing the general term $\frac{\sqrt{n^2-n+1}}{n\ln n}$ with $\frac{1}{\ln n}$:

$$\lim_{n\to\infty}\frac{\frac{\sqrt{n^2-n+1}}{n\ln n}}{\frac{1}{\ln n}} = \lim_{n\to\infty}\frac{\sqrt{n^2-n+1}}{n} = \lim_{n\to\infty}\sqrt{1-\frac{1}{n}+\frac{1}{n^2}} = 1,$$

we conclude that the series $\sum_{n=2}^{\infty}\frac{\sqrt{n^2-n+1}}{n\ln n}$ diverges since the simpler series $\sum_{n=2}^{\infty}\frac{1}{\ln n}$ diverges.

In the case $\alpha \neq 2m\pi$, we apply Abel's test. Choose the sequence $a_n = \frac{\sqrt{n^2-n+1}}{n}$ and show that it is monotone and bounded. Indeed, the comparison between $a_n = \frac{\sqrt{n^2-n+1}}{n}$ and $a_{n+1} = \frac{\sqrt{(n+1)^2-(n+1)+1}}{n+1}$ is equivalent to the comparison between $\sqrt{n^2 - n + 1} \cdot (n+1)$ and $\sqrt{(n+1)^2 - n} \cdot n$, which, in turn, is equivalent to the comparison between the corresponding squares $(n^2 - n + 1) \cdot (n+1)^2$ and $((n+1)^2 - n) \cdot n^2$. The last is reduced to the comparison between $n^4 + n^3 + n + 1$ and $n^4 + n^3 + n^2$. Since $n + 1 < n^2$, $\forall n \geq 2$, it follows that $n^4 + n^3 + n + 1 < n^4 + n^3 + n^2$, $\forall n \geq 2$, and consequently, $a_n = \frac{\sqrt{n^2-n+1}}{n} < \frac{\sqrt{(n+1)^2-(n+1)+1}}{n+1} = a_{n+1}$, $\forall n \geq 2$. Hence, a_n is increasing starting from the index $n = 2$. Since its limit is 1, we have the upper bound of a_n: $a_n \leq 1$, $\forall n$. On the other hand, $a_n > 0$, $\forall n$, and consequently, the sequence a_n is bounded.

To verify the second condition of Abel's test—the convergence of the series $\sum_{n=2}^{\infty} b_n = \sum_{n=2}^{\infty} \frac{\cos \alpha n}{\ln n}$, we use Dirichlet's test and the techniques employed in Example 3D. Choose the sequence $\tilde{a}_n = \frac{1}{\ln n}$, that convergence monotonically to 0, and evaluate the partial sums $\tilde{B}_n$ of the series $\sum_{n=1}^{\infty} \tilde{b}_n = \sum_{n=1}^{\infty} \cos \alpha n$, bearing in mind that $\sin \frac{\alpha}{2} \neq 0$ for $\alpha \neq 2m\pi$:

$$\tilde{B}_n = \sum_{k=1}^{n} \cos \alpha k = \frac{1}{2 \sin \frac{\alpha}{2}} \left(\sin \frac{2n+1}{2}\alpha - \sin \frac{1}{2}\alpha \right),$$

and consequently,

$$|\tilde{B}_n| \leq \frac{1}{|\sin \frac{\alpha}{2}|}.$$

Hence, the conditions of Dirichlet's test are satisfied for the series $\sum_{n=2}^{\infty} b_n = \sum_{n=2}^{\infty} \frac{\cos \alpha n}{\ln n}$ that guarantees its convergence.

Returning to the original series, we see that all the conditions of Abel's test hold, and therefore, the original series converges for $\alpha \neq 2m\pi$.

Historical Remarks The first version of his test Abel published in 1826 in the paper on the binomial series. In the same paper he derived the summation by parts formula. The Dirichlet's test was published posthumously in 1862.

4.3   Absolute and Conditional Convergence

▶ **Definition** (**Absolute Convergence of a Series**) A *series* $\sum a_n$ *converges absolutely* if the series $\sum |a_n|$ *converges*.

Theorem (**Relationship Between Absolute Convergence and Convergence**) *If a series* $\sum a_n$ *converges absolutely, then it converges.*

Proof If the series $\sum |a_n|$ converges, according to the Cauchy criterion we have: for $\forall \varepsilon > 0$ there exists N such that for $\forall m > n > N$ it follows $\left| \sum_{k=n+1}^{m} |a_k| \right| = \sum_{k=n+1}^{n+p} |a_k| < \varepsilon$. Then, for the same indices $m > n > N$ we have $\left| \sum_{k=n+1}^{m} a_k \right| \leq \sum_{k=n+1}^{m} |a_k| < \varepsilon$, that is, the Cauchy condition for the series $\sum a_n$ holds, which means that the original series converges. $\qquad \square$

Remark The converse is not true, that can be seen from the example of the alternating harmonic series $\sum_{n=1}^{\infty} \frac{(-1)^n}{n}$.

The last Theorem shows that there are series which converge, but do not converge absolutely. Therefore, it is natural to name this type of series.

▶ **Definition** (**Conditional Convergence of a Series**) A *series* $\sum a_n$ *converges conditionally* if it is convergent, but not absolutely convergent.

Tests of Absolute Convergence

Since the series of the absolute values $\sum |a_n|$ are the series of non-negative (or positive if $a_n \neq 0$) terms, all the discussed tests for non-negative and positive series (which allow $a_n = 0$) are applicable to investigation of the absolute convergence of an arbitrary series.

Notice additionally that the verification of the divergence of the series $\sum |a_n|$ by D'Alembert's or Cauchy's test implies the divergence of the original series $\sum a_n$, since in both tests the divergence follows from the violation of the necessary condition of the convergence: $|a_n| \underset{n \to \infty}{\not\to} 0$, and the last is equivalent to the condition $a_n \underset{n \to \infty}{\not\to} 0$.

4.4 Product of Two Series

With the notions of absolute and conditional convergence at hand, we return to the discussion of the *product of two series*. Recall that in Sect. 2.1 this product was introduced in the following manner:

▶ **Definition** (**Cauchy Product**) The *Cauchy product of two series* $\sum_{n=0}^{\infty} a_n$ and $\sum_{n=0}^{\infty} b_n$ is a series $\sum_{n=0}^{\infty} c_n$ with the general term defined by the formula

$$c_n = \sum_{k=0}^{n} a_k b_{n-k} = a_0 b_n + a_1 b_{n-1} + \ldots + a_{n-1} b_1 + a_n b_0.$$

For this product, the convergence of each of the original series does not guarantee the convergence of the product, as it follows from the following example.

Example Consider the series

$$\sum_{n=0}^{\infty} a_n = \sum_{n=0}^{\infty} \frac{(-1)^n}{\sqrt{n+1}} = 1 - \frac{1}{\sqrt{2}} + \frac{1}{\sqrt{3}} - \frac{1}{\sqrt{4}} + \cdots$$

convergent (conditionally) according to Leibniz's test. Consider the Cauchy product of this series with itself, that is, take $b_n = a_n$:

$$\sum_{n=0}^{\infty} c_n = 1 - \left(\frac{1}{\sqrt{2}} + \frac{1}{\sqrt{2}}\right) + \left(\frac{1}{\sqrt{3}} + \frac{1}{\sqrt{2}\sqrt{2}} + \frac{1}{\sqrt{3}}\right) - \left(\frac{1}{\sqrt{4}} + \frac{1}{\sqrt{3}\sqrt{2}} + \frac{1}{\sqrt{2}\sqrt{3}} + \frac{1}{\sqrt{4}}\right) + \cdots.$$

The general term of the product is found by the formula

$$c_n = (-1)^n \sum_{k=0}^{n} \frac{1}{\sqrt{k+1}} \frac{1}{\sqrt{n-k+1}}.$$

Since

$$(k+1)(n-k+1) = \left(\frac{n}{2}+1\right)^2 - \left(\frac{n}{2}-k\right)^2 \le \left(\frac{n}{2}+1\right)^2,$$

the general term can be evaluated in the form

$$|c_n| \ge \sum_{k=0}^{n} \frac{2}{n+2} = \frac{2(n+1)}{n+2},$$

that shows that c_n does not approach 0, and consequently, the series of the product is divergent.

However, a modest increase in the requirements to the original series—the condition that one of these series converges absolutely—is already sufficient to ensure the convergence of the product. Moreover, the sum of the product is $c = ab$, although the dependence of the partial sums $C_n = \sum_{k=0}^{n} c_k$ from the original partial sums $A_n = \sum_{k=0}^{n} a_k$ and $B_n = \sum_{k=0}^{n} b_k$ has rather complicated form, which does not satisfy the relation $C_n = A_n B_n$. This property is formulated in exact form and demonstrated below.

Property of the product *Let $\sum_{n=0}^{\infty} a_n = a$ and $\sum_{n=0}^{\infty} b_n = b$ be two convergent series. If one of these series converges absolutely, then their Cauchy product $\sum_{n=0}^{\infty} c_n$, $c_n = \sum_{k=0}^{n} a_k b_{n-k}$ converges to the sum $c = ab$.*

Proof Due to the symmetry of the coefficients a_n and b_n in the formula of the general term c_n, we can suppose, without loss of generality, that the series $\sum_{n=0}^{\infty} a_n$ converges absolutely. Denoting $A_n = \sum_{k=0}^{n} a_k$, $B_n = \sum_{k=0}^{n} b_k$, $C_n = \sum_{k=0}^{n} c_k$ and $\beta_n = B_n - b$, we obtain

$$C_n = \sum_{k=0}^{n} c_k = a_0 b_0 + (a_0 b_1 + a_1 b_0) + \ldots + (a_0 b_n + a_1 b_{n-1} + \ldots a_n b_0)$$

$$= a_0 B_n + a_1 B_{n-1} + \ldots + a_n B_0 = a_0 (b + \beta_n) + a_1 (b + \beta_{n-1}) + \ldots + a_n (b + \beta_0)$$

$$= A_n b + a_0 \beta_n + a_1 \beta_{n-1} + \ldots + a_n \beta_0.$$

Since $A_n b \underset{n \to \infty}{\to} ab$, we have to show that $\gamma_n = a_0 \beta_n + a_1 \beta_{n-1} + \ldots + a_n \beta_0 \underset{n \to \infty}{\to} 0$.

The convergence of $\sum_{n=0}^{\infty} b_n$ implies that $\beta_n \underset{n \to \infty}{\to} 0$. Then, for any $\varepsilon_1 > 0$ there exists N such that $|\beta_n| < \varepsilon_1$ whenever $n > N$. In particular, we can choose $\varepsilon_1 = \frac{\varepsilon}{2(\alpha+1)}$, where $\varepsilon > 0$ is an arbitrary positive number and $\alpha = \sum_{n=0}^{\infty} |a_n|$. For any $n > N$, we divide γ_n into two parts:

$$\gamma_n = a_n \beta_0 + \ldots + a_0 \beta_n = g_n + h_n, \quad g_n = a_n \beta_0 + \ldots + a_{n-N} \beta_N, \quad h_n = a_{n-N-1} \beta_{N+1} + \ldots + a_0 \beta_n.$$

The part h_n, where $|\beta_n|$ are sufficiently small quantities, can be evaluated as follows:

$$|h_n| = |a_{n-N-1} \beta_{N+1} + \ldots + a_0 \beta_n| \leq |a_{n-N-1}||\beta_{N+1}| + \ldots + |a_0||\beta_n|$$

$$< \frac{\varepsilon}{2(\alpha+1)} (|a_{n-N-1}| + \ldots + |a_0|) < \frac{\varepsilon}{2}.$$

Now we keep N fixed and evaluate the part g_n. First, we find $\beta = \max\{|\beta_0|, \ldots, |\beta_N|\}$. The convergence of $\sum_{n=0}^{\infty} a_n$ implies that $a_n \underset{n \to \infty}{\to} 0$. Then, for any $\varepsilon_2 > 0$, in particular, for $\varepsilon_2 = \frac{\varepsilon}{2(\beta+1)(N+1)}$, it can be chosen N_2 such that $|a_n| < \varepsilon_2$ whenever $n > N_2$. Taking then $N_3 = N + N_2$, for all the indices $n > N_3$ we have $n > n - 1 > \ldots > n - N > N_2$. Therefore,

$$|g_n| = |a_n \beta_0 + \ldots + a_{n-N} \beta_N| \leq |a_n||\beta_0| + \ldots + |a_{n-N}||\beta_N|$$

$$\leq (|a_n| + \ldots + |a_{n-N}|)\beta < \frac{\varepsilon}{2(\beta+1)(N+1)} (N+1)\beta < \frac{\varepsilon}{2}.$$

Since $n > N_3 = N + N_2$ implies $n > N$, the estimate for h_n is also satisfied. Joining the obtained evaluations for h_n and g_n, we arrive at the following inequality

$$|\gamma_n| \le |g_n| + |h_n| < \frac{\varepsilon}{2} + \frac{\varepsilon}{2} = \varepsilon \,,$$

that means, due to the arbitrariness of $\varepsilon > 0$, that $\gamma_n \underset{n \to \infty}{\to} 0$.

This completes the proof of the Theorem. $\square$

5   Associative and Commutative Properties of Series

5.1   Positive and Negative Parts of Series

Let us denote a given series $\sum_{n=1}^{\infty} a_n$ by (A). It may happen that this series has a finite number of negative terms. Recall that elimination or addition of a finite number of elements does not affect the convergence/divergence of a series. Then, the series A behaves qualitatively in the same manner as the non-negative series (B), which contains all the non-negative elements of (A), but all the negative elements are excluded. More specifically, the series (A) converges if and only if the series (B) converges, and their sums A and B are related by the formula $A = B - C$, where C is the sum of all the negative elements of (A) taken with opposite sign. Using again the result on the invariance of the convergence of a series with respect to a change of a finite number of its terms, we can conclude that the series of the absolute values $\sum_{n=1}^{\infty} |a_n|$ (denote it by (A^*)) is also convergent (if the series (A) converges), because it is obtained from the series (A) by substituting a finite number of negative elements by their opposites. Hence, if the series (A) converges and has a finite number of negative terms, it converges absolutely. Besides, for the sum A^* of the series (A^*) we have $A^* = B + C$.

In the same manner, if a series (A) contains a finite number of positive terms, it behaves in the same way as the series of its non-positive part (the series with all the non-positive elements of (A), but without all its positive elements). For convenience, we compose the series of the non-positive part of (A) using the non-positive elements with opposite sign in order to obtain the non-negative series (C). In particular, the series (A) converges if and only if the series (C) converges, and the sums A and C are related by the formula $A = B - C$, where B is the finite sum of all the positive elements of (A). Besides, the series of the absolute values (A^*) converges, which means that the series (A) converges absolutely, and $A^* = B + C$.

Thus, using the above discussion, we arrive at the following simple result.

Theorem 1 *If a convergent series $\sum_{n=1}^{\infty} a_n$ has a finite number of positive or negative terms, then it converges absolutely and the following relations among the sums of the four series (A), (A^*), (B) and (C) hold: $A = B - C$ and $A^* = B + C$.*

Remark Notice that the converse with respect to the number of positive and negative elements is not true: a series may converge absolutely and still contain an infinite number of both positive and negative terms. An elementary example of this type is the alternating $p = 2$-series $\sum_{n=1}^{\infty} \frac{(-1)^n}{n^2}$. However, the relations between the sums—$A = B - C$ and $A^* = B + C$—continue to be true for any absolutely convergent series, which will be shown below.

Consider now the situation when the series (A) contains an infinite number of both positive and negative elements. Notice that, according to Theorem 1, this is certainly the case of conditionally convergent series, but it may also happen with absolutely convergent series (see example $\sum_{n=1}^{\infty} \frac{(-1)^n}{n^2}$ of the last Remark). In this case, we construct the two auxiliary series: the series $\sum_{k=1}^{\infty} b_k$ (denoted by (B)) composed of all the positive elements of the series (A), which follow the same order as in the series (A), and the series $\sum_{m=1}^{\infty} c_m$ (denoted by (C)) composed of all the negative elements of (A) taken with opposite sign and in the same order as in the original series. The relationships between convergence of these series and between their sums are explained in the following Theorem.

Theorem 2 *Suppose that a series $\sum_{n=1}^{\infty} a_n$ (series (A)) has an infinite number of both positive and negative elements, and series $\sum_{k=1}^{\infty} b_k$ and $\sum_{m=1}^{\infty} c_m$ (series (B) and (C)) are its positive and negative parts, respectively (the latter contains the negative elements of (A) with opposite sign). In this case:*

(1) if the series (A) converges absolutely, then both series (B) and (C) converge and their sums are related by the formula $A = B - C$, where $A = \sum_{n=1}^{\infty} a_n$, $B = \sum_{k=1}^{\infty} b_k$ and $C = \sum_{m=1}^{\infty} c_m$; besides, $A^* = B + C$, where $A^* = \sum_{n=1}^{\infty} |a_n|$;

(2) if the series (A) converges conditionally, then both series (B) and (C) diverge.

Proof By the definition of the series (A), (B), (C) and (A^*), we have the following relations among their partial sums $A_n = \sum_{i=1}^{n} a_i$, $A_n^* = \sum_{i=1}^{n} |a_i|$, $B_k = \sum_{j=1}^{k} b_j$ and $C_m = \sum_{j=1}^{m} c_j$:

$$A_n = B_k - C_m, \quad A_n^* = B_k + C_m, \quad n = k + m.$$

By hypothesis, the series (A) contains an infinite number of positive and negative elements, which implies that $n \to \infty$ if and only if both $k \to \infty$ and $m \to \infty$.

Consider the first situation, when the series (A) converges absolutely, that is, the series (A^*) converges. In this case, there exists the number M^* such that for $\forall n \in \mathbb{N}$ it follows $A_n^* \leq M^*$. By construction, $B_k \leq A_n^* \leq M^*$ and $C_m \leq A_n^* \leq M^*$, $\forall k, m \in \mathbb{N}$. Since the partial sums B_k and C_m are bounded above, by the criterion for convergence of positive series, the series (B) and (C) converge, that is, there exist the finite limits $\lim_{k \to \infty} B_k = B$

and $\lim_{m \to \infty} C_m = C$. Then, passing to the limit in the two relations for the partial sums, we obtain

$$A = \lim_{n \to \infty} A_n = \lim_{k \to \infty} B_k - \lim_{m \to \infty} C_m = B - C, \quad A^* = \lim_{n \to \infty} A_n^* = \lim_{k \to \infty} B_k + \lim_{m \to \infty} C_m = B + C.$$

In the second case, when the series (A) converges conditionally (that is, the series (A^*) diverges), we notice first that $\lim_{n \to \infty} A_n^* = +\infty$. Then, from the equality $A_n^* = B_k + C_m$ it follows that at least one of the limits of the partial sums $\lim_{k \to \infty} B_k$ or $\lim_{m \to \infty} C_m$ is infinite. If we suppose, by absurd, that only one of these two limits is infinite, for instance, $\lim_{k \to \infty} B_k = +\infty$ and $\lim_{m \to \infty} C_m = C$, then we have $\lim_{n \to \infty} A_n = \lim_{n \to \infty} (B_k - C_m) = +\infty$, that contradicts the convergence of the series (A) (in the case $\lim_{k \to \infty} B_k = B$ and $\lim_{m \to \infty} C_m = +\infty$ we arrive at the same contradiction). Therefore, both limits should be infinite, that is, both series (B) and (C) should diverge.   □

5.2   Associative Property of Convergent Series

It is an elementary fact that finite sums possess the associative property, that is, the summands of a finite sum can be grouped in any manner without changing the value of the sum; in other words, we can introduce or drop parentheses among any group of terms (as usual, parentheses indicate the precedence of the operations among the terms within them), and it has no effect on the value of the sum. Let us analyze if there is an analogous property for convergent series.

We denote a given series $\sum_{n=1}^{\infty} a_n$ by (A) and compose a new series, denoted $(\tilde{A})$, by introducing some parentheses in the series (A) (under this operation, we do not either include new elements or drop the elements of the original series, and we do not change the order in which the terms appear in the series (A)). Then, we obtain the series $\sum_{k=1}^{\infty} \tilde{a}_k$ with the elements

$$\tilde{a}_1 = a_1 + a_2 + \ldots + a_{n_1}, \ \tilde{a}_2 = a_{n_1+1} + \ldots + a_{n_2}, \ \ldots, \ \tilde{a}_k = a_{n_{k-1}+1} + a_{n_{k-1}+2} + \ldots + a_{n_k}, \ \ldots.$$

Theorem 1 *If a series (A) converges, then the series $(\tilde{A})$ also converges and its sum $\tilde{A}$ coincides with the sum A of the original series.*

Proof To demonstrate the convergence of the series $(\tilde{A})$ we use the definition of convergence of a series. Consider the partial sums of the series $(\tilde{A})$, taking into account that finite sums satisfy the associative property

$$\tilde{A}_k = \sum_{m=1}^{k} \tilde{a}_m = \tilde{a}_1 + \tilde{a}_2 + \ldots + \tilde{a}_k = (a_1 + \ldots + a_{n_1}) + (a_{n_1+1} + \ldots + a_{n_2}) + \ldots + (a_{n_{k-1}+1} + \ldots + a_{n_k})$$

$$= a_1 + a_2 + \ldots + a_{n_k} = A_{n_k}.$$

Since the series (A) converges, the sequence of its partial sums has a finite limit and any subsequence of this sequence has the same limit. Therefore, $\lim\limits_{k\to\infty} \tilde{A}_k = \lim\limits_{k\to\infty} A_{n_k} = A$. Hence, the series $(\tilde{A})$ converges to $\tilde{A} = A$. $\square$

Remark 1 Notice that the associative property does not work in the inverse direction: the convergence of the series $(\tilde{A})$ does not imply the convergence of (A). Indeed, the series

$$\sum_{k=1}^{\infty} \tilde{a}_k = \sum_{k=1}^{\infty} (1-1) = (1-1) + (1-1) + \ldots + (1-1) + \ldots$$

converges because $\tilde{a}_k = 1 - 1 = 0$, and consequently $\tilde{A}_k = 0$ and $\tilde{A} = \lim\limits_{k\to\infty} \tilde{A}_k = 0$. However, the series (A), obtained from $(\tilde{A})$ by disarranging its terms (dropping the parentheses), $1 - 1 + 1 - 1 + \ldots = \sum_{n=1}^{\infty} (-1)^{n+1}$ diverges, because its general term does not approach zero.

Remark 2 If the series (A), obtained by dropping the parentheses in the series $(\tilde{A})$, is convergent, then the sums of these two series coincide. This follows immediately from Theorem 1.

Corollary *If a series (A) converges absolutely, then the series $(\tilde{A})$ also converges absolutely.*

Proof In fact, the absolute value of the general term of the series $(\tilde{A})$ satisfies the inequality

$$|\tilde{a}_k| = |a_{n_{k-1}+1} + \ldots + a_{n_k}| \le |a_{n_{k-1}+1}| + \ldots + |a_{n_k}| = \tilde{a}_k^*.$$

The series $\sum_{k=1}^{\infty} \tilde{a}_k^*$ converges (according to Theorem 1), since its terms represent the group of the elements of the convergent series $\sum_{n=1}^{\infty} |a_n|$. Then, by the Comparison test, the series $\sum_{k=1}^{\infty} |\tilde{a}_k|$ also converges, that is, $(\tilde{A})$ converges absolutely. $\square$

Theorem 2 *If the grouped elements of the convergent series $(\tilde{A})$ are such that within each parentheses the terms have the same sign (while in different parentheses the signs can be different), then the series (A), obtained from $(\tilde{A})$ by disarranging its terms (dropping the parentheses), is also convergent and has the same sum. Notice that zero elements can be considered as both negative and positive.*

Proof Take any number $m \in \mathbb{N}$ and evaluate the partial sum A_m of the series (A). Since for any m there exists $n_k, n_{k-1} \in \mathbb{N}$ such that $n_{k-1} < m \le n_k$, we get

$$\tilde{A}_{k-1} = A_{n_{k-1}} \le A_m \le A_{n_k} = \tilde{A}_k,$$

if all the elements in the k-th parentheses are non-negative, or

$$\tilde{A}_{k-1} = A_{n_{k-1}} \geq A_m \geq A_{n_k} = \tilde{A}_k,$$

if all the elements in the k-th parentheses are non-positive. The convergence of the series $(\tilde{A})$ means that $\lim\limits_{k \to \infty} \tilde{A}_k = \lim\limits_{k \to \infty} \tilde{A}_{k-1} = \tilde{A}$. Passing to the limit in the double inequalities obtained for the partial sums A_m, taking into account that the tendency $m \to \infty$ implies $k \to \infty$, and utilizing the squeeze Theorem, we obtain: $A = \lim\limits_{m \to \infty} A_m = \lim\limits_{k \to \infty} \tilde{A}_k = \tilde{A}$, that is, the series (A) converges and its sum A coincide with the sum $\tilde{A}$ of the series $(\tilde{A})$.

$\square$

Joining this result with that of Theorem 1, we conclude that under the conditions of Theorem 2, the associative property holds in the two directions (both for grouping and ungrouping of the terms).

Corollary 1 *For a non-negative series the associative property works in both directions, that is, the series (A) of separated terms and the series $(\tilde{A})$ of grouped terms converge or diverge simultaneously. In the case of convergence, both series have the same sum. This statement follows immediately from Theorems 1 and 2.*

Corollary 2 *If a series, obtained from $\sum_{k=1}^{\infty} |a_k|$ by some grouping of its terms, is convergent, then the original series (A) converges absolutely and the sums of the two series (A) and $(\tilde{A})$ are equal (as well as the sums of the series of the absolute values).*

Proof Indeed, the series $\sum_{k=1}^{\infty} |a_k|$ is non-negative and, by Corollary 1, the convergence of the series of grouped elements $\sum_{k=1}^{\infty} |\tilde{a}_k|$, where $|\tilde{a}_k| = |a_{n_{k-1}+1}| + |a_{n_{k-1}+2}| + \ldots + |a_{n_k}|$, is equivalent to the convergence of the series $\sum_{n=1}^{\infty} |a_n|$, which means that the original series (A) converges absolutely. Then, by Theorem 1, the sum of (A) is equal to the sum of $(\tilde{A})$.

$\square$

5.3   Commutative Property of Absolutely Convergent Series

One of the elementary properties of finite sums is the commutative property, which allows us to change the order of the terms any way we wanted keeping the same value of the sum. Let us see if a similar property is true for infinite sums.

We denote a given series $\sum_{n=1}^{\infty} a_n$ by (A) and consider a new series (A') with the elements a'_k, obtained by a rearrangement (permutation) of the elements of (A) (without including new elements or dropping any elements of the original series). More precisely, the new series $\sum_{k=1}^{\infty} a_k'$ is such that for any k there exists n_k such that $a_k' = a_{n_k}$ and the converse is true, that is, for any index n of the original series there exists k_n such that $a_n = a_{k_n}'$.

The Dirichlet Theorem 1 *If a series* (A) *is positive and convergent, then it satisfies the commutative property, that is, the rearranged series* (A') *converges and its sum coincides with the sum of the series* (A).

Proof Consider a partial sum of the series (A') (evidently, the series (A') is also positive):

$$A_k' = a_1' + a_2' + \ldots + a_k' = a_{n_1} + a_{n_2} + \ldots + a_{n_k}.$$

Compare k indices $n_1, n_2, \ldots, n_k$ and choose the maximum among them: $m = \max\{n_1, n_2, \ldots, n_k\}$. Since all the terms of the series (A) are positive, we have the following inequality between the partial sums A_k' and A_m:

$$A_k' = a_{n_1} + a_{n_2} + \ldots + a_{n_k} \leq a_1 + a_2 + \ldots + a_m = A_m.$$

Since (A) is a positive convergent series, its partial sums are less than or equal to the sum A. Therefore, for $\forall k \in \mathbb{N}$ the partial sums of the positive series (A') are bounded above: $A_k' \leq A_m \leq A$, which implies the convergence of (A'). Besides, passing to the limit in the inequality $A_k' \leq A$, we have $A' = \lim_{k \to \infty} A_k' \leq A$. On the other hand, the series (A) can be obtained from (A') using the reverse rearrangement (reordering) of its elements. Then knowing that (A') converges and considering it as the original series, we can apply the same arguments to arrive at the reverse inequality $A \leq A'$. Therefore, the two obtained inequalities between A' and A amount to $A' = A$.   □

The Dirichlet Theorem 2 *If a series* (A) *converges absolutely, then it satisfies the commutative property, that is, the rearranged series* (A') *converges absolutely and its sum coincides with the sum of the series* (A).

Proof Since the series (A) converges absolutely, the series (D) of the absolute values $\sum_{n=1}^{\infty} |a_n|$ converges. A rearrangement (permutation) of the terms of the series (A) causes the corresponding rearrangement of the terms of the series (D), that is, we obtain the series (D'): $\sum_{k=1}^{\infty} |a_k'|$. Since the series (D) converges, by the Theorem 1, the series (D') also converges and $D' = D$. This means that the series (A') converges absolutely. To find the sum of the series (A'), we consider the following two auxiliary series: the series $\sum_{k=1}^{\infty} b_k$

(denoted by (B)) composed of all the positive elements of the series (A) placed in the same order as they appear in (A); and the series $\sum_{m=1}^{\infty} c_m$ (denoted by (C)) composed of all the negative terms of the series (A) taken with opposite sign and following their original order. A rearrangement of the terms of the series (A) induces the corresponding rearrangement of the terms of the series (B) and (C), that results in the series (B') and (C'). Since the series (A) converges absolutely, the positive series (B) and (C) also converge (see Theorem 2 in Sect. 5.1). Then, by the Dirichlet Theorem 1, the series (B') and (C') also converge and $B' = B$, $C' = C$. Therefore, $A' = B' - C' = B - C = A$, that is, the sum of the series (A') coincides with the sum of the series (A).   □

5.4   Commutative Property of Conditionally Convergent Series

If a series $\sum_{n=1}^{\infty} a_n$ converges conditionally, then the commutative property does not hold. Let us start with the famous *example given by Dirichlet*, which illustrates the situation nicely.

The Dirichlet Example Consider the alternating harmonic series (conditionally convergent)

$$\sum_{n=1}^{\infty} a_n = \sum_{n=1}^{\infty} \frac{(-1)^{n+1}}{n} = 1 - \frac{1}{2} + \frac{1}{3} - \frac{1}{4} + \frac{1}{5} - \frac{1}{6} + \frac{1}{7} - \frac{1}{8} + \ldots = \ln 2.$$

The specific value of the sum does not matter (it could be denoted just by the letter A), but since it is well-known we use this information. Let us perform some operations with this series admissible for convergent series. First we multiply this series by $\frac{1}{2}$:

$$\sum_{n=1}^{\infty} b_n = \frac{1}{2} \sum_{n=1}^{\infty} \frac{(-1)^{n+1}}{n} = \sum_{n=1}^{\infty} \frac{1}{2} \frac{(-1)^{n+1}}{n} = \frac{1}{2} - \frac{1}{4} + \frac{1}{6} - \frac{1}{8} + \ldots = \frac{1}{2} \ln 2.$$

Next, we add the series $\sum_{n=1}^{\infty} b_n$ to the series of zeros $\sum_{n=1}^{\infty} 0$:

$$\sum_{n=1}^{\infty} c_n = \sum_{n=1}^{\infty} 0 + \sum_{n=1}^{\infty} \frac{(-1)^{n+1}}{2n} = \sum_{n=1}^{\infty} \left(0 + \frac{(-1)^{n+1}}{2n}\right) = 0 + \frac{1}{2} + 0 - \frac{1}{4} + 0 + \frac{1}{6} + 0 - \frac{1}{8} + \ldots = \frac{1}{2} \ln 2.$$

Notice that the elements of the last series are defined by the formula

$$c_n = \begin{cases} 0, & n = 2k - 1 \\ \frac{(-1)^{k+1}}{2k}, & n = 2k \end{cases}, \quad k \in \mathbb{N}.$$

Finally, we add the original and the last series and obtain:

$$\sum_{n=1}^{\infty} d_n = \sum_{n=1}^{\infty} a_n + \sum_{n=1}^{\infty} c_n = \sum_{n=1}^{\infty}(a_n + c_n)$$

$$= (1+0)+\left(-\frac{1}{2}+\frac{1}{2}\right)+\left(\frac{1}{3}+0\right)+\left(-\frac{1}{4}-\frac{1}{4}\right)+\left(\frac{1}{5}+0\right)+\left(-\frac{1}{6}+\frac{1}{6}\right)+\left(\frac{1}{7}+0\right)+\left(-\frac{1}{8}-\frac{1}{8}\right)+\dots$$

$$= 1 + 0 + \frac{1}{3} - \frac{1}{2} + \frac{1}{5} + 0 + \frac{1}{7} - \frac{1}{4} + \dots = \ln 2 + \frac{1}{2}\ln 2 = \frac{3}{2}\ln 2.$$

In the general form, the terms of the last series are defined by the formula

$$d_n = \begin{cases} \frac{1}{4k-3}, & n = 4k-3 \\ 0, & n = 4k-2 \\ \frac{1}{4k-1}, & n = 4k-1 \\ -\frac{1}{2k}, & n = 4k \end{cases}, \quad k \in \mathbb{N}.$$

We notice that the last series has the same terms as the original series, but they appear in a different order, and because of this the sum of the last series is different from the sum of the original series.

This property of the specific series is a reflection of the general property of rearrangement (permutation) of the elements of conditionally convergent series. The general statement, frequently called the Riemann theorem (or the Riemann rearrangement theorem or still the Riemann series theorem), is presented below.

The Riemann Theorem *If a series converges conditionally, then there is a rearrangement of its terms (without adding or dropping any element) such that the new series converges to any chosen beforehand constant, or tends to $\pm\infty$, or even diverges without any specific tendency.*

Proof Recall first that a conditionally convergent series has an infinite number of both positive and negative terms (see Theorem 1 in Sect. 5.1) and that the series of positive and negative parts diverge (see Theorem 2 in Sect. 5.1).

In what follows, we denote the original series $\sum_{n=1}^{\infty} a_n$ by (A), the reordered series (the series with the rearranged elements) by (A') and use the two auxiliary series already introduced in the previous sections: the series $\sum_{k=1}^{\infty} b_k$ (denoted by (B)) composed of all the positive terms of the series (A) placed in the same order as in the series (A); and the series $\sum_{m=1}^{\infty} c_m$ (denoted by (C)) composed of all the negative elements of (A) which are taken with opposite sign and follow the same order as in the original series.

Part 1 In the first part we prove that for any chosen beforehand constant M the elements of (A) can be rearranged in such a way that the reordered series (A') converges to the chosen constant M. Since the series (B) and (C) are positive, their divergence means that their partial sums (and also the remainders of these series) tend to $+\infty$. Then, it follows that for an arbitrary number $E \in \mathbb{R}$ we can choose such number of consecutive summands of the series (B) and (C) (starting with any element of these series) that the obtained partial sum will be greater than the number E.

Then, at the first step of the construction of the series (A'), we choose the number of the first consecutive terms of (B) sufficient for their sum to exceed M:

$$b_1 + b_2 + \ldots + b_{n_1} > M . \tag{2.7}$$

At the same time we choose the number n_1 to be the smallest number with this property, that is, without the last summand the sum is still less than or equal to M:

$$b_1 + b_2 + \ldots + b_{n_1-1} \le M . \tag{2.8}$$

From the last two inequalities it follows that

$$M < b_1 + b_2 + \ldots + b_{n_1} \le M + b_{n_1}. \tag{2.9}$$

The sum of the chosen elements $\tilde{a}_1' = b_1 + b_2 + \ldots + b_{n_1}$ will be the first term of the auxiliary series $(\tilde{A}')$.

At the second step, we choose the number of the first negative terms of (A) (that is, the first terms of the series (C) with their original negative sign) just enough to make their sum with $\tilde{a}_1'$ less than M:

$$b_1 + b_2 + \ldots + b_{n_1} - c_1 - c_2 - \ldots - c_{n_2} < M . \tag{2.10}$$

As in the first step, this number n_2 should the smallest possible of this kind, that is, without the last element the sum is greater than or equal to M:

$$b_1 + b_2 + \ldots + b_{n_1} - c_1 - c_2 - \ldots - c_{n_2-1} \ge M . \tag{2.11}$$

From (2.10) and (2.11) we have

$$M - c_{n_2} \le \left(b_1 + b_2 + \ldots + b_{n_1}\right) + \left(-c_1 - c_2 - \ldots - c_{n_2}\right) < M . \tag{2.12}$$

The sum of the chosen negative elements $\tilde{a}_2' = -c_1 - c_2 - \ldots - c_{n_2}$ represents the second element of the auxiliary series $(\tilde{A}')$.

At the next (third) step, we return to the positive elements and choose such number of them (starting from b_{n_1+1}) which is necessary to satisfy the double inequality

$$M < (b_1+\ldots+b_{n_1})+(-c_1-\ldots-c_{n_2})+(b_{n_1+1}+b_{n_1+2}+\ldots+b_{n_3}) \leq M+b_{n_3}. \quad (2.13)$$

The sum of these elements $\tilde{a}_3' = b_{n_1+1}+b_{n_1+2}+\ldots+b_{n_3}$ is the third term of the auxiliary series $(\tilde{A}')$.

At the fourth step, we choose such number of negative terms (starting from c_{n_2+1}) that the following double inequality is satisfied:

$$M - c_{n_4} \leq (b_1 + \ldots + b_{n_1}) + (-c_1 - \ldots - c_{n_2}) + (b_{n_1+1} + \ldots + b_{n_3})$$
$$+ (-c_{n_2+1} - c_{n_2+2} - \ldots - c_{n_4}) < M, \quad (2.14)$$

The sum $\tilde{a}_4' = -c_{n_2+1} - c_{n_2+2} - \ldots - c_{n_4}$ is the fourth element of $(\tilde{A}')$.

And so on: at each odd step we add the next group of the positive elements $b_{n_{2k-1}+1}, b_{n_{2k-1}+2}, \ldots, b_{n_{2k+1}}$ such that the following double inequality is satisfied

$$M < \left(b_1 + \ldots + b_{n_1}\right) + \left(-c_1 - \ldots - c_{n_2}\right) + \ldots + \left(b_{n_{2k-1}+1} + b_{n_{2k-1}+2} + \ldots + b_{n_{2k+1}}\right)$$
$$\leq M + b_{n_{2k+1}}, \quad (2.15)$$

and define the $(2k + 1)$-th element $\tilde{a}_{2k+1}' = b_{n_{2k-1}+1} + b_{n_{2k-1}+2} + \ldots + b_{n_{2k+1}}$ of the series $(\tilde{A}')$; while at each even step we take the next group of the negative elements $c_{n_{2k}+1}, c_{n_{2k}+2}, \ldots, c_{n_{2k+2}}$, whose sum with the previous elements satisfies the double inequality

$$M - c_{n_{2k+2}} \leq (b_1 + \ldots + b_{n_1}) + (-c_1 - \ldots - c_{n_2}) + \ldots + \left(-c_{n_{2k}+1} - c_{n_{2k}+2} - \ldots - c_{n_{2k+2}}\right) < M, \quad (2.16)$$

and define the $(2k + 2)$-th element $\tilde{a}_{2k+2}' = -c_{n_{2k}+1} - c_{n_{2k}+2} - \ldots - c_{n_{2k+2}}$ of the series $(\tilde{A}')$.

In this construction of the auxiliary series $(\tilde{A}')$ we have applied both a specific rearrangement of the terms of the original series (A) and a specific grouping of these terms (without omitting or adding any element). The obtained series $(\tilde{A}')$ has the form

$$\tilde{a}_1' + \tilde{a}_2' + \tilde{a}_3' + \ldots + \tilde{a}_{2k+1}' + \tilde{a}_{2k+2}' + \ldots$$
$$= \left(b_1 + b_2 + \ldots + b_{n_1}\right) + \left(-c_1 - c_2 - \ldots - c_{n_2}\right) + \left(b_{n_1+1} + b_{n_1+2} \ldots + b_{n_3}\right)$$
$$+ \ldots + \left(b_{n_{2k-1}+1} + b_{n_{2k-1}+2} + \ldots + b_{n_{2k+1}}\right) + \left(-c_{n_{2k}+1} - c_{n_{2k}+2} - \ldots - c_{n_{2k+2}}\right) + \ldots . \quad (2.17)$$

Inequalities (2.15) and (2.16) can be rewritten for the partial sums of the series $(\tilde{A}')$ in the form

$$M < \tilde{A}'_{2k+1} \leq M + b_{n_{2k+1}} \tag{2.18}$$

and

$$M - c_{n_{2k+2}} \leq \tilde{A}'_{2k+2} < M. \tag{2.19}$$

Since the series (A) converges, its general term tends to 0, which means that the general terms of the series (B) and (C) also tend to 0. Then, passing to the limit in inequalities (2.18), (2.19) and using the squeeze sequence Theorem, we obtain: $\lim\limits_{k\to\infty} \tilde{A}'_{2k+1} = M$ and $\lim\limits_{k\to\infty} \tilde{A}'_{2k+2} = M$, that is, the series $(\tilde{A}')$ converges and its sum $\tilde{A}'$ is equal to M. Since each element of $(\tilde{A}')$ represents a group of the terms of the series (A) with the same sign, by Theorem 2 of Sect. 5.2, the series of ungrouped terms (with all the parentheses dropped) also converges to the same sum. The last is the desired series (A') obtained from the original series (A) by a specific rearrangement (permutation) of its elements:

$$\sum_{i=1}^{+\infty} a'_i = b_1 + b_2 + \ldots + b_{n_1} - c_1 - c_2 - \ldots - c_{n_2} + b_{n_1+1} + b_{n_1+2} \ldots + b_{n_3}$$

$$+ \ldots + b_{n_{2k-1}+1} + b_{n_{2k-1}+2} + \ldots + b_{n_{2k+1}} - c_{n_{2k}+1} - c_{n_{2k}+2} - \ldots - c_{n_{2k+2}} + \ldots . \tag{2.20}$$

This series converges to a chosen in advance number M.

Part 2 Consider now a rearrangement which results in the tendency to infinity. We analyze the case of $+\infty$ in detail and a similar construction for $-\infty$ is left for the reader. Again we use the fact that the partial sums of the series (B) and (C) (as well as the partial sums of their remainders) tend to $+\infty$. Initially, we choose the first group of the consecutive elements of (B) (starting with b_1) such that their sum exceeds 1:

$$\tilde{a}'_1 = b_1 + b_2 + \ldots + b_{n_1} > 1.$$

This sum is defined to be the first element $\tilde{a}'_1$ of the auxiliary series $(\tilde{A}')$. Then, we add the first negative element of (C) (just to take into account all the elements of (A)), which we consider as $\tilde{a}'_2 = -c_1$:

$$\tilde{a}'_1 + \tilde{a}'_2 = b_1 + b_2 + \ldots + b_{n_1} - c_1.$$

At the next step, we choose the second group of the consecutive elements of (B) (starting from b_{n_1+1}), which makes the total current sum to be greater than 2, and define $\tilde{a}'_3 = b_{n_1+1} + b_{n_1+2} + \ldots + b_{n_3}$:

$$\tilde{a}'_1 + \tilde{a}'_2 + \tilde{a}'_3 = (b_1 + b_2 + \ldots + b_{n_1}) - c_1 + (b_{n_1+1} + b_{n_1+2} + \ldots + b_{n_3}) > 2.$$

Then, we take the next (second) element of (C) considered as $\tilde{a}'_4 = -c_2$:

$$\tilde{a}'_1 + \tilde{a}'_2 + \tilde{a}'_3 + \tilde{a}'_4 = (b_1 + b_2 + \ldots + b_{n_1}) - c_1 + (b_{n_1+1} + b_{n_1+2} + \ldots + b_{n_3}) - c_2.$$

And so on. At each odd step we add the next group of the positive elements of (B), whose sum determines the $(2k-1)$-th element of the series $(\tilde{A}')$: $\tilde{a}'_{2k-1} = b_{n_{2k-3}+1} + b_{n_{2k-3}+2} + \ldots + b_{n_{2k-1}}$. The number of the chosen elements should be such that the current partial sum exceeds k:

$$\left(b_1 + \ldots + b_{n_1}\right) + (-c_1) + \left(b_{n_1+1} + \ldots + b_{n_3}\right) + (-c_2) + \ldots + \left(b_{n_{2k-3}+1} + \ldots + b_{n_{2k-1}}\right) > k.$$

At each even step we add to the obtained partial sum the current negative element of (C), which defines $\tilde{a}'_{2k} = -c_k$, and the new partial sum takes the form

$$\left(b_1 + \ldots + b_{n_1}\right) + (-c_1) + \left(b_{n_1+1} + \ldots + b_{n_3}\right) + (-c_2) + \ldots + \left(b_{n_{2k-3}+1} + \ldots + b_{n_{2k-1}}\right) + (-c_k).$$

In this manner, we construct the auxiliary series $(\tilde{A}')$ by rearranging and grouping the elements of the series (A) (without omitting or adding any element). For every $k \in \mathbb{N}$, the odd partial sums $\tilde{A}'_{2k-1}$ of this auxiliary series satisfy the inequality

$$\tilde{A}'_{2k-1} = \tilde{a}'_1 + \tilde{a}'_2 + \ldots + \tilde{a}'_{2k-1} = \left(b_1 + \ldots + b_{n_1}\right) + (-c_1) + \ldots + \left(b_{n_{2k-3}+1} + \ldots + b_{n_{2k-1}}\right) > k,$$

which shows that $\tilde{A}'_{2k-1} \underset{k\to\infty}{\to} +\infty$. For the even partial sums of $(\tilde{A}')$, we use the tendency of the general term of (A) to 0 (due to the convergence of (A)), which implies that $c_k \underset{k\to\infty}{\to} 0$, and consequently $\tilde{A}'_{2k} = \tilde{A}'_{2k-1} + \tilde{a}'_{2k} = \tilde{A}'_{2k-1} + (-c_k) \underset{k\to\infty}{\to} +\infty$. Thus, the series $(\tilde{A}')$ diverges in such a manner that its partial sums tend to infinity: $\tilde{A}'_m \underset{m\to\infty}{\to} +\infty$. It remains to drop parentheses in the series $(\tilde{A}')$ (ungroup its terms) to arrive at the desired series (A') obtained form (A) by a rearrangement of its terms:

$$b_1 + b_2 + \ldots + b_{n_1} - c_1 + b_{n_1+1} + \ldots + b_{n_3} - c_2 + \ldots + b_{n_{2k-3}+1} + \ldots + b_{n_{2k-1}} - c_k + \ldots.$$

To evaluate the partial sums A'_m of the series (A'), notice that for any index $m \in \mathbb{N}$ there exists $k \in \mathbb{N}$ such that $n_k + k \le m \le n_{k+1}$, and consequently $\tilde{A}'_{2k} \le A'_m \le \tilde{A}'_{2k+1}$. Then, by the squeeze sequence Theorem, we have $A'_m \underset{m\to\infty}{\to} +\infty$.

Part 3 Finally, using a rearrangement of the elements of the series (A), we build a divergent series without any specific tendency. Again we use repeatedly the fact that the partial sums of the series (B) and (C) as well as their remainders tend to $+\infty$. Initially we choose the number of the first consecutive terms of (B) sufficient for their sum to exceed 1:

$$b_1 + b_2 + \ldots + b_{n_1} > 1.$$

At the second step we add as many consecutive negative terms of (C) as necessary to obtain the second partial sum smaller than -1:

$$b_1 + b_2 + \ldots + b_{n_1} - c_1 - c_2 - \ldots - c_{n_2} < -1 .$$

At the next step, we return to the positive terms of (B) and choose the consecutive elements $b_{n_1+1}, b_{n_1+2}, \ldots, b_{n_3}$ which make the next partial sum greater than 1

$$(b_1 + \ldots + b_{n_1}) + (-c_1 - \ldots - c_{n_2}) + (b_{n_1+1} + b_{n_1+2} + \ldots + b_{n_3}) > 1.$$

Then again we use the number of the negative terms of (C) which is necessary to obtain the fourth partial sum smaller than -1:

$$(b_1+\ldots+b_{n_1})+(-c_1-\ldots-c_{n_2})+(b_{n_1+1}+\ldots+b_{n_3})+(-c_{n_2+1}-c_{n_2+2}-\ldots-c_{n_4}) < -1.$$

And so on: at each odd step we take the next group of the positive elements $b_{n_{2k-1}+1}, b_{n_{2k-1}+2}, \ldots, b_{n_{2k+1}}$ to make the partial sum be greater than 1

$$\left(b_1 + \ldots + b_{n_1}\right) + \left(-c_1 - \ldots - c_{n_2}\right) + \ldots + \left(b_{n_{2k-1}+1} + b_{n_{2k-1}+2} + \ldots + b_{n_{2k+1}}\right) > 1,$$

and in the following even step we add as many negative terms $c_{n_{2k}+1}, c_{n_{2k}+2}, \ldots, c_{n_{2k+2}}$ of (C) as necessary to obtain the partial sum smaller than -1

$$(b_1 + \ldots + b_{n_1}) + (-c_1 - \ldots - c_{n_2}) + \ldots + \left(-c_{n_{2k}+1} - c_{n_{2k}+2} - \ldots - c_{n_{2k+2}}\right) < -1.$$

In this manner, we construct the auxiliary series $(\tilde{A}')$ obtained from (A) by rearranging and grouping its elements (without omitting or adding any element):

$$\left(b_1 + b_2 + \ldots + b_{n_1}\right) + \left(-c_1 - c_2 - \ldots - c_{n_2}\right) + \left(b_{n_1+1} + b_{n_1+2} \ldots + b_{n_3}\right)$$

$$+ \ldots + \left(b_{n_{2k-1}+1} + b_{n_{2k-1}+2} + \ldots + b_{n_{2k+1}}\right) + \left(-c_{n_{2k}+1} - c_{n_{2k}+2} - \ldots - c_{n_{2k+2}}\right) + \ldots .$$

$$(2.21)$$

All the odd partial sums of the series $(\tilde{A}')$ satisfies the inequality $\tilde{A}'_{2k+1} > 1$, while the even partial sums have the evaluation $\tilde{A}'_{2k} < -1$, which means that the series $(\tilde{A}')$ diverges without having any specific limit (finite or infinite).

Ungrouping the terms of the series $(\tilde{A}')$, we obtain the desired series (A') built from (A) through a rearrangement of its terms:

$$b_1 + b_2 + \ldots + b_{n_1} - c_1 - c_2 - \ldots - c_{n_2} + b_{n_1+1} + b_{n_1+2} \ldots + b_{n_3}$$

$$+ \ldots + b_{n_{2k-1}+1} + b_{n_{2k-1}+2} + \ldots + b_{n_{2k}+1} - c_{n_{2k}+1} - c_{n_{2k}+2} - \ldots - c_{n_{2k}+2} + \ldots \ . \tag{2.22}$$

Notice that the same partial sums $\tilde{A}'_{2k+1} > 1$ and $\tilde{A}'_{2k} < -1$, $k \in \mathbb{N}$ form the two subsequences of the partial sums of the series (A'). Therefore, the series (A') diverges and does not have any specific limit.

The proof is complete.

$\square$

Historical Remarks Probably, Goldbach, in correspondence with Euler in 1742, was the first to notice that the terms of the alternating harmonic series can be rearranged so that the new series converges to a different sum.

Dirichlet published the first example of this type in the paper of 1827, and a few years later, a similar example was presented by Cauchy. Studying the question why this is happening, Dirichlet did not obtain a general answer, but he did formulate and prove in 1837 the theorem on the possibility to change the order of terms without changing the sum in the case of absolutely convergent series. A general answer was left to be discovered by Riemann.

In 1852 Riemann began the work on the convergence of Fourier series and sought the advice of Dirichlet, one of the main authorities on the subject. Dirichlet draw his attention to a strange behavior of rearrangements of conditionally convergent series, and Riemann soon found the exact explanation for this phenomenon, which he included in the habilitation paper on real analysis submitted to the faculty of the University of Göttingen in 1854. Unfortunately, this work, which contains solutions to many fundamental problems of real analysis, was published only in 1868, after Riemann death.

6 Complement: Double and Repeated Series

The subject of this section will be used in the demonstration of some results for the power series (composition of series and change of the central point).

Consider an infinite matrix of real numbers $(a_{ij})_{i,j=1}^{\infty}$, or in detailed form

$$
\begin{array}{l}
a_{11}\ a_{12}\ a_{13}\ \ldots\ a_{1j}\ \ldots \\
a_{21}\ a_{22}\ a_{23}\ \ldots\ a_{2j}\ \ldots \\
a_{31}\ a_{32}\ a_{33}\ \ldots\ a_{3j}\ \ldots \\
\qquad\ \ldots \\
a_{i1}\ a_{i2}\ a_{i3}\ \ldots\ a_{ij}\ \ldots \\
\qquad\ \ldots
\end{array}
\tag{2.23}
$$

There are different ways to order the elements of this matrix with the goal to form an infinite sequence (with the unique index). For example, we can choose the first element $u_1 = a_{11}$, then the elements whose indices sum to 3: $u_2 = a_{12}, u_3 = a_{21}$ (ordering these elements by the increase of the row index), then we take those whose sum of the indices is 4: $u_4 = a_{13}, u_5 = a_{22}, u_6 = a_{31}$; etc. In this manner, we exhaust eventually all the elements of the infinite matrix. Summing all the elements of the sequence u_n, we obtain the *double series*

$$\sum_{n=1}^{\infty} u_n. \tag{2.24}$$

Another form to sum the elements of an infinite matrix is trying first to find the sums of the row sequences

$$A_i = \sum_{j=1}^{\infty} a_{ij}, i = 1, 2, \ldots, \infty \tag{2.25}$$

and then add up all these sums

$$\sum_{i=1}^{\infty} A_i. \tag{2.26}$$

In this case, we obtain the *repeated row series*

$$\sum_{i=1}^{\infty} \left(\sum_{j=1}^{\infty} a_{ij} \right). \tag{2.27}$$

This series is called convergent if, first, all the row series (2.25) converge, and second, the resulting series (2.26) also converges.

In a similar mode, we can start with the sums of the column sequences and then form the series of their sums, obtaining another *repeated series—column series*:

$$\sum_{j=1}^{\infty} \left(\sum_{i=1}^{\infty} a_{ij} \right). \tag{2.28}$$

A natural question that arises is if the convergence and sums of the series (2.24), (2.27) and (2.28) are connected or not. The following results answer this question.

Theorem 1 (Double Series) *If for some specific ordering of the elements* (2.23) *in an infinite sequence* u_n, *the double series* (2.24) *converges absolutely to the sum* U, *then the repeated series* (2.27) *converges absolutely to the same sum.*

Proof By the hypothesis, the series of the absolute values converges: $\sum_{n=1}^{\infty} |u_n| = \tilde{U}$. Then, for any fixed i we have $\sum_{j=1}^{n} |a_{ij}| \leq \tilde{U}$, $\forall n \in \mathbb{N}$, which implies the absolute convergence of the series (2.25) for any fixed i.

Since $\sum_{n=1}^{\infty} |u_n|$ converges, the sequence of its remainders approaches 0, that is, for any $\varepsilon > 0$ there exists N such that

$$|\tilde{r}_n| = \sum_{k=n+1}^{\infty} |u_k| = \left| \tilde{U} - \sum_{k=1}^{n} |u_k| \right| < \varepsilon, \forall n \geq N. \qquad (2.29)$$

Choosing the indices i and j large enough, we can guarantee that all the elements $u_1, \ldots, u_N$ are contained in the first I_0 rows and J_0 columns of the matrix (2.23). Then, the inequality (2.29) implies that

$$\left| \tilde{U} - \sum_{i=1}^{I} \left(\sum_{j=1}^{J} |a_{ij}| \right) \right| < \varepsilon$$

for any $I \geq I_0$ and $J \geq J_0$. Fixing now $I \geq I_0$ and using the convergence (already proved) of the series $\sum_{j=1}^{\infty} |a_{ij}| = \tilde{A}_i$, $\forall i$, we take the limit in the last inequality as $J \to +\infty$ and obtain $\left| \tilde{U} - \sum_{i=1}^{I} \tilde{A}_i \right| < \varepsilon$, $\forall I \geq I_0$. Finally, the arbitrariness of ε implies that $\lim_{I \to +\infty} \sum_{i=1}^{I} \tilde{A}_i = \tilde{U}$. Hence, the repeated row series (2.27) converges absolutely to $\tilde{U}$.

Finally, let us show that the sum of the series (2.27) is U. Since the series (2.24) converges absolutely, it satisfies the associative and commutative properties, that is, we can group and reorder the elements of this series in any way, keeping the convergence of the new series to U. In particular, we can consider as the partial sums of the series (2.24) the sums of the elements of the finite submatrices $(a_{ij})_{i,j=1}^{I,J}$ of the matrix (2.23):

$$\sum_{i=1}^{I} \sum_{j=1}^{J} a_{ij}. \qquad (2.30)$$

Since the sum of the series (2.24) under a specific ordering of its elements is equal to U, by reordering with the use of the partial sums (2.30) we obtain the series convergent to the same sum, that is,

$$\lim_{I,J \to +\infty} \sum_{i=1}^{I} \sum_{j=1}^{J} a_{ij} = U.$$

Then, by definition, for $\forall \varepsilon > 0$ (choose for convenience $\frac{\varepsilon}{2}$) there exist I_0, J_0 such that

$$\left| \sum_{i=1}^{I} \sum_{j=1}^{J} a_{ij} - U \right| < \frac{\varepsilon}{2}, \forall I > I_0, J > J_0.$$

Fixing $I \geq I_0$ in this evaluation and using the convergence (already proved) of the series (2.25), we pass to the limit in the last inequality as $J \rightarrow +\infty$ and obtain $\left| U - \sum_{i=1}^{I} A_i \right| \leq \frac{\varepsilon}{2} < \varepsilon, \forall I \geq I_0$. Due to the arbitrariness of ε, this means that the series $\sum_{i=1}^{\infty} A_i$ converges and its sum is equal to U. Thus, the repeated series (2.27) converges to U.

This completes the proof.   □

Remark For the repeated column series (2.28) the statement and proof are the same as in Theorem 1.

Theorem 2 (Double Series) *If the repeated row series (2.27) converges absolutely to the sum U, then the double series (2.24) converges absolutely to the same sum U, whatever ordering of the elements of the matrix (2.23) is used.*

Proof The series $\sum_{i=1}^{\infty} \left(\sum_{j=1}^{\infty} |a_{ij}| \right)$ converges by the hypothesis. Denote its sum by $\tilde{U}$. Then, for any I and J we have $\sum_{i=1}^{I} \left(\sum_{j=1}^{J} |a_{ij}| \right) < \tilde{U}$. Consider now the sequence u_n generated by an arbitrary ordering of the elements of the matrix (2.23), and analyze the series of its absolute values $\sum_{n=1}^{\infty} |u_n|$. Choose a partial sum $\sum_{k=1}^{N} |u_k|$ of this series and notice that there are the indices I and J large enough to guarantee that all the elements $u_1, \ldots, u_N$ are contained in the first I rows and J columns of the matrix (2.23). For these I, J and N, we have

$$\sum_{k=1}^{N} |u_k| \leq \sum_{i=1}^{I} \left(\sum_{j=1}^{J} |a_{ij}| \right) < \tilde{U},$$

which proves the convergence of the series $\sum_{n=1}^{\infty} |u_n|$, that is, the double series $\sum_{n=1}^{\infty} u_n$ converges absolutely.

It remains to apply Theorem 1 to see that the series (2.27) and (2.24) converge to the same sum.   □

Remark For the repeated column series (2.28) the statement and proof follow those presented in Theorem 2.

Corollary *The repeated row series (2.27) converges absolutely to the sum U if and only if the repeated column series (2.28) converges absolutely to the same sum. This statement is an immediate consequence of the results of Theorems 1 and 2.*

Exercises

1. Given two series $(A) \sum a_n$ and $(B) \sum b_n$, what can be said about the series $\sum (a_n + b_n)$ if:
   (1) both series (A) and (B) converge;
   (2) both series (A) and (B) diverge;
   (3) the series (A) converges, while the series (B) diverges.

2. Given two series $(A) \sum a_n$ and $(B) \sum b_n$, what can be said about the series $\sum (a_n b_n)$ if:
   (1) both series (A) and (B) converge;
   (2) both series (A) and (B) diverge;
   (3) the series (A) converges, while the series (B) diverges.

3. The series $\sum a_n^2$ and $\sum b_n^2$ converge. Prove that the following series converge:
   (1) $\sum |a_n b_n|$;
   (2) $\sum (a_n + b_n)^2$;
   (3) $\sum \frac{|a_n|}{n}$;
   (4) $\sum \frac{|a_n| + |b_n|}{n}$.

4. Demonstrate that if $\lim\limits_{n \to +\infty} n a_n = a \neq 0$ then the series $\sum a_n$ diverges. What happens if the condition is changed to $\lim\limits_{n \to +\infty} n a_n \neq 0$?

5*. Demonstrate that if a positive series $\sum a_n$ converges and a_n is decreasing, then $\lim\limits_{n \to +\infty} n a_n = 0$. What happens if a series is arbitrary (non-positive)? What happens if the second condition is dropped (a_n is non-decreasing)?

6. Given three series $(A) \sum a_n$, $(B) \sum b_n$ and $(C) \sum c_n$, such that $a_n \leq c_n \leq b_n$. What can be said about the series $\sum c_n$ if:
   (1) both series (A) and (B) converge;
   (2) both series (A) and (B) diverge;
   (3) the series (A) converges, while the series (B) diverges.

7. Investigate the convergence or divergence of a given series by definition. In the case of convergence, find the sum of the series:
   (1) $\sum_{n=0}^{+\infty} \frac{(-1)^n}{3^n}$;
   (2) $\sum_{n=1}^{+\infty} \frac{2}{n^2 + 2n}$;
   (3) $\sum_{n=0}^{+\infty} \sin^n \frac{\pi}{3}$;
   (4) $\sum_{n=1}^{+\infty} (-1)^{\frac{n(n+1)}{2}}$;
   (5) $\sum_{n=1}^{+\infty} \ln \frac{2n-1}{2n+1}$;

(6) $\sum_{n=1}^{+\infty} \frac{1}{(3n-1)(3n+2)}$;

(7) $\sum_{n=0}^{+\infty} \sin \frac{n\pi}{4}$;

(8) $\sum_{n=1}^{+\infty} \frac{2n+1}{n^2(n+1)^2}$;

(9) $\sum_{n=1}^{+\infty} \frac{3^{n+2}}{5^n}$;

(10) $\sum_{n=0}^{+\infty} \cos \frac{n\pi}{2}$;

(11) $\sum_{n=1}^{+\infty} \left(\sqrt{n+2} - 2\sqrt{n+1} + \sqrt{n}\right)$.

8. Investigate the convergence or divergence of a given series using the Cauchy criterion:

(1) $\sum_{n=1}^{+\infty} \frac{1}{\sqrt{n}}$;

(2) $\sum_{n=1}^{+\infty} \frac{1}{\sqrt{(2n-1)(2n+1)}}$;

(3) $\sum_{n=1}^{+\infty} \frac{1}{n^2}$;

(4) $1 + \frac{1}{2} - \frac{1}{3} + \frac{1}{4} + \frac{1}{5} - \frac{1}{6} + \ldots$;

(5) $\sum_{n=1}^{+\infty} \frac{\cos n - \cos(n+1)}{n}$.

9. Examine the convergence or divergence of a positive series using the Integral test:

(1) $\sum_{n=1}^{+\infty} \frac{1}{n^2+4}$;

(2) $\sum_{n=0}^{+\infty} \frac{1}{\sqrt[3]{(2n+3)^2}}$;

(3) $\sum_{n=1}^{+\infty} n(1 + n^2)^p$;

(4) $\sum_{n=1}^{+\infty} \frac{n}{n^4+1}$;

(5) $\sum_{n=1}^{+\infty} \frac{1}{\sqrt{n^4+2n^3+1}}$;

(6) $\sum_{n=2}^{+\infty} \frac{e^{1/n}}{n^2}$;

(7) $\sum_{n=1}^{+\infty} \frac{n^2}{e^n}$;

(8) $\sum_{n=1}^{+\infty} \frac{1}{e^{n^2}}$;

(9) $\sum_{n=3}^{+\infty} \frac{1}{n \ln n \ln(\ln n)}$;

(10) $\sum_{n=1}^{+\infty} \frac{\ln n}{n^p}$.

10. Determine whether a positive series is convergent or divergent using the Comparison tests:

(1) $\sum_{n=3}^{+\infty} \frac{1}{\sqrt{(n-1)(n-2)}}$;

(2) $\sum_{n=1}^{+\infty} \frac{1}{\sqrt{(n+1)(n+2)}}$;

(3) $\sum_{n=1}^{+\infty} \frac{n}{(n+2)3^n}$;

(4*) $\sum_{n=1}^{+\infty} \frac{n^2}{2\sqrt{n}}$;

(5) $\sum_{n=2}^{+\infty} \frac{1}{\ln n}$;

(6) $\sum_{n=2}^{+\infty} \frac{1}{\sqrt[3]{n} \ln n}$;

(7*) $\sum_{n=2}^{+\infty} \frac{1}{(\ln n)^{\ln n}}$;

(8*) $\sum_{n=1}^{+\infty} \arctan \frac{n+1}{\sqrt{n^5+2n^3+4}}$;

(9) $\sum_{n=1}^{+\infty} (\sqrt{n} - \sqrt{n-1})$;

(10) $\sum_{n=1}^{+\infty} \frac{1}{n}(\sqrt{n+1} - \sqrt{n-1})$.

11. Examine the convergence or divergence of a positive series using the Cauchy condensation test:

(1) $\sum_{n=2}^{\infty} \frac{\ln n}{n}$;

(2) $\sum_{n=1}^{\infty} \frac{1}{n^{1+\frac{1}{n}}}$;

(3) $\sum_{n=2}^{\infty} (\ln n)^{-\ln n}$;

(4) $\sum_{n=3}^{\infty} \frac{1}{n \ln n (\ln \ln n)^2}$.

12. Determine whether a positive series is convergent or divergent using D'Alembert's or Cauchy's test:

(1) $\sum_{n=1}^{+\infty} \frac{(2n-1)!!}{n!}$;

(2) $\sum_{n=1}^{+\infty} \frac{4 \cdot 7 \cdot 10 \cdot \ldots \cdot (3n+4)}{3 \cdot 7 \cdot 11 \cdot \ldots \cdot (4n+3)}$;

(3) $\sum_{n=1}^{+\infty} \frac{3 \cdot 6 \cdot 9 \cdot \ldots \cdot (3n)}{(n+1)!} \arctan \frac{1}{2^n}$;

(4) $\sum_{n=1}^{+\infty} \frac{n^n \sin \frac{\pi}{2^n}}{n!}$;

(5) $\sum_{n=1}^{+\infty} \frac{(5-(-1)^n)^n}{n^2 7^n}$;

(6) $\sum_{n=1}^{+\infty} \frac{(5-(-1)^n)^n}{n^2 3^n}$;

(7) $\sum_{n=2}^{+\infty} \left(\frac{n^2+3}{n^2+4}\right)^{n^3+1}$;

(8) $\sum_{n=1}^{+\infty} \arctan^n \frac{\sqrt{3n+2}}{\sqrt{n+1}}$.

13. Investigate the convergence/divergence of a positive series applying a suitable test:

(1) $\sum_{n=2}^{+\infty} \frac{1}{\sqrt[n]{\ln n}}$;

(2) $\sum_{n=1}^{+\infty} \frac{2^n n!}{n^n}$;

(3) $\sum_{n=1}^{+\infty} \frac{3^n n!}{n^n}$;

(4) $\sum_{n=1}^{+\infty} \frac{(n!)^2}{(2n)!}$;

(5) $\sum_{n=2}^{+\infty} \left(\frac{n-1}{n+1}\right)^{n(n-1)}$;

(6) $\sum_{n=1}^{+\infty} \frac{n^{n-1}}{(2n^2+n+1)^{\frac{n+1}{2}}}$;

(7) $\sum_{n=1}^{+\infty} \frac{n^{n+\frac{1}{n}}}{\left(n+\frac{1}{n}\right)^n}$;

(8) $\sum_{n=1}^{+\infty} (\sqrt{2} - \sqrt[3]{2})(\sqrt{2} - \sqrt[5]{2}) \cdot \ldots \cdot (\sqrt{2} - \sqrt[2n+1]{2})$;

(9) $\sum_{n=1}^{+\infty} n e^{-n^2}$;

(10) $\sum_{n=1}^{+\infty} n^2 e^{-n}$;

(11) $\sum_{n=1}^{+\infty} \frac{\arctan n}{n^2+1}$;

(12) $\sum_{n=1}^{+\infty} \operatorname{arccot} n$;

(13) $\sum_{n=1}^{+\infty} \arctan^n \frac{1}{n}$;

(14) $\sum_{n=1}^{+\infty} \sin \frac{\pi}{2n}$;

(15) $\sum_{n=1}^{+\infty} \frac{n^2}{4n^3+1}$;

(16) $\sum_{n=1}^{+\infty} \frac{1}{3^n - \cos n}$;

(17) $\sum_{n=3}^{+\infty} \frac{1}{n \ln n (\ln \ln n)^p}$, $p > 0$;

(18) $\sum_{n=1}^{+\infty} \frac{\ln 2 \cdot \ln 3 \cdot \ldots \cdot \ln(n+1)}{\ln(2+p) \cdot \ln(3+p) \cdot \ldots \cdot \ln(n+1+p)}$, $p > 0$;

(19*) $\sum_{n=1}^{+\infty} a_n$, $a_1 = \sqrt{2}, a_2 = \sqrt{2-\sqrt{2}}, a_3 = \sqrt{2-\sqrt{2+\sqrt{2}}}, \ldots,$

$$a_n = \underbrace{\sqrt{2-\sqrt{2+\ldots+\sqrt{2}}}}_{n};$$

(20) $\sum_{n=1}^{+\infty} \frac{\sqrt[3]{n}}{2n+3}$;

(21) $\sum_{n=1}^{+\infty} \frac{n^4}{2^n+3^n}$;

(22) $\sum_{n=1}^{+\infty} \frac{\ln 2 \cdot \ln 3 \cdot \ldots \cdot \ln(n+1)}{\ln(2^p+1) \cdot \ln(3^p+1) \cdot \ldots \cdot \ln((n+1)^p+1)}$, $p > 0$;

(23) $\sum_{n=2}^{+\infty} \frac{\sqrt[n]{n}}{n \ln^2 n}$;

(24) $\sum_{n=1}^{+\infty} \left(1 + \frac{1}{n}\right)^{2n} \cdot \frac{1}{n^{3/2}}$.

14. Investigate the convergence/divergence of an alternating series applying Leibniz's test:

(1) $\sum_{n=1}^{+\infty} \frac{(-1)^{n+1}}{\sqrt[3]{2n}}$;

(2) $\sum_{n=1}^{+\infty} (-1)^n \frac{3n}{\sqrt{4n^2+1}}$;

(3) $\sum_{n=1}^{+\infty} \frac{(-1)^{n-1} \ln^2 n}{\sqrt{n}}$;

(4) $\sum_{n=1}^{+\infty} (-1)^n \frac{\sqrt{2n+1}}{n+2}$;

(5*) $\sum_{n=1}^{+\infty} \frac{(-1)^n n^3}{2\sqrt{n}}$;

(6*) $\sum_{n=1}^{+\infty} \sin(\pi \sqrt{n^2 + 1})$.

15. Examine the convergence/divergence of a series using Dirichlet's or Abel's test:

(1) $\sum_{n=1}^{+\infty} \frac{\sin(n\alpha)}{n}$, $\alpha \in \mathbb{R}$;

(2) $\sum_{n=1}^{+\infty} (-1)^n \frac{\arctan^2 n}{n}$;

(3) $\sum_{n=1}^{+\infty} \frac{\cos(2n-1)\alpha}{2n-1}$, $\alpha \in \mathbb{R}$;

(4) $\sum_{n=1}^{+\infty} \frac{\sin(n\alpha)\cos(n^2\alpha)}{\sqrt{n}}$, $\alpha \in \mathbb{R}$;

(5*) $\sum_{n=1}^{+\infty} (-1)^n \frac{\sin^2 n}{n}$;

(6*) $\sum_{n=2}^{+\infty} \frac{\cos \frac{\pi n^2}{n+1}}{\ln^2 n}$.

16. Analyze if a given series is absolutely convergent, conditionally convergent or divergent using a suitable test:

(1) $\sum_{n=1}^{+\infty} \frac{(-1)^{n+1}}{2n+3}$;

(2) $\sum_{n=1}^{+\infty} (-1)^{n+1} \sin \frac{\pi}{2n}$;

(3) $\sum_{n=1}^{+\infty} (-1)^{n+1} \sin \frac{\pi}{2n^2}$;

(4) $\sum_{n=1}^{+\infty} (-1)^n \frac{\ln n}{n}$;

(5) $\sum_{n=1}^{+\infty} \frac{(-1)^{n+1}}{(n+1) \ln(n+1)}$;

(6) $\sum_{n=1}^{+\infty} \frac{(-1)^n}{\sqrt[n]{n}}$;

(7) $\sum_{n=1}^{+\infty} (-1)^n \frac{3^n}{n!}$;

(8) $\sum_{n=1}^{+\infty} \frac{(-1)^n}{(n+1)^p}$;

(9) $\sum_{n=2}^{+\infty} \frac{(-1)^n}{\ln n}$;

(10) $\sum_{n=2}^{+\infty} \frac{(-1)^n}{n(\ln n)^p}$;

(11) $\sum_{n=1}^{+\infty} \frac{(-1)^{\frac{n(n+1)}{2}}}{2^n}$;

(12) $\sum_{n=1}^{+\infty} \frac{(-1)^{\frac{n(n+1)}{2}}}{n^p}$;

(13) $1 + \frac{1}{2} + \frac{1}{3} - \frac{1}{4} - \frac{1}{5} - \frac{1}{6} + \frac{1}{7} + \frac{1}{8} + \frac{1}{9} - \ldots$;

(14*) $\sum_{n=1}^{+\infty} \frac{\sin \frac{n\pi}{4}}{n + \sin \frac{n\pi}{4}}$;

(15*) $\sum_{n=2}^{+\infty} \frac{(-1)^n}{n + (-1)^n}$;

(16*) $\sum_{n=1}^{+\infty} \frac{\cos n}{(2n+3)^p}$;

(17*) $\sum_{n=1}^{+\infty} \sin(\pi \sqrt{n^2 + p^2})$, $p \in \mathbb{R}$;

(18*) $\sum_{n=2}^{+\infty} \frac{\sin \frac{n\pi}{6}}{n} \ln^{10} n$;

(19*) $\sum_{n=1}^{+\infty} (-1)^n \frac{\cos^2 n}{\sqrt[3]{n}}$;

(20) $\sum_{n=3}^{+\infty} \frac{1}{\ln^2 n} \cos \frac{\pi n^2}{n+1}$;

(21) $\sum_{n=2}^{+\infty} \frac{\sin \frac{\pi n}{12}}{\ln n}$;

(22) $\sum_{n=2}^{+\infty} \frac{(-1)^n \sqrt[n]{n}}{\ln n}$.

17. Investigate the convergence of two series and of their Cauchy product:

(1) $\sum_{n=1}^{+\infty} \frac{(-1)^{n-1}}{\sqrt[4]{n}}$ and $\sum_{n=1}^{+\infty} \frac{(-1)^{n-1}}{\sqrt{n}}$;

(2) $1 - \sum_{n=1}^{+\infty} \left(\frac{3}{2}\right)^n$ and $1 + \sum_{n=1}^{+\infty} \left(\frac{3}{2}\right)^{n-1} \left(2^n + \frac{1}{2^{n+1}}\right)$.

18. Consider the double and repeated series defined by a given infinite matrix and analyze their convergence:

(1)
$$\begin{matrix} \frac{1}{1\cdot 2} & \frac{1}{2\cdot 3} & \frac{1}{3\cdot 4} & \frac{1}{4\cdot 5} & \cdots \\ 0 & \frac{1}{2\cdot 3} & \frac{1}{3\cdot 4} & \frac{1}{4\cdot 5} & \cdots \\ 0 & 0 & \frac{1}{3\cdot 4} & \frac{1}{4\cdot 5} & \cdots \\ 0 & 0 & 0 & \frac{1}{4\cdot 5} & \cdots \end{matrix}$$
$\cdots$

(2)
$$\begin{matrix} a & -a^2 & a^2 & -a^3 & a^3 & \cdots \\ a(1-a) & -a^2(1-a^2) & a^2(1-a^2) & -a^3(1-a^3) & a^3(1-a^3) & \cdots \\ a(1-a)^2 & -a^2(1-a^2)^2 & a^2(1-a^2)^2 & -a^3(1-a^3)^2 & a^3(1-a^3)^2 & \cdots \\ a(1-a)^3 & -a^2(1-a^2)^3 & a^2(1-a^2)^3 & -a^3(1-a^3)^3 & a^3(1-a^3)^3 & \cdots \end{matrix}$$
, $0 < a < 1$.

$\cdots$

19*. Verify whether the following statement is true or false:

(1) if $\lim_{n \to +\infty} a_n = 0$ and the sequence of the partial sums is bounded, then $\sum a_n$ converges;

(2) if $\sum a_n$ converges and $\sum b_n$ diverges, then $|a_n| < |b_n|$, $\forall n$;

(3) if $\sum (-1)^n a_n$ converges and $0 \le b_n \le a_n$, $\forall n$, then $\sum (-1)^n b_n$ also converges;

(4) if $\sum a_n$ converges and $\lim_{n \to +\infty} \frac{b_n}{a_n} = 1$, then $\sum b_n$ also converges;

(5) if $f(x)$ is continuous and positive on $[1, +\infty)$ function, whose improper integral $\int_1^{+\infty} f(x)dx$ converges, then the series $\sum_{n=1}^{+\infty} f(n)$ also converges;

(6) if Cauchy's and D'Alembet's tests, as well as the Raabe and integral tests are inconclusive with respect to a positive series, then any other test is also inconclusive.

Sequences of Functions

3

Newton considered the sequence of functions ... whereas Wallis had only considered the sequence of numbers.
Charles Henry Edwards, 1979

1 Pointwise Convergence and Introductory Examples

▶ **Definition (Sequence of Functions).** A *sequence of functions* defined on a set $X \subset \mathbb{R}$ is a relationship between the set of natural numbers $\mathbb{N}$ and the set of functions F such that with each element (index) $n \in \mathbb{N}$ there is associated only one function $f_n(x) \in F$ whose domain is X. The set X is called the *domain of the sequence.*

As it was made for sequences of numbers, we extend this definition to the set of indices n such that $n \in \mathbb{N}_0 = \{k_1, k_2 = k_1 + 1, k_3 = k_2 + 1, \dots, k_{i+1} = k_i + 1, \dots\}$, where $k_1 \in \mathbb{Z}$.

The standard notations are $\{f_n(x)\}_{n=1}^{\infty}$ or $\{f_n(x)\}$ or simply $f_n(x)$. If the domain X is not explicitly indicated, it is considered such a set X where all the functions of F are defined.

Electronic Supplementary Material The online version of this article (https://doi.org/10.1007/978-3-030-79431-6_3) contains supplementary material, which is available to authorized users.

▶ **Definition (Convergence at a Point).** Pick an arbitrary point $x_0 \in X$ and fix it. Then the sequence $f_n(x)$ becomes a sequence of numbers $f_n(x_0)$. If the last sequence converges, we say that the *sequence $f_n(x)$ converges at the point x_0.* Otherwise, the *sequence $f_n(x)$ diverges at x_0*.

▶ **Definition (Pointwise Convergence on a Set)** If a sequence $f_n(x)$ converges (as a sequence of numbers) at every fixed point $x \in X$, then we say that $f_n(x)$ *converges pointwise on X* (or it is *pointwise convergent on X*). Otherwise, $f_n(x)$ *diverges on X*. The set X where $f_n(x)$ converges pointwise is called the *set of convergence*.

When it is clear from the context that we consider the pointwise convergence, we can say shortly that $f_n(x)$ converges on X.

If $f_n(x)$ converges on X, it is evident that the value of the limit depends on the chosen point $x \in X$ in such a way that we obtain the *limit function* $f(x)$: $\lim_{n \to \infty} f_n(x) = f(x)$.

The usual notation for the pointwise convergence on X is $f_n(x) \xrightarrow[n \to \infty]{X} f(x)$.

Consider some introductory examples of convergent and divergent sequences of functions.

Example 1a $f_n(x) = x^n$, $X = \mathbb{R}$

For every fixed x we have a geometric sequence, which converges to 0 for $x \in (-1, 1)$, converges to 1 at $x = 1$ and diverges when $x \leq -1$ and $x > 1$ (see details in Example 7a, Sect. 1.2 of Chap. 1). Hence, the set of convergence is the interval $(-1, 1]$.

Example 2a $f_n(x) = \frac{x^n}{n}$, $X = \mathbb{R}$

For every $|x| \leq 1$ this sequence converges to 0, but for every $|x| > 1$ it diverges (the indeterminate form in the latter case is easy to handle by L'Hospital's rule). Hence, the set of convergence is the interval $[-1, 1]$.

Example 3a $f_n(x) = \frac{x^n}{n!}$, $X = \mathbb{R}$

In this case, the sequence converges to 0 on the set of real numbers $\mathbb{R}$, because $n!$ increases much faster than x^n for any fixed x. Indeed, take any $x \neq 0$ and fix it. Choose $n > |x|$ and represent it in the form $n = m + k$, where $m = [|x|]$ is the entire part of $|x|$. Then, $\frac{|x|^n}{n!} = \frac{|x|^m}{m!} \cdot \frac{|x|}{m+1} \cdot \ldots \cdot \frac{|x|}{m+k} < C_m \cdot q^k$, where $C_m = \frac{|x|^m}{m!}$ is a constant, since x, and consequently m, are fixed numbers, and $q = \frac{|x|}{m+1} < 1$. Therefore, we have the evaluation through the geometric sequence convergent to 0.

Example 4a $f_n(x) = (xn)^n$, $X = \mathbb{R}$

This sequence converges only at the point $x = 0$. Indeed, for any fixed $x \neq 0$, we have $|x|n > 1$ for n large enough—it is sufficient to choose $n > \frac{1}{|x|}$, and for these values of n we have the evaluation through the divergent geometric sequence $(|x|n)^n > q^n, q > 1$.

Example 5a $f_n(x) = (1 + x^2)n$, $X = \mathbb{R}$

This sequence diverges at any real point $x \in \mathbb{R}$ (evidently, $(1+x^2)n \geq n \underset{n\to\infty}{\to} \infty, \forall x \in \mathbb{R}$).

Let us consider now different examples that bring rather surprising results, at least at first glance: certain properties inherent to each function of a sequence are not satisfied by the limit function.

Example 1b $f_n(x) = \frac{n}{nx+1}$, $X = (0, 1)$

Rewriting $f_n(x)$ in the form $f_n(x) = \frac{1}{x+\frac{1}{n}}$, we immediately conclude that $f_n(x) \underset{n\to\infty}{\overset{(0,1)}{\to}} f(x) = \frac{1}{x}$. Notice that each function $f_n(x)$ is bounded on $(0, 1)$, but the limit function $f(x)$ is unbounded on the same interval (see Fig. 3.1).

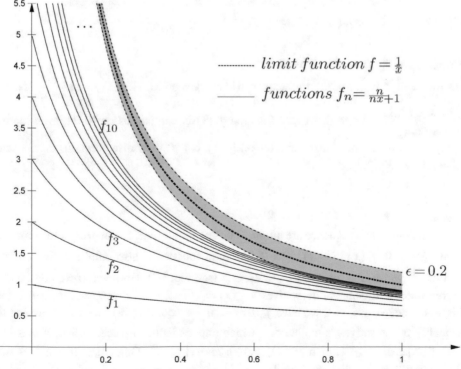

Fig. 3.1 Example 1b: $f_n(x) = \frac{n}{nx+1}$, $X = (0, 1)$

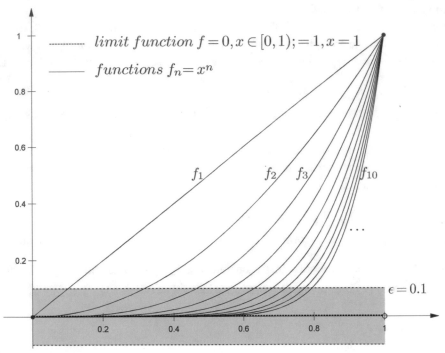

Fig. 3.2 Example 2b: $f_n(x) = x^n$, $X = [0, 1]$

Example 2b $f_n(x) = x^n$, $X = [0, 1]$

As was seen in Example 1a, this sequence converges on $[0, 1]$: $f_n(x) = x^n \overset{[0,1]}{\underset{n \to \infty}{\to}} f(x) =$
$\begin{cases} 0, & x \in [0, 1) \\ 1, & x = 1 \end{cases}$. Notice that every function $f_n(x)$ is continuous on $[0, 1]$, even more—
every $f_n(x)$ is infinitely differentiable on $[0, 1]$, but $f(x)$ is discontinuous on the same
interval (see Fig. 3.2).

Example 3b $f_n(x) = \frac{\sin nx}{\sqrt{n}}$, $X = \mathbb{R}$

The functions of this sequence are infinitely differentiable on $\mathbb{R}$ and at every x the sequence
converges to 0: $f_n(x) \overset{\mathbb{R}}{\underset{n \to \infty}{\to}} f(x) \equiv 0$. The limit function is also infinitely differentiable
and its derivative is zero: $f'(x) = 0$, $\forall x \in \mathbb{R}$ (see Fig. 3.3). However, if we consider the
corresponding sequence of derivatives—$f_n'(x) = \sqrt{n} \cos nx$, it diverges at any $x \in \mathbb{R}$ (see
Fig. 3.4). Indeed, at $x = 0$ we have $f_n'(0) = \sqrt{n} \underset{n \to \infty}{\to} \infty$. If $x \neq 0$, we can show that the
sequence of derivatives is unbounded, which implies its divergence. In fact, if $n \in \mathbb{N}$ is
such that $|\cos nx| \geq \frac{1}{2}$, then $|f_n'(x)| = |\sqrt{n} \cos nx| \geq \frac{\sqrt{n}}{2}$. Otherwise (if $n \in \mathbb{N}$ is such
that $|\cos nx| < \frac{1}{2}$), we have $|\cos 2nx| = |2\cos^2 nx - 1| = 1 - 2\cos^2 nx > \frac{1}{2}$, which
means that the subsequence $f_{2n}'(x)$ is unbounded: $|f_{2n}'(x)| = \sqrt{2n}|\cos 2nx| > \frac{\sqrt{2n}}{2}$.

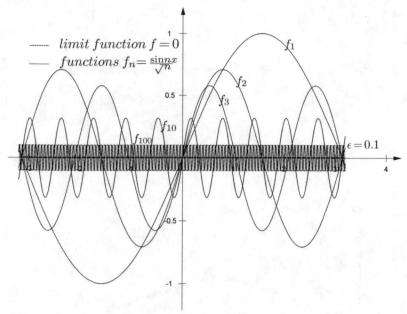

Fig. 3.3 Example 3b: $f_n(x) = \frac{\sin nx}{\sqrt{n}}$, $X = \mathbb{R}$

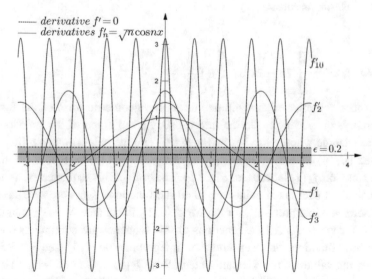

Fig. 3.4 Example 3b: $f_n'(x) = \sqrt{n}\cos nx$, $X = \mathbb{R}$

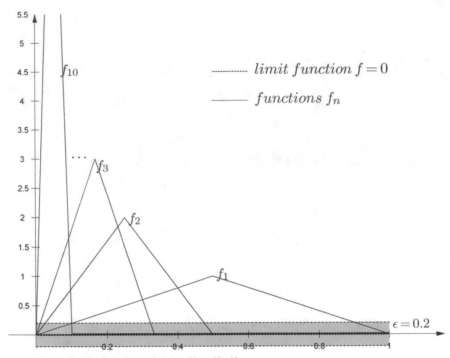

Fig. 3.5 Example 4b: functions $f_n(x)$, $X = [0, 1]$

Example 4b $f_n(x) = \begin{cases} 2n^2x, & x \in [0, \frac{1}{2n}] \\ 2n - 2n^2x, & x \in [\frac{1}{2n}, \frac{1}{n}] \\ 0, & x \in [\frac{1}{n}, 1] \end{cases}$

Despite a more complicated analytic definition, the geometric form of the functions of this sequence is quite simple. On the interval $[0, \frac{1}{n}]$, the graph of the function $f_n(x)$ is an isosceles triangle whose base lies on the x-axis and the height equals n (it is achieved at the vertex $(\frac{1}{2n}, n)$). On the remaining part of the interval $[0, 1]$ (that is, on the interval $[\frac{1}{n}, 1]$) the function $f_n(x)$ is zero (see Fig. 3.5). The limit of this sequence exists and equals $f(x) \equiv 0$, since at each fixed point $x_0 > 0$ all the functions $f_n(x)$ are zero starting from a certain index n. More precisely, for any $x_0 \in (0, 1]$, the choice $N > \frac{1}{x_0}$ guarantees that $f_n(x_0) = 0$ for $\forall n > N$. All the functions of this sequence are continuous on $[0, 1]$, as well as the limit function, which ensures their integrability on $[0, 1]$ (see Fig. 3.5).

However, the values of the Riemann integrals of $f_n(x)$ and $f(x)$ (which represent in this case the areas of the figures between the x-axis and function graph) are very different. On the one hand, the area of each triangle is equal to $\frac{1}{2}n \cdot \frac{1}{n} = \frac{1}{2}$, that can be confirmed by the calculation of the integral:

$$\int_0^1 f_n(x)dx = \int_0^{\frac{1}{2n}} 2n^2x\,dx + \int_{\frac{1}{2n}}^{\frac{1}{n}} (2n - 2n^2x)dx = \frac{1}{4} + \frac{1}{4} = \frac{1}{2}.$$

But, on the other hand, $\int_0^1 f(x)dx = \int_0^1 0dx = 0$, that is, there is no connection between the integrals of the sequence functions and the integral of the limit function. This means that there is no geometric similarity between the graphs of $f_n(x)$ and that of $f(x)$: the property of the isosceles triangles with constant area inherent to the sequence functions is lost for the limit function.

Let us discuss the results of these examples. The convergence of a sequence $f_n(x)$ to the limit function $f(x)$ means that, in a sense, the functions $f_n(x)$ approximate better the function $f(x)$ each time as n increases. Then, it is natural to expect that the limit function inherits, in a certain degree, the analytic properties shared by all the functions of the sequence. However, it is not observed in the above examples.

Evidently, it does not mean that a relation between the properties of $f_n(x)$ and $f(x)$ is never observed. There are a number of examples when $f(x)$ has the same properties as $f_n(x)$. For the sake of illustration, we provide some examples of this kind: the sequence $f_n(x) = 1$ converges on any set $X \subset \mathbb{R}$ to $f(x) = 1$; the sequence $f_n(x) = \frac{\arctan x}{n}$ of bounded and infinitely differentiable functions converges on any set $X \subset \mathbb{R}$ to the bounded and infinitely differentiable function $f(x) \equiv 0$; the sequence $f_n(x) = \frac{n}{2n+1} \sin x$ of bounded, infinitely differentiable and 2π-periodic functions, whose Riemann integral on the interval $[0, \pi]$ equals $\frac{2n}{2n+1}$ and on the interval $[0, 2\pi]$ equals 0, converges on any set $X \subset \mathbb{R}$ to $f(x) = \frac{1}{2} \sin x$, which is a bounded, infinitely differentiable and 2π-periodic function with the Riemann integral equal 1 on $[0, \pi]$ and 0 on $[0, 2\pi]$ (note that $\frac{2n}{2n+1} \underset{n\to\infty}{\to} 1$).

Let us clarify what is "wrong" with the sequences in Examples 1b–4b, when the limit function does not keep the relevant properties of the sequence functions $f_n(x)$. We start with Example 4b which has a more direct geometric interpretation, because the studied properties are related to the behavior of the graphs of functions. Looking at the graphs of $f_n(x)$ and $f(x)$, we notice that at each fixed point $x_0 \in (0, 1]$ the points of the sequence approach 0 (the x-axis), because all the functions $f_n(x)$ have zero value at x_0 starting from some elevated index n (this is true for any $x_0 \in (0, 1]$ if we choose the indices $n > N > \frac{1}{x_0}$). However, this approaching to zero does not occur with the same rate at all the points of $[0, 1]$: it is necessary to choose larger and larger N as x is chosen closer and closer to 0, and no one N works for all x at once (see Fig. 3.5). Therefore, there is no N that could be used as the rate of convergence of $f_n(x)$ simultaneously for all points $x \in [0, 1]$. If we represent geometrically the degree of approximation of the limit function $f(x) \equiv 0$ by the functions $f_n(x)$ using a strip of width 2ε centralized around the limit function (this strip corresponds to the value ε in the definition of convergence), then we can notice that the graphs of $f_n(x)$ do not lie completely within this strip, no matter how large n we choose. Actually, near the origin the graphs of $f_n(x)$ have nothing to do with the graph of $f(x)$ (for the fixed but large values of n), and even worse: in a sense, the parts of the graphs (near the vertices $P_n = (\frac{1}{2n}, n)$ whose heights increase as n grows), move away from the x-axis as n increases (see Fig. 3.5).

Looking at Figs. 3.1, 3.2, and 3.4, we can notice that a quite similar situation is observed in each of the last four examples, although it can be not so evident geometrically as for Example 4b (in the case of Example 3b it happens with the derivatives). As a contrast to this, Fig. 3.3 shows a "correct", desired behavior of the sequence of Example 3b. Indeed, with the growth of n, the values of $f_n(x) = \frac{\sin nx}{\sqrt{n}}$, despite the oscillations, approximate closer and closer the limit function $f(x) \equiv 0$: for sufficiently large n the graphs of $f_n(x)$ lie completely within the ε-strip. It is easy to check this analytically: for any $\varepsilon > 0$ we can choose $N = \left[\frac{1}{\varepsilon^2}\right]$ such that $\left|\frac{\sin nx}{\sqrt{n}}\right| \leq \frac{1}{\sqrt{n}} < \varepsilon$ for all $n > N$ and simultaneously for all $x \in \mathbb{R}$. Figure 3.3 illustrates this behavior for the value $\varepsilon = 0.1$ and the corresponding index $N = 100$. Apparently, in this case, there is a good agreement between the properties of the sequence functions $f_n(x)$ and those of the limit function: all these functions are bounded, continuous (even infinitely differentiable) and the numerical sequence of integrals $\int_a^b f_n(x)dx$ converges to $\int_a^b f(x)dx = 0$ for any interval $[a, b]$:

$$\int_a^b f_n(x)\,dx = \int_a^b \frac{\sin nx}{\sqrt{n}}dx = -\frac{1}{\sqrt{n}}\frac{1}{n}\cos nx\big|_a^b = \frac{1}{n^{3/2}}(\cos na - \cos nb) \underset{n \to \infty}{\to} 0\,.$$

Nevertheless, even these "good" properties of the original sequence $f_n(x)$ are not sufficient to guarantee a desired correspondence between the sequence of derivatives $f_n'(x)$ and the derivative of the limit function $f'(x) \equiv 0$ (see Fig. 3.4).

Returning to the properties of the sequences themselves (in Examples 1b, 2b and 4b), we can make an important note that in each of these cases (contrary to Example 3b) it does not exist a unique N in the definition of convergence, which could be used for all x of the set X at once. At the moment, we arrive to this conclusion through geometric illustrations, but in the next sections we substantiate this result analytically. This leads to the supposition that the concept of the pointwise convergence can be not strict enough to ensure the transfer of relevant properties of the sequence functions to the limit function, and consequently, we need another stronger notion of convergence.

In the next sections we will see that the relationship between the properties of $f_n(x)$ and $f(x)$ can be really established if we will use a special type of convergence, stronger than pointwise convergence, called the uniform convergence. We will focus on the properties of *boundedness, continuity, differentiability and integrability of limit functions*, that is, on the properties which play the primary role in analysis, and other parts of mathematics. It is important to find the conditions which guarantee the transfer of these properties from the sequence functions $f_n(x)$ to the limit function $f(x)$. We will solve this problem in the next sections using the following development of topics. In Sect. 2 we introduce the concept of uniform convergence and discuss some techniques to investigate the type of convergence. In Sect. 3 we introduce a sufficient condition for uniform convergence, and in Sect. 4 we present the results clarifying when the limit function inherits the properties of the sequence functions.

2   Uniform and Non-uniform Convergence

2.1   Concept of the Uniform and Non-uniform Convergence

Recall the definition of the pointwise convergence given in the previous section.

▶ **Definition (Pointwise Convergence on a Set)** If a sequence $f_n(x)$ converges (as a sequence of numbers) at every fixed point $x \in X$, then $f_n(x)$ converges pointwise on X to a limit function $f(x)$: $\lim\limits_{n \to \infty} f_n(x) = f(x)$.

In the detailed $\varepsilon - N$ form this definition goes as follows:

$$\forall x \in X, \forall \varepsilon > 0 \; \exists N_\varepsilon(x) \text{ such that } \forall n > N_\varepsilon(x) \text{ it follows } |f_n(x) - f(x)| < \varepsilon.$$
(3.1)

Notice that, in general mode, the number N_ε depends on both ε and the chosen point x in the set X. Then, we actually have a set of natural numbers $\{N_\varepsilon(x), \forall x \in X\}$.

It may happen that for a fixed ε there exists an upper bound $\tilde{N}_\varepsilon \in \mathbb{N}$ of the set of numbers $N_\varepsilon(x)$, $\forall x \in X$, that is, a constant $\tilde{N}_\varepsilon$ such that $N_\varepsilon(x) \leq \tilde{N}_\varepsilon$, $\forall x \in X$ (in particular, we can take the supremum of the set $N_\varepsilon(x)$—if this set is bounded, its supremum exists). If this occurs for any ε, then in the definition of the pointwise convergence of a sequence $f_n(x)$ we can choose the same index N_ε for all points $x \in X$ (in what follows we will drop the tilde). In this case, we arrive at the notion of uniform convergence, which is a special case of pointwise convergence. The formal definition of uniform convergence is as follows.

▶ **Definition (Uniform Convergence on a Set)** A *sequence* $f_n(x)$ *converges uniformly on* X (or is *uniformly convergent on* X) if

$$\forall \varepsilon > 0 \; \exists N_\varepsilon \text{ (that does not depend on } x) \text{ such that}$$

$$\forall n > N_\varepsilon \text{ (and simultaneously for all } x \in X) \text{ it follows } |f_n(x) - f(x)| < \varepsilon.$$

(continued)

Due to the fundamental importance of this definition, we introduce some pertinent explanations in parentheses, but in what follows we use a shorter form of this definition:

$$\forall \varepsilon > 0 \ \exists N_\varepsilon \ \text{such that} \ \forall n > N_\varepsilon \ \text{and} \ \forall x \in X \ \text{it follows} \ |f_n(x) - f(x)| < \varepsilon. \tag{3.2}$$

The principal feature differentiating this definition from pointwise convergence is that the *number N_ε in uniform convergence depends only on ε and does not depend on the choice of points of the set X*, that is, the same N_ε serves for all points $x \in X$ at once. This kind of convergence is usually denoted as follows: $f_n(x) \underset{n\to\infty}{\overset{X}{\rightrightarrows}} f(x)$.

As we have seen in Examples 1b–4b, geometrically speaking, uniform convergence of $f_n(x)$ to the limit function $f(x)$ on a set X can be visualized by constructing a strip of width 2ε around $f(x)$. In the case of uniform convergence there exists an index N of the sequence after which every $f_n(x)$ has its graph completely contained in the ε-strip.

Evidently, as was noted, uniform convergence is a stronger requirement than pointwise convergence: if a sequence $f_n(x)$ converges uniformly, it also converges pointwise, but the converse is not true (see Examples 1u–6u below).

We say that the *convergence of a sequence $f_n(x)$ is non-uniform* if it converges pointwise, but not uniformly. In this case, it does not exist an upper bound $\tilde{N}_\varepsilon$ for the set $N_\varepsilon(x)$. This situation is described in detail in the following definition.

▶ **Definition (Non-uniform Convergence on a Set)** A *sequence $f_n(x)$ converges non-uniformly on X* if it converges pointwise, but

$$\exists \varepsilon_0 > 0 \ \text{such that} \ \forall N \in \mathbb{N} \ \exists n_N > N \ \text{and} \ \exists x_N \in X \ \text{such that} \ \left| f_{n_N}(x_N) - f(x_N) \right| \geq \varepsilon_0. \tag{3.3}$$

The definitions of uniform and non-uniform convergence given in this section can be written in other forms useful both in theoretical studies and solution of exercises. These forms are presented in the following two Remarks.

Remark 1 Condition (3.2) in the definition of uniform convergence is equivalent to the following condition:

$$\sup_{\forall x \in X} |f_n(x) - f(x)| \underset{n\to\infty}{\to} 0. \tag{3.4}$$

Indeed, choose an arbitrary $\varepsilon > 0$ and fix it for a moment. If condition (3.2) holds, then the inequality $|f_n(x) - f(x)| < \varepsilon$ is satisfied for all points $x \in X$ simultaneously (whenever $n > N_\varepsilon$), that is, the number ε represents an upper bound of the set $|f_n(x) - f(x)|$, $\forall x \in X$, and consequently, $\sup_{\forall x \in X} |f_n(x) - f(x)| \leq \varepsilon$ (for the same $n > N_\varepsilon$), since the least upper bound (supremum) is less than or equal to any other upper bound. The sequence $\sup_{\forall x \in X} |f_n(x) - f(x)|$ is a sequence of numbers (there exists the only number $\sup_{\forall x \in X} |f_n(x) - f(x)|$ associated with each n) and its last evaluation through ε means, due to the arbitrariness of ε, that $\sup_{\forall x \in X} |f_n(x) - f(x)| \underset{n \to \infty}{\to} 0$.

In the opposite direction, if $\sup_{\forall x \in X} |f_n(x) - f(x)| \underset{n \to \infty}{\to} 0$, then, by the definition of the limit, for $\forall \varepsilon > 0 \, \exists N_\varepsilon$ such that $\sup_{\forall x \in X} |f_n(x) - f(x)| < \varepsilon$ whenever $n > N_\varepsilon$. Therefore, for the same $n > N_\varepsilon$ and simultaneously for all $x \in X$, the inequality $|f_n(x) - f(x)| \leq \sup_{x \in X} |f_n(x) - f(x)| < \varepsilon$ is satisfied, which means that condition (3.2) holds.

Correspondingly, the condition of non-uniform convergence can be formulated in the form

$$\sup_{\forall x \in X} |f_n(x) - f(x)| \underset{n \to \infty}{\not\to} 0. \tag{3.5}$$

Remark 2 In many cases, condition (3.3) in the definition of non-uniform convergence can be substituted by a stronger restriction (which implies condition (3.3)) that has a simpler form:

$$\exists \varepsilon_0 > 0 \text{ such that } \forall n \in \mathbb{N} \, \exists x_n \in X \text{ such that } |f_n(x_n) - f(x_n)| \geq \varepsilon_0. \tag{3.6}$$

Remark 3 Notice that non-uniform convergence on a part of a given set implies immediately that the convergence is non-uniform on all the given set.

Remark 4 The type of convergence (uniform or not) of a sequence $f_n(x)$ detected on a set X does not change if we add to X (or remove from X) a finite number of points at which the sequence $f_n(x)$ converges. To prove this, it is sufficient to consider the case when one point is added to the original set X (an extension to the case of adding or eliminating a finite number of points is elementary). First, if $f_n(x)$ converges non-uniformly on X, it also converges non-uniformly on any broader set. In particular, if we add a point $c \notin X$ to a set X, then the convergence continue to be non-uniform on $X_c = X \cup \{c\}$. On the other hand, if $f_n(x)$ converges uniformly on X and $c \notin X$ is a convergence point of $f_n(x)$, then the convergence continue to be uniform on X_c. Indeed, for any given $\varepsilon > 0$ we can choose the same value of $N_X(\varepsilon)$ for all points $x \in X$ according to the definition of uniform convergence on X. Besides, for the same ε we can find a specific value N_c in the definition of convergence at the point c. Taking now $N_m = \max\{N_X, N_c\}$, we define the unique number N_m that works for all points $x \in X_c$ at once in the definition of uniform convergence on X_c, which means that $f_n(x)$ converges uniformly on X_c.

Now we return to the four Examples 1b–4b of Sect. 1 and analyze their behavior from the point of view of uniform convergence.

Example 1u $f_n(x) = \frac{n}{nx+1}$, $X = (0, 1)$
It was shown in Sect. 1 (Example 1b) that this sequence converges pointwise in $(0, 1)$ to the limit function $f(x) = \frac{1}{x}$. Let us prove that this convergence is not uniform, that causes the discrepancy between boundedness of $f_n(x)$ and the lack of this property for $f(x)$. It is intuitively clear that the most probable region that causes the problem is a neighborhood of the origin, where the limit function tends to $+\infty$ and the sequence functions are more distant from $f(x)$ (see illustration in Fig. 3.1). Then, we can try to use the sequence of points $x_n = \frac{1}{n} \in (0, 1)$ approaching 0 to show the non-uniformity of convergence. Following Remark 2, we can evaluate the difference $|f_n(x_n) - f(x_n)|$ in the form:

$$|f_n(x_n) - f(x_n)| = \left| \frac{n}{n \cdot \frac{1}{n} + 1} - \frac{1}{\frac{1}{n}} \right| = \left| \frac{n}{2} - n \right| = \frac{n}{2} \geq \frac{1}{2},$$

which confirms that the sequence converges non-uniformly on $(0, 1)$.

Example 2u $f_n(x) = x^n$, $X = [0, 1]$
As was proved in Sect. 1 (Example 2b), this sequence converges pointwise on $[0, 1]$:
$f_n(x) = x^n \overset{[0,1]}{\underset{n \to \infty}{\to}} f(x) = \begin{cases} 0, & x \in [0, 1) \\ 1, & x = 1 \end{cases}$. The discrepancy between continuity of the functions $f_n(x)$ and discontinuity of $f(x)$ on $[0, 1]$ (noted in Sect. 1) might be induced by the non-uniform character of convergence on $[0, 1]$. To investigate this, notice first that the region of the problem is located near the point 1, according to illustration in Fig. 3.2. Then, with intention to apply Remark 2, we try to use the sequence of points $x_n = 1 - \frac{1}{n}$ which tend to 1 and lie in $[0, 1]$. This leads to the following evaluation:

$$|f_n(x_n) - f(x_n)| = \left(1 - \frac{1}{n} \right)^n \underset{n \to \infty}{\to} e^{-1},$$

which means that the sequence converges non-uniformly on $[0, 1]$.
Another manner to achieve the same conclusion is by using Remark 1. In this case, we have the evaluation:

$$\sup_{\forall x \in [0,1]} |f_n(x) - f(x)| = \sup_{\forall x \in [0,1)} |f_n(x) - f(x)| = \sup_{\forall x \in [0,1)} |f_n(x)| = \sup_{\forall x \in [0,1)} x^n = 1 \underset{n \to \infty}{\not\to} 0,$$

which confirms the non-uniform character of convergence on $[0, 1]$.

Notice that the non-uniform convergence on the interval $[0, 1]$ implies immediately the non-uniform convergence on any set containing this interval, in particular, on the interval $(-1, 1]$, which is the convergence set of the sequence x^n. On the other hand, we can remove a finite number of points without changing the character of convergence. This means that the sequence x^n converges non-uniformly on the intervals $(0, 1)$ and $(-1, 1)$.

Notice also that the convergence of $f_n(x) = x^n$ is uniform on any interval $[-a, a]$, $0 < a < 1$ due to the following evaluation:

$$|f_n(x) - f(x)| = |x^n| \leq a^n \underset{n \to \infty}{\to} 0$$

valid for all x in $[-a, a]$ at once.

Example 3u $f_n(x) = \frac{\sin nx}{\sqrt{n}}$, $X = \mathbb{R}$

This is the sequence of Example 3b of Sect. 1. It was shown there that the sequence converges pointwise to $f(x) \equiv 0$ on $\mathbb{R}$ and later it was also provided the evaluation which ensures the uniform convergence on $\mathbb{R}$: for any $\varepsilon > 0$ we can find $N = \left[\frac{1}{\varepsilon^2}\right]$ such that $\left|\frac{\sin nx}{\sqrt{n}}\right| \leq \frac{1}{\sqrt{n}} < \varepsilon$ whenever $n > N$ for all $x \in \mathbb{R}$ at once. Hence, this sequence converges uniformly.

Example 4u $f_n(x) = \begin{cases} 2n^2x, & x \in [0, \frac{1}{2n}] \\ 2n - 2n^2x, & x \in [\frac{1}{2n}, \frac{1}{n}] \\ 0, & x \in [\frac{1}{n}, 1] \end{cases}$, $X = [0, 1]$.

In Example 4b of Sect. 1 it was shown that this sequence converges pointwise to $f(x) \equiv 0$ on $[0, 1]$, and later, in geometric interpretation, it was noted that the rate of its convergence slows down significantly in a neighnborhood of the points $x = \frac{1}{2n}$, especially near the origin, where the vertices $P_n = (\frac{1}{2n}, n)$ of the graphs move away from the x-axis. Now we can show analytically, using the rigorous definition that there is no unique bound for all $x \in [0, 1]$ in the evaluation of the proximity between $f_n(x)$ and $f(x)$, which means that the convergence is not uniform on $[0, 1]$. Indeed, choosing the sequence of points approaching 0 from the right $x_n = \frac{1}{2n} \in [0, 1]$, we have

$$|f_n(x_n) - f(x_n)| = |n - 0| = n > 1, \forall n \in \mathbb{N},$$

that is, according to Remark 2, the sequence $f_n(x)$ converges non-uniformly.

Consider two more examples to train techniques of investigation of uniform convergence.

Example 5u $f_n(x) = \frac{x}{1+n^2x^2}$, $X = [0, 1]$

We start with finding $\lim_{n\to\infty} f_n(x)$. If $x = 0$, then $f_n(0) = 0, \forall n \in \mathbb{N}$, and consequently $\lim_{n\to\infty} f_n(0) = 0$; if $x \neq 0$, $\forall x \in (0, 1]$, then $\lim_{n\to\infty} f_n(x) = \lim_{n\to\infty} \frac{x}{1+n^2x^2} = 0$. Therefore, the limit function $f(x)$ exists and equals 0: $\lim_{n\to\infty} f_n(x) = f(x) \equiv 0, \forall x \in [0, 1]$. Next, we analyze if the convergence of $f_n(x)$ to its limit $f(x) = 0$ is uniform or not. To do this, we evaluate the difference

$$|f_n(x) - f(x)| = \frac{|x|}{1 + n^2x^2} = \frac{2n|x|}{1 + n^2x^2} \cdot \frac{1}{2n} \leq \frac{1}{2n} \tag{3.7}$$

(here, we use the inequality $|2ab| \leq a^2 + b^2$). Take $\forall \varepsilon > 0$ and find the natural numbers n such that the inequality $\frac{1}{2n} < \varepsilon$ is true. This inequality is equivalent to $n > \frac{1}{2\varepsilon}$. Denote $N_\varepsilon = \left[\frac{1}{2\varepsilon}\right]$ and notice that the evaluation (3.7) holds simultaneously for all x of the interval $[0, 1]$. Then, for all the indices $n > N_\varepsilon$ and for all $x \in [0, 1]$ we have

$$|f_n(x) - f(x)| \leq \frac{1}{2n} < \varepsilon,$$

that is, according to the definition of uniform convergence, the sequence converges to the limit function $f(x) \equiv 0$ uniformly on $[0, 1]$.

Notice that the type of convergence can also be analyzed using the result of Remark 1. To this end, we have to find the extrema of the functions $f_n(x) - f(x) = f_n(x)$ on the interval $[0, 1]$. Using the derivative

$$f_n'(x) = \frac{1 + n^2x^2 - 2n^2x^2}{\left(1 + n^2x^2\right)^2} = \frac{1 - n^2x^2}{\left(1 + n^2x^2\right)^2},$$

we find the critical points from the equation $f_n'(x) = 0$, which is reduced to $1 - n^2x^2 = 0$. The last equation has the two solutions $x = \pm\frac{1}{n}$, but only one of them $x_n = \frac{1}{n}$ belongs to the interval $[0, 1]$. Since $f_n'(x) > 0$ for $0 \leq x < \frac{1}{n}$ and $f_n'(x) < 0$ when $\frac{1}{n} < x \leq 1$, it follows that $x_n = \frac{1}{n}$ is a local maximum of the function $f_n(x)$. Taking into account that $f_n(x)$ is non-negative on $[0, 1]$ and x_n is its unique local extremum in this interval, we conclude that $x_n = \frac{1}{n}$ is a global maximum of $f_n(x)$ on $[0, 1]$. Therefore,

$$\sup_{\forall x \in [0,1]} |f_n(x) - f(x)| = f_n(x_n) = \frac{1}{2n} \underset{n\to\infty}{\to} 0.$$

Then, by Remark 1, f_n converges uniformly on $[0, 1]$: $f_n(x) \underset{n\to\infty}{\overset{[0,1]}{\rightrightarrows}} f(x) \equiv 0$.

Example 6u $f_n(x) = \frac{nx}{1+n^2x^2}$, $X = [0, 1]$

Here, we follow the same line of reasoning as in the previous example. Initially, we find the limit function: if $x = 0$, then $f_n(0) = 0$, $\forall n \in \mathbb{N}$, and consequently $\lim\limits_{n\to\infty} f_n(0) = 0$; if $x \neq 0$ then for $\forall x \in (0, 1]$ we have $\lim\limits_{n\to\infty} f_n(x) = \lim\limits_{n\to\infty} \frac{nx}{1+n^2x^2} = 0$. Hence, the sequence converges to 0: $\lim\limits_{n\to\infty} f_n(x) = f(x) \equiv 0$, $\forall x \in [0, 1]$. To investigate the character of this convergence, we employ Remark 2. It is intuitively clear that problems with convergence may arise for small values of x, which reduce the values of x^2n^2 in the denominator, and this term is the main one which determines the rate of convergence. Therefore, we try to use the sequence of points $x_n = \frac{1}{n} \in [0, 1]$ that approaches 0 and obtain

$$|f_n(x_n) - f(x_n)| = \left| \frac{n \cdot \frac{1}{n}}{1 + n^2 \cdot \frac{1}{n^2}} - 0 \right| = \frac{1}{2} \geq \frac{1}{2} = \varepsilon_0.$$

This shows that the convergence is not uniform.

An alternative way is to employ Remark 1. Since the sequence of this example differs from that of the previous example just by the factor n (constant with respect to x), it is clear that the critical point of the function $f_n(x) - f(x) = f_n(x)$ is the same—$x_n = \frac{1}{n}$, and this point is a global maximum on $[0, 1]$. Taking into account that $f_n(x) \geq 0$ on $[0, 1]$, we have

$$\sup_{[0,1]} |f_n(x) - f(x)| = \sup_{[0,1]} |f_n(x)| = f_n(x_n) = \frac{1}{2} \underset{n\to\infty}{\not\to} 0.$$

According to Remark 1, this means that the sequence $f_n(x) = \frac{nx}{1+n^2x^2}$ converges to the function $f(x) \equiv 0$ non-uniformly on $[0, 1]$.

2.2   Arithmetic Properties of Uniform Convergence

Property 1 *If $f_n(x)$ and $g_n(x)$ converge uniformly on X to $f(x)$ and $g(x)$, respectively, then $f_n(x) + g_n(x)$ converges uniformly on X to $f(x) + g(x)$.*

Proof This demonstration follows the definition. The following statements of the uniform convergence of the sequences $f_n(x)$ and $g_n(x)$ are satisfied simultaneously for all $x \in X$:

$$\forall \varepsilon_f > 0 \; \exists N_{\varepsilon_f} \text{ such that } \forall n > N_{\varepsilon_f} \text{ it follows } |f_n(x) - f(x)| < \varepsilon_f$$

and

$$\forall \varepsilon_g > 0 \; \exists N_{\varepsilon_g} \text{ such that } \forall n > N_{\varepsilon_g} \text{ it follows } |g_n(x) - g(x)| < \varepsilon_g.$$

Then, for $N_\varepsilon = \max\{N_{\varepsilon_f}, N_{\varepsilon_g}\}$, both inequalities for $f_n(x)$ and $g_n(x)$ hold, and consequently, choosing $\varepsilon_f = \varepsilon_g = \frac{\varepsilon}{2}$, we arrive at the following statement:

$$\forall \varepsilon > 0 \; \exists N_\varepsilon \text{ such that } \forall n > N_\varepsilon \text{ and } \forall x \in X \text{ it follows}$$

$$|f_n(x) + g_n(x) - (f(x) + g(x))| \le |f_n(x) - f(x)| + |g_n(x) - g(x)| < \frac{\varepsilon}{2} + \frac{\varepsilon}{2} = \varepsilon.$$

The last inequality is valid for all $x \in X$ at once, which means that the last statement is the definition of uniform convergence of the sequence $f_n(x) + g_n(x)$ on X.   □

Property 2 *If $f_n(x)$ converges uniformly on X to $f(x)$, and $g(x)$ is bounded on X, then $g(x) f_n(x)$ converges uniformly on X to $g(x) f(x)$.*

Proof In this demonstration we follow the definition. The uniform convergence of $f_n(x)$ on X means that simultaneously for all $x \in X$ we have

$$\forall \varepsilon_f > 0 \; \exists N_{\varepsilon_f} \text{ such that } |f_n(x) - f(x)| < \varepsilon_f \text{ whenever } n > N_{\varepsilon_f}.$$

The boundedness of $g(x)$ means that there exists a constant $M > 0$ such that $|g(x)| \le M$ for all $x \in X$. Then, choosing $\varepsilon_f = \frac{\varepsilon}{M}$ and $N_\varepsilon = N_{\varepsilon_f}$, we obtain the following statement valid for all $x \in X$ at once:

$$\forall \varepsilon > 0 \; \exists N_\varepsilon \text{ such that } |g(x) f_n(x) - g(x) f(x)| = |g(x)| \cdot |f_n(x) - f(x)|$$

$$< M \frac{\varepsilon}{M} = \varepsilon \text{ whenever } n > N_\varepsilon.$$

Therefore, the definition of the uniform convergence of the sequence $g(x) f_n(x)$ on X is satisfied.   □

Remark Notice that without the requirement of boundedness of $g(x)$ the property does not hold. Consider, for example, the sequence $f_n(x) = \frac{\cos nx}{n}$ that converges uniformly on $(0, \infty)$ to $f(x) \equiv 0$ as is shown by the inequality $\frac{|\cos nx|}{n} \le \frac{1}{n}$ valid simultaneously for all $x \in (0, \infty)$. Choosing the unbounded function $g(x) = \frac{1}{x}$, we obtain the sequence $g(x) f_n(x) = \frac{1}{x} \frac{\cos nx}{n}$ that converges to 0 on $(0, \infty)$, but this convergence is not uniform: for the sequence of points $x_n = \frac{1}{n}$ we have $g(x_n) f_n(x_n) = n \frac{\cos 1}{n} = \cos 1 \underset{n \to \infty}{\nrightarrow} 0$.

Complement: Product of Uniformly Convergent Sequences

The product property is trivially valid for pointwise convergence: at each point of the convergence set it is reduced to the corresponding property of the product of two sequences of numbers. However, it is not applicable for uniform convergence: in general, the uniform convergence of $f_n(x)$ and $g_n(x)$ on X does not guarantee the uniform convergence of

$f_n(x)g_n(x)$ on X. To see this, consider the following example: $f_n(x) = g_n(x) = x + \frac{1}{n}$ on $X = \mathbb{R}$. The sequence $f_n(x) = x + \frac{1}{n}$ converges uniformly on $\mathbb{R}$ to $f(x) = x$ according to the elementary evaluation

$$|f_n(x) - f(x)| = \frac{1}{n} < \varepsilon$$

valid for all $n > N = \left[\frac{1}{\varepsilon}\right]$ and for all real x. It is evident that the product $f_n(x)g_n(x) = f_n^2(x) = x^2 + \frac{2x}{n} + \frac{1}{n^2}$ converges pointwise on $\mathbb{R}$ to $f^2(x) = x^2$, but this convergence is not uniform due to the following evaluation at the points $x_n = n$:

$$\left| f_n^2(x_n) - f^2(x_n) \right| = \frac{2x_n}{n} + \frac{1}{n^2} = \frac{2n}{n} + \frac{1}{n^2} > 2 \underset{n\to\infty}{\nrightarrow} 0.$$

The product property can be recovered if we add the condition that $f_n(x)$ and $g_n(x)$ are bounded functions on X for any fixed n. Then this property can be formulated as follows.

Product Property *If $f_n(x)$ and $g_n(x)$ converge uniformly on X to $f(x)$ and $g(x)$, respectively, and every function $f_n(x)$ and $g_n(x)$ is bounded on X, then $f_n(x)g_n(x)$ converges uniformly on X to $f(x)g(x)$.*

Proof First, we show that $f_n(x)$ and $g_n(x)$ are bounded uniformly on X, that is, for the sequence $f_n(x)$ there exists a constant M_f such that $|f_n(x)| \le M_f$ simultaneously for all $x \in X$ and all n, and the same is true for $g_n(x)$. We demonstrate this property only for $f_n(x)$, since for $g_n(x)$ the arguments are the same. Let us take $\varepsilon = 1$ in the definition of uniform convergence and find N such that $|f_n(x) - f(x)| < 1$ for all $n > N$ and for all $x \in X$ at once. Each of the first $N + 1$ functions $f_1(x), \ldots, f_N(x), f_{N+1}(x)$ is bounded on X by its proper constant:

$$|f_1(x)| \le M_1, \ldots, |f_N(x)| \le M_N, |f_{N+1}(x)| \le M_{N+1}.$$

The last function satisfies also the inequality $|f_{N+1}(x) - f(x)| < 1$. Then, for any $n > N$ and all $x \in X$ we have

$$|f_n(x) - f_{N+1}(x)| \le |f_n(x) - f(x)| + |f(x) - f_{N+1}| < 2,$$

and consequently,

$$|f_n| \le |f_n(x) - f_{N+1}(x)| + |f_{N+1}(x)| < M_{N+1} + 2.$$

Defining $M_f = \max\{M_1, \ldots, M_N, M_{N+1} + 2\}$, we guarantee that $|f_n(x)| \le M_f$ for all indices n and for all $x \in X$. Besides, it follows from the inequality $|f_{N+1}(x) - f(x)| < 1$

that $|f(x)| < M_{N+1} + 1 < M_f$. The same is true for $g_n(x)$ and $g(x)$: there exists a constant M_g such that $|g_n(x)| \le M_g$ and $|g(x)| \le M_g$ for all n and all $x \in X$. Now we can prove that the definition of uniform convergence of the product $f_n(x)g_n(x)$ is satisfied. Denote $M = \max\{M_f, M_g\}$ and choose the quantity $\frac{\varepsilon}{2M}$ in the definition of uniform convergence of $f_n(x)$ and $g_n(x)$ to obtain the inequalities $|f_n(x) - f(x)| < \frac{\varepsilon}{2M}$ and $|g_n(x) - g(x)| < \frac{\varepsilon}{2M}$ satisfied for all $n > N$ and all $x \in X$. Then

$$|f_n(x)g_n(x) - f(x)g(x)| \le |f_n(x)g_n(x) - f(x)g_n(x)| + |f(x)g_n(x) - f(x)g(x)|$$

$$= |g_n(x)||f_n(x) - f(x)| + |f(x)||g_n(x) - g(x)| < M\frac{\varepsilon}{2M} + M\frac{\varepsilon}{2M} = \varepsilon,$$

that is, the definition of uniform convergence of $f_n(x)g_n(x)$ on X is satisfied. $\square$

Remark As a by-product of this proof, we have shown that uniformly convergent sequence of bounded functions has a bounded limit function on the convergence set. This simple result will also appear later among the analytic properties of the limit functions of uniformly convergent sequences.

2.3   Cauchy Criterion for Uniform Convergence

Consider now the *Cauchy criterion for uniform convergence*.

> ▶ **Definition** (**Uniform Cauchy Sequence**) Let $f_n(x)$ be a sequence of functions defined on a set $X \subset \mathbb{R}$. The sequence $f_n(x)$ is a *uniform Cauchy sequence on X* if the following Cauchy condition holds:
>
> $\forall \varepsilon > 0 \; \exists N_\varepsilon$ such that for $\forall n, m > N_\varepsilon$ (and for $\forall x \in X$) it follows $|f_n(x) - f_m(x)| < \varepsilon$.
> $$(3.8)$$

We emphasize that in this definition N_ε depends only on ε and does not depend on points of the set X, that is, the last inequality is true for all $n, m > N_\varepsilon$ and for all points $x \in X$ at once.

> **Theorem** (**Cauchy Criterion for Uniform Convergence**) *Let $f_n(x)$ be a sequence of functions defined on a set $X \subset \mathbb{R}$. The sequence $f_n(x)$ converges uniformly on X if and only if $f_n(x)$ is a uniform Cauchy sequence on X.*

Proof *Necessity.* By the condition, we have that $f_n(x) \underset{n \to \infty}{\overset{X}{\rightrightarrows}} f(x)$ where $f(x)$ is a limit function, that is, for $\forall \varepsilon > 0$ (take for convenience $\frac{\varepsilon}{2}$) there exists N_ε (that depends only on ε) such that for all $n > N_\varepsilon$ and simultaneously for all $x \in X$ it follows that $|f_n(x) - f(x)| < \frac{\varepsilon}{2}$. Analogously, if we take $\forall m > N_\varepsilon$, then for $\forall x \in X$ we have $|f_m(x) - f(x)| < \frac{\varepsilon}{2}$. Evaluating the difference $|f_n(x) - f_m(x)|$ for all $n, m > N_\varepsilon$ and for all $x \in X$, we obtain:

$$|f_n(x) - f_m(x)| = |f_n(x) - f(x) + f(x) - f_m(x)|$$

$$\leq |f_n(x) - f(x)| + |f_m(x) - f(x)| < \frac{\varepsilon}{2} + \frac{\varepsilon}{2} = \varepsilon.$$

Sufficiency. If $f_n(x)$ is a uniform Cauchy sequence, then, by definition, the Cauchy condition (3.8) holds. First we show that it guarantees pointwise convergence of $f_n(x)$ on X. Pick an arbitrary point $x \in X$ and fix it. At this point, the sequence of functions becomes the sequence of numbers, which satisfies the Cauchy condition. Then, from the Cauchy criterion for a sequence of numbers it follows that there exists a finite limit $\lim\limits_{n \to \infty} f_n(x)$ for any fixed $x \in X$. Therefore, $f_n(x)$ converges pointwise on X to a limit function: $\lim\limits_{n \to \infty} f_n(x) = f(x), \forall x \in X$.

Now we prove that the convergence of $f_n(x)$ to $f(x)$ is uniform on X. Let us choose, for convenience, the quantity $\frac{\varepsilon}{2}$ instead of ε in inequality (3.8) and write this inequality in the form

$$f_m(x) - \frac{\varepsilon}{2} < f_n(x) < f_m(x) + \frac{\varepsilon}{2}.$$

Passing to the limit in the last inequality as $m \to \infty$ and using the fact that $\lim\limits_{m \to \infty} f_m(x) = f(x), \forall x \in X$ (which was already proved), we obtain for $\forall n > N_\varepsilon$ and simultaneously for all $\forall x \in X$:

$$f(x) - \frac{\varepsilon}{2} \leq f_n(x) \leq f(x) + \frac{\varepsilon}{2}$$

or

$$|f_n(x) - f(x)| \leq \frac{\varepsilon}{2} < \varepsilon.$$

Since the last inequality is true for $\forall n > N_\varepsilon$ and for all $x \in X$ at once, by definition, we have the uniform convergence of $f_n(x)$ on X: $f_n(x) \underset{n \to \infty}{\overset{X}{\rightrightarrows}} f(x)$.

This completes the proof. $\qquad\square$

3   Dini's Theorem

In this section we consider Dini's Theorem which offers sufficient conditions for uniform convergence of a sequence of functions and, for this reason, can be considered as a test for uniform convergence.

Dini's Theorem *Let $f_n(x)$ be a sequence of functions defined on a compact set $X \subset \mathbb{R}$. If*

(1) for any fixed $x \in X$ the sequence of numbers $f_n(x)$ is monotone,
(2) all the functions $f_n(x)$ are continuous on X,
(3) the sequence $f_n(x)$ converges pointwise on X to $f(x)$,
(4) the limit function $f(x)$ is continuous on X,

then the sequence $f_n(x)$ converges uniformly on X.

Proof Just to specify the situation, let us suppose that the sequence $f_n(x)$ is increasing in n (for a decreasing sequence the proof is the same). Then, for $\forall n \in N$ we have $f_n(x) \leq f(x)$ for any fixed x in X.

Let us use the method of contradiction. Assume that $f_n(x)$ converges non-uniformly on X, that is, according to definition, there exists $\varepsilon_0 > 0$ such that for any $N \in \mathbb{N}$ it can be found $n_N > N$ and $x_N \in X$ such that

$$\left| f_{n_N}(x_N) - f(x_N) \right| \geq \varepsilon_0. \tag{3.9}$$

In this way, we obtain a sequence of points $\{x_N\}_{N=1}^{\infty}$, $x_N \in X$. Since X is compact (in particular, it is bounded), by the Bolzano-Weierstrass Theorem, the sequence x_N has a convergent subsequence x_k with a finite limit: $\lim_{k \to \infty} x_k = x_0$. Due to the compactness of X this limit point x_0 belongs to X.

Now we show that the conditions obtained so far lead to the conclusion that $f_n(x)$ does not converge at the point x_0. First, we evaluate $|f_n(x_0) - f(x_0)|$ in the form

$$|f_n(x_0) - f(x_0)| = |f_n(x_0) - f_n(x_k) + f_n(x_k) - f(x_k) + f(x_k) - f(x_0)|$$
$$\geq |f_n(x_k) - f(x_k)| - |f_n(x_0) - f_n(x_k)| - |f(x_k) - f(x_0)| \tag{3.10}$$

Since x_k is a subsequence of x_N, it follows from (3.9) that (take for convenience $2\varepsilon_0$)

$$\left| f_{n_k}(x_k) - f(x_k) \right| \geq 2\varepsilon_0.$$

The growth of the sequence $f_n(x)$ allows us to choose $n_k > n$ such that

$$f_n(x) \leq f_{n_k}(x) \leq f(x)$$

for every fixed x in X. Using this inequality at the points x_k, we can evaluate the first absolute value on the right-hand side of (3.10) as follows:

$$|f_n(x_k) - f(x_k)| = f(x_k) - f_n(x_k) \geq f(x_k) - f_{n_k}(x_k) = |f_{n_k}(x_k) - f(x_k)| \geq 2\varepsilon_0.$$

The second and third absolute values on the right-hand side of (3.10) are small quantities due to the continuity of $f_n(x)$ and $f(x)$ at x_0, namely: for $\forall \varepsilon > 0$ (choose here $\varepsilon = \frac{\varepsilon_0}{2}$) the following two inequalities

$$|f_n(x) - f_n(x_0)| < \frac{\varepsilon_0}{2}, \quad |f(x) - f(x_0)| < \frac{\varepsilon_0}{2}$$

are satisfied for all $x \in X$ such that $|x - x_0| < \delta$. Since $x_k \underset{k \to \infty}{\to} x_0$, for sufficiently large values of k we guarantee that $|x_k - x_0| < \delta$ (whatever small $\delta > 0$ is chosen). Bringing the obtained inequalities for the three absolute values in (3.10), we conclude that

$$|f_n(x_0) - f(x_0)| \geq |f_n(x_k) - f(x_k)| - |f_n(x_0) - f_n(x_k)| \tag{3.11}$$

$$- |f(x_k) - f(x_0)| > 2\varepsilon_0 - \frac{\varepsilon_0}{2} - \frac{\varepsilon_0}{2} = \varepsilon_0,$$

which means that $f_n(x)$ does not converge to $f(x)$ at the point $x_0 \in X$. This contradicts the hypothesis of pointwise convergence of the sequence $f_n(x)$ on X. Hence, the obtained contradiction indicates that our assumption about non-uniform convergence on X is false. $\qquad \square$

4 Properties of Limit Functions Under Uniform Convergence

4.1 Boundedness of Limit Function

Theorem of Boundedness *Let $f_n(x)$ be a sequence of functions defined on a set $X \subset \mathbb{R}$. If each function $f_n(x)$ is bounded on X and the sequence converges uniformly on X, then the limit function $f(x)$ is bounded on X.*

Proof Choose $\varepsilon = 1$ in the definition of uniform convergence. Then, there exists N_1 such that for $\forall n > N_1$ and simultaneously for all $x \in X$ we have

$$|f_n(x) - f(x)| < 1. \tag{3.12}$$

Pick now some specific $n > N_1$ and consider the corresponding function $f_n(x)$. By the condition of the Theorem, this function is bounded on X, that is, there exists a constant

M_n such that

$$|f_n(x)| < M_n \tag{3.13}$$

for all $x \in X$. From inequalities (3.12) and (3.13) it follows that

$$|f(x)| = |f(x) - f_n(x) + f_n(x)| \leq |f(x) - f_n(x)| + |f_n(x)| < 1 + M_n.$$

This means that $f(x)$ is bounded on X. $\qquad\qquad\qquad\qquad\qquad\square$

Remark 1 There is a useful consequence of this Theorem: if a sequence of bounded functions converges to an unbounded function, then this convergence is not uniform.

Remark 2 In this Theorem we do not assume that the sequence $f_n(x)$ is uniformly bounded, that is, all the functions $f_n(x)$ are bounded by the same constant. If such an assumption would be made, the boundedness of the limit function would follow directly from pointwise convergence of $f_n(x)$ on X.

Remark 3 The condition of uniform convergence is essential for this result. At the same time, this is a sufficient, but not necessary condition. The examples below illustrate these properties.

Example 1s $f_n(x) = \frac{x^n}{n}$, $X = [-1, 1]$
As was shown in Sect. 1, this sequence converges pointwise on $[-1, 1]$ to the limit function $f(x) \equiv 0$. Notice that every function $f_n(x)$ is bounded on the considered interval: $\left|\frac{x^n}{n}\right| \leq \frac{1}{n}, \forall x \in [-1, 1]$. Although both the sequence functions and the limit function are bounded, it does not guarantee that the convergence is uniform on $[-1, 1]$. Therefore, we make the evaluation by definition:

$$|f_n(x) - f(x)| = \left|\frac{x^n}{n}\right| \leq \frac{1}{n} < \varepsilon \,,$$

which is true for $\forall n > N = \left[\frac{1}{\varepsilon}\right]$ and for all $x \in [-1, 1]$ at once. Thus, the convergence is uniform on $[-1, 1]$.

Example 2s $f_n(x) = \begin{cases} \min(n, \frac{1}{x}), & x \in (0, 1] \,; \\ 0, & x = 0 \end{cases}$, $X = [0, 1]$.
Each function $f_n(x)$ is non-negative and bounded above on $[0, 1]$ by the constant n. The sequence $f_n(x)$ converges on $[0, 1]$ to $f(x) = \begin{cases} \frac{1}{x}, & x \in (0, 1] \\ 0, & x = 0 \end{cases}$, but the limit function is unbounded. Consequently, using Remark 1, we conclude that the convergence is not

uniform. We can also prove it by definition. Evidently, the problem occurs when x approaches 0. Therefore, we can choose the sequence of points $x_n = \frac{1}{2n} \in [0, 1]$ such that

$$|f_n(x_n) - f(x_n)| = |n - 2n| = n \to \infty ,$$

which shows that the convergence is non-uniform.

Example 3s $f_n(x) = \frac{nx}{1+n^2x^2}$, $X = [0, 1]$
This sequence was considered in Example 6u of Sect. 2, where it was shown that the sequence converges non-uniformly on $[0, 1]$ to the limit function $f(x) \equiv 0$. It remains to note that the limit function $f(x)$ as well as the sequence functions $f_n(x)$ are bounded on $[0, 1]$: $|f_n(x)| = \frac{nx}{1+n^2x^2} \le \frac{1}{2}$ (this is a well-known evaluation that follows from the inequality $(1 - |nx|)^2 \ge 0$).

Example 4s $f_n(x) = \begin{cases} \min(1, \frac{1}{nx}), & x \in (0, 1] ; \\ 0, & x = 0 \end{cases}$, $X = [0, 1]$.

This example illustrates the same situation as Example 3s. Each function $f_n(x)$ is non-negative and bounded above by 1 on $[0, 1]$. The sequence $f_n(x)$ converges to the bounded function $f(x) \equiv 0$. However, this convergence is non-uniform, that can be verified by definition. Indeed, the problem occurs when x approaches 0. Therefore, choosing the sequence of points $x_n = \frac{1}{n} \in [0, 1]$ we obtain

$$|f_n(x_n) - f(x_n)| = 1 \not\to 0 ,$$

which reveals that the convergence is non-uniform.

4.2   Limit of the Limit Function

Theorem of Limit *Let $f_n(x)$ be a sequence of functions defined on a set $X \subset \mathbb{R}$ and a be an accumulation point of X. If for each fixed n there exists the limit $\lim_{x \to a} f_n(x) = A_n$ and the sequence $f_n(x)$ converges uniformly on X, then the limit function $f(x)$ has the limit at the point a, the sequence of numbers A_n converges, and $\lim_{x \to a} f(x) = \lim_{n \to \infty} A_n$.*

In other words,

$$\lim_{x \to a} \lim_{n \to \infty} f_n(x) = \lim_{n \to \infty} \lim_{x \to a} f_n(x), \tag{3.14}$$

that is, the order of the limits is interchangeable.

Proof First we demonstrate that the sequence of numbers A_n converges by showing that A_n is a Cauchy sequence. Indeed, from the Cauchy criterion for uniform convergence we have that for $\forall \varepsilon > 0$ (choose for convenience $\varepsilon_1 = \frac{\varepsilon}{3}$) it can be found N_ε such that for all $n, m > N_\varepsilon$ and simultaneously for all $x \in X$ it follows that

$$|f_n(x) - f_m(x)| < \frac{\varepsilon}{3}. \tag{3.15}$$

Fixing two values m and n such that $n, m > N_\varepsilon$ and using the hypothesis of the Theorem that $\lim_{x \to a} f_n(x) = A_n$ and $\lim_{x \to a} f_m(x) = A_m$, we obtain the estimates

$$|f_n(x) - A_n| < \frac{\varepsilon}{3}, \quad |f_m(x) - A_m| < \frac{\varepsilon}{3} \tag{3.16}$$

for all points $x \in X$ close to a. Joining the last three inequalities (3.15)–(3.16), we get

$$|A_n - A_m| = |A_n - f_n(x) + f_n(x) - f_m(x) + f_m(x) - A_m|$$

$$\leq |A_n - f_n(x)| + |f_n(x) - f_m(x)| + |f_m(x) - A_m| < \frac{\varepsilon}{3} + \frac{\varepsilon}{3} + \frac{\varepsilon}{3} = \varepsilon,$$

that is, the sequence of numbers A_n is a Cauchy sequence, and consequently, it converges to a limit which we denote by A.

Second we prove that $\lim_{x \to a} f(x) = A$. In the definition of the uniform convergence of $f_n(x)$ on X we find N_f that corresponds to $\frac{\varepsilon}{3}$, and in the definition of the convergence of A_n we pick N_A corresponding to the same $\frac{\varepsilon}{3}$. Then, for any $n > N_m = \max\{N_f, N_A\}$ we have the following two inequalities:

$$|f_n(x) - f(x)| < \frac{\varepsilon}{3}, \quad |A_n - A| < \frac{\varepsilon}{3}, \tag{3.17}$$

where the first inequality is valid for all $x \in X$. Now we fix one of these indices $n > N_m$ and use the fact that $\lim_{x \to a} f_n(x) = A_n$, that is, for the same value $\frac{\varepsilon}{3}$ there exists $\delta > 0$ such that for all $x \in X, 0 < |x - a| < \delta$ it follows that

$$|f_n(x) - A_n| < \frac{\varepsilon}{3}. \tag{3.18}$$

Then, for the same x chosen in the last definition, employing the three inequalities (3.17)–(3.18), we obtain:

$$|f(x) - A| = |f(x) - f_n(x) + f_n(x) - A_n + A_n - A|$$

$$\leq |f(x) - f_n(x)| + |f_n(x) - A_n| + |A_n - A| < \frac{\varepsilon}{3} + \frac{\varepsilon}{3} + \frac{\varepsilon}{3} = \varepsilon,$$

which means that the definition of the limit $\lim_{x \to a} f(x) = A$ is satisfied. $\square$

Example 11 $f_n(x) = \frac{nx}{n^4x^2+2}$, $X = (0, 1]$, $a = 0$

It is evident that this sequence converges to $f(x) \equiv 0$:

$$f_n(x) = \frac{nx}{n^4x^2+2} = \frac{\frac{x}{n^3}}{x^2 + \frac{2}{n^4}} \underset{n\to\infty}{\to} \frac{0}{x^2+0} = 0, \forall x \in (0, 1].$$

Moreover, this convergence is uniform on $(0, 1]$ due to the following evaluation

$$|f_n(x) - f(x)| = \frac{nx}{n^4x^2+2} = \frac{1}{2n}\frac{2n^2x}{n^4x^2+2} < \frac{1}{2n} \underset{n\to\infty}{\to} 0$$

valid for all $x \in (0, 1]$ at once.

Besides, an application of the arithmetic rules of limits gives $\lim\limits_{x\to 0} f_n(x) = \frac{0}{0+2} = 0 = A_n$ for every fixed n (notice that x approaches 0 from the right according to the problem conditions). Therefore, all the conditions of the Theorem of limit are satisfied, and consequently $\lim\limits_{x\to 0}\lim\limits_{n\to\infty} f_n(x) = \lim\limits_{n\to\infty}\lim\limits_{x\to 0} f_n(x)$. The last formula can be verified directly for this sequence, since $\lim\limits_{x\to 0}\lim\limits_{n\to\infty} f_n(x) = \lim\limits_{x\to 0} f(x) = \lim\limits_{x\to 0} 0 = 0$ and $\lim\limits_{n\to\infty}\lim\limits_{x\to 0} f_n(x) = \lim\limits_{n\to\infty} A_n = \lim\limits_{n\to\infty} 0 = 0$.

Remark The next example shows that the condition of uniform convergence is essential for the Theorem, but is not necessary.

Example 21 $f_n(x) = \frac{n^2x}{n^4x^2+2}$, $X = (0, 1]$, $a = 0$

It is evident that this sequence converges to $f(x) \equiv 0$:

$$f_n(x) = \frac{n^2x}{n^4x^2+2} = \frac{\frac{x}{n^2}}{x^2 + \frac{2}{n^4}} \underset{n\to\infty}{\to} \frac{0}{x^2+0} = 0, \forall x \in (0, 1].$$

However, this convergence is not uniform on $(0, 1]$, that can be seen from the following evaluation: choosing the sequence of points $x_n = \frac{1}{n^2} \in (0, 1]$, we obtain

$$|f_n(x_n) - f(x_n)| = \frac{1}{1+2} = \frac{1}{3} \underset{n\to\infty}{\not\to} 0.$$

Nevertheless, the two iterated limits coincide. In fact, for the first iterated limit we have $\lim\limits_{x\to 0}\lim\limits_{n\to\infty} f_n(x) = \lim\limits_{x\to 0} f(x) = 0$, and for the second one we get the same result $\lim\limits_{n\to\infty}\lim\limits_{x\to 0} f_n(x) = \lim\limits_{n\to\infty} 0 = 0$.

4.3   Continuity of the Limit Function

Theorem of Continuity *Let $f_n(x)$ be a sequence of functions defined on a set $X \subset$ $\mathbb{R}$. If each function $f_n(x)$ is continuous on the set X and the sequence converges uniformly on X, then the limit function $f(x)$ is continuous on the set X.*
We can express this result in the following manner:

$$\lim_{x \to x_0} \lim_{n \to \infty} f_n(x) = \lim_{x \to x_0} f(x) = f(x_0) = \lim_{n \to \infty} f_n(x_0) = \lim_{n \to \infty} \lim_{x \to x_0} f_n(x), \ \forall x_0 \in X.$$
$$(3.19)$$

Proof We pick an arbitrary point $x_0 \in X$ and show that $f(x)$ is continuous at this point. The uniform convergence of $f_n(x)$ to $f(x)$ means that for $\forall \varepsilon > 0$ (choose for convenience $\varepsilon_1 = \frac{\varepsilon}{3}$) there exists N_ε (depending only on ε) such that for $\forall n > N_\varepsilon$ and simultaneously for $\forall x \in X$ we have

$$|f_n(x) - f(x)| < \frac{\varepsilon}{3}. \qquad (3.20)$$

In particular, the same relation (with the same values $\frac{\varepsilon}{3}$ and N_ε) holds at the point x_0:

$$|f_n(x_0) - f(x_0)| < \frac{\varepsilon}{3}. \qquad (3.21)$$

On the other hand, each function $f_n(x)$ is continuous on X (by the hypothesis) and, in particular, at the point $x_0 \in X$. Considering any $n > N_\varepsilon$ (where N_ε is from the definition of uniform convergence), fixing this n and using the definition of continuity of $f_n(x)$ at the point x_0, we get: $\forall \varepsilon > 0$ (again take $\varepsilon_1 = \frac{\varepsilon}{3}$) there exists $\delta = \delta(\varepsilon)$ such that for $\forall x \in X$, $|x - x_0| < \delta$ it follows

$$|f_n(x) - f_n(x_0)| < \frac{\varepsilon}{3}. \qquad (3.22)$$

Then, for x chosen in the definition of continuity (3.22) and n chosen in the definition of uniform convergence (3.20), we obtain

$$|f(x) - f(x_0)| = |f(x) - f_n(x) + f_n(x) - f_n(x_0) + f_n(x_0) - f(x_0)|$$

$$\leq |f(x) - f_n(x)| + |f_n(x) - f_n(x_0)| + |f_n(x_0) - f(x_0)| < \frac{\varepsilon}{3} + \frac{\varepsilon}{3} + \frac{\varepsilon}{3} = \varepsilon.$$

This means that the function $f(x)$ is continuous at the point x_0. Since x_0 is an arbitrary point of the set X, we have the continuity of the limit function $f(x)$ on X. □

Example 1c $f_n(x) = \frac{x^n}{n}$, $X = [-1, 1]$

It was shown before (see Example 1s of the boundedness property) that this sequence converges uniformly on $[-1, 1]$. Since each function $f_n(x)$ is continuous on $[-1, 1]$, by the last Theorem, the limit function $f(x)$ must be a continuous function on $[-1, 1]$. It was already found that the limit function is $f(x) \equiv 0$ and it is continuous on $[-1, 1]$.

Remark 1 There is a useful consequence of this Theorem: if a sequence of continuous on a set X functions converges on X to a discontinuous function, then this convergence is not uniform on X.

This conclusion can be extended to the following situation which can be found in practice. Let us suppose that a sequence of continuous functions $f_n(x)$ converges on X to a function $f(x)$ continuous on X. Then we cannot make any conclusion about the character of this convergence. However, it may happen that adding one more point b to X (where b is an accumulation point of X) and considering the set $X_b = X \cup b$, we are able to prove that the sequence $f_n(x)$ is still continuous on X_b and it converges on X_b to a discontinuous function (evidently, this discontinuity occurs at the point b). Under these conditions we can conclude that the convergence is not uniform both on X_b and on X. (Recall that removing or adding a point to a given set does not change the type of convergence on this set.)

In practice, this situation frequently occurs when the original set X is an open (or half-open) interval and the extended set X_b is a half-open (or closed) interval. A typical example of this is the following one.

Example 2c $f_n(x) = x^n$, $X = (0, 1)$

All the functions $f_n(x) = x^n$ are continuous on the interval $X = (0, 1)$. It was shown in Sects. 1 and 2 (Examples 2b and 2u, respectively), that the sequence $f_n(x)$ converges on $(0, 1)$ to the continuous function $f(x) \equiv 0$, but this convergence is not uniform.

Now we have another option to analyze the character of convergence by adding the point $b = 1$ to the interval X and considering the same sequence on the interval $X_b = (0, 1]$. In this case, the limit function becomes discontinuous at 1 (and consequently on $(0, 1]$):

$$f(x) = \begin{cases} 0, & 0 < x < 1 \\ 1, & x = 1 \end{cases}.$$

Therefore, according to the last Theorem, the convergence cannot be uniform on $(0, 1]$, and consequently, according to Remark 1, it is also non-uniform on the original interval $(0, 1)$. For this sequence, the investigation by definition was quite simple, but in more complicated cases, the second approach can be more efficient (less technically involved).

Remark 2 Uniform convergence is an important condition in the last Theorem, without this condition the statement is false. At the same time, uniform convergence is a sufficient, but not necessary condition for continuity of the limit function. We will see it in the following examples.

Example 3c We use the sequence of Example 2c, but consider it on the closed interval: $f_n(x) = x^n$, $X = [0, 1]$

In the Sects. 1 and 2 it was shown that the sequence $f_n(x)$ converges on $[0, 1]$:

$$\lim_{n \to \infty} f_n(x) = \lim_{n \to \infty} x^n = \begin{cases} 0, & 0 \le x < 1; \\ 1, & x = 1 \end{cases} = f(x),$$

but this convergence is not uniform on $[0, 1]$. The last conclusion can be obtained immediately (according to Remark 1), noting that all the functions $f_n(x) = x^n$ are continuous on the interval $[0, 1]$, while the limit function $f(x)$ is discontinuous at the point 1 (and consequently on $[0, 1]$). This shows that uniform convergence is an important condition to ensure the continuity of the limit function.

Example 4c In Sect. 2 (Example 6u) we studied the sequence $f_n(x) = \frac{nx}{1+n^2x^2}$ and showed that it converges non-uniformly to the limit function $f(x) \equiv 0$ on the interval $X = [0, 1]$. Nevertheless, in this case, the limit function is continuous on the interval $[0, 1]$.

Remark 3 A sequence of discontinuous on X functions may converge uniformly on X to a continuous function (of course, it may also happen under non-uniform convergence). An example of this type is given below.

Example 5c $f_n(x) = \frac{1}{n}D(x)$ where $D(x)$ is the Dirichlet function—$D(x) = \begin{cases} 1, & x \in \mathbb{Q}; \\ 0, & x \in \mathbb{I} \equiv \mathbb{R} \backslash \mathbb{Q} \end{cases}$, and the set X can be any interval, for instance, $X = \mathbb{R}$

Evidently, each function $f_n(x)$ is discontinuous at every point of its domain, but the sequence $f_n(x)$ converges on $\mathbb{R}$ to $f(x) \equiv 0$ and the limit function is continuous on $\mathbb{R}$. Moreover, this convergence is uniform, because for all x we have

$$|f_n(x) - f(x)| = \frac{1}{n}D(x) \le \frac{1}{n},$$

and consequently, for $\forall \varepsilon > 0$ choosing the same $N = \left[\frac{1}{\varepsilon}\right]$ for all $x \in \mathbb{R}$, we guarantee that

$$|f_n(x) - f(x)| \le \frac{1}{n} < \varepsilon$$

whenever $n > N$. This means that the convergence is uniform on $\mathbb{R}$.

4.4 Integrability of the Limit Function (Integration by Parameter)

Theorem of Integration *If each function $f_n(x)$ is Riemann integrable on $[a, b]$ and the sequence $f_n(x)$ converges uniformly on $[a, b]$ to the limit function $f(x)$, which is also Riemann integrable on $[a, b]$, then*

$$\lim_{n \to \infty} \int_a^b f_n(x)\,dx = \int_a^b f(x)\,dx = \int_a^b \lim_{n \to \infty} f_n(x)\,dx , \qquad (3.23)$$

that is, the order of limit and integral can be interchanged.

Proof By definition, the uniform convergence of the sequence $f_n(x)$ to $f(x)$ on $[a, b]$ means that for $\forall \varepsilon > 0$ (take for convenience $\varepsilon_1 = \frac{\varepsilon}{b-a}$) there exists N_ε (depending only on ε) such that for $\forall n > N_\varepsilon$ and simultaneously for $\forall x \in [a, b]$ we have

$$|f_n(x) - f(x)| < \frac{\varepsilon}{b-a}. \qquad (3.24)$$

Then, applying (3.24), for $\forall n > N_\varepsilon$ we obtain

$$\left| \int_a^b f_n(x)\,dx - \int_a^b f(x)\,dx \right| = \left| \int_a^b (f_n(x) - f(x))\,dx \right|$$

$$\leq \int_a^b |f_n(x) - f(x)|\,dx < \frac{\varepsilon}{b-a} \int_a^b dx = \frac{\varepsilon}{b-a}(b-a) = \varepsilon.$$

This means that the sequence of numbers $y_n = \int_a^b f_n(x)\,dx$ converges to the number $A = \int_a^b f(x)\,dx$: $\lim_{n \to \infty} y_n = A$. In other words, we have proved that $\lim_{n \to \infty} \int_a^b f_n(x)\,dx = \int_a^b f(x)\,dx$. The last relation can also we written in the form $\lim_{n \to \infty} \int_a^b f_n(x)\,dx = \int_a^b \lim_{n \to \infty} f_n(x)\,dx$, which shows that, under the conditions of the Theorem, it is permissible to change the order of two operations—calculation of the limit and integration. $\quad \square$

Corollary 1 *We can formulate a weaker form of this result, imposing more restrictive conditions on the functions $f_n(x)$, but omitting any requirement on the limit function $f(x)$:*

If each function $f_n(x)$ is continuous on $[a, b]$ and the sequence $f_n(x)$ converges uniformly on $[a, b]$ to the limit function $f(x)$, then

$$\lim_{n \to \infty} \int_a^b f_n(x)\,dx = \int_a^b f(x)\,dx = \int_a^b \lim_{n \to \infty} f_n(x)\,dx .$$

Proof In this case, the integrability of each $f_n(x)$ follows from its continuity, and the integrability of $f(x)$ follows from the Theorem on continuity of the limit function, whose conditions are satisfied in this formulation. Hence, the conditions of the weaker form imply the conditions of the original Theorem.   □

The advantage of this weaker result is that it does not require any information about limit function. This leads to its wide applicability in practice.

Corollary 2 *Let a sequence $f_n(x)$ be defined on an interval $[a, b]$. If*

(1) at every fixed point $x \in [a, b]$ the sequence $f_n(x)$ is monotone,
(2) each function $f_n(x)$ is continuous on $[a, b]$,
(3) the sequence $f_n(x)$ converges pointwise to $f(x)$ on $[a, b]$,
(4) the limit function $f(x)$ is continuous on $[a, b]$,

then

$$\lim_{n \to \infty} \int_a^b f_n(x) \, dx = \int_a^b f(x) \, dx = \int_a^b \lim_{n \to \infty} f_n(x) \, dx.$$

Proof The sequence $f_n(x)$ satisfies all the conditions of Dini's Theorems, and consequently, this sequence converges uniformly to its limit function on $[a, b]$. Therefore, the statement of this Corollary follows from Corollary 1.   □

Example 1i $f_n(x) = \frac{\sin nx}{\sqrt{n}}$, $X = \mathbb{R}$
In Example 3u of Sect. 2, it was shown that this sequence converges uniformly on $\mathbb{R}$ to $f(x) \equiv 0$. Since the functions $f_n(x)$ are continuous on $\mathbb{R}$, we can apply Corollary 1 to obtain

$$\lim_{n \to \infty} \int_a^b f_n(x) \, dx = \int_a^b \lim_{n \to \infty} f_n(x) \, dx = \int_a^b f(x) \, dx = \int_a^b 0 \, dx = 0 .$$

Changing the order of operations we arrive at the same result:

$$\int_a^b f_n(x) \, dx = \int_a^b \frac{\sin nx}{\sqrt{n}} dx = -\frac{1}{\sqrt{n}} \frac{1}{n} \cos nx \Big|_a^b = \frac{1}{n^{3/2}} (\cos na - \cos nb) \underset{n \to \infty}{\to} 0 .$$

Example 2i $f_n(x) = \begin{cases} 2n^2 x, & x \in [0, \frac{1}{2n}] \\ 2n - 2n^2 x, & x \in [\frac{1}{2n}, \frac{1}{n}] \\ 0, & x \in [\frac{1}{n}, 1] \end{cases}$, $X = [0, 1]$.

This sequence was considered in Example 4b of Sect. 1 and Example 4u of Sect. 2. In particular, it was shown there that the sequence converges non-uniformly on $[0, 1]$ to

$f(x) \equiv 0$. It was also verified that the limit of the integrals $\int_0^1 f_n(x)dx$ is not equal to the integral of the limit function $\int_0^1 f(x)dx$:

$$\int_0^1 f_n(x)dx = \int_0^{\frac{1}{2n}} 2n^2x\,dx + \int_{\frac{1}{2n}}^{\frac{1}{n}} (2n - 2n^2x)dx = \frac{1}{4} + \frac{1}{4} = \frac{1}{2} \underset{n \to \infty}{\nrightarrow} 0$$

$$= \int_0^1 0\,dx = \int_0^1 f(x)dx \ .$$

Now we can say that this result is a consequence of the non-uniformity of convergence.

Example 3i $f_n(x) = 2(n + 1)x(1 - x^2)^n$, $X = [0, 1]$

Since $n^k q^n \underset{n \to \infty}{\to} 0$ for $\forall |q| < 1$ and $\forall k \in \mathbb{N}$ (this indeterminate form is easily calculated by L'Hospital's rule) and at the points $x = 0$ and $x = 1$ we have $f_n(0) = f_n(1) = 0$, it follows that this sequence converges to 0 at every point $x \in [0, 1]$: $f_n(x) \overset{[0,1]}{\underset{n \to \infty}{\to}} f(x) \equiv 0$. All the functions of this sequence as well as the limit function are infinitely differentiable on $[0, 1]$. Let us see what happens with the Riemann integrals on $[0, 1]$. On the one hand,

$$\int_0^1 f_n(x)dx = \int_0^1 2(n + 1)x(1 - x^2)^n dx = -(1 - x^2)^{n+1}|_0^1 = 1,$$

but on the other hand, $\int_0^1 f(x)dx = \int_0^1 0\,dx = 0$, that is, there is no relation between the integrals of the sequence functions and the integral of the limit function. Therefore, there is no similarity between the graphs of $f_n(x)$ and the graph of $f(x)$.

Since the functions $f_n(x)$ and the limit function $f(x)$ are continuous on $[0, 1]$, the different results for the integrals indicate that the convergence is not uniform: if it would be uniform, then, according to the Theorem or Corollary 1, equality (3.23) between the integrals should be satisfied. We can also prove this directly: choosing $x_n = \frac{1}{\sqrt{n}}$, we obtain the evaluation

$$|f_n(x_n) - f(x_n)| = 2(n + 1)\frac{1}{\sqrt{n}}\left(1 - \frac{1}{n}\right)^n \underset{n \to \infty}{\to} \infty,$$

which confirm the non-uniform character of convergence.

Remark Uniform convergence is an important condition to guarantee equality (3.23) between the integrals, but it is not necessary condition as it is illustrated in the next examples.

Example 4i $f_n(x) = n^2 x e^{-n^2 x^2}$, $X = [0, 1]$

If $x = 0$, then $f_n(0) = 0$, $\forall n \in \mathbb{N}$ and $\lim\limits_{n \to \infty} f_n(0) = 0$. If $x \neq 0$, then

$$\lim_{n \to \infty} f_n(x) = \lim_{n \to \infty} n^2 x \cdot e^{-n^2 x^2} = \lim_{n \to \infty} \frac{n^2 x}{e^{n^2 x^2}} = 0.$$

Hence, $f(x) \equiv 0$, $\forall x \in [0, 1]$.

Let us investigate the character of convergence of the sequence $f_n(x)$ to the limit function $f(x) \equiv 0$. To do this, we find $\sup\limits_{\forall x \in [0,1]} |f_n(x) - f(x)|$ for $\forall n \in \mathbb{N}$. Since

$$(f_n(x) - f(x))' = f_n'(x) = n^2 e^{-n^2 x^2} - n^2 x \cdot 2n^2 x e^{-n^2 x^2}$$

$$= n^2 e^{-n^2 x^2} \left(1 - 2n^2 x^2 \right) = 0,$$

the critical points are $x = \pm \frac{1}{n\sqrt{2}}$ and only one of them belongs to the given interval: $x_n = \frac{1}{n\sqrt{2}} \in [0, 1]$. Since the derivative changes its sign from positive to negative passing through this point, the critical point $x_n = \frac{1}{n\sqrt{2}}$ is a local maximum of the function $f_n(x)$. Taking into account that $f_n(x)$ is non-negative on $[0, 1]$ and x_n is its unique local extremum in this interval, we conclude that $x_n = \frac{1}{n\sqrt{2}}$ is a global maximum of $f_n(x)$ on $[0, 1]$. Therefore,

$$\sup_{\forall x \in [0,1]} |f_n(x) - f(x)| = f_n(x_n) = f_n\left(\frac{1}{n\sqrt{2}} \right) = n^2 \cdot \frac{1}{n\sqrt{2}} \cdot e^{-n^2 \cdot \frac{1}{2n^2}} = \frac{n}{\sqrt{2}} \cdot e^{-\frac{1}{2}} \not\to 0.$$

This shows that the sequence $f_n(x)$ converges non-uniformly on $[0, 1]$.

Let us see if it is possible to change the order of limit and integral in this example:

$$\lim_{n \to \infty} \int_0^1 f_n(x)\, dx = \lim_{n \to \infty} \int_0^1 n^2 x e^{-n^2 x^2}\, dx = \lim_{n \to \infty} \left(-\frac{1}{2} e^{-n^2 x^2} \Big|_0^1 \right)$$

$$= \lim_{n \to \infty} \left(-\frac{1}{2} e^{-n^2} + \frac{1}{2} \right) = \frac{1}{2} \neq 0 = \int_0^1 0 \cdot dx = \int_0^1 f(x)\, dx.$$

Hence, in this case, the lack of uniform convergence makes it impossible to calculate the limit inside the integral.

Example 5i $f_n(x) = \frac{nx}{1+n^2x^2}$, $X = [0, 1]$

As was shown (see Example 6u in Sect. 2), the sequence $f_n(x)$ converges to $f(x) \equiv 0$ on $[0, 1]$, but this convergence is not uniform. Nevertheless,

$$\lim_{n\to\infty} \int_0^1 f_n(x)\,dx = \lim_{n\to\infty} \int_0^1 \frac{nx}{1+n^2x^2}dx = \lim_{n\to\infty} \frac{1}{2n} \ln\left(1+n^2x^2\right) \Big|_0^1$$

$$= \lim_{n\to\infty} \frac{\ln\left(1+n^2\right)}{2n} = 0 = \int_0^1 0 \cdot dx = \int_0^1 f(x)\,dx,$$

that is, the uniform convergence is a sufficient but not necessary condition for passing to the limit inside the integral.

Example 6i $f_n(x) = \begin{cases} \min(1, \frac{1}{nx}), & x \in (0, 1] \,; \\ 0, & x = 0 \end{cases}$, $X = [0, 1]$

This is one more example similar to Example 5i. In Example 4s (Sect. 4.1) it was shown that this sequence converges non-uniformly on $[0, 1]$ to the function $f(x) \equiv 0$. Calculating the integral of each $f_n(x)$, we get

$$\int_0^1 f_n(x)\,dx = \int_0^{1/n} 1dx + \int_{1/n}^1 \frac{1}{nx}dx = \frac{1}{n} + \frac{1}{n}(\ln 1 - \ln \frac{1}{n}) = \frac{1}{n} + \frac{\ln n}{n}.$$

Therefore,

$$\lim_{n\to\infty} \int_0^1 f_n(x)\,dx = \lim_{n\to\infty} \frac{1+\ln n}{n} = 0 = \int_0^1 0 \cdot dx = \int_0^1 f(x)\,dx.$$

Hence, in this case, it is possible to change the order of the limit and integral, even though the convergence is non-uniform.

Complement: Improper Integral

Although the problem of improper integral is out of scope of this text, as a matter of warning and curiosity, we present the following example that shows that under the same conditions of the Theorem of integrability, the change of the order of the limit and integral is not justified for an improper integral. Consider the sequence $f_n(x) = \begin{cases} \frac{1}{n}, & 0 \le x \le n \\ 0, & x > n \end{cases}$

on the set $X = [0, +\infty)$. Evidently, each function $f_n(x)$ is Riemann integrable on $[0, +\infty)$ (that is, Riemann integrable on any interval $[0, b]$, $\forall b > 0$), because $f_n(x)$ is a continuous on $[0, +\infty)$ function except at the point of jump $x = n$. Also the improper integral of

$f_n(x)$ exists (converges):

$$\int_0^{+\infty} f_n(x)dx = \lim_{b\to\infty} \left(\int_0^n \frac{1}{n}dx + \int_n^b 0dx\right) = \lim_{b\to\infty}(1+0) = 1.$$

Besides, the sequence converges uniformly on $[0, +\infty)$ to $f(x) \equiv 0$, the limit function is Riemann integrable on $[0, +\infty)$ and its improper integral exists. Nevertheless,

$$\lim_{n\to\infty}\int_0^{+\infty} f_n(x)dx = \lim_{n\to\infty} 1 = 1 \neq 0 = \int_0^{+\infty} f(x)dx = \int_0^{+\infty} \lim_{n\to\infty} f_n(x)dx .$$

The problem in this case is caused by non-uniform convergence of the improper integral: to guarantee the validity of interchanging the order of the limit and improper integral, besides the conditions required in the Theorem of integration (the uniform convergence of the sequence $f_n(x)$ to the limit function $f(x)$ on $X = \mathbb{R}$ or $X = [0, +\infty)$ and the integrability of $f_n(x)$ and $f(x)$ on X), some additional conditions on the improper integrals should be satisfied. More specifically: the improper integrals of the sequence functions $f_n(x)$ and of the limit function $f(x)$ should exist, and the convergence of the improper integrals $\int_0^{+\infty} f_n(x)dx$ should be uniform with respect to the set of indices n ($\mathbb{N}$ or $\mathbb{N}_0$). The former condition is satisfied in the considered example, but the latter is not. Indeed, the latter condition means that for $\forall \varepsilon > 0$ there exists the number $A_\varepsilon > 0$ such that for $\forall A > A_\varepsilon$ it follows $\left|\int_A^{+\infty} f_n(x)dx\right| < \varepsilon$ simultaneously for all indices $n \in \mathbb{N}$. It happens that this condition is not satisfied in our example: for $\varepsilon = \frac{1}{2}$ whatever $A > 1$ is used, by choosing the index $n_A = [2A]$ (consequently, $n_A \leq 2A < n_A + 1$), we obtain

$$\int_A^{+\infty} f_{n_A}(x)dx = \int_A^{n_A} \frac{1}{n_A}dx = \frac{n_A - A}{n_A} > \frac{2A - 1 - A}{2A} = \frac{A-1}{2A} \underset{A\to+\infty}{\to} \frac{1}{2} \neq 0.$$

Complement: Integrability with a Stronger Formulation and More Involved Proof

Theorem *If each function $f_n(x)$ is Riemann integrable on $[a, b]$ and the sequence $f_n(x)$ converges uniformly on $[a, b]$, then the limit function $f(x)$ is Riemann integrable on $[a, b]$ and*

$$\lim_{n\to\infty}\int_a^b f_n(x)\,dx = \int_a^b f(x)\,dx = \int_a^b \lim_{n\to\infty} f_n(x)dx .$$

Proof We divide the demonstration into two parts. First we prove that the sequence of numbers $y_n = \int_a^b f_n(x)dx$ is convergent. Indeed, the sequence $f_n(x)$ converges uniformly on $[a, b]$, which means that it is a uniform Cauchy sequence, that is, for $\forall \varepsilon > 0$ (take for convenience $\varepsilon_1 = \frac{\varepsilon}{b-a}$) there exists N_ε (that depends only on ε) such that for $\forall n, m > N_\varepsilon$ we have

$$|f_n(x) - f_m(x)| < \frac{\varepsilon}{b-a}$$

for all $x \in [a, b]$ at once. Then, for the same values of N and $n, m > N$, we obtain

$$|y_n - y_m| = \left| \int_a^b f_n(x)dx - \int_a^b f_m(x)dx \right| \leq \int_a^b |f_n(x) - f_m(x)|\, dx < \int_a^b \frac{\varepsilon}{b-a} dx = \varepsilon,$$

which shows that y_n is a Cauchy sequence. The last condition is equivalent to the convergence of y_n, whose limit we denote by y: $\lim_{n \to \infty} y_n = y$.

In the second part, we show that the limit function $f(x)$ is integrable on $[a, b]$ and that $y = \int_a^b f(x)dx$. To do this, we demonstrate that there exists the limit of the Riemann sums of the function $f(x)$ and this limit is y. Initially, we set the three auxiliary evaluations. First, from the convergence of y_n it follows that for $\forall \varepsilon > 0$ (take for convenience $\varepsilon_1 = \frac{\varepsilon}{3}$) there exists N_{ε_1} such that for $\forall n > N_{\varepsilon_1}$ we have

$$|y_n - y| < \frac{\varepsilon}{3}.$$

Second, we employ the uniform convergence of $f_n(x)$ to $f(x)$ on $[a, b]$: for $\forall \varepsilon > 0$ (choose for convenience $\varepsilon_2 = \frac{\varepsilon}{3(b-a)}$) there exists N_{ε_2} (that depends only on ε_2) such that for $\forall n > N_{\varepsilon_2}$ and simultaneously for all $x \in [a, b]$ we get

$$|f_n(x) - f(x)| < \frac{\varepsilon}{3(b-a)}.$$

Third, we choose some function $f_n(x)$ with $n > N_{\varepsilon_2}$ from the last inequality and consider such partition $P \equiv \{a = x_0 < x_1 < \ldots < x_k = b\}$ of the interval $[a, b]$ that the corresponding Riemann sum $S(P, f_n) \equiv \sum_{i=1}^k f_n(x_i^*)\Delta x_i$ (recall that $\Delta x_i \equiv x_i - x_{i-1}$ and $x_i^* \in [x_{i-1}, x_i]$) is close enough to the integral of $f_n(x)$, namely:

$$\left| S(P, f_n) - \int_a^b f_n(x)dx \right| < \frac{\varepsilon}{3}$$

(this is true for a sufficiently fine partition due to the integrability of $f_n(x)$).

Consider now the Riemann sum of $f(x)$ that corresponds to the same partition P. Employing the last three evaluations, we obtain for $\forall n > \max\{N_{\varepsilon_1}, N_{\varepsilon_2}\}$:

$$|S(P, f) - y| \leq |S(P, f) - S(P, f_n)| + \left| S(P, f_n) - \int_a^b f_n(x)dx \right| + \left| \int_a^b f_n(x)dx - y \right|$$

$$\leq \sum_{i=1}^k \left| f(x_i^*) - f_n(x_i^*) \right| \Delta x_i + \left| S(P, f_n) - \int_a^b f_n(x)dx \right| + |y_n - y|$$

$$< \sum_{i=1}^k \frac{\varepsilon}{3(b-a)} \Delta x_i + \frac{\varepsilon}{3} + \frac{\varepsilon}{3} = \frac{\varepsilon}{3} + \frac{\varepsilon}{3} + \frac{\varepsilon}{3} = \varepsilon.$$

From this evaluation it follows that the Riemann sums $S(P, f)$ tend to y as partitions are refined (as the diameter of partitions approaches 0), which means that $f(x)$ is integrable on $[a, b]$ and y is its Riemann integral. In other words, we have proved that $f(x)$ is integrable and $\lim_{n\to\infty} \int_a^b f_n(x)\, dx = \lim_{n\to\infty} y_n = y = \int_a^b f(x)\, dx$. The last relation can also be written in the form $\lim_{n\to\infty} \int_a^b f_n(x)\, dx = \int_a^b \lim_{n\to\infty} f_n(x)\, dx$, which evidences that the order of the limit and integral is interchangeable under the conditions of the Theorem.

This completes the proof.   □

4.5   Differentiability of the Limit Function (Differentiation by Parameter)

Theorem of Differentiation *Let $f_n(x)$ be a sequence of functions defined on an interval $[a, b]$. Suppose that each $f_n(x)$ is differentiable on $[a, b]$ and its derivative $f_n'(x)$ is Riemann integrable on $[a, b]$. Suppose also that the sequence $f_n(x)$ converges pointwise on $[a, b]$ to $f(x)$, and the sequence $f_n'(x)$ converges uniformly on $[a, b]$ to a continuous function $g(x)$. Then, $f(x)$ is differentiable on $[a, b]$ and*

$$f'(x) = g(x), \forall x \in [a, b].$$

Another manner to express this result is as follows:

$$(\lim_{n\to\infty} f_n(x))' = \lim_{n\to\infty} f_n'(x), \forall x \in [a, b], \tag{3.25}$$

that is, the order of limit and differentiation is interchangeable.

Proof Consider the interval $[a, x]$, where $\forall x \in [a, b]$. On this interval, the functions $f_n'(t)$ satisfy all the conditions of the Theorem of integrability of a sequence of functions (from the previous section). Then, according to the result of that Theorem, we have

$$\int_a^x g(t)\, dt = \int_a^x \lim_{n\to\infty} f_n'(t)\, dt = \lim_{n\to\infty} \int_a^x f_n'(t)\, dt$$

$$= \lim_{n\to\infty} (f_n(x) - f_n(a)) = f(x) - f(a).$$

Since $g(x)$ is a continuous function, its integral with a variable upper limit $\int_a^x g(t)\, dt$ is a differentiable function on $[a, b]$, and, due to the last equality, the function $f(x)$ is also differentiable on $[a, b]$. Then, differentiating both sides of the last formula, we obtain

$$\left(\int_a^x g(t)\, dt \right)' = g(x) = (f(x) - f(a))' = f'(x).$$

Hence, the function $f(x)$ is differentiable and $f'(x) = g(x)$, $\forall x \in [a,b]$. This formula can also be written in the form $\lim\limits_{n\to\infty} f_n'(x) = \left(\lim\limits_{n\to\infty} f_n(x)\right)'$, which states that it is permissible to pass to the limit inside the derivative under the Theorem conditions. $\quad\square$

Corollary *We can formulate a weaker version of this result, which imposes stronger restrictions on the functions $f_n(x)$, but does not need any information about the limit function $g(x)$:*

Suppose that each $f_n(x)$ is continuously differentiable on $[a,b]$, that the sequence $f_n(x)$ converges pointwise on $[a,b]$ to $f(x)$, and the sequence $f_n'(x)$ converges uniformly on $[a,b]$ to $g(x)$. Then, $f(x)$ is differentiable on $[a,b]$ and

$$f'(x) = g(x), \forall x \in [a,b].$$

Proof Indeed, since the functions $f_n'(x)$ are continuous on $[a,b]$ and the sequence $f_n'(x)$ converges uniformly on $[a,b]$, its limit function $g(x)$ is continuous on $[a,b]$ according to the Theorem of continuity. In this way, the condition of continuity of $g(x)$ is immediately recovered and we return to the original formulation. $\quad\square$

Since the weaker result does not require any information about limit function, it is very useful in practice.

Consider some examples which illustrate the result of the Theorem of differentiability and show what may happen when the conditions of the Theorem are not satisfied.

Example 1d $f_n(x) = \frac{\sin nx}{\sqrt{n}}$, $X = \mathbb{R}$

This sequence was analyzed in Sects. 1 and 2 (Examples 3b and 3u, respectively). It was demonstrated that $f_n(x)$ converges uniformly on $\mathbb{R}$ to $f(x) \equiv 0$, and that the sequence of derivatives $f_n'(x) = \sqrt{n}\cos nx$ diverges at every point $x \in \mathbb{R}$. This reveals that the conditions of the uniform convergence of the sequence $f_n(x)$ and differentiability of the functions $f_n(x)$ and $f(x)$ have no effect on the properties of the sequence $f_n'(x)$.

Example 2d $f_n(x) = \sqrt{\frac{1}{n^2} + x^2}$, $X = (-1, 1)$

Evidently, this sequence converges (monotonically) to $f(x) = \sqrt{x^2} = |x|$ at every fixed x, in particular, on $(-1, 1)$. Moreover, this convergence is uniform on $(-1, 1)$ due to the following evaluation:

$$|f_n(x) - f(x)| = \left|\sqrt{\frac{1}{n^2} + x^2} - |x|\right| = \sqrt{\frac{1}{n^2} + x^2} - |x| \le \frac{1}{n} + |x| - |x| = \frac{1}{n} < \varepsilon,$$

which is valid for $\forall n > N = \left[\frac{1}{\varepsilon}\right]$ and for all $x \in (-1, 1)$ at once. Notice that all the functions $f_n(x)$ are infinitely differentiable on $(-1, 1)$, but the limit function does not have a derivative at 0. Hence, the uniform convergence of differentiable funcitons does not guarantee the differentiability of limit function.

Example 3d $f_n(x) = \frac{x^n}{n}, X = [0, 1]$

This is the sequence of Example 2a in Sect. 1 and Example 1s in Sect. 4.1. It was shown there that $f_n(x)$ converges uniformly on $[0, 1]$ to $f(x) \equiv 0$, that follows from the evaluation

$$|f_n(x) - f(x)| = \left|\frac{x^n}{n}\right| \le \frac{1}{n} < \varepsilon,$$

for $\forall n > N = \left[\frac{1}{\varepsilon}\right]$ and simultaneously for all $x \in [0, 1]$. The sequence of derivatives $f_n'(x) = x^{n-1}$, $X = [0, 1]$ (with n instead of $n - 1$) was considered in Sects. 1 and 2 (Examples 2b and 2u, respectively). It was demonstrated that $f_n'(x)$ converges to $g(x) = \begin{cases} 0, & 0 \le x < 1 \\ 1, & x = 1 \end{cases}$, but this convergence is not uniform, which is evidenced by the fact that $f_n'(x)$ are continuous, while $g(x)$ is discontinuous on $[0, 1]$. One of the consequences of this is that at the point $x = 1$ we obtain different values of $f'(x)$ and $g(x)$: $f'(1) = 0 \ne 1 = g(1)$. Thus, in this case, we have the sequence of differentiable functions $f_n(x)$ convergent uniformly to the differentiable function $f(x) \equiv 0$; besides, the sequence of derivatives $f_n'(x)$ also converges (albeit non-uniformly) to the function $g(x)$. However, $f'(x) \ne g(x)$ on the given set.

Example 4d $f_n(x) = \frac{x^n}{n^2}, X = [0, 1]$

This sequence converges uniformly on $[0, 1]$ to $f(x) \equiv 0$ according to the following evaluation:

$$|f_n(x) - f(x)| = \left|\frac{x^n}{n^2}\right| \le \frac{1}{n^2} < \varepsilon,$$

for $\forall n > N = \left[\frac{1}{\sqrt{\varepsilon}}\right]$ and for all $x \in [0, 1]$ at once. The sequence of derivatives $f_n'(x) = \frac{x^{n-1}}{n}$, $X = [0, 1]$ is virtually the same sequence $g_n(x) = \frac{x^n}{n}$ (with $n - 1$ substituted by n) considered in the previous example. Therefore, $f_n'(x)$ converges uniformly on $[0, 1]$ to $g(x) \equiv 0$. Besides, $f_n'(x)$ are continuous functions on $[0, 1]$. Thus, the conditions of the Corollary are satisfied, and consequently $(\lim_{n \to \infty} f_n(x))' = f'(x) = g(x) \equiv 0, \forall x \in [0, 1]$.

Remark Uniform convergence is an important condition to guarantee the Theorem result. At the same time, just like in the previous properties, this condition is sufficient, but not necessary for the possibility to change the order of the limit and derivative. This point is illustrated in the next two examples.

Example 5d $f_n(x) = e^{-n^2x^2}$, $X = [0, 1]$
First we establish pointwise convergence of the sequences $f_n(x)$ and $f_n'(x) = -2n^2x \cdot e^{-n^2x^2}$ and their limit functions:

$$\lim_{n\to\infty} f_n(x) = \lim_{n\to\infty} e^{-n^2x^2} = \left\{ \begin{array}{l} 1, \ x = 0 \\ 0, \ x \in (0, 1] \end{array} \right\} = f(x),$$

$$\lim_{n\to\infty} f_n'(x) = \lim_{n\to\infty} \left(-2n^2x \cdot e^{-n^2x^2} \right) = 0 \equiv g(x).$$

The limit function $f(x)$ has discontinuity at $x = 0$, and consequently, it is not differentiable at this point. Since all the remaining hypotheses of the Theorem of differentiability are satisfied, this behavior of $f(x)$ at $x = 0$ can only be caused by the non-uniformity of convergence of the sequence $f_n'(x)$. Indeed, this fact was already verified in Example 4i of Sect. 4.4:

$$\sup_{\forall x \in [0,1]} \left| f_n'(x) - g(x) \right| = \sup_{\forall x \in [0,1]} 2n^2x \cdot e^{-n^2x^2} \geq 2n^2x \cdot e^{-n^2x^2} \Big|_{x=\frac{1}{n}} = 2n \cdot e^{-1} > \frac{2}{e}.$$

Example 6d $f_n(x) = \frac{1}{n} \ln\left(1 + x^2n^2\right)$, $X = [0, 1]$
Calculating the limits of the sequences $f_n(x)$ and $f_n'(x) = \frac{2nx}{1+n^2x^2}$, we obtain:

$$\lim_{n\to\infty} f_n(x) = \lim_{n\to\infty} \frac{1}{n} \ln\left(1 + x^2n^2\right) = 0 \equiv f(x),$$

$$\lim_{n\to\infty} f_n'(x) = \lim_{n\to\infty} \frac{2nx}{1 + n^2x^2} = 0 \equiv g(x).$$

In this case, for all $\forall x \in [0, 1]$ we have $f'(x) = g(x)$, although the sequence $f_n'(x) = \frac{2nx}{1+n^2x^2}$ converges non-uniformly on $[0, 1]$ (see Example 6u in Sect. 2). This shows that uniform convergence is a sufficient, but not necessary condition for differentiation inside the limit.

Complement: Differentiation with Stronger Formulation and More Involved Proof

Theorem *Let $f_n(x)$ be a sequence of functions defined on an interval $[a, b]$. Suppose that each $f_n(x)$ is differentiable on $[a, b]$, the sequence $f_n'(x)$ converges uniformly on $[a, b]$ to $g(x)$ and the sequence $f_n(x)$ converges at a point $x_0 \in [a, b]$. Then, the sequence $f_n(x)$ converges uniformly on $[a, b]$ to the differentiable function $f(x)$ and $f'(x) = g(x)$, $\forall x \in [a, b]$.*

Proof We divide the proof into the two parts. First we demonstrate that the sequence $f_n(x)$ converges uniformly on $[a, b]$. Since the sequence of numbers $f_n(x_0)$ converges, it is a Cauchy sequence: for $\forall \varepsilon > 0$ (take for convenience $\varepsilon_1 = \frac{\varepsilon}{2}$) there exists N_{ε_1} such that for $\forall n, m > N_{\varepsilon_1}$ it follows

$$|f_n(x_0) - f_m(x_0)| < \frac{\varepsilon}{2}.$$

At the same time, the uniform convergence of the sequence $f_n'(x)$ means that this sequence is a uniform Cauchy sequence: for $\forall \varepsilon > 0$ (choose for convenience $\varepsilon_2 = \frac{\varepsilon}{2(b-a)}$) there exists N_{ε_2} such that for $\forall n, m > N_{\varepsilon_2}$ we have

$$\left| f_n'(x) - f_m'(x) \right| < \frac{\varepsilon}{2(b-a)}$$

for all $x \in [a, b]$ at once. Notice also that the differentiability of $f_n(x)$ on $[a, b]$ implies that (by the mean value theorem applied to the difference $f_n(x) - f_m(x)$) for any two points $t, p \in [a, b]$ there exists a point $s \in (a, b)$ such that

$$(f_n(t) - f_m(t)) - (f_n(p) - f_m(p)) = \left(f_n'(s) - f_m'(s) \right) \cdot (t - p).$$

Employing these three formulas, for $\forall n, m > \max\{N_{\varepsilon_1}, N_{\varepsilon_2}\}$ and for all $x \in [a, b]$, we obtain

$$|f_n(x) - f_m(x)| \leq |(f_n(x) - f_m(x)) - (f_n(x_0) - f_m(x_0))| + |f_n(x_0) - f_m(x_0)|$$

$$= \left| f_n'(s) - f_m'(s) \right| \cdot |x - x_0| + |f_n(x_0) - f_m(x_0)| < \frac{\varepsilon}{2(b-a)} \cdot |x - x_0| + \frac{\varepsilon}{2} \leq \frac{\varepsilon}{2} + \frac{\varepsilon}{2} = \varepsilon.$$

Hence, $f_n(x)$ is a uniform Cauchy sequence, and consequently, it converges uniformly on $[a, b]$: $f_n(x) \underset{n \to \infty}{\overset{[a,b]}{\rightrightarrows}} f(x)$.

In the second part, we prove the statements on differentiability of $f(x)$. In this procedure we use the following plan. Fix an arbitrary point $x \in [a, b]$ and define the sequence of functions $F_n(y)$ on the interval $[a, b]$ by the following formula: $F_n(y) = \begin{cases} \frac{f_n(y) - f_n(x)}{y - x}, & y \neq x \\ f_n'(x), & y = x \end{cases}$. Also define the function $F(y)$ on the interval $[a, b]$ as follows:

$F(y) = \begin{cases} \frac{f(y) - f(x)}{y - x}, & y \neq x \\ g(x), & y = x \end{cases}$. Now our goal is to demonstrate that the functions $F_n(y)$ are

continuous on $[a, b]$ and that the sequence $F_n(y)$ converges uniformly on $[a, b]$ to $F(y)$. This will imply that $F(y)$ is continuous at x and, according to definition of $F(y)$, this will lead to conclusion that $f(x)$ is differentiable at x and $f'(x) = g(x)$.

Following this plan, we note initially some properties of $F_n(y)$ which follow directly from its definition and the properties of $f_n(y)$. First, the differentiability of $f_n(y)$ on

$[a, b]$ implies that $F_n(y)$ is continuous at $\forall y \in [a, b]$, $y \neq x$. Besides, since $f_n(y)$ is differentiable at x, it follows that

$$\lim_{y \to x} F_n(y) = \lim_{y \to x} \frac{f_n(y) - f_n(x)}{y - x} = f_n'(x) = F_n(x),$$

that is, $F_n(y)$ is also continuous at x. Hence, $F_n(y)$ is continuous on $[a, b]$. Second, due to the uniform convergence of $f_n(y)$ to $f(y)$ on $[a, b]$, the sequence $F_n(y)$ converges, at least pointwise, to $F(y)$ at $\forall y \in [a, b]$, $y \neq x$. Besides, since $f_n'(x)$ converges at x to $g(x)$, the sequence $F_n(y)$ also converges at the point x. Therefore, $F_n(y)$ converges, at least pointwise, to $F(y)$ on $[a, b]$.

Now we will show that this convergence is uniform, that is, $F_n(y)$ converges uniformly on $[a, b]$ to $F(y)$. Since $f_n'(x)$ converges uniformly on $[a, b]$, it is a uniform Cauchy sequence, that is, for $\forall \varepsilon > 0$ (take for convenience $\frac{\varepsilon}{2}$) there exists N_ε such that for $\forall n, m > N_\varepsilon$ it follows

$$\left| f_n'(x) - f_m'(x) \right| < \frac{\varepsilon}{2}$$

for all $x \in [a, b]$ at once. Using the last inequality together with the mean value theorem, we obtain for the same N_ε and $n, m > N_\varepsilon$, and simultaneously for all $y \in [a, b]$, $y \neq x$:

$$|F_n(y) - F_m(y)| = \left| \frac{f_n(y) - f_n(x)}{y - x} - \frac{f_m(y) - f_m(x)}{y - x} \right|$$

$$= \left| \frac{1}{y - x} \right| \cdot |(f_n(y) - f_m(y)) - (f_n(x) - f_m(x))|$$

$$= \left| \frac{1}{y - x} \right| \cdot |(f_n'(t) - f_m'(t)) \cdot (y - x)| \leq \left| \frac{1}{y - x} \right| \cdot \frac{\varepsilon}{2} \cdot |y - x| < \varepsilon.$$

Besides, at the point x itself, we have (for the same N_ε and $n, m > N_\varepsilon$)

$$|F_n(x) - F_m(x)| = \left| f_n'(x) - f_m'(x) \right| < \frac{\varepsilon}{2} < \varepsilon.$$

From these evaluations, we conclude that $F_n(y)$ is a uniform Cauchy sequence on $[a, b]$, and consequently, $F_n(y)$ converges uniformly on $[a, b]$. Then, according to the Theorem of continuity, the conditions of continuity of $F_n(y)$ and uniform convergence on $[a, b]$ guarantee that the limit function $F(y)$ is continuous on $[a, b]$ and, in particular, at x. By the definition of $F(y)$, its continuity at x means that

$$f'(x) = \lim_{y \to x} \frac{f(y) - f(x)}{y - x} = \lim_{y \to x} F(y) = F(x) = g(x),$$

that is, $f(x)$ is differentiable at x and $f'(x) = g(x)$.

Recalling that x is an arbitrary point in $[a, b]$, the proof is complete.   □

5   Complement: The Weierstrass Approximation Theorem

The basis of the theory of approximation of functions of a real
variable is a theorem discovered by Weierstrass which was of great
importance in the development of the whole mathematical analysis.
Aleksandr Tilman, 1960

**The Weierstrass Approximation Theorem (Fundamental Theorem of Approx-
imation Theory).** *If $f(x)$ is a continuous on $[a, b]$ function, then there exists a
sequence of polynomials that converges uniformly on $[a, b]$ to $f(x)$.*

Proof We start with the demonstration of the statement for the interval $[0, 1]$ and then
perform a simple generalization to the case of the interval $[a, b]$. To construct the desired
sequence of polynomials we use the Bernstein polynomials $B_n(x)$ defined for a given
function $f(x)$ as follows:

$$B_n(x) = \sum_{k=0}^{n} \binom{n}{k} x^k (1 - x)^{n-k} f\left(\frac{k}{n}\right), x \in [0, 1], n \in \mathbb{N}, \qquad (3.26)$$

where $\binom{n}{k}$ are the binomial coefficients (the number of combinations from n given elements
in unordered groups of k elements):

$$\binom{n}{k} = \frac{n!}{k!(n - k)!}.$$

Our goal is to show that for $\forall \varepsilon > 0$ there exists N such that $|f(x) - B_n(x)| < \varepsilon$
simultaneously for all $x \in [0, 1]$ and for all $n > N$.

We will need a considerable amount of preliminary computations. Recall that, by the
Newton binomial theorem, for any two real numbers p and q we have the following
equality:

$$\sum_{k=0}^{n} \binom{n}{k} p^k q^{n-k} = (p + q)^n, n \in \mathbb{N}. \qquad (3.27)$$

Considering that p is a variable and differentiating the last formula with respect to p, we
obtain

$$\sum_{k=0}^{n} \binom{n}{k} k p^{k-1} q^{n-k} = n(p + q)^{n-1}, n \in \mathbb{N}$$

or

$$\sum_{k=0}^{n} \frac{k}{n}\binom{n}{k}p^k q^{n-k} = p(p+q)^{n-1}. \tag{3.28}$$

Differentiating the last expression one more time, we get

$$\sum_{k=0}^{n} \frac{k^2}{n}\binom{n}{k}p^{k-1}q^{n-k} = p(n-1)(p+q)^{n-2} + (p+q)^{n-1}$$

or

$$\sum_{k=0}^{n} \frac{k^2}{n^2}\binom{n}{k}p^k q^{n-k} = p^2\left(1-\frac{1}{n}\right)(p+q)^{n-2} + \frac{p}{n}(p+q)^{n-1}. \tag{3.29}$$

Now, using $p = x, q = 1 - x$ in (3.27), (3.28), and (3.29) we have

$$\sum_{k=0}^{n}\binom{n}{k}x^k(1-x)^{n-k} = 1, \tag{3.30}$$

$$\sum_{k=0}^{n}\frac{k}{n}\binom{n}{k}x^k(1-x)^{n-k} = x,$$

$$\sum_{k=0}^{n}\frac{k^2}{n^2}\binom{n}{k}x^k(1-x)^{n-k} = x^2\left(1-\frac{1}{n}\right) + \frac{x}{n}.$$

Multiplying the first equality by x^2, the second by $-2x$, and summing these two results with the third equality, we obtain

$$\sum_{k=0}^{n}\left(\frac{k}{n}-x\right)^2\binom{n}{k}x^k(1-x)^{n-k} = \frac{x(1-x)}{n}, \forall x \in [0, 1]. \tag{3.31}$$

Since $f(x)$ is continuous on the compact $[0, 1]$, it is bounded on this interval: $|f(x)| \leq M, \forall x \in [0, 1]$. Moreover, by the Cantor theorem, $f(x)$ is uniformly continuous on $[0, 1]$, which means that for $\forall \varepsilon > 0$ there exists $\delta > 0$ such that $|f(x) - f(y)| < \varepsilon$ for all $|x - y| < \delta$, $x, y \in [0, 1]$. Let us choose N such that $\frac{1}{\sqrt[4]{N}} < \delta$ and $\frac{1}{\sqrt{N}} < \frac{\varepsilon}{4M}$, that is, $N > \max\{\left[\frac{1}{\delta^4}\right], \left[\frac{(4M)^2}{\varepsilon^2}\right]\}$. Now we pick an arbitrary x in $[0, 1]$ and fix it. Multiplying (3.30) by $f(x)$ and subtracting (3.26), we obtain for $\forall n \in \mathbb{N}$

$$f(x) - B_n(x) = \sum_{k=0}^{n}\left(f(x) - f\left(\frac{k}{n}\right)\right)\binom{n}{k}x^k(1-x)^{n-k} = \Sigma_1 + \Sigma_2, \tag{3.32}$$

where Σ_1 is the sum over those values of k that satisfy the inequality

$$\left| \frac{k}{n} - x \right| < \frac{1}{\sqrt[4]{n}} \tag{3.33}$$

(denote these indices by k_1) and Σ_2 is the sum over all the remaining indices k (denoted k_2). If k does not satisfy (3.33), that is, $\left(\frac{k}{n} - x\right)^2 \geq \frac{1}{\sqrt{n}}$, then

$$|\Sigma_2| = \left| \sum_{k_2} \left(f(x) - f\left(\frac{k}{n}\right) \right) \binom{n}{k} x^k (1-x)^{n-k} \right|$$

$$\leq \sum_{k_2} \left(|f(x)| + \left| f\left(\frac{k}{n}\right) \right| \right) \binom{n}{k} x^k (1-x)^{n-k}$$

$$\leq 2M \sum_{k_2} \binom{n}{k} x^k (1-x)^{n-k} = 2M\sqrt{n} \sum_{k_2} \frac{1}{\sqrt{n}} \binom{n}{k} x^k (1-x)^{n-k}$$

$$\leq 2M\sqrt{n} \sum_{k_2} \left(\frac{k}{n} - x\right)^2 \binom{n}{k} x^k (1-x)^{n-k}$$

$$\leq 2M\sqrt{n} \sum_{k=0}^{n} \left(\frac{k}{n} - x\right)^2 \binom{n}{k} x^k (1-x)^{n-k} = 2M\sqrt{n}\frac{x(1-x)}{n} \leq 2M\frac{1}{\sqrt{n}} .$$

If $n > N$, it follows from the inequality $\frac{1}{\sqrt{N}} < \frac{\varepsilon}{4M}$ that $\frac{2M}{\sqrt{n}} < \frac{\varepsilon}{2}$, and consequently, $|\Sigma_2| < \frac{\varepsilon}{2}$.

Besides, if $n \geq N$ and k satisfies (3.33), then, together with $\frac{1}{\sqrt[4]{N}} < \delta$, it results in $\left|\frac{k}{n} - x\right| < \delta$, and consequently, $\left|f(x) - f\left(\frac{k}{n}\right)\right| < \frac{\varepsilon}{2}$. Therefore, we obtain the following evaluation of Σ_1:

$$|\Sigma_1| = \left| \sum_{k_1} \left(f(x) - f\left(\frac{k}{n}\right) \right) \binom{n}{k} x^k (1-x)^{n-k} \right| < \frac{\varepsilon}{2} \sum_{k=0}^{n} \binom{n}{k} x^k (1-x)^{n-k}.$$

Using (3.30) to calculate the last sum, we get $|\Sigma_1| < \frac{\varepsilon}{2}$.

Bringing the estimates of Σ_1 and Σ_2 in formula (3.32), we obtain

$$|f(x) - B_n(x)| \leq |\Sigma_1| + |\Sigma_2| < \frac{\varepsilon}{2} + \frac{\varepsilon}{2} = \varepsilon.$$

Since we have chosen an arbitrary point x in $[0, 1]$ and the last evaluation does not depend on x, this shows that

$$|f(x) - B_n(x)| < \varepsilon$$

simultaneously for all $x \in [0, 1]$ and for all $n > N$, which means the uniform convergence of the sequence $B_n(x)$ to $f(x)$ on $[0, 1]$.

Finally, we generalize the result obtained for $[0, 1]$ to an arbitrary interval $[a, b]$. If $g(x)$ is a continuous function on $[a, b]$, then using the change of the variable $x = a + (b - a)t$, $t \in [0, 1]$ that maps (bijectively) $[0, 1]$ onto $[a, b]$, we transform $g(x)$ into the function $f(t) = g(a + (b - a)t) = g(x)$ continuous on $[0, 1]$. Since for $f(t)$ the Theorem was already proved, for sufficiently large indices n we have $|f(t) - B_n(t)| < \varepsilon$ for all $t \in [0, 1]$ at once. Correspondingly, for $g(x)$ we obtain $\left| g(x) - B_n\left(\frac{x-a}{b-a}\right) \right| < \varepsilon$ simultaneously for all $x \in [a, b]$. Noting that each $B_n\left(\frac{x-a}{b-a}\right)$ is a polynomial (of degree n) of the variable x, the last inequality means the uniform convergence of these polynomials to $g(x)$ on $[a, b]$.

Thus, the Theorem is proved. $\square$

Historical Remarks The approximation theorem was published by Weierstrass in 1885 when he was 70 years old. It seems that the significance of the paper was immediately recognized, since it was translated to French already one year later. Initially Weierstrass and other authors often formulated the approximation theorem in terms of infinite series, but after some years this form was not used anymore. In 1891 Picard presented an alternative proof of Weierstrass' theorems and showed how to extend the results to functions of several variables. Besides the original proof by Weierstrass and approach by Picard, a wide circle of mathematicians were attracted to this theorem and proposed different proofs even in the first three decades after the original Weierstarss' publication. Among these mathematicians one can find such names as Volterra, Lebesgue, de la Vallée Poussin, Mittag-Leffler, and Bernstein, which clearly evidences the importance of the approximation theorem. The version of the proof presented in our text follows the approach proposed by Bernstein in 1912. This proof is relatively simple, does not require any additional knowledge and, at the same time, it is constructive, allowing us to generate the form of polynomials that achieve the desired approximation.

Exercises

1. Determine whether a given sequence of functions is convergent or divergent. If it converges, find the set of convergence and the limit function:
   (1) $f_n(x) = \frac{nx}{1+n+|x|}$;
   (2*) $f_n(x) = \sin nx$;

(3*) $f_n(x) = \cos nx$;

(4) $f_n(x) = x^n - x^{2n}$;

(5) $f_n(x) = \frac{1}{n^2 + x^2}$;

(6) $f_n(x) = \frac{nx}{n^2 + x^2}$;

(7) $f_n(x) = \frac{x^{2n}}{1 + x^{2n}}$;

(8) $f_n(x) = e^{nx}$;

(9) $f_n(x) = \left(1 + \frac{x}{n}\right)^n$;

(10) $f_n(x) = \frac{1}{nx} \ln \frac{1}{nx}$;

(11) $f_n(x) = x \arctan nx$;

(12) $f_n(x) = \sqrt{n^2 x + n} - \sqrt{n^2 x}$;

(13) $f_n(x) = \frac{n^2 x^2}{n^2 + x^2}$.

2. Investigate the character of convergence of a given sequence on different intervals:

(1) $f_n(x) = \frac{x^2 n^2}{x^4 n^4 + 9}$, (a) $X_1 = [0, 1]$, (b) $X_2 = [1, +\infty)$;

(2) $f_n(x) = \frac{x^3 n^3}{x^6 + n^6}$, (a) $X_1 = [0, 1]$, (b) $X_2 = [-10, 10]$, (c) $X_3 = [1, +\infty)$;

(3) $f_n(x) = nx e^{-n^2 x^2}$, (a) $X_1 = [0, 1]$, (b) $X_2 = [-1, 0]$, (c) $X_3 = [1, +\infty)$;

(4) $f_n(x) = \frac{2n}{x} e^{-\frac{2n}{x}}$, (a) $X_1 = (0, 10]$, (b) $X_2 = [10, +\infty)$;

(5) $f_n(x) = \frac{x}{\sqrt{n}} e^{-\frac{x}{\sqrt{n}}}$, (a) $X_1 = [0, 1]$, (b) $X_2 = [1, +\infty)$;

(6) $f_n(x) = \frac{x}{n} \ln \frac{x}{n}$, (a) $X_1 = (0, 2]$, (b) $X_2 = [2, +\infty)$;

(7) $f_n(x) = \frac{1}{nx} \ln \frac{1}{nx}$, (a) $X_1 = (0, 2)$, (b) $X_2 = [2 + \infty)$;

(8) $f_n(x) = \frac{1}{n} \ln(1 + n^2 x^2)$, (a) $X_1 = [0, 1]$, (b) $X_2 = [1, +\infty)$;

(9) $f_n(x) = \frac{1}{\sqrt{n+1}} \ln(2 + nx^2)$, (a) $X_1 = [0, 1]$, (b) $X_2 = [1, +\infty)$;

(10) $f_n(x) = \arctan xn$, (a) $X_1 = (0, 5)$, (b) $X_2 = [5, +\infty)$;

(11) $f_n(x) = \frac{1}{n} \arctan nx$, (a) $X_1 = [-1, 1]$, (b) $X_2 = [1, +\infty)$, (c)$X_3 = \mathbb{R}$;

(12*) $f_n(x) = x \arctan xn$, (a) $X_1 = [0, +\infty)$, (b) $X_2 = \mathbb{R}$;

(13) $f_n(x) = \sin \frac{x}{n}$, (a) $X_1 = [-1, 1]$, (b) $X_2 = [1, +\infty)$;

(14*) $f_n(x) = \sqrt{n^2 x + n} - \sqrt{n^2 x}$, (a) $X_1 = (0, 2)$, (b) $X_2 = [2, +\infty)$;

(15) $f_n(x) = x^n - x^{n+1}$, (a) $X_1 = [0, 1]$, (b) $X_2 = (-1, 0]$;

(16) $f_n(x) = x^n - x^{2n}$, (a) $X_1 = [0, 1]$, (b) $X_2 = (-1, 0]$.

3. What can we say about the sequence $h_n(x) = f_n(x) + g_n(x)$ if:

(1) $f_n(x)$ and $g_n(x)$ converge uniformly on X;

(2) $f_n(x)$ converges uniformly on X and $g_n(x)$ converges non-uniformly on X;

(3) $f_n(x)$ and $g_n(x)$ converge non-uniformly on X.

4*. What can we say about the sequence $h_n(x) = f_n(x)g_n(x)$ if:

(1) $f_n(x)$ and $g_n(x)$ converge uniformly on X;

(2) $f_n(x)$ converges uniformly on X and $g_n(x)$ converges non-uniformly on X;

(3) $f_n(x)$ and $g_n(x)$ converge non-uniformly on X;

Hint: try to construct different examples; if it would be hard, try solve some (or all) examples of Exercise 5.

5. Investigate the character of convergence of the sequences $f_n(x)$, $g_n(x)$ and $h_n(x) = f_n(x)g_n(x)$ on the indicated sets:

(1) $f_n(x) = g_n(x) = \frac{x}{1+n^2x^2}$, $X = \mathbb{R}$;

(2) $f_n(x) = \frac{x}{1+n^2x^2}$, $g_n(x) = \frac{x}{n^2+x^2}$, $X = \mathbb{R}$;

(3) $f_n(x) = g_n(x) = x^2 + \frac{1}{n^2}$, $X = \mathbb{R}$;

(4) $f_n(x) = x^2 + \frac{1}{n^2}$, $g_n(x) = x + \frac{1}{n}$, $X = \mathbb{R}$;

(5) $f_n(x) = \frac{x}{1+n^2x^2}$, $g_n(x) = \frac{nx}{1+n^2x^2}$, $X = [0, 1]$;

(6) $f_n(x) = \frac{x^2}{1+n^4x^4}$, $g_n(x) = \frac{n^3x}{1+n^4x^4}$, $X = [0, 1]$;

(7) $f_n(x) = \frac{nx}{1+n^2x^2}$, $g_n(x) = \frac{1}{nx}$, $X = (0, 1]$;

(8) $f_n(x) = \frac{n^2x^2}{1+n^2x^2}$, $g_n(x) = \frac{1}{n^2x}$, $X = (0, 1]$;

(9) $f_n(x) = x \sin \frac{x}{n}$, $g_n(x) = \frac{1}{nx}$, $X = (0, +\infty)$;

(10) $f_n(x) = \sin \frac{x}{n}$, $g_n(x) = \frac{x}{n}$, $X = \mathbb{R}$;

(11) $f_n(x) = \frac{x}{1+n^4x^2}$, $g_n(x) = \frac{n^2x}{1+n^4x^2}$, $X = \mathbb{R}$;

(12) $f_n(x) = \frac{nx}{1+n^4x^2}$, $g_n(x) = \frac{n^3x}{1+n^4x^2}$, $X = \mathbb{R}$;

(13) $f_n(x) = x \arctan \frac{x}{n}$, $g_n(x) = \frac{1}{nx}$, $X = (0, +\infty)$;

(14) $f_n(x) = nx \arctan \frac{1}{nx}$, $g_n(x) = \frac{1}{n^2x}$, $X = (0, 1]$.

6. Verify if the conditions of Dini's theorem are satisfied and investigate the character of convergence of the sequence $f_n(x)$ on X:

    (1) $f_n(x) = (1 - x^2)^n$, $X = [0, 1]$;

    (2) $f_n(x) = (1 - x^2)^n$, $X = (0, 1]$;

    (3) $f_n(x) = \frac{(-1)^n}{n} \operatorname{sgn} x$, $X = \mathbb{R}$;

    (4) $f_n(x) = \frac{2n+3}{n} \operatorname{sgn} x$, $X = [-1, 1]$;

    (5) $f_n(x) = \cot^n x$, $X = [\frac{\pi}{4}, \frac{\pi}{2}]$;

    (6) $f_n(x) = \cot^n x$, $X = (\frac{\pi}{4}, \frac{\pi}{2}]$;

    (7) $f_n(x) = x(1 - x)^n$, $X = [0, 1]$;

    (8) $f_n(x) = x^n - x^{n+1}$, $X = [0, 1]$;

    (9) $f_n(x) = x^n - x^{n+1}$, $X = (-1, 0]$;

    (10) $f_n(x) = x^n - x^{2n}$, $X = [0, 1]$;

    (11) $f_n(x) = \frac{2n^2x^2}{1+n^4x^4}$, $X = [-1, 1]$;

(12*) $f_n(x) = \frac{nx^2}{1+n^4x^4}$, $X = \mathbb{R}$.

7. Verify if the functions $f_n(x)$ and the corresponding limit functions of the sequence are bounded on the set X. Investigate the character of convergence of the sequence $f_n(x)$ on X:

    (1) $f_n(x) = \frac{1}{nx}$, $X = (0, +\infty)$;

    (2) $f_n(x) = \sqrt{n^2x + n} - \sqrt{n^2x}$, $X = (0, +\infty)$;

    (3) $f_n(x) = n \arctan \frac{1}{nx}$, $X = (0, +\infty)$;

    (4) $f_n(x) = x^n - x^{2n}$, $X = [0, 1]$;

    (5) $f_n(x) = \frac{n^4x}{1+n^4x^2}$, $X = [0, 1]$;

    (6) $f_n(x) = \frac{n^2x}{1+n^4x^2}$, $X = [0, 1]$;

    (7) $f_n(x) = n \sin \frac{1}{n^2x^2}$, $X = (0, 1]$;

(8) $f_n(x) = n^2 \sin \frac{1}{n^2 x^2}$, $X = (0, 1]$;

(9) $f_n(x) = \frac{1}{n^2} + \frac{1}{x^2}$, $X = (0, +\infty)$;

(10) $f_n(x) = \frac{1}{n^2 x^2} + \frac{1}{x^2}$, $X = (0, +\infty)$;

(11) $f_n(x) = \frac{1}{n} \sin \frac{1}{n^2 x^2}$, $X = (0, 1]$.

8. Investigate the character of convergence of $f_n(x)$ on the given set X and verify if the iterated limits $\lim_{x \to a} \lim_{n \to \infty} f_n(x)$ and $\lim_{n \to \infty} \lim_{x \to a} f_n(x)$ are equal:

(1) $f_n(x) = \frac{n^4 x^2 - 5}{n^4 x^3 + 1}$, $X = (0, 1]$, $a = 0$;

(2) $f_n(x) = \frac{n^2 x^2 + 2}{n^4 x^4 + 1}$, $X = (0, +\infty)$, $a = 0$;

(3) $f_n(x) = \frac{2n^2 x^2 + (-1)^n}{n^4 x^4 + 1}$, $X = (0, +\infty)$, $a = 0$;

(4) $f_n(x) = \frac{n^2 x - 3}{n^4 x^2 + 2}$, $X = (0, 1]$, $a = 0$;

(5) $f_n(x) = \frac{n^2 x^2 + 2}{n^4 x^4 + 1} \operatorname{sgn} x$, $X = [-1, 1]$, $a = 0$;

(6) $f_n(x) = \frac{n}{x} \sin \frac{x}{n}$, $X = (0, +\infty)$, $a = 0$;

(7) $f_n(x) = nx \sin \frac{1}{nx}$, $X = (0, 1]$, $a = 0$;

(8) $f_n(x) = \sin \frac{1}{nx}$, $X = (0, 1]$, $a = 0$;

(9) $f_n(x) = nx \arctan \frac{1}{nx}$, $X = (0, 1]$, $a = 0$;

(10) $f_n(x) = (-1)^n \arctan \frac{1}{nx}$, $X = (0, 1]$, $a = 0$;

(11) $f_n(x) = \begin{cases} 0, & x = 0 \\ n|x| \arctan \frac{1}{nx}, & 0 < |x| \le 1 \end{cases}$, $X = [-1, 1]$, $a = 0$;

(12) $f_n(x) = nx \arctan \frac{1}{n^2 x^2}$, $X = (0, 1]$, $a = 0$.

9. Verify if the functions $f_n(x)$ and the corresponding limit functions of the sequence are continuous on the set X. Investigate the character of convergence of the sequence $f_n(x)$ on X:

(1) $f_n(x) = \frac{nx^2}{1 + n^4 x^4}$, $X = \mathbb{R}$;

(2) $f_n(x) = \frac{nx}{1 + n^4 x^4}$, $X = \mathbb{R}$;

(3) $f_n(x) = \frac{1}{n^2} + \frac{1}{x^2}$, $X = (0, +\infty)$;

(4) $f_n(x) = \frac{1}{n^2 x^2} + \frac{1}{x^2}$, $X = (0, +\infty)$;

(5) $f_n(x) = \begin{cases} 0, & x = 0 \\ \frac{1}{n^2} + \frac{1}{x^2}, & x > 0 \end{cases}$, $X = [0, +\infty)$;

(6) $f_n(x) = \begin{cases} 0, & x = 0 \\ \frac{1}{n^2 x^2} + \frac{1}{x^2}, & x > 0 \end{cases}$, $X = [0, +\infty)$;

(7) $f_n(x) = x^n - x^{2n}$, $X = [0, 1]$;

(8) $f_n(x) = \arctan nx$, $X = \mathbb{R}$;

(9) $f_n(x) = \frac{1}{n} \operatorname{sgn} x$, $X = \mathbb{R}$;

(10) $f_n(x) = \frac{|x|}{n}$, $X = \mathbb{R}$;

(11) $f_n(x) = \frac{1}{n} \widetilde{D}(x) = \frac{1}{n} \cdot \begin{cases} 1, & x \in \mathbb{Q} \\ -1, & x \in \mathbb{I} \end{cases}$, $X = \mathbb{R}$;

(12) $f_n(x) = \frac{x}{n}\tilde{D}(x) = \frac{x}{n} \cdot \begin{cases} 1, & x \in \mathbb{Q} \\ -1, & x \in \mathbb{I} \end{cases}$, $X = \mathbb{R}$;

(13) $f_n(x) = \sin\frac{\pi n^2 x^2}{2n^2 x^2 + 4}$, $X = \mathbb{R}$.

10. Verify the character of convergence of the sequence $f_n(x)$ on X and check the possibility to interchange the limit of the sequence and the integration:

(1) $f_n(x) = \frac{nx}{1+n^4 x^2}$, $X = [0, 1]$;

(2) $f_n(x) = \frac{n^2 x}{1+n^4 x^2}$, $X = [0, 1]$;

(3) $f_n(x) = \frac{n^2 x}{1+n^4 x^4}$, $X = [0, 1]$;

(4) $f_n(x) = \frac{nx}{1+n^4 x^4}$, $X = [0, 1]$;

(5) $f_n(x) = \arctan nx$, $X = [0, 1]$;

(6) $f_n(x) = \frac{1}{\sqrt{n}}\arctan nx$, $X = [0, 1]$;

(7) $f_n(x) = x\sqrt{n}e^{-n^2 x^2}$, $X = [0, 1]$;

(8) $f_n(x) = 2nxe^{-n^2 x^2}$, $X = [0, 1]$;

(9) $f_n(x) = 2n^2 xe^{-n^2 x^2}$, $X = [0, 1]$.

11*. Demonstrate that the functions $f_n(x) = \frac{1}{n}\tilde{D}(x) = \frac{1}{n} \cdot \begin{cases} 1, & x \in \mathbb{Q} \\ -1, & x \in \mathbb{I} \end{cases}$ and $g_n(x) =$

$\frac{x}{n}\tilde{D}(x) = \frac{x}{n} \cdot \begin{cases} 1, & x \in \mathbb{Q} \\ -1, & x \in \mathbb{I} \end{cases}$ are nowhere differentiable, but their limit functions are everywhere differentiable. Analyze the character of convergence of these two sequences on $\mathbb{R}$.

12. Verify the character of convergence of the sequences $f_n(x)$ and $f_n'(x)$ on X and check the possibility to interchange the limit of the sequence and the differentiation.

(1) $f_n(x) = x \arctan nx$, $X = \mathbb{R}$;

(2) $f_n(x) = \arctan nx$, $X = \mathbb{R}$;

(3) $f_n(x) = \frac{x}{n}\arctan nx$, $X = \mathbb{R}$;

(4) $f_n(x) = \frac{x}{n}\arctan nx$, $X = [-1, 1]$;

(5) $f_n(x) = \frac{1}{n}\arctan nx$, $X = \mathbb{R}$;

(6) $f_n(x) = \frac{\cos nx}{n}$, $X = \mathbb{R}$;

(7) $f_n(x) = \frac{\sin nx}{\sqrt{n}}$, $X = \mathbb{R}$;

(8) $f_n(x) = e^{-x^4 n^4}$, $X = [-1, 1]$;

(9) $f_n(x) = e^{-x^4 n^4}$, $X = (0, +\infty)$;

(10) $f_n(x) = \frac{1}{n}e^{-x^2 n^2}$, $X = \mathbb{R}$;

(11) $f_n(x) = xe^{-\frac{x^2}{n^4}}$, $X = \mathbb{R}$;

(12*) $f_n(x) = x^n - x^{n+1}$, $X = [0, 1]$;

(13) $f_n(x) = x^n - x^{2n}$, $X = [0, 1]$;

(14) $f_n(x) = \frac{x}{1+n^2 x^2}$, $X = [-1, 1]$;

(15) $f_n(x) = (1 - x^2)^n$, $X = [-1, 1]$;

(16) $f_n(x) = \frac{(1-x)^n}{n}$, $X = [0, 1]$;

(17) $f_n(x) = n \arctan \frac{x}{n}$, $X = \mathbb{R}$.

13*. Construct an example of the sequence with the following property:

(1) a sequence $f_n(x)$ converges non-uniformly on X, but its subsequence converges uniformly on X;

(2) a sequence $f_n(x)$ converges on X, but it does not converge uniformly on any subinterval of X;

(3) a sequence $f_n(x)$ converges uniformly on X to $f(x)$, but $f_n^2(x)$ does not converge uniformly on X;

(4) a sequence of continuous functions converges on X to a continuous function, but this convergence is non-uniform on X.

14*. Verify whether the following statement is true or false:

(1) if $f_n(x)$ and $g_n(x)$ converge uniformly on X to bounded functions $f(x)$ and $g(x)$, then $f_n(x)g_n(x)$ converges uniformly on X to $f(x)g(x)$; what happens if the condition of boundedness is dropped?

(2) if a sequence of unbounded and discontinuous functions converges on X to unbounded and discontinuous function, then the convergence is uniform on X; can we state that this convergence is non-uniform?

(3) if a sequence of uniformly continuous on X functions converges on X, then the limit function is continuous on X;

(4) if functions $f_n(x)$ are continuously differentiable on X and the sequence $f_n'(x)$ converges uniformly on X to a continuous function, then $f_n(x)$ converges at least at one point;

(5) if a sequence of non-integrable on X functions converges uniformly on X, then the limit function is also non-integrable on X.

Series of Functions

4

The cases of action at a distance are becoming, in a physical point of view, daily more and more important. Sound, light, electricity, magnetism, gravitation, present them as a series.
Michael Faraday, 1857

1   Pointwise Convergence and Introductory Examples

Newton understood by analysis the investigation of equations by means of infinite series. In other words, Newton's basic discovery was that everything had to be expanded in infinite series.
Vladimir Arnold, 1989

In the same way as the theory of series of numbers is based on the theory of sequences of numbers, the methods and results of sequences of functions set the stage for development of the theory of series of functions. This logic connection between sequences and series follows from the fact that the initial definitions and fundamental concepts of series are introduced by means of sequences: a series is usually defined as a sum of all the elements of a given sequence and the convergence (of any kind) of a series is reduced to the convergence (of the corresponding kind) of the sequence of its partial sums. We have already seen how it works for series of numbers and this is also true for series of functions.

Electronic Supplementary Material The online version of this article (https://doi.org/10.1007/978-3-030-79431-6_4) contains supplementary material, which is available to authorized users.

For this reason, the theory of series is intimately tied to the theory of sequence, and many fundamental results about series have their counterparts in statements on sequences.

Another natural root of the theory of series of functions is the series of numbers, especially when the question is the behavior of a series of functions at individual/specific points. However, in more important topics, such as the type of convergence on sets and analytic properties of sums of series, this theory goes far beyond the results of series of numbers.

The relationship between series and sequences of functions determines a natural way for developing the theory of series: in a large part, it follows the topics of the theory of sequences. In the major part of this chapter, and especially in the first section, we adopt this natural line of presentation of material on series of functions and use systematically the results of sequences of functions to formulate (and prove) the corresponding results regarding series. It will be clear from the context (and from the notations utilized) what kind of series (numbers or functions) we are considering. For this reason, frequently we will shorten the name to a series, implying its type implicitly.

▶ **Definition** (**Series of Functions**) Let $u_n(x)$, $n \in \mathbb{N}$ be a sequence of function defined on a set $X \subset \mathbb{R}$. The corresponding *series of functions* is an infinite sum of all the terms of the sequence $u_n(x)$ in the order prescribed by the sequence:

$$u_1(x) + u_2(x) + \ldots + u_n(x) + \ldots.$$

The common notation is similar to that used for series of numbers:

$$u_1(x) + u_2(x) + \ldots + u_n(x) + \ldots \equiv \sum_{n=1}^{\infty} u_n(x).$$

The *domain of a series* is X (the domain of the original sequence). If the domain X is not indicated explicitly, it is considered the largest set where all the functions $u_n(x)$ are defined. A function $u_n(x)$ is called the *general term of a series*.

If the original sequence is defined on the extended set of indices $\mathbb{N}_0 = \{k_1, k_2 = k_1 + 1, \ldots, k_{i+1} = k_i + 1, \ldots\}$, $k_1 \in \mathbb{Z}$, then the series follows the same variation of indices:

$$u_{k_1}(x) + u_{k_1+1}(x) + \ldots + u_n(x) + \ldots \equiv \sum_{n=k_1}^{\infty} u_n(x).$$

In what follows, for simplicity of notation and exposition of material, we will use series with natural indices, but the same concepts and results are valid for series with indices varying in $\mathbb{N}_0$. For this reason, we will also use the notations without any specification of the variation of indices.

> ▶ **Definition** (**Partial Sum and Remainder**) The sum of the first n elements of a series $\sum_{n=1}^{\infty} u_n(x)$ is called a *partial sum*:
>
> $$f_n(x) = \sum_{k=1}^{n} u_k(x)$$
>
> and the sum of the remaining terms is called a *remainder of the series*:
>
> $$r_n(x) = \sum_{k=n+1}^{\infty} u_k(x).$$

> ▶ **Definition** (**Convergence at a Point and Sum of a Series**) Take an arbitrary point $x_0 \in X$ and fix it. If the corresponding numerical sequence of the partial sums $f_n(x_0)$ converges, then we say that the *series $\sum_{n=1}^{\infty} u_n(x)$ is convergent at the point x_0* and the value of the limit is called the *sum of the series at the point x_0*:
>
> $$\sum_{n=1}^{\infty} u_n(x_0) = \lim_{n \to \infty} f_n(x_0).$$
>
> Otherwise, the *series is said to be divergent at x_0*. Notice that this concept can be equivalently defined employing the convergence of series of numbers: if the series of numbers $\sum_{n=1}^{\infty} u_n(x_0)$ converges, we say that the series $\sum_{n=1}^{\infty} u_n(x)$ converges at the point x_0.

> ▶ **Definition** (**Pointwise Convergence on a Set**) If a series $\sum_{n=1}^{\infty} u_n(x)$ converges (as a series of numbers) at every fixed point $x \in X$, then we say that the *series $\sum_{n=1}^{\infty} u_n(x)$ converges pointwise on X* (or it is *pointwise convergent on X*). Otherwise, the *series is divergent on X*. The set X on which the series is pointwise convergent is called the *set of convergence*. When it is clear that we are considering the pointwise convergence, we can simply say that a series converges on X.

If a series $\sum_{n=1}^{\infty} u_n(x)$ converges on X, it is evident that the value of its sum (the value of the limit of the sequence of partial sums $f_n(x)$) depends on the chosen point $x \in X$ in a way that the sum represents a function of x:

$$\sum_{n=1}^{\infty} u_n(x) = \lim_{n \to \infty} f_n(x) = f(x).$$

Remark Just as in the case of a series of numbers, we can give a formal rigorous definition of series and its convergence as follows. A series of functions $\sum_{n=1}^{\infty} u_n(x)$ is a pair of two sequences of functions $u_n(x)$ and $f_n(x)$, where the former is an original sequence with the domain X and the latter (called the sequence of partial sums) is defined by the formula $f_n(x) = \sum_{k=1}^{n} u_k(x)$. If at the point $x_0 \in X$ the numerical sequence $f_n(x_0)$ converges to $f(x_0)$, we say that the series $\sum_{n=1}^{\infty} u_n(x)$ converges at x_0 to the sum $f(x_0)$ and write $\sum_{n=1}^{\infty} u_n(x_0) = f(x_0)$. If the numerical sequence $f_n(x)$ converges at every $x \in X$ to $f(x)$, we say that the series $\sum_{n=1}^{\infty} u_n(x)$ converges pointwise on X to the sum $f(x)$ and write $\sum_{n=1}^{\infty} u_n(x) = f(x)$.

Nevertheless, we prefer more intuitive and comprehensible definitions of a series and its convergence given above. It does not affect in any way the development of the theory.

▶ **Definition (Absolute Convergence at a Point and on a Set)** If the series $\sum_{n=1}^{\infty} |u_n(x)|$ converges at a point $x_0 \in X$, then the original *series* $\sum_{n=1}^{\infty} u_n(x)$ *converges absolutely at* x_0. If the series $\sum_{n=1}^{\infty} |u_n(x)|$ converges at every point $x \in X$, then the original *series* $\sum_{n=1}^{\infty} u_n(x)$ *converges absolutely on* X.

Consider now some introductory examples of convergent and divergent series of functions. We make use of sequences of functions already studied in the previous Chapter.

Example 1a $\sum_{n=0}^{\infty} x^n$, $X = \mathbb{R}$
For any fixed x this is a geometric series that converges (absolutely) to $\frac{1}{1-x}$ when $|x| < 1$, and diverges when $|x| \geq 1$. Therefore, its set of convergence is the interval $(-1, 1)$.

Example 2a $\sum_{n=1}^{\infty} \frac{x^n}{n}$, $X = \mathbb{R}$
For any $|x| < 1$ this series converges (absolutely) according to the Comparison test for series of numbers: $\frac{|x|^n}{n} \leq |x|^n$, $\forall n \in \mathbb{N}$. For $x = -1$ this is the alternating harmonic series $\sum_{n=1}^{\infty} \frac{(-1)^n}{n}$ convergent (conditionally) by Leibniz's test. For $x = 1$ this is the divergent harmonic series $\sum_{n=1}^{\infty} \frac{1}{n}$. Finally, if $|x| > 1$, the general term does not approach 0: $\frac{x^n}{n} \underset{n \to \infty}{\not\to} 0$ and, by the Divergence test, the series diverges. Hence, the set of convergence is the interval $[-1, 1)$.

Example 3a $\sum_{n=0}^{\infty} \frac{x^n}{n!}$, $X = \mathbb{R}$

This series converges (absolutely) on all the real axis $\mathbb{R}$. Indeed, using D'Alembert's test,

we have: $\left| \frac{u_{n+1}}{u_n} \right| = \left| \frac{\frac{x^{n+1}}{(n+1)!}}{\frac{x^n}{n!}} \right| = |x| \cdot \frac{1}{n+1} \underset{n \to \infty}{\to} 0$ for every $x \in \mathbb{R}$.

Example 4a $\sum_{n=1}^{\infty} (xn)^n$, $X = \mathbb{R}$

For any $x \neq 0$, the general term of this series does not tend to 0 (more precisely, the sequence $(xn)^n$ is divergent), and, by the Divergence test, the series diverges. Hence, the set of convergence is the only point $x = 0$.

Example 5a $\sum_{n=1}^{\infty} (1 + x^2)n$, $X = \mathbb{R}$

This series diverges at any $x \in \mathbb{R}$, because the sequence of the general term $(1 + x^2)n$ diverges at any $x \in \mathbb{R}$.

Now we turn to the series that lead to some unexpected results, similar to those we have seen in the previous Chapter.

Example 1b A series of bounded functions converges to unbounded sum: $\sum_{n=0}^{\infty} x^n$, $X = (-1, 1)$.

The general term $u_n(x) = x^n$ is a bounded function on $(-1, 1)$, but the sum of the series $f(x) = \frac{1}{1-x}$ is unbounded on the same interval.

Example 2b A series of continuous functions has a discontinuous sum: $\sum_{n=1}^{\infty} (1 - x)x^n$, $X = [0, 1]$.

Observing that $1 - x$ is a factor independent from n, we can immediately reduce this series to the geometric one: $\sum_{n=1}^{\infty} (1 - x)x^n = (1 - x) \sum_{n=1}^{\infty} x^n = (1 - x)\frac{x}{1-x} = x$, $\forall x \in [0, 1)$. The original series also converges at the point $x = 1$, since all its terms

vanish. Therefore, the series converges on $[0, 1]$ to the sum $f(x) = \begin{cases} x, & x \in [0, 1) \\ 0, & x = 1 \end{cases}$. It

remains to note that each function $(1 - x)x^n$ is continuous on $[0, 1]$, but the sum of the series is discontinuous at $x = 1$.

Example 3b A convergent series has a differentiable general term, but the series of derivatives diverges: $\sum_{n=1}^{\infty} \frac{\sin n^2 x}{n^2}$, $X = \mathbb{R}$.

For the absolute value of the general term we have the elementary estimate $\left| \frac{\sin n^2 x}{n^2} \right| \leq \frac{1}{n^2}$, which shows that the series converges absolutely at every $x \in \mathbb{R}$ according to the Comparison test for numerical series (recall that the $p = 2$-series $\sum_{n=1}^{\infty} \frac{1}{n^2}$ is convergent). The functions $u_n(x) = \frac{\sin n^2 x}{n^2}$ are infinitely differentiable on $\mathbb{R}$, but the series of derivatives $\sum_{n=1}^{\infty} u_n'(x) = \sum_{n=1}^{\infty} \cos n^2 x$ diverges at any x, since the general term $\cos n^2 x$ does not approaches 0. Indeed, for $x = 2m\pi, m \in \mathbb{Z}$ this is evident, and for all other x this can be demonstrated using the contradiction method. Let us suppose, by absurd, that $\cos n^2 x$

tends to 0. Then, for $\forall n > N$ we have $|\cos n^2 x| < \frac{1}{4}$. Using the trigonometric formula of the double argument $\cos 4a = 2(2\cos^2 a - 1)^2 - 1$, for indices $2n > N$ we obtain:

$$|\cos(2n)^2 x| = |\cos 4n^2 x| = |2(2\cos^2 n^2 x - 1)^2 - 1| = |8\cos^2 n^2 x(\cos^2 n^2 x - 1) + 1|$$

$$= 1 - 8\cos^2 n^2 x(1 - \cos^2 n^2 x) \geq 1 - 8\cos^2 n^2 x > 1 - 8 \cdot \frac{1}{16} = \frac{1}{2}$$

that contradicts the inequality $|\cos n^2 x| < \frac{1}{4}$ which should be valid for $\forall n > N$, in particular, for $2n$. Therefore, the assumption that $\cos n^2 x$ converges to 0 is false.

Example 4b Integral of the sum of a series is different from the series of integrals of the general term: $\sum_{n=1}^{\infty} u_n(x)$, $u_1(x) = f_1(x)$, $u_n(x) = f_n(x) - f_{n-1}(x)$, $f_n(x) =$

$$\begin{cases} 2n^2 x, & x \in [0, \frac{1}{2n}] \\ 2n - 2n^2 x, & x \in [\frac{1}{2n}, \frac{1}{n}], n \in \mathbb{N}, \ X = [0, 1]. \\ 0, & x \in [\frac{1}{n}, 1] \end{cases}$$

In this Example we make use of the functions $f_n(x)$ of Example 4b in Sect. 1, Chap. 3. These functions have the areas of figures between their graphs and the x-axis equal to $\frac{1}{2}$: $\int_0^1 f_n(x)dx = \frac{1}{2}$, and their sequence converges on $[0, 1]$ to $f(x) \equiv 0$ (the geometric form of these functions is simple and is illustrated in Fig. 3.5 of Examples 4b in Chap. 3). For the general term we have $u_n(x) = f_n(x) - f_{n-1}(x)$, and consequently, the n-th partial sum of this series is exactly $f_n(x)$. Therefore, the series converges to the sum $f(x) \equiv 0$ on $[0, 1]$ and its integral is 0:

$$\int_0^1 f(x)dx = \int_0^1 \lim_{n\to\infty} f_n(x)dx = \int_0^1 \sum_{n=1}^{\infty} u_n(x)dx = 0.$$

On the other hand, each function $u_n(x)$ is continuous (and consequently integrable) on $[0, 1]$ and the corresponding integrals are

$$\int_0^1 u_1(x)dx = \int_0^1 f_1(x)dx = \frac{1}{2}, \ \int_0^1 u_n(x)dx = \int_0^1 (f_n(x) - f_{n-1}(x))dx = 0, \ n = 2, 3, \ldots.$$

Then, $\sum_{n=1}^{\infty} \int_0^1 u_n(x)dx = \frac{1}{2}$, and consequently, $\sum_{n=1}^{\infty} \int_0^1 u_n(x)dx \neq \int_0^1 \sum_{n=1}^{\infty} u_n(x)dx$, that is, there is no relation between the integral of the sum of the series and the corresponding series of integrals.

Remark In general, having an example of a sequence of functions $f_n(x)$, it is easy to construct the corresponding example of the series $\sum_{n=1}^{\infty} u_n(x)$ by considering the series whose partial sums are $f_n(x)$, that is, by defining $u_1(x) = f_1(x)$ and $u_n(x) = f_n(x) - f_{n-1}(x)$, $n = 2, 3, \ldots$. This technique was used in Example 4b and will be used later in this Chapter. It allows us to make use of different examples of sequences of the previous Chapter for construction of series with similar properties of convergence.

2 Uniform and Non-uniform Convergence

> *After a scientific meeting at which Cauchy presented his theory on*
> *the convergence of series Laplace hastened home and remained*
> *there in reclusion until he had examined the series in his Mécanique*
> *céleste. Luckily every one was found to be convergent. When*
> *Weierstrass's work [on uniform convergence] became known*
> *through his lectures, the effect was even more noticeable.*
> *Morris Kline, 1972*

2.1 Concept of Uniform and Non-uniform Convergence

First, we recall the definition of pointwise convergence given in the previous section and write it in detailed form using ε-N notation.

▶ **Definition** (**Pointwise Convergence on a Set**) A series $\sum_{n=1}^{\infty} u_n(x)$ converges pointwise on X to the sum $f(x)$ if it converges at every fixed point $x \in X$. This means that the sequence of partial sums $f_n(x) = \sum_{k=1}^{n} u_k(x)$ converges at every point $x \in X$ to $f(x)$. In ε-N terms, for each fixed $x \in X$ we have:

$$\forall \varepsilon > 0 \; \exists N_\varepsilon(x) \; \text{ such that } \forall n > N_\varepsilon(x) \; \text{ it follows } |f_n(x) - f(x)| = |r_n(x)| = \left| \sum_{k=n+1}^{\infty} u_k(x) \right| < \varepsilon.$$

$$(4.1)$$

Like in the case of pointwise convergence of a sequence, the number N_ε in this definition depends on both parameter ε and the chosen point x of the set X.

▶ **Definition** (**Uniform Convergence on a Set**) A *series* $\sum_{n=1}^{\infty} u_n(x)$ *converges uniformly on* X to the sum $f(x)$ if the sequence of its partial sums $f_n(x) = \sum_{k=1}^{n} u_k(x)$ converges uniformly on X to $f(x)$, that is, for all $x \in X$ at once we have:

$$\forall \varepsilon > 0 \; \exists N_\varepsilon \; \text{ such that } \forall n > N_\varepsilon \; \text{ it follows } |f_n(x) - f(x)| = |r_n(x)| = \left| \sum_{k=n+1}^{\infty} u_k(x) \right| < \varepsilon.$$

$$(4.2)$$

Here, the *number N_ε is the same for all $x \in X$*. Evidently, uniform convergence is a stronger requirement on a series than pointwise convergence: a series that converges uniformly also converges pointwise, but the converse is not true.

▶ **Definition** (**Non-uniform Convergence on a Set**) *A series* $\sum_{n=1}^{\infty} u_n(x)$ *converges non-uniformly on* X *to the sum* $f(x)$ *if it converges pointwise, but not uniformly on* X, *that is,*

$$\exists \varepsilon_0 > 0 \text{ such that for } \forall N \in \mathbb{N} \; \exists n_N > N \text{ and } \exists x_N \in X \text{ such that}$$

$$\left| f_{n_N}(x_N) - f(x_N) \right| = \left| r_{n_N}(x_N) \right| = \left| \sum_{k=n_N+1}^{\infty} u_k(x_N) \right| \geq \varepsilon_0. \tag{4.3}$$

Remark 1 From the theory of sequences of functions we know that condition (4.2) in the definition of uniform convergence is equivalent to the following condition:

$$\sup_{\forall x \in X} |f_n(x) - f(x)| = \sup_{\forall x \in X} |r_n(x)| = \sup_{\forall x \in X} \left| \sum_{k=n+1}^{\infty} u_k(x) \right| \xrightarrow[n \to \infty]{} 0. \tag{4.4}$$

Therefore, the condition of non-uniform convergence can be formulated as follows:

$$\sup_{\forall x \in X} |f_n(x) - f(x)| = \sup_{\forall x \in X} |r_n(x)| = \sup_{\forall x \in X} \left| \sum_{k=n+1}^{\infty} u_k(x) \right| \not\xrightarrow[n \to \infty]{} 0. \tag{4.5}$$

Remark 2 In the same manner as for the sequences of functions, condition (4.3) of non-uniform convergence can be changed to more restrictive condition, which allows a simpler verification:

$$\exists \varepsilon_0 > 0 \text{ such that } \forall n \in \mathbb{N} \; \exists x_n \in X \text{ such that } |f_n(x_n) - f(x_n)| = |r_n(x_n)| = \left| \sum_{k=n+1}^{\infty} u_k(x_n) \right| \geq \varepsilon_0. \tag{4.6}$$

Theorem (**Test for Non-uniform Convergence, Necessary Condition of Uniform Convergence**) *In the same way as the convergence of a series of numbers implies the tendency of its general term to* 0, *the uniform convergence of a series of functions implies the uniform convergence of its general term to* 0, *namely: if a series* $\sum_{n=1}^{\infty} u_n(x)$ *converges uniformly on a set* X, *then its general term* $u_n(x)$ *converges uniformly to* 0 *on* X.

Proof If $\sum_{n=1}^{\infty} u_n(x)$ converges uniformly on X, then, by definition, the sequence of partial sums $f_n(x) = \sum_{k=1}^{n} u_k(x)$ converges uniformly on X. Since $u_n(x) = f_n(x) - f_{n-1}(x)$, we conclude that $u_n(x) = f_n(x) - f_{n-1}(x) \underset{n \to \infty}{\overset{X}{\rightrightarrows}} f(x) - f(x) = 0$. □

Remark The uniform convergence of the general term to 0 is a necessary, but not sufficient condition for the uniform convergence of the corresponding series. An example of this type is the series $\sum_{n=1}^{\infty} \frac{x^n}{n}$ on the interval $(-1, 1)$. Its general term $\frac{x^n}{n}$ tends uniformly to 0 on $(-1, 1)$ according to the evaluation $\left| \frac{x^n}{n} \right| \leq \frac{1}{n} \underset{n \to \infty}{\to} 0$ satisfied for all $x \in (-1, 1)$ at once. However, the series does not converge uniformly on $(-1, 1)$ as shown in Example 5u below.

Examples Let us analyze the character of convergence of the four series considered in Examples 1b–4b of Sect. 1.

Example 1u $\sum_{n=0}^{\infty} x^n$, $X = (-1, 1)$

The partial sums of this series are easily found: $f_n(x) = \sum_{k=0}^{n} x^k = \frac{1-x^{n+1}}{1-x}$, $\forall x \in (-1, 1)$, as well as the sum of the series $\sum_{n=0}^{\infty} x^n = f(x) = \frac{1}{1-x}$, $\forall x \in (-1, 1)$. Therefore, the series converges on $(-1, 1)$. However, this convergence is non-uniform that can be seen from the evaluation of remainders at the points $x_n = 1 - \frac{1}{n+1} \in (-1, 1)$:

$$|f_n(x_n) - f(x_n)| = \left| \frac{1-x_n^{n+1}}{1-x_n} - \frac{1}{1-x_n} \right| = \frac{\left(1-\frac{1}{n+1}\right)^{n+1}}{1-1+\frac{1}{n+1}} = (n+1)\left(1-\frac{1}{n+1}\right)^{n+1} \underset{n \to \infty}{\to} \infty.$$

Hence, according to Remark 2 on the definition of non-uniform convergence, the series converges non-uniformly on $(-1, 1)$.

Another way to show that the convergence is non-uniform on $(-1, 1)$ is by the application of the Test for non-uniform convergence: recalling that the general term x^n does not converge uniformly on $(-1, 1)$ (see Example 2u in Sect. 2.1 of Chap. 3), we conclude immediately that the series converges non-uniformly on $(-1, 1)$.

Notice that the convergence of this series is uniform on any interval $[-a, a]$, where $0 < a < 1$. In fact, using the definition of uniform convergence with partial sums, we obtain

$$|f_n(x) - f(x)| = \left| \frac{1-x^{n+1}}{1-x} - \frac{1}{1-x} \right| = \frac{|x|^{n+1}}{1-x} \leq \frac{a^{n+1}}{1-a} < \varepsilon$$

for all $x \in [-a, a]$ at once and for all indices n large enough (note that $\frac{a^{n+1}}{1-a} \underset{n \to \infty}{\to} 0$).

Example 2u $\sum_{n=1}^{\infty} (1-x)x^n$, $X = [0, 1]$

This case is similar to the previous Example. For $x \neq 1$ the partial sums are found in the form $f_n(x) = \sum_{k=1}^{n} (1-x)x^k = (1-x)x\frac{1-x^n}{1-x} = x(1-x^n)$, $\forall x \in [0, 1)$. Notice that the value at $x = 1$ can also be included in the expression on the right-hand side, since $f_n(1) = 0$. Therefore, the sum of this series is $f(x) = \lim_{n \to \infty} f_n(x) = \begin{cases} x, & x \in [0, 1) \\ 0, & x = 1 \end{cases}$

and the remainders are found in the form $r_n(x) = f(x) - f_n(x) = \begin{cases} x^{n+1}, & x \in [0, 1) \\ 0, & x = 1 \end{cases}$.

Since the sequence x^{n+1} converges non-uniformly on $[0, 1)$ (see Example 2u in Sect. 2.1 of Chap. 3), the sequence $r_n(x)$ converges non-uniformly on $[0, 1]$, which means the non-uniform convergence of the series on $[0, 1]$. In particular, this causes a discontinuity of the sum $f(x)$ at $x = 1$.

Example 3u $\sum_{n=1}^{\infty} \frac{\sin n^2 x}{n^2}$, $X = \mathbb{R}$
This series converges uniformly on $\mathbb{R}$ according to the following evaluation of remainders:

$$\sup_{\forall x \in \mathbb{R}} |r_n(x)| = \sup_{\forall x \in \mathbb{R}} \left| \sum_{k=n+1}^{\infty} \frac{\sin k^2 x}{k^2} \right| < \sum_{k=n+1}^{\infty} \frac{1}{k^2} \xrightarrow[n \to \infty]{} 0$$

(the last tendency fóllows from the convergence of the p-series $\sum_{n=1}^{\infty} \frac{1}{n^2}$). The functions $u_n(x) = \frac{\sin n^2 x}{n^2}$ are infinitely differentiable on $\mathbb{R}$, but the series of derivatives $\sum_{n=1}^{\infty} u'_n(x) = \sum_{n=1}^{\infty} \cos n^2 x$ diverges at any x, as was shown in Sect. 1.

Example 4u $\sum_{n=1}^{\infty} u_n(x)$, $u_1(x) = f_1(x)$, $u_n(x) = f_n(x) - f_{n-1}(x)$, $f_n(x) = \begin{cases} 2n^2 x, & x \in [0, \frac{1}{2n}] \\ 2n - 2n^2 x, & x \in [\frac{1}{2n}, \frac{1}{n}], n \in \mathbb{N}, X = [0, 1]. \\ 0, & x \in [\frac{1}{n}, 1] \end{cases}$
   The partial sums of this series form the sequence considered in Example 4u of Sect. 2.1, Chap. 3, where it was shown that the sequence converges non-uniformly on $[0, 1]$. For this reason, the integral of the sum of the series and the corresponding series of integrals are different.

Let us analyze two more series that have already appeared in Sect. 1.

Example 5u $\sum_{n=1}^{\infty} \frac{x^n}{n}$, $X = (-1, 1)$
In Example 2a it was shown that this series converges absolutely on the interval $(-1, 1)$ and conditionally at the point $x = -1$. To analyze the type of convergence on the interval $(-1, 1)$, notice that the problem (if it exists) might occur near the point 1. Then, we restrict our attention to the positive values of x and evaluate the remainders at the points $x_n = 1 - \frac{1}{n} \in (-1, 1)$:

$$|r_n(x_n)| = \left| \sum_{k=n+1}^{\infty} \frac{x_n^k}{k} \right| = \sum_{k=n+1}^{\infty} \frac{\left(1 - \frac{1}{n}\right)^k}{k} > \sum_{k=n+1}^{2n} \frac{\left(1 - \frac{1}{n}\right)^k}{k} > \sum_{k=n+1}^{2n} \frac{\left(1 - \frac{1}{n}\right)^{2n}}{2n}$$

$$= \frac{1}{2} \left(1 - \frac{1}{n}\right)^{2n} \xrightarrow[n \to \infty]{} \frac{1}{2} e^{-2} \neq 0.$$

Hence, by definition (4.3), the series does not converge uniformly on $(-1, 1)$.

Example 6u $\sum_{n=0}^{\infty} \frac{x^n}{n!}$, $X = \mathbb{R}$

This series was considered in Example 3a and it was proved there that it converges (absolutely) on all the real axis $\mathbb{R}$. If a problem with non-uniform convergence occurs on this set, it is most likely to happen for large values of x that give the large values of x^n. Bearing this in mind, we evaluate the remainders r_n at the points $x_n = n$:

$$\sup_{\forall x \in \mathbb{R}} |r_n(x)| = \sup_{\forall x \in \mathbb{R}} \left| \sum_{k=n+1}^{\infty} \frac{x^k}{k!} \right| \geq \sum_{k=n+1}^{\infty} \frac{x_n^k}{k!}$$

$$= \sum_{k=n+1}^{\infty} \frac{n^k}{k!} > \frac{n^{n+1}}{(n+1)!} = \frac{n}{n+1} \frac{n^n}{n!} \geq \frac{n}{n+1} \geq \frac{1}{2} \underset{n \to \infty}{\nrightarrow} 0.$$

Therefore, by definition (4.5), the series does not converge uniformly on $\mathbb{R}$.

The same result can be obtained using the Test for non-uniform convergence. Indeed, evaluating the sequence $u_n(x) = \frac{x^n}{n!}$ at the points $x_n = n$, we have $u_n(x_n) = \frac{n^n}{n!} \underset{n \to \infty}{\to} \infty \neq 0$, that is, the general term does not converge uniformly to 0.

Notice that this series converges uniformly on any interval $[-a, a], a > 0$. Indeed, evaluating the remainders $r_n(x)$, simultaneously for all $x \in [-a, a]$ we obtain

$$\sup_{\forall x \in [-a,a]} |r_n(x)| = \sup_{\forall x \in [-a,a]} \left| \sum_{k=n+1}^{\infty} \frac{x^k}{k!} \right| \leq \sum_{k=n+1}^{\infty} \frac{a^k}{k!} \underset{n \to \infty}{\to} 0.$$

The last tendency follows from the convergence of the series of numbers $\sum_{n=0}^{\infty} \frac{a^n}{n!}$ for any fixed a.

The next example shows that uniform convergence does not imply absolute convergence.

Example 7u $\sum_{n=1}^{\infty} (-1)^n \frac{x^2+n}{n^2}$, $X = [-a, a], a > 0$

First, it is clear that for every fixed $x \in \mathbb{R}$ this alternating series converges according to Leibniz's test: the sequence $\frac{x^2+n}{n^2}$ converges monotonically to 0 for any fixed x. To show that the convergence is uniform on $[-a, a]$, recall that the remainder of convergent alternating series satisfies the estimate $|r_n(x)| \leq |u_{n+1}(x)| = \frac{x^2+n+1}{(n+1)^2}$. Then, for all $x \in [-a, a]$ we have

$$|r_n(x)| \leq \frac{x^2 + n + 1}{(n+1)^2} \leq \frac{a^2 + n + 1}{(n+1)^2} \underset{n \to \infty}{\to} 0,$$

which implies the uniform convergence on $[-a, a]$.

However, the series does not converge absolutely on $[-a, a]$. Actually, the series of absolute values $\sum_{n=1}^{\infty} \frac{x^2+n}{n^2}$ diverges at every $x \in \mathbb{R}$ according to the Comparison test: $\frac{x^2+n}{n^2} \geq \frac{1}{n}$ and the harmonic series $\sum_{n=1}^{\infty} \frac{1}{n}$ is divergent.

2.2 Arithmetic Properties of Uniform Convergence

The *arithmetic properties of the uniformly convergent series of functions* are replicas of the corresponding properties of sequences.

Property 1 *If series $\sum u_n(x)$ and $\sum v_n(x)$ converge uniformly on X, then $\sum(u_n(x) + v_n(x))$ converges uniformly on X.*

Proof The uniform convergence of the series $\sum u_n(x)$ and $\sum v_n(x)$ means the uniform convergence of the sequences $f_n(x)$ and $g_n(x)$ of their partial sums. Then, the analogous property of sequences of functions (Property 1 in Sect. 2.2 of Chap. 3) guarantees that $f_n(x) + g_n(x)$ converges uniformly on X. Since the last is the sequence of the partial sums of $\sum(u_n(x) + v_n(x))$, this series converges uniformly on X. $\qquad\square$

Property 2 *If $\sum u_n(x)$ converges uniformly on X, and $g(x)$ is bounded on X, then $\sum g(x)u_n(x)$ converges uniformly on X.*

Proof The uniform convergence of the series $\sum u_n(x)$ means the uniform convergence of the sequence of its partial sums $f_n(x)$. Then, the analogous property of the sequences (Property 2 in Sect. 2.2 of Chap. 3) ensures that $g(x) f_n(x)$ converges uniformly on X, and the last is the sequence of the partial sums of $\sum g(x)u_n(x)$. $\qquad\square$

2.3 The Cauchy Criterion for Uniform Convergence

The *Cauchy criterion for uniform convergence of series* follows directly from the Cauchy criterion for uniform convergence of sequences.

Theorem (Cauchy Criterion for Uniform Convergence) *Let $\sum_{n=1}^{\infty} u_n(x)$ be a series of functions defined on a set $X \subset \mathbb{R}$. The series converges uniformly on X if and only if the sequence of its partial sums is a uniform Cauchy sequence on X, that is,*

$$\forall \varepsilon > 0 \, \exists N_\varepsilon \text{ such that for } \forall m > n > N_\varepsilon \text{ (and for } \forall x \in X \text{ at once) it follows } \left| \sum_{k=n+1}^{m} u_k(x) \right| < \varepsilon.$$

$$(4.7)$$

Proof The corresponding Cauchy criterion for uniform convergence of sequences (Sect. 2.3 of Chap. 3), and the equivalence between uniform convergence of a series and uniform convergence of the sequence of its partial sums dismiss any necessity of proof.   □

2.4   Uniform and Absolute Convergence

Theorem of Uniform and Absolute Convergence *If a series $\sum_{n=1}^{\infty} |u_n(x)|$ converges uniformly on X, then the series $\sum_{n=1}^{\infty} u_n(x)$ converges uniformly and absolutely on X. The converse is not true.*

Proof If a series $\sum_{n=1}^{\infty} |u_n(x)|$ converges uniformly on X, the absolute convergence of $\sum_{n=1}^{\infty} u_n(x)$ at every fixed point of X follows directly from the properties of series of numbers. To show the uniform convergence we use the Cauchy criterion:

$$\left| \sum_{k=n+1}^{m} u_k(x) \right| \le \sum_{k=n+1}^{m} |u_k(x)| < \varepsilon$$

for all $x \in X$ at once and for $\forall m > n > N$, where N is from the Cauchy criterion for the series $\sum_{n=1}^{\infty} |u_n(x)|$.

A counterexample for the converse statement can be provided by the series $\sum_{n=1}^{\infty} (-1)^n x^n (1 - x)$ on $[0, 1]$. First, this series converges absolutely at every point of $[0, 1]$. Indeed, as was seen before (Example 2b in Sect. 1), the series $\sum_{n=1}^{\infty} x^n (1 - x)$ converges on $[0, 1]$ to the sum $g(x) = \begin{cases} x, & x \in [0, 1) \\ 0, & x = 1 \end{cases}$. This guarantees also the pointwise convergence of the original series, which can alternatively be proved using the formula for its partial sums:

$$f_n(x) = \sum_{k=1}^{n} (-x)^k (1 - x) = (1 - x) \cdot (-x) \frac{1 - (-x)^n}{1 - (-x)}$$

$$= \frac{x(x - 1)}{x + 1} (1 - (-x)^n) \xrightarrow[n \to \infty]{} f(x) = \begin{cases} \frac{x(x-1)}{x+1}, & x \in [0, 1) \\ 0, & x = 1 \end{cases} = \frac{x(x - 1)}{x + 1}.$$

To demonstrate that the series $\sum_{n=1}^{\infty} (-1)^n x^n (1 - x)$ converges uniformly on $[0, 1]$, we use formulation (4.4). Start with the evaluation of $f_n(x) - f(x)$:

$$f_n(x) - f(x) = \frac{x(x - 1)}{x + 1} (1 - (-x)^n) - \frac{x(x - 1)}{x + 1} = \frac{(x - 1)}{x + 1} (-x)^{n+1}.$$

Then,

$$|f_n(x) - f(x)| = \frac{(1-x)}{1+x} x^{n+1} \leq (1-x)x^{n+1} \equiv \varphi(x), \forall x \in [0, 1].$$

Now we find the extrema of $\varphi(x)$ on $[0, 1]$: $\varphi'(x) = (1-x)(n+1)x^n - x^{n+1} = x^n(n + 1 - (n+2)x)$ and the condition of critical point $\varphi'(x) = 0$ gives the unique solution $x_n = \frac{n+1}{n+2} \in (0, 1)$. The derivative changes the sign from positive to negative passing through the point x_n, which indicates that x_n is a local maximum. Since $\varphi(x)$ is non-negative and has the only local extremum on $(0, 1)$, the point x_n is its global maximum. Therefore, we arrive at the following evaluation:

$$\sup_{x\in[0,1]} |f_n(x) - f(x)| \leq \sup_{x\in[0,1]} \varphi(x) = \max_{x\in[0,1]} \varphi(x) = \varphi(x_n) = \frac{1}{n+2} \cdot \left(\frac{n+1}{n+2}\right)^{n+1} < \frac{1}{n+2} \underset{n\to\infty}{\to} 0,$$

that demonstrates the uniform convergence of $\sum_{n=1}^{\infty}(-1)^n x^n (1 - x)$ on $[0, 1]$.

Nevertheless, the series of absolute values $\sum_{n=1}^{\infty} x^n(1-x)$ does not converge uniformly on $[0, 1]$, as was shown in Example 2u of Sect. 2.1.   □

Historical Remarks The necessity of some additional conditions to guarantee the continuity of the sum of a convergent series of functions was first noted by Abel in discussion of Cauchy's problematic theorem on the sum of a series of continuous functions. In 1826 Abel constructed a counterexample to Cauchy's theorem using trigonometric series and also indicated a safe part of the domain where a special case of the theorem holds, but still did not arrived at a new concept of convergence.

The uniform convergence was probably first introduced by Gudermann in the 1838 paper on elliptic functions, where he considered the mode of convergence of a series independent of the involved parameters. While he used the phrase "convergence in a uniform way", he did not give an exact definition of a new type of convergence, nor did use the property in the proof of any theorem.

The proper concept of uniform convergence was defined independently (and in different ways) by Weierstrass in 1841, Stokes in 1847 and Seidel in 1848. However, the definitions by Stokes and Seidel were not quite precise, and were introduced to study particular series. Besides, Stokes and Seidel considered a uniform and quasi-uniform convergence, respectively, in a neighborhood of a point, while Weierstrass introduced the concept of uniform convergence on an interval. Comparing the three definitions, Hardy made the following comment, which became currently a general opinion on the subject: "Weierstrass's discovery was the earliest, and he alone fully realized its far-reaching importance as one of the fundamental ideas of analysis".

Weierstrass probably learned about the new concept of convergence from his teacher Gudermann, and he used it in a critical way in the paper of 1841, where he showed that the sum of uniformly convergent series of analytic functions is analytic and can be differentiated term by term. However, this paper remained unpublished until 1894, and the importance of uniform convergence continued to be concealed from mathematical community until Weierstrass began to lecture at the University of Berlin in 1856, where he included the exposition of uniform convergence in the course of the theory of analytic functions. Although the content of this course was not published during his lifetime, but the main ideas were fast disseminated by many German and foreign students gathered in Berlin to attend Weierstrass's lectures.

3   Sufficient Conditions for Uniform Convergence of Series

3.1   Comparison Tests

General Comparison Test *Let $\sum_{n=1}^{\infty} u_n(x)$ and $\sum_{n=1}^{\infty} v_n(x)$ be series of functions defined on a set $X \subset \mathbb{R}$. If $|u_n(x)| \leq v_n(x)$ for $\forall n$ and $\forall x \in X$, then the uniform convergence of the series $\sum_{n=1}^{\infty} v_n(x)$ on X implies the uniform convergence of the series $\sum_{n=1}^{\infty} |u_n(x)|$, which in turn guarantees the uniform (and absolute) convergence of the series $\sum_{n=1}^{\infty} u_n(x)$ on X.*

Proof If $\sum_{n=1}^{\infty} v_n(x)$ converges uniformly on X, then, according to the Cauchy criterion, its partial sums form a uniform Cauchy sequence on X. Using this property and the inequality $|u_n(x)| \leq v_n(x)$, we arrive at the following evaluation valid for all $x \in X$:

$$\left| \sum_{k=n+1}^{m} u_k(x) \right| \leq \sum_{k=n+1}^{m} |u_k(x)| \leq \sum_{k=n+1}^{m} v_k(x) = \left| \sum_{k=n+1}^{m} v_k(x) \right| < \varepsilon.$$

This shows that the partial sums of the series $\sum_{n=1}^{\infty} |u_n(x)|$ also constitute a uniform Cauchy sequence on X, and consequently, the series $\sum_{n=1}^{\infty} |u_n(x)|$ converges uniformly on X. Therefore, by the Theorem of Sect. 2.4, the series $\sum_{n=1}^{\infty} u_n(x)$ converges uniformly and absolutely on X.   □

Remark As it always happens with sequences and series, this and other statements continue to be valid if the hypotheses are satisfied starting from a certain index n.

▶ **Definition** Let $\sum_{n=1}^{\infty} u_n(x)$ be a series of functions defined on a set $X \subset \mathbb{R}$. If there exists a series of numbers $\sum_{n=1}^{\infty} M_n$ such that $|u_n(x)| \leq M_n$, $\forall n$ and $\forall x \in X$, then it is called the *majorant (dominant) series* for $\sum_{n=1}^{\infty} u_n(x)$.

> **Weierstrass Test (Weierstrass M-Test)** *Let $\sum_{n=1}^{\infty} u_n(x)$ be a series of functions defined on a set $X \subset \mathbb{R}$. If it has a convergent majorant series $\sum_{n=1}^{\infty} M_n$, then the series $\sum_{n=1}^{\infty} |u_n(x)|$ converges uniformly on X, and consequently, the series $\sum_{n=1}^{\infty} u_n(x)$ converges uniformly and absolutely on X.*

Proof Taking $v_n(x) = M_n$ and noting that convergence of $\sum_{n=1}^{\infty} M_n$ means the uniform convergence of $\sum_{n=1}^{\infty} v_n(x)$ on X, we confirm that the hypotheses of the General Comparison test are satisfied, and consequently, the series $\sum_{n=1}^{\infty} |u_n(x)|$ converges uniformly on X, which implies the uniform and absolute convergence of the series $\sum_{n=1}^{\infty} u_n(x)$ on X.   □

Remark Frequently, the main goal is to investigate the behavior of the original series $\sum_{n=1}^{\infty} u_n(x)$ instead of the series of its absolute values. Therefore, the Weierstrass test is usually formulated in a simpler manner: if there exists a convergent numerical series $\sum_{n=1}^{\infty} M_n$ such that $|u_n(x)| \leq M_n$, for $\forall n$ and $\forall x \in X$, then the series $\sum_{n=1}^{\infty} u_n(x)$ converges uniformly (and absolutely) on X.

Let us apply the Weierstrass test to verify uniform convergence of some series.

Example 1W $\sum_{n=1}^{\infty} x^n$, $X = [-a, a], 0 < a < 1$
In Example 1u, Sect. 2.1, using definition we have shown that this series converges uniformly on $[-a, a], a < 1$. Let us confirm this result applying the Weierstrass test. We define the general term of a majorant series in the following way: $M_n = \max\limits_{x \in [-a,a]} |x^n| = a^n$.
Then, $\sum_{n=1}^{\infty} M_n = \sum_{n=1}^{\infty} a^n$ is a convergent geometric series and, by the Weierstrass test, the original series converges uniformly and absolutely on $[-a, a]$.

Example 2W $\sum_{n=1}^{\infty} \frac{x}{1+n^4 x^2}$, $X = (0, +\infty)$
Using the well-known inequality $1 + n^4 x^2 \geq 2n^2 x$, we evaluate the (positive) general term in the form $\frac{x}{1+n^4 x^2} = \frac{2n^2 x}{1+n^4 x^2} \cdot \frac{1}{2n^2} \leq \frac{1}{2n^2} \equiv M_n$, $\forall x > 0$. Since the series of numbers $\sum_{n=1}^{\infty} M_n = \sum_{n=1}^{\infty} \frac{1}{2n^2}$ converges, by the Weierstrass test, the original series converges uniformly on $(0, \infty)$.

Example 3W $\sum_{n=1}^{\infty} x^2 e^{-nx}$, $X = [0, +\infty)$
To evaluate the general term $u_n(x) = x^2 e^{-nx}$, we find its extrema. Differentiating $u'_n(x) = 2xe^{-nx} - x^2 n e^{-nx} = xe^{-nx}(2-nx)$ and applying the condition of critical point $u'_n(x) = 0$ we obtain the only point $x_n = \frac{2}{n}$ that belongs to the given interval. The derivative is positive on the left and negative on the right of this point, which means that x_n is a local maximum. (The point $x = 0$ can be dismissed because $u_n(0) = 0$ and $u_n(x) > 0$ for $\forall x > 0$). Since $u_n(x)$ is positive and does not have any other critical point, it follows that x_n is a global maximum of $u_n(x)$, and we arrive at the following estimate

$$u_n(x) = x^2 e^{-nx} \leq \max_{x \geq 0} u_n(x) = u_n(x_n) = \frac{4}{n^2} e^{-2} < \frac{1}{n^2} \equiv M_n.$$

Since the series of numbers $\sum_{n=1}^{\infty} M_n = \sum_{n=1}^{\infty} \frac{1}{n^2}$ converges, by the Weierstrass test, the original series converges uniformly on $[0, \infty)$.

The next example shows that even if a series converges uniformly, the corresponding convergent majorant series may not exist.

Example 4W $\sum_{n=1}^{\infty}(-1)^n x^n(1-x)$, $X = [0,1]$

In Sect. 2.4 it was shown that this series converges uniformly (and absolutely) on $[0,1]$. Let us see what we can say about a majorant series. For each fixed n the function $v_n(x) = |u_n(x)| = x^n(1-x)$ is continuous on $[0,1]$, positive on $(0,1)$ and equal to zero at the endpoints—$v_n(0) = v_n(1) = 0$. Under these conditions, $v_n(x)$ attains its global maximum inside the interval $(0,1)$ and this global maximum is also the local one. Therefore, it is sufficient to find critical points on $(0,1)$, by solving the corresponding equation for the derivative:

$$v_n'(x) = (x^n - x^{n+1})' = nx^{n-1} - (n+1)x^n = (n+1)x^{n-1}\left(\frac{n}{n+1} - x\right) = 0.$$

Then, the only critical point in $(0,1)$ is $x_n = \frac{n}{n+1}$, and consequently,

$$|u_n(x)| = v_n(x) \leq \max_{x\in[0,1]} v_n(x) = |u_n(x_n)| = \frac{1}{n+1}\left(1 - \frac{1}{n+1}\right)^n \equiv M_n.$$

Since $\lim\limits_{n\to\infty}\left(1 - \frac{1}{n+1}\right)^n = e^{-1}$ and the series $\sum_{n=1}^{\infty}\frac{1}{n+1}$ diverges, the numerical series $\sum_{n=1}^{\infty} M_n = \sum_{n=1}^{\infty}\frac{1}{n+1}\left(1 - \frac{1}{n+1}\right)^n$ diverges according to the Comparison test for positive series. On the other hand, the evaluation of $|u_n(x)|$ by $M_n = \frac{1}{n+1}\left(1 - \frac{1}{n+1}\right)^n$ is the best possible in the sense that the numbers M_n are the minimum possible for each n because $M_n = \max\limits_{x\in[0,1]} |u_n(x)|$. Thus, in this case, the uniformly convergent series does not have a convergent majorant series.

Historical Remarks Since many of the Weierstrass results were first widely disseminated by the students attending his lectures in the University of Berlin and appeared published in his works many years later, it is difficult to establish even approximate date when Weierstrass discovered the famous M-test. It is well known that he provided an explicit reference to this test in the treatise of 1880. However, historical studies of the notes of the Weierstrass lectures on analytic functions made by his students and papers on series of functions published by his pupils infer that Weierstrass had formulated the M-test before 1866.

3.2   Dirihlet's and Abel's Tests

In this section we prove the two tests for uniform convergence of series $\sum_{n=1}^{\infty} u_n(x)$ with the general term in the form $u_n(x) = a_n(x)b_n(x)$. These results are similar to Dirichlet's and Abel's tests for series of numbers (Sect. 4.2, Chap. 2) and we adapt the line of reasoning used in the proofs of those tests to the context of uniform convergence.

To avoid repetitive indications that all the elements depend on x and shorten notations, in the great part of this section we omit x in the formulas.

Abel's Summation by Parts Formula *Let $\sum_{n=1}^{\infty} u_n(x)$ be a series of functions with the general term in the form $u_n(x) = a_n(x)b_n(x)$ and $B_n(x) = \sum_{k=1}^{n} b_k(x)$ be the partial sums of the series of functions $\sum_{n=1}^{\infty} b_n(x)$, all the functions defined on a set $X \subset \mathbb{R}$. Then for any $m > n \geq 1$ and for each $x \in X$ the following formula is true*

$$\sum_{k=n+1}^{m} u_k = \sum_{k=n+1}^{m} a_k b_k = \sum_{k=n+1}^{m-1} B_k (a_k - a_{k+1}) + B_m a_m - B_n a_{n+1}.$$

Proof This result is an exact repetition of Abel's summation formula for series of numbers (given in Sect. 4.2 of Chap. 2), since it is formulated separately for each $x \in X$.   □

We use this result to demonstrate the next two tests.

Dirichlet's Test *Let $\sum_{n=1}^{\infty} u_n(x)$, $u_n(x) = a_n(x)b_n(x)$ be a series of functions defined on a set $X \subset \mathbb{R}$. If*

(1) the partial sums $B_n(x)$ of the series $\sum_{n=1}^{\infty} b_n(x)$ are uniformly bounded on X, that is, there exists a constant M such that $|B_n(x)| = \left| \sum_{k=1}^{n} b_k(x) \right| \leq M$ for $\forall n$ and $\forall x \in X$,
(2) the sequence $a_n(x)$ is monotone at every fixed $x \in X$,
(3) the sequence $a_n(x)$ converges uniformly to 0 on X,

then the series $\sum_{n=1}^{\infty} a_n(x)b_n(x)$ is uniformly convergent on X.

Proof Let us show that the series $\sum_{n=1}^{\infty} u_n(x)$ satisfies the conditions of the Cauchy criterion for uniform convergence. To do this, take $m > n$ and use Abel's summation formula to evaluate the sums $S \equiv \sum_{k=n+1}^{m} u_k(x) = \sum_{k=n+1}^{m} a_k(x) \cdot b_k(x)$, in the form

$$|S| = \left| \sum_{k=n+1}^{m} a_k b_k \right| = \left| \sum_{k=n+1}^{m-1} B_k (a_k - a_{k+1}) + B_m a_m - B_n a_{n+1} \right|$$

$$\leq \sum_{k=n+1}^{m-1} |B_k| \cdot |a_k - a_{k+1}| + |B_m| \cdot |a_m| + |B_n| \cdot |a_{n+1}| \equiv S_1.$$

By the condition (1) of the Theorem, the partial sums B_n are uniformly bounded, that is, there exists $M > 0$ such that $|B_n| \leq M$ for $\forall n \in \mathbb{N}$ and for all $x \in X$. Then, we can proceed with the evaluation of S_1 in the following way:

$$S_1 \le M \left(\sum_{k=n+1}^{m-1} |a_k - a_{k+1}| + |a_m| + |a_{n+1}| \right) \equiv S_2.$$

Since the sequence a_n is monotone at each $x \in X$, all the differences $a_k - a_{k+1}$ have the same sign, and consequently, the sum of the absolute values is equal to the absolute value of the sum:

$$S_2 = M \left(\left| \sum_{k=n+1}^{m-1} (a_k - a_{k+1}) \right| + |a_m| + |a_{n+1}| \right)$$

$$= M \left(|a_{n+1} - a_{n+2} + a_{n+2} - a_{n+3} + \ldots + a_{m-1} - a_m| + |a_m| + |a_{n+1}| \right)$$

$$= M \left(|a_{n+1} - a_m| + |a_m| + |a_{n+1}| \right) \le 2M \left(|a_{n+1}| + |a_m| \right).$$

Now we use the condition of the uniform convergence of a_n: $a_n \overset{X}{\underset{n \to \infty}{\rightrightarrows}} 0$. By definition, this means that for $\forall \varepsilon > 0$ (take for convenience $\varepsilon_1 = \frac{\varepsilon}{4M}$) there exists N_ε such that for $\forall n > N_\varepsilon$ and simultaneously for all $x \in X$ it follows that $|a_n| < \frac{\varepsilon}{4M}$. Choosing for the evaluation of the sum S the indices $\forall m > n > N_\varepsilon$, we have

$$|S| = \left| \sum_{k=n+1}^{m} a_k b_k \right| \le 2M \left(|a_{n+1}| + |a_m| \right) < 2M \left(\frac{\varepsilon}{4M} + \frac{\varepsilon}{4M} \right) = \varepsilon.$$

Hence, according to the Cauchy criterion for uniform convergence, the series $\sum_{n=1}^{\infty} a_n(x) b_n(x)$ converges uniformly on X. $\square$

Remark 1 It can be noted, that the proof follows the same line of reasoning as in Dirichlet's test for series of numbers, slightly adjusted for uniform convergence.

Remark 2 Dirichlet's test for series of numbers guarantees immediately the pointwise convergence of the series $\sum_{n=1}^{\infty} a_n(x) b_n(x)$, but to show the uniform convergence we are forced to perform again similar transformations, just taking into account the uniformity of the properties in the hypotheses and result of the Theorem.

Abel's Test *Let $\sum_{n=1}^{\infty} u_n(x)$, $u_n(x) = a_n(x) b_n(x)$ be a series of functions defined on a set $X \subset \mathbb{R}$. If*

(1) the series $\sum_{n=1}^{\infty} b_n(x)$ converges uniformly on X,
(2) the sequence $a_n(x)$ is monotone at every fixed $x \in X$,
(3) the sequence $a_n(x)$ is uniformly bounded on X,

then the series $\sum_{n=1}^{\infty} a_n(x) b_n(x)$ converges uniformly on X.

Proof Let us show that the series $\sum_{n=1}^{\infty} a_n(x)b_n(x)$ satisfies the conditions of the Cauchy criterion for uniform convergence. First we use Abel's summation by parts formula and the telescopic representation $a_{n+1} = \sum_{k=n+1}^{m-1}(a_k - a_{k+1}) + a_m$ in order to express the sum $S \equiv \sum_{k=n+1}^{m} u_k(x) = \sum_{k=n+1}^{m} a_k(x)b_k(x)$, $m > n$ in the following manner:

$$S = \sum_{k=n+1}^{m} a_k b_k = \sum_{k=n+1}^{m-1} B_k\,(a_k - a_{k+1}) + B_m a_m - B_n a_{n+1}$$

$$= \sum_{k=n+1}^{m-1} B_k\,(a_k - a_{k+1}) + B_m a_m - B_n \left(\sum_{k=n+1}^{m-1} (a_k - a_{k+1}) + a_m \right)$$

$$= \sum_{k=n+1}^{m-1} (B_k - B_n)\,(a_k - a_{k+1}) + (B_m - B_n)\,a_m.$$

From this formula we get the following evaluation:

$$|S| \le \sum_{k=n+1}^{m-1} |B_k - B_n| \cdot |a_k - a_{k+1}| + |B_m - B_n| \cdot |a_m| \equiv S_1.$$

Since the series $\sum_{n=1}^{\infty} b_n(x)$ converges uniformly on X, it satisfies the Cauchy criterion for uniform convergence, that is, $\forall \varepsilon_1 > 0$ there exists N_{ε_1} such that for $\forall m > n > N_{\varepsilon_1}$ and simultaneously for all $x \in X$ we have

$$\left| \sum_{k=n+1}^{m} b_k \right| = |B_m - B_n| < \varepsilon_1.$$

Applying this property in the evaluation of S_1, we obtain

$$S_1 < \varepsilon_1 \left(\sum_{k=n+1}^{m-1} |a_k - a_{k+1}| + |a_m| \right) \equiv S_2.$$

Since the sequence a_n is monotone at each $x \in X$, all the differences $a_k - a_{k+1}$ have the same sign, and consequently, the sum of the absolute values is equal to the absolute value of the sum:

$$S_2 = \varepsilon_1 \left(\left| \sum_{k=n+1}^{m-1} (a_k - a_{k+1}) \right| + |a_m| \right)$$

$$= \varepsilon_1\,(|a_{n+1} - a_{n+2} + a_{n+2} - a_{n+3} + \ldots + a_{m-1} - a_m| + |a_m|)$$

$$= \varepsilon_1\,(|a_{n+1} - a_m| + |a_m|) \le \varepsilon_1\,(|a_{n+1}| + 2\,|a_m|).$$

Finally, we employ the condition of the uniform boundedness of a_n: $|a_n| \leq M$, for $\forall n \in \mathbb{N}$ and for all $x \in X$, and specify $\varepsilon_1 = \frac{\varepsilon}{3M}$ to arrive at the inequality

$$|S| = \left| \sum_{k=n+1}^{m} a_k b_k \right| < \frac{\varepsilon}{3M} \left(|a_{n+1}| + 2\,|a_m| \right) \leq \frac{\varepsilon}{3M} \cdot 3M = \varepsilon.$$

This shows that the series $\sum_{n=1}^{\infty} a_n(x) b_n(x)$ satisfies the conditions of the Cauchy criterion for uniform convergence, and consequently, this series converges uniformly on X. □

Remark 1 Frequently Dirichlet's and Abel's tests are used in a simplified form when a_n or b_n does not depend on x and represents a sequence of numbers. In this case, the condition of the uniform convergence of the series $\sum_{n=1}^{\infty} b_n(x)$ or that of the uniform boundedness of its partial sums is changed to the convergence of the series of numbers $\sum_{n=1}^{\infty} b_n$ or boundedness of its partial sums. Similarly, the uniform convergence or uniform boundedness of the sequence $a_n(x)$ is substituted by the convergence or boundedness of the sequence of numbers. To exemplify this kind of application, we consider the following test, which represents a generalization of Leibniz's test for series of numbers.

Generalized Leibniz's Test *If a sequence* $a_n(x)$ *converges uniformly to* 0 *on* X, *and for any fixed* $x \in X$ *the sequence of numbers* $a_n(x)$ *is non-negative and decreasing (that is,* $0 \leq a_{n+1}(x) \leq a_n(x)$ *for* $\forall n \in \mathbb{N}$ *and* $\forall x \in X$*), then the series* $\sum_{n=1}^{\infty}(-1)^n a_n(x)$ *converges uniformly on* X.

Proof To prove this statement, we apply Dirichlet's test with the sequence $a_n(x)$ determined in the conditions of the Generalized Leibniz's test and with $b_n(x) = (-1)^n$. In this case, the second and third conditions of Dirichlet's test are satisfied, and the partial sums B_n of the series $\sum_{n=1}^{\infty}(-1)^n$ are uniformly bounded: $|B_n| = \left| \sum_{k=1}^{n}(-1)^k \right| \leq 1$, for $\forall n$ and $\forall x \in X$. □

Remark 2 The proof of Abel's test for uniform convergence follows the line of arguments of Abel's test for numerical series, with the only adjustment for the uniformity of the properties in the hypotheses and result.

Remark 3 Abel's test for series of numbers implies immediately the pointwise convergence of the series $\sum_{n=1}^{\infty} a_n(x) b_n(x)$, but to show the uniform convergence we are forced to perform once more similar transformations.

Remark 4 It was discussed before that for series of numbers Abel's test can be deduced from Dirichlet's test. More specifically: if the conditions of Abel's test are satisfied (the series $\sum_{n=1}^{\infty} b_n$ converges, the sequence a_n is monotone and bounded) then the partial sums B_n are bounded and the sequence a_n converges to a, and consequently, representing

$\sum_{n=1}^{\infty} a_n b_n = \sum_{n=1}^{\infty} (a_n - a) b_n + \sum_{n=1}^{\infty} a b_n$, we can apply Dirichlet's test to the first series and the supposition of Abel's test to the second one. This kind of reducing does not work for uniform convergence tests: the pointwise monotonicity and uniform boundedness of the sequence $a_n(x)$ do not necessary imply the uniform convergence of $a_n(x)$ on X (albeit it implies the pointwise convergence). A simple example of such situation is the sequence $a_n(x) = x^n$ on the interval $(0, 1)$: this sequence is uniformly bounded ($|x^n| \leq 1$, for $\forall x \in (0, 1)$ and $\forall n \in \mathbb{N}$) and decreasing with respect to n (for each fixed $x \in (0, 1)$ we have $x^n > x^{n+1}$, $\forall n \in \mathbb{N}$), but it was shown before (Example 2u in Sect. 2.1, Chap. 3) that this sequence does not converge uniformly on $(0, 1)$.

Let us solve some examples with application of Dirichlet's and Abel's tests.

Example 1D $\sum_{n=1}^{\infty} (-1)^n \frac{x^n}{n}$, $X = [0, 1]$
First, let us apply Dirichlet's test. For this purpose, choose $a_n = \frac{1}{n}$ and $b_n(x) = (-1)^n x^n$. Then, the series $\sum_{n=1}^{\infty} b_n(x) = \sum_{n=1}^{\infty} (-1)^n x^n$ has the partial sums bounded uniformly on $[0, 1]$:

$$|B_n(x)| = \left| \sum_{k=1}^{n} (-1)^k x^k \right| = \left| -x \frac{1 - (-x)^n}{1 + x} \right| \leq 2 \frac{x}{1 + x} < 2, \ \forall n, \ \forall x \in [0, 1].$$

The sequence $a_n = \frac{1}{n}$ is monotone and uniformly convergent to 0 (since it does not depend on x). Thus, all the conditions of Dirichlet's test are satisfied, and consequently, the original series converges uniformly on $[0, 1]$.

To use Abel's test, we choose $a_n = x^n$ and $b_n(x) = \frac{(-1)^n}{n}$. Then, the alternating harmonic series $\sum_{n=1}^{\infty} b_n(x) = \sum_{n=1}^{\infty} \frac{(-1)^n}{n}$ converges uniformly on $[0, 1]$ (since $b_n(x)$ does not depend on x). The sequence $a_n = x^n$ is decreasing for every fixed $x \in [0, 1]$ and uniformly bounded by 1 on $[0, 1]$. Thus, all the conditions of Abel's test hold, and consequently, the original series converges uniformly on $[0, 1]$.

Example 2D $\sum_{n=1}^{\infty} \frac{(-1)^{n-1}}{n} \frac{x^n}{1+x^n}$, $X = (0, 1)$
We can apply Abel's test with $a_n = \frac{x^n}{1+x^n}$ and $b_n(x) = \frac{(-1)^{n-1}}{n}$. Indeed, the alternating harmonic series $\sum_{n=1}^{\infty} b_n(x) = \sum_{n=1}^{\infty} \frac{(-1)^{n-1}}{n}$ converges uniformly on $(0, 1)$ (since $b_n(x)$ does not depend on x). The sequence $a_n = \frac{x^n}{1+x^n}$ is uniformly bounded on $(0, 1)$ by 1 and decreasing for every fixed $x \in (0, 1)$:

$$a_n - a_{n+1} = \frac{x^n}{1 + x^n} - \frac{x^{n+1}}{1 + x^{n+1}} = \frac{x^n(1 - x)}{(1 + x^n)(1 + x^{n+1})} > 0.$$

Thus, all the conditions of Abel's test hold, and consequently, the series converges uniformly on $(0, 1)$.

Example 3D $\sum_{n=1}^{\infty} \frac{\sin nx}{\sqrt{n}}$, $X = [a, 2\pi - a], 0 < a < \pi$

To use Dirichlet's test, we consider the sequence $a_n(x) = \frac{1}{\sqrt{n}}$ (which actually does not depend on x) and the trigonometric series $\sum_{n=1}^{\infty} b_n(x)$ with $b_n(x) = \sin nx$. The sequence $a_n = \frac{1}{\sqrt{n}}$ converges uniformly and monotonically to 0. The series $\sum_{n=1}^{\infty} b_n(x)$ has the partial sums $B_n(x) = \sum_{k=1}^{n} \sin kx$ that can be calculated in the following way:

$$\sum_{k=1}^{n} \sin kx = \frac{1}{2 \sin \frac{x}{2}} \sum_{k=1}^{n} 2 \sin \frac{x}{2} \sin kx = \frac{1}{2 \sin \frac{x}{2}} \sum_{k=1}^{n} \left(\cos \left(k - \frac{1}{2} \right) x - \cos \left(k + \frac{1}{2} \right) x \right)$$

$$= \frac{1}{2 \sin \frac{x}{2}} \left(\cos \frac{x}{2} - \cos \left(n + \frac{1}{2} \right) x \right) , \forall x \neq 2k\pi, k \in \mathbb{Z}.$$

Then, $B_n(x)$ are uniformly bounded on $[a, 2\pi - a]$ according to the evaluation

$$|B_n(x)| \leq \frac{1}{|\sin \frac{x}{2}|} \leq \frac{1}{|\sin \frac{a}{2}|}, \forall x \in [a, 2\pi - a].$$

Thus, all the conditions of Dirichlet's test are satisfied, and consequently, the series converges uniformly on $[a, 2\pi - a]$.

Notice that on the interval $[0, 2\pi]$ this series converges, but not uniformly. Its pointwise convergence follows from Dirichlet's test for series of numbers: the sequence $a_n = \frac{1}{\sqrt{n}}$ converges monotonically to 0 and the sequence of the partial sums $B_n = \sum_{k=1}^{n} \sin kx$ is bounded for every fixed $x \in [0, 2\pi]$ (for $x = m\pi, m \in \mathbb{Z}$ we have $B_n = 0$ and for all other values of x we use the representation $B_n = \frac{\cos \frac{x}{2} - \cos(nx + \frac{x}{2})}{2 \sin \frac{x}{2}}$). The fact that the convergence on $[0, 2\pi]$ is not uniform can be verified by applying the Cauchy criterion. We choose $\varepsilon_0 = \frac{1}{2}$ and show that for any N it can be found $m > n > N$ and $x_N \in [0, 2\pi]$ such that $\left| \sum_{k=n+1}^{m} \frac{\sin kx_N}{\sqrt{k}} \right| > \frac{1}{2}$. Indeed, for an arbitrary N we take $n = N + 1, m = 5N$ and $x_N = \frac{\pi}{6N}$. In this case, the argument $kx_N = k\frac{\pi}{6N}$ lies between $\frac{\pi}{6}$ and $\frac{5\pi}{6}$ for all $k = N + 1, \ldots, 5N$, and consequently, $\sin kx_N \geq \frac{1}{2}$. Then,

$$\left| \sum_{k=N+1}^{5N} \frac{\sin kx_N}{\sqrt{k}} \right| \geq \left| \sum_{k=N+1}^{5N} \frac{1/2}{\sqrt{k}} \right| > \frac{1}{2} \cdot \frac{1}{\sqrt{5N}} \cdot 4N > \frac{1}{2}.$$

Since the Cauchy condition is violated, the series converges non-uniformly on $[0, 2\pi]$.

3.3   Dini's Theorem

Dini's Theorem *Let $\sum_{n=1}^{\infty} u_n(x)$ be a series of functions defined on a compact set $X \subset \mathbb{R}$. If*

(1) all the functions $u_n(x)$ have the same sign em X ($u_n(x) \geq 0$ or $u_n(x) \leq 0$ for $\forall n$ and $\forall x \in X$),
(2) the functions $u_n(x)$ are continuous on X,
(3) the series $\sum_{n=1}^{\infty} u_n(x)$ converges pontwise on X to $f(x)$,
(4) the sum of the series $f(x)$ is continuous on X,

then the series $\sum_{n=1}^{\infty} u_n(x)$ converges uniformly on X.

Proof We show that the partial sums $f_n(x) = \sum_{k=1}^{n} u_k(x)$ of series $\sum_{n=1}^{\infty} u_n(x)$ satisfy the conditions of Dini's Theorem for sequences. To specify the situation, suppose that $u_n(x) \geq 0$ on X for $\forall n$ (the case with the negative sign has the same demonstration). In this case, $f_n(x)$ is an increasing sequence at every fixed point of X. Second, each $f_n(x)$ is continuous on X as the sum of a finite number of continuous functions $u_n(x)$. Third, the pointwise convergence of the series $\sum_{n=1}^{\infty} u_n(x)$ on X to the sum $f(x)$ is equivalent (by definition) to pointwise convergence of the sequence of the partial sums $f_n(x)$ to $f(x)$. Finally, the limit function $f(x)$ is continuous on X. Thus, all the conditions of Dini's Theorem for sequences are satisfied, which guarantees the uniform convergence of the sequence $f_n(x)$ on X, that is, the series $\sum_{n=1}^{\infty} u_n(x)$ converges uniformly on X.   □

We illustrate the application of Dini's Theorem using the following series.

Example D $\sum_{n=1}^{\infty} (1-x)^2 x^n$, $X = [0, 1]$
Notice that $[0, 1]$ is a compact set and verify the conditions of Dini's Theorem. First, all the functions $u_n(x) = (1-x)^2 x^n$ are non-negative on $[0, 1]$. Second, all $u_n(x)$ are continuous on $[0, 1]$. Third, for $x \neq 0$, $x \neq 1$ the series is reduced to a geometric series whose sum is easily calculated:

$$f(x) = \sum_{n=1}^{\infty} (1-x)^2 x^n = (1-x)^2 \sum_{n=1}^{\infty} x^n = (1-x)^2 \frac{x}{1-x} = x(1-x).$$

For $x = 0$ and $x = 1$ the sum is equal to 0, and these two values can be included in the general formula for $f(x)$. Finally, the sum $f(x) = x(1-x)$ is a continuous function on $[0, 1]$. Hence, all the conditions of Dini's Theorem are satisfied, and consequently, the series converges uniformly on $[0, 1]$.

We can confirm this result by using Remark 1 on the definition of uniform convergence. First we calculate the remainders

$$r_n(x) = \sum_{k=n+1}^{\infty} (1-x)^2 x^k = (1-x)^2 \frac{x^{n+1}}{1-x} = x^{n+1}(1-x)$$

(the results for $x = 0$ and $x = 1$ can be included in this expression) and evaluate their extrema. The equation of critical points

$$r'_n(x) = (n+1)x^n - (n+2)x^{n+1} = x^n(n+1-(n+2)x) = 0$$

has only one solution $x_n = \frac{n+1}{n+2}$ in the interval $(0, 1)$. Together with the properties that $r_n(x)$ is continuous on $[0, 1]$, that $r_n(x) \geq 0, \forall x \in [0, 1]$ and at the endpoints $r_n(0) = r_n(1) = 0$, this gives the following evaluation of the remainders valid for all $x \in [0, 1]$:

$$\sup_{x\in[0,1]} |r_n(x)| = \max_{x\in[0,1]} r_n(x) = r_n(x_n) = \left(1 - \frac{n+1}{n+2}\right)\left(\frac{n+1}{n+2}\right)^{n+1}$$

$$= \frac{1}{n+2}\left(1 - \frac{1}{n+2}\right)^{n+1} \underset{n\to\infty}{\to} 0 \cdot e^{-1} = 0.$$

Therefore, by definition, the series converges uniformly on $[0, 1]$.

---

4   Properties of the Sum of Uniformly Convergent Series

> *Until now the theory of infinite series in general has been very badly grounded. One applies all the operations to infinite series as if they were finite; but is that permissible? I think not. Where is it demonstrated that one obtains the differential of an infinite series by taking the differential of each term? Nothing is easier than to give instances where this is not so.*
> Niels Henrik Abel, 1826

The results of this section are straightforward consequences of the corresponding properties of uniformly convergent sequences (see Sect. 4 of Chap. 3). Therefore, in the proofs of these properties we systematically use reducing to the conditions of the properties for sequences and make use of their conclusions.

4.1   Boundedness of a Sum

Theorem *If a series $\sum_{n=1}^{\infty} u_n(x)$ of bounded on X functions $u_n(x)$ converges uniformly on X, then its sum $f(x)$ is a bounded function on X.*

Proof The partial sums $f_n(x) = \sum_{k=1}^{n} u_k(x)$ are bounded on X functions that form the sequence uniformly convergent on X to their limit function $f(x)$. Therefore, according to the property of boundedness for uniformly convergent sequences (Sect. 4.1 in Chap. 3), the function $f(x)$ is bounded on X.   □

Remark This result can be used as an indicator of non-uniform convergence dismissing a direct investigation of the type of convergence: if a series of bounded on X functions converges on X to an unbounded sum, then this convergence is non-uniform on X.

The following examples show the relevance of uniform convergence for boundedness of the sum of a series.

Example 1s $\sum_{n=0}^{\infty} x^n$, $X = (-1, 1)$
This is the series of Examples 1b (Sect. 1) and 1u (Sect. 2.1), where it was proved that the series converges pointwise, but not uniformly on $(-1, 1)$. The functions $u_n(x) = x^n$ are bounded (even uniformly bounded) on $(-1, 1)$, but the sum of the series $f(x) = \frac{1}{1-x}$ is unbounded on this interval.

It was also shown that the convergence of this sequence is uniform on any interval $[-a, a]$, where $0 < a < 1$, and the property of boundedness states that the sum of the series should be a bounded function on $[-a, a]$, which can be checked directly:

$$|f(x)| = \left| \frac{1}{1-x} \right| \le \frac{1}{1-a}, \forall x \in [-a, a].$$

Example 2s $\sum_{n=1}^{\infty} \frac{x}{(1+x^2)^n}$, $X = \mathbb{R}$
For $x = 0$ the sum of this series is 0, and for any $x \neq 0$ this series can be reduced to a geometric series, and in this way we can both calculate its sum and show the pointwise convergence on $\mathbb{R}$:

$$\sum_{n=1}^{\infty} \frac{x}{(1+x^2)^n} = x \sum_{n=1}^{\infty} \left(\frac{1}{1+x^2} \right)^n = x \frac{\frac{1}{1+x^2}}{1 - \frac{1}{1+x^2}} = \frac{1}{x}, \forall x \neq 0.$$

Notice that all the functions $u_n(x) = \frac{x}{(1+x^2)^n}$ are bounded (even uniformly) on $\mathbb{R}$:

$$|u_n(x)| = \frac{|x|}{1+x^2} \cdot \frac{1}{(1+x^2)^{n-1}} \le \frac{1}{2}, \forall n \in \mathbb{N}, \forall x \in \mathbb{R}. \text{ However, the sum } f(x) = \begin{cases} \frac{1}{x}, & x \neq 0 \\ 0, & x = 0 \end{cases}$$

is an unbounded function on $\mathbb{R}$, and consequently, the convergence is not uniform on $\mathbb{R}$.

The last statement can be confirmed by analyzing the remainders of this series at the points $x \neq 0$:

$$|r_n(x)| = \left| \sum_{k=n+1}^{\infty} \frac{x}{(1+x^2)^k} \right| = |x| \frac{\frac{1}{(1+x^2)^{n+1}}}{1 - \frac{1}{1+x^2}} = \frac{|x|}{x^2} \cdot \frac{1}{(1+x^2)^n}.$$

Choosing the points $x_n = \frac{1}{\sqrt{n}}$, we obtain

$$|r_n(x_n)| = \sqrt{n}\left(1 + \frac{1}{n}\right)^{-n} \underset{n\to\infty}{\to} \infty,$$

that indicates that the convergence is non-uniform on $\mathbb{R}$.

At the same time, uniform convergence is a sufficient, but not necessary condition for boundedness of the sum of a series. The next two examples illustrate this point.

Example 3s $\sum_{n=1}^{\infty} u_n(x)$, $u_1(x) = f_1(x)$, $u_n(x) = f_n(x) - f_{n-1}(x)$, $f_n(x) =$
$$\begin{cases} 2n^2x, & x \in [0, \frac{1}{2n}] \\ 2n - 2n^2x, & x \in [\frac{1}{2n}, \frac{1}{n}], n \in \mathbb{N}, \ X = [0, 1]. \\ 0, & x \in [\frac{1}{n}, 1] \end{cases}$$
This series was studied in Examples 4b (Sect. 1) and 4u (Sect. 2.1), and it was found that the series converges to the sum $f(x) \equiv 0$ on $[0, 1]$, but this convergence is not uniform. However, each function $u_n(x)$ is bounded on $[0, 1]$, as well as the sum of the series.

Example 4s $\sum_{n=1}^{\infty} \frac{x^2}{(1+x^2)^n}$, $X = \mathbb{R}$
For $x = 0$ the sum of the series is 0, and for any $x \neq 0$ this series can be reduced to a geometric series:

$$\sum_{n=1}^{\infty} \frac{x^2}{(1+x^2)^n} = x^2 \sum_{n=1}^{\infty} \left(\frac{1}{1+x^2}\right)^n = x^2 \frac{\frac{1}{1+x^2}}{1 - \frac{1}{1+x^2}} = 1, \ \forall x \neq 0.$$

The functions $u_n(x) = \frac{x^2}{(1+x^2)^n}$ are bounded (even uniformly) on $\mathbb{R}$: $|u_n(x)| = \frac{x^2}{(1+x^2)^n} \leq 1$, $\forall n \in \mathbb{N}, \forall x \in \mathbb{R}$. The sum $f(x) = \begin{cases} 1, x \neq 0 \\ 0, x = 0 \end{cases}$ is also a bounded function on $\mathbb{R}$.

However, the convergence is not uniform on $\mathbb{R}$ (and on any interval containing 0). Indeed, for any $x \neq 0$, the remainders of the series can be calculated as follows

$$|r_n(x)| = \left| \sum_{k=n+1}^{\infty} \frac{x^2}{(1+x^2)^k} \right| = x^2 \frac{\frac{1}{(1+x^2)^{n+1}}}{1 - \frac{1}{1+x^2}} = \frac{1}{(1+x^2)^n}.$$

Evaluating them at the points $x_n = \frac{1}{\sqrt{n}}$, we obtain

$$|r_n(x_n)| = \left(1 + \frac{1}{n}\right)^{-n} \underset{n\to\infty}{\to} e^{-1},$$

that shows non-uniformity of the convergence on $\mathbb{R}$.

4.2   Limit of a Sum

Theorem *Let $\sum_{n=1}^{\infty} u_n(x)$ be a series defined on a set X and a be an accumulation point of X. If for every fixed n there exists the limit $\lim_{x \to a} u_n(x) = a_n$, and the series $\sum_{n=1}^{\infty} u_n(x)$ converges uniformly on X, then its sum $f(x)$ has the limit at the point a, and the series of numbers $\sum_{n=1}^{\infty} a_n$ is convergent, and $\lim_{x \to a} f(x) = \sum_{n=1}^{\infty} a_n$.*

Proof By the arithmetic properties of limits, there exist a limit of each partial sum $f_n(x)$:

$$\lim_{x \to a} f_n(x) = \lim_{x \to a} \sum_{k=1}^{n} u_k(x) = \sum_{k=1}^{n} \lim_{x \to a} u_k(x) = \sum_{k=1}^{n} a_n \equiv A_n.$$

By the condition of the Theorem, the sequence of the partial sums $f_n(x)$ converges uniformly on X. Then this sequence satisfies the conditions of the property of limit of uniformly convergent sequences (Sect. 4.2, Chap. 3), and consequently, the following results of this property are valid: the limit function $f(x)$ has a limit at the point a, the sequence of numbers A_n is convergent and $\lim_{x \to a} f(x) = \lim_{n \to \infty} A_n$. Rewording in the terms of series, we have: the sum $f(x)$ of the series $\sum_{n=1}^{\infty} u_n(x)$ has a limit at the point a, the series of numbers $\sum_{n=1}^{\infty} a_n$ (with partial sums A_n) is convergent and $\lim_{x \to a} f(x) = \lim_{n \to \infty} A_n = \sum_{n=1}^{\infty} a_n$, that is, all the results of the Theorem are proved.   □

The demonstrated formula can be written in the form

$$\lim_{x \to a} \sum_{n=1}^{\infty} u_n(x) = \sum_{n=1}^{\infty} \lim_{x \to a} u_n(x), \tag{4.8}$$

that is, the order of the limit and the summation is interchangeable, or equivalently, it is permissible to pass to the limit inside the summation sign.

Remark If the limits of $u_n(x)$ exist and the series $\sum_{n=1}^{\infty} u_n(x)$ converges on X, but the order of operations is not interchangeable, then the convergence is not uniform on X.

The following example shows that the uniform convergence is an important condition to ensure that the order of operations is interchangeable.

Example 11 $\sum_{n=0}^{\infty} (-1)^n (1 - x) x^{2n}$, $X = (0, 1)$, $a = 1$
At every point $x \in (0, 1)$ the series is geometric convergent to $f(x) = \frac{1-x}{1+x^2}$. Therefore, $\lim_{x \to 1} f(x) = 0$. For every function $u_n(x)$ we have $\lim_{x \to 1} u_n(x) = 0$. Consequently,

$$\lim_{x \to 1} \sum_{n=0}^{\infty} u_n(x) = \lim_{x \to 1} f(x) = 0 = \sum_{n=0}^{\infty} 0 = \sum_{n=0}^{\infty} \lim_{x \to 1} u_n(x),$$

that is, the order of the operations is interchangeable. Let us see that this is guaranteed by the uniformity of convergence on $(0, 1)$. For any $x \in (0, 1)$ the remainders of this series are found by the formula

$$r_n(x) = \sum_{k=n+1}^{\infty} (-1)^n (1 - x) x^{2n} = (-1)^{n+1} (1 - x) x^{2n+2} \frac{1}{1 + x^2}.$$

Then,

$$|r_n(x)| = \frac{(1 - x) x^{2n+2}}{1 + x^2} \le (1 - x) x^{2n+2} \equiv g(x).$$

The derivative $g'(x) = x^{2n+1}(2n + 2 - (2n + 3)x)$ gives the only critical point $x_n = \frac{2n+2}{2n+3} \in (0, 1)$, which is the maximum of $g(x)$ since the derivative changes the sign from positive to negative at x_n. Consequently, for $\forall x \in (0, 1)$ we obtain the evaluation

$$|r_n(x)| \le g(x) \le \max_{(0,1)} g(x) = g(x_n) = \left(1 - \frac{2n + 2}{2n + 3}\right) \left(\frac{2n + 2}{2n + 3}\right)^{2n+2}$$

$$= \frac{1}{2n + 3} \left(1 + \frac{1}{2n + 2}\right)^{-(2n+2)} \underset{n \to \infty}{\to} 0 \cdot e^{-1} = 0,$$

which means that the series converges uniformly on $(0, 1)$.

   At the same time, the uniform convergence is a sufficient but not necessarily condition of the possibility to interchange the order of the operations, as is shown in the next example.

Example 21 $\sum_{n=1}^{\infty} \frac{x}{(x+n)(x+n+1)}$, $X = (0, +\infty)$, $a = 0$
At every point $x \in (0, +\infty)$ the convergence can be proved by representing the series in a telescopic form. For the partial sums we have

$$f_n(x) = \sum_{k=1}^{n} \frac{x}{(x + k)(x + k + 1)} = \sum_{k=1}^{n} x \left(\frac{1}{x + k} - \frac{1}{x + k + 1}\right)$$

$$= x \left(\frac{1}{x + 1} - \frac{1}{x + 2} + \frac{1}{x + 2} - \frac{1}{x + 3} + \ldots + \frac{1}{x + n} - \frac{1}{x + n + 1}\right) = \frac{x}{x + 1} - \frac{x}{x + n + 1},$$

whence $f(x) = \lim_{n \to \infty} f_n(x) = \frac{x}{x+1}$, $\forall x \in (0, +\infty)$. For every function $u_n(x)$ we have $\lim_{x \to 0} u_n(x) = 0$ and for $f(x)$ we get $\lim_{x \to 0} f(x) = 0$. Therefore,

$$\lim_{x \to 0} \sum_{n=1}^{\infty} u_n(x) = \lim_{x \to 0} f(x) = 0 = \sum_{n=1}^{\infty} 0 = \sum_{n=1}^{\infty} \lim_{x \to 0} u_n(x),$$

that is, the order of the operations is interchangeable. Nevertheless, the series converges non-uniformly on $(0, +\infty)$: at the points $x_n = n \in (0, +\infty)$ the remainders take the form

$$|r_n(x_n)| = |f(x_n) - f_n(x_n)| = \frac{x_n}{x_n + n + 1} = \frac{n}{2n + 1} \underset{n \to \infty}{\to} \frac{1}{2} \neq 0.$$

4.3 Continuity of a Sum

Theorem *If a series $\sum_{n=1}^{\infty} u_n(x)$ of continuous on X functions $u_n(x)$ converges uniformly on X, then its sum $f(x)$ is continuous on X.*

This result can be expressed in the following manner:

$$\lim_{x \to a} \sum_{n=1}^{\infty} u_n(x) = \lim_{x \to a} f(x) = f(a) = \sum_{n=1}^{\infty} u_n(a) = \sum_{n=1}^{\infty} \lim_{x \to a} u_n(x), \ \forall a \in X.$$

$$(4.9)$$

Proof The continuity of $u_n(x)$ implies the continuity of the partial sums $f_n(x)$, and the uniform convergence of the series means the uniform convergence of the sequence $f_n(x)$. Hence, the conditions of the property of continuity of uniformly convergent sequences are satisfied (see Sect. 4.3 in Chap. 3), and therefore, the limit function $f(x)$ (that is, the sum of the series) is a continuous function on X.   □

Remark This result can be used as an indicator of non-uniform convergence dismissing a direct investigation of the type of convergence: if a series of continuous on X functions converges on X to a discontinuous sum, then this convergence is non-uniform on X.

The following examples show that the uniform convergence is an indispensable condition to guarantee the continuity of the sum of a series.

Example 1c $\sum_{n=1}^{\infty} (1 - x) x^n$, $X = [0, 1]$
In Examples 2b (Sect. 1) and 2u (Sect. 2.1) it was shown that this series converges pointwise, but not uniformly on $[0, 1]$. In the consequence of this, the functions $u_n(x) = (1 - x) x^n$ are continuous on $[0, 1]$, but the sum $f(x) = \begin{cases} x, & x \in [0, 1) \\ 0, & x = 1 \end{cases}$ is discontinuous at $x = 1$.

Example 2c $\sum_{n=1}^{\infty} (1 - \sin x) \sin^n x$, $X = [0, \pi]$

At the points $x = 0, \frac{\pi}{2}, \pi$ the general term of this series vanishes, and consequently, the sum of the series is 0. At all other points of the interval $[0, \pi]$ we can use the formula of a geometric series to calculate the sum:

$$\sum_{n=1}^{\infty} (1 - \sin x) \sin^n x = (1 - \sin x) \sum_{n=1}^{\infty} \sin^n x = (1 - \sin x) \frac{\sin x}{1 - \sin x} = \sin x.$$

Therefore, the series converges on $[0, \pi]$ to the function $f(x) = \begin{cases} \sin x, & x \neq \frac{\pi}{2} \\ 0, & x = \frac{\pi}{2} \end{cases}$. Since all the functions $u_n(x) = (1 - \sin x) \sin^n x$ are continuous on $[0, \pi]$, but the sum $f(x)$ is discontinuous at $\frac{\pi}{2}$, we can conclude that the convergence is non-uniform on $[0, \pi]$.

At the same time, the uniform convergence is a sufficient but not necessary condition for the continuity of the sum of a series. The next series exemplifies this situation.

Example 3c $\sum_{n=1}^{\infty} x^n$, $X = (-1, 1)$
It was found in Examples 1b (Sect. 1) and 1u (Sect. 2.1) that this series converges non-uniformly on $(-1, 1)$ to the sum $f(x) = \frac{1}{1-x}$. Even so, all the functions $u_n(x) = x^n$ and the sum $f(x)$ are continuous on $(-1, 1)$.

4.4   Integrability of a Sum (Integration Term by Term)

Theorem *If a series $\sum_{n=1}^{\infty} u_n(x)$ of Riemann integrable on $[a, b]$ functions converges uniformly on $[a, b]$ to the sum $f(x)$, which is also Riemann integrable on $[a, b]$, then*

$$\sum_{n=1}^{\infty} \int_a^b u_n(x) dx = \int_a^b f(x) dx = \int_a^b \sum_{n=1}^{\infty} u_n(x) dx . \tag{4.10}$$

This means that the order of summation and integration can be interchanged, or equivalently, the series can be integrated term by term.

Proof The integrability of $u_n(x)$ on $[a, b]$ guarantees the integrability of the partial sums $f_n(x)$ on $[a, b]$, and the uniform convergence of the series means the uniform convergence of the sequence $f_n(x)$. Hence, the conditions of the property of integrability of uniformly convergent sequences are satisfied (Sect. 4.4, Chap. 3), and consequently, the change of the order of the summation and integration in formula (4.10) is justified.   □

Corollary *A weaker formulation of the Theorem, which has a broad practical application, is as follows:*

> If a series $\sum_{n=1}^{\infty} u_n(x)$ of continuous on $[a, b]$ functions $u_n(x)$ converges uniformly on $[a, b]$ to the sum $f(x)$, then $f(x)$ is Riemann integrable on $[a, b]$ and formula (4.10) is true.

The main advantage of this formulation is that it omits any requirement to the sum of a series. This result follows straightforward from Corollary 1 of the Theorem of integrability of uniformly convergent sequences (Sect. 4.4, Chap. 3).

As the Theorem states, the uniform convergence is an essential condition for the term-by-term integration. The next example shows that formula (4.10) can be false if the convergence is non-uniform.

Example 1i $\sum_{n=1}^{\infty} u_n(x)$, $u_1(x) = f_1(x)$, $u_n(x) = f_n(x) - f_{n-1}(x)$, $f_n(x) =$
$$\begin{cases} 2n^2x, & x \in [0, \frac{1}{2n}] \\ 2n - 2n^2x, & x \in [\frac{1}{2n}, \frac{1}{n}], n \in \mathbb{N}, \ X = [0, 1]. \\ 0, & x \in [\frac{1}{n}, 1] \end{cases}$$
This series was considered in Examples 4b (Sect. 1) and 4u (Sect. 2.1), where it was demonstrated that the series converges to $f(x) \equiv 0$ non-uniformly on $[0, 1]$. For this reason, the integral of the sum of this series and the corresponding series of integrals have different values: $\int_0^1 f(x)dx = \int_0^1 0dx = 0$, but

$$\int_0^1 u_1(x)dx = \int_0^1 f_1(x)dx = \frac{1}{2}, \ \int_0^1 u_n(x)dx = \int_0^1 (f_n(x) - f_{n-1}(x))dx = 0, \ n = 2, 3, \ldots,$$

and consequently, $\sum_{n=1}^{\infty} \int_0^1 u_n(x)dx = \frac{1}{2}$. Hence,

$$\sum_{n=1}^{\infty} \int_0^1 u_n(x)dx \neq \int_0^1 \sum_{n=1}^{\infty} u_n(x)dx.$$

The next example evidences that the uniform convergence is a sufficient but not necessary condition for term-by-term integration.

Example 2i $\sum_{n=1}^{\infty} (1 - x)x^n$, $X = [0, 1]$
This series was already studied in Examples 2b, 2u, and, 1c (Sects. 1, 2.1, and 4.3). It converges pointwise but not uniformly on $[0, 1]$ to the sum $f(x) = \begin{cases} x, & x \in [0, 1) \\ 0, & x = 1 \end{cases}$. One of the indicators of non-uniform convergence is that the functions $u_n(x) = (1 - x)x^n$ are

continuous, while $f(x)$ is discontinuous on $[0, 1]$. Nevertheless, in this case, the lack of uniform convergence does not impede the use of formula (4.10). Indeed, the integral of the sum is $\int_0^1 f(x)dx = \int_0^1 x dx = \frac{1}{2}$ and the integral of the general term is

$$
\int_0^1 u_n(x)dx = \int_0^1 (1-x)x^n dx = \left(\frac{x^{n+1}}{n+1} - \frac{x^{n+2}}{n+2} \right)\Big|_0^1 = \frac{1}{n+1} - \frac{1}{n+2}.
$$

The sum of the last integrals gives a telescopic series with the partial sums $a_n = \sum_{k=1}^n (\frac{1}{k+1} - \frac{1}{k+2}) = \frac{1}{2} - \frac{1}{n+2}$. Then,

$$
\sum_{n=1}^\infty \int_0^1 u_n(x)dx = \sum_{n=1}^\infty \left(\frac{1}{n+1} - \frac{1}{n+2} \right) = \lim_{n\to\infty} a_n = \lim_{n\to\infty} \left(\frac{1}{2} - \frac{1}{n+2} \right) = \frac{1}{2}
$$

$$
= \int_0^1 f(x)dx = \int_0^1 \sum_{n=1}^\infty u_n(x)dx.
$$

Complement: Theorem of Integrability with a Stronger Formulation *If a series $\sum_{n=1}^\infty u_n(x)$ of Riemann integrable on $[a, b]$ functions $u_n(x)$ converges uniformly on $[a, b]$, then its sum $f(x)$ is also Riemann integrable on $[a, b]$ and*

$$
\sum_{n=1}^\infty \int_a^b u_n(x)dx = \int_a^b f(x)dx = \int_a^b \sum_{n=1}^\infty u_n(x)dx.
$$

Proof This result is a direct consequence of the Complement of the property of integrability in Sect. 4.4 of Chap. 3. □

4.5 Differentiability of a Sum (Differentiation Term by Term)

Theorem *Let $\sum_{n=1}^\infty u_n(x)$ be a series of functions defined on an interval $[a, b]$. Suppose that each $u_n(x)$ is differentiable on $[a, b]$ and its derivative $u_n'(x)$ is Riemann integrable on $[a, b]$. If the series $\sum_{n=1}^\infty u_n(x)$ converges pointwise on $[a, b]$ to $f(x)$, and the series of derivatives $\sum_{n=1}^\infty u_n'(x)$ converges uniformly on $[a, b]$ to a continuous function $g(x)$, then $f(x)$ is differentiable on $[a, b]$ and*

$$
f'(x) = g(x), \forall x \in [a, b].
$$

Another mode to express this result is as follows:

$$\left(\sum_{n=1}^{\infty} u_n(x)\right)' = \sum_{n=1}^{\infty} u_n'(x) \, , \forall x \in [a, b] \, , \tag{4.11}$$

that is, the order of summation and differentiation is interchangeable, or equivalently, the series is differentiable term by term.

Proof The differentiability of $u_n(x)$ implies the differentiability of the partial sums $f_n(x) = \sum_{k=1}^{n} u_k(x)$, and the integrability of $u_n'(x)$ on $[a, b]$ results in the integrability of the partial sums $f_n'(x) = \sum_{k=1}^{n} u_k'(x)$ in this interval. The convergence of the series $\sum_{n=1}^{\infty} u_n(x)$ means the convergence of the sequence $f_n(x)$, and the uniform convergence of $\sum_{n=1}^{\infty} u_n'(x)$ to a continuous sum $g(x)$ means the uniform convergence of the sequence $f_n'(x)$ to $g(x)$. Hence, all the conditions of the property of differentiability of sequences are satisfied (see Sect. 4.5 in Chap. 3), and consequently, the result of this Theorem and formula (4.11) are proved.   □

Corollary *The following weaker formulation of the Theorem is quite useful since it does not require any information about properties of the function $g(x)$:*

Suppose that each function $u_n(x)$ is continuously differentiable on $[a, b]$. If the series $\sum_{n=1}^{\infty} u_n(x)$ converges pointwise on $[a, b]$ to $f(x)$, and the series of derivatives $\sum_{n=1}^{\infty} u_n'(x)$ converges uniformly on $[a, b]$ to $g(x)$, then $f(x)$ is differentiable on $[a, b]$ and $f'(x) = g(x), \forall x \in [a, b]$.

This result follows directly from the Corollary to the Theorem of differentiation of sequences (see Sect. 4.5 in Chap. 3).

The next two examples show that the uniform convergence of the series of derivatives is an essential condition and that "good" properties of the original series do not guarantee the result of the Theorem.

Example 1d $\sum_{n=1}^{\infty} \frac{\sin n^2 x}{n^2}$, $X = \mathbb{R}$
In Example 3b of Sect. 1 and Example 3u of Sect. 2.1 it was shown that this series converges uniformly on $\mathbb{R}$, but the series of derivatives $\sum_{n=1}^{\infty} u_n'(x) = \sum_{n=1}^{\infty} \cos n^2 x$ diverges at every x.

Example 2d $\sum_{n=1}^{\infty} \frac{\sin nx}{\sqrt{n}}$, $X = \mathbb{R}$
This series was considered in Example 3D (Sect. 3.2), where it was proved that it converges uniformly on $[a, 2\pi - a], 0 < a < \pi$. However, the series of derivatives $\sum_{n=1}^{\infty} u_n'(x) = \sum_{n=1}^{\infty} \sqrt{n} \cos nx$ diverges at every x, since the general term $u_n'(x)$ does not approach 0 (in

Example 3b of Sect. 1, Chap. 3 it was shown that the sequence $\sqrt{n}\cos nx$ is unbounded at every real x).

Complement: Example of a Series Convergent Non-uniformly, Whose Series of Derivatives Also Converges Non-uniformly, But Term-by-Term Differentiation Still Can Be Used

The next example reveals that the uniform convergence of the series of derivatives is an important but not necessary condition for term-by-term differentiation.

Example 3d $\sum_{n=1}^{\infty}\left(nx^2e^{-n^2x^4}-(n-1)x^2e^{-(n-1)^2x^4}\right)$, $X=\mathbb{R}$

This series is telescopic and it is easy to find its partial sums:

$$f_n(x)=\sum_{k=1}^{n}\left(kx^2e^{-k^2x^4}-(k-1)x^2e^{-(k-1)^2x^4}\right)$$

$$=x^2e^{-x^4}+2x^2e^{-2^2x^4}-x^2e^{-x^4}+\ldots+nx^2e^{-n^2x^4}-(n-1)x^2e^{-(n-1)^2x^4}=nx^2e^{-n^2x^4}.$$

Then, at the point 0 we have $f(0)=\lim_{n\to\infty}f_n(0)=0$, and at any other point $x\neq 0$ we can calculate the limit applying L'Hospital's rule with the change of the variable $t=nx^2$:

$$f(x)=\lim_{n\to\infty}f_n(x)=\lim_{n\to\infty}\frac{nx^2}{e^{n^2x^4}}=\lim_{t\to+\infty}\frac{t}{e^{t^2}}=\lim_{t\to+\infty}\frac{1}{2te^{t^2}}=0.$$

Hence, the sum of the series is $f(x)\equiv 0,\ \forall x\in\mathbb{R}$.

The functions $u_n=nx^2e^{-n^2x^4}-(n-1)x^2e^{-(n-1)^2x^4}$ are differentiable on $\mathbb{R}$ and the series of derivatives has the form

$$\sum_{n=1}^{\infty}u_n'(x)=\sum_{n=1}^{\infty}\left(2nxe^{-n^2x^4}-4n^3x^5e^{-n^2x^4}-2(n-1)xe^{-(n-1)^2x^4}+4(n-1)^3x^5e^{-(n-1)^2x^4}\right).$$

This is one more telescopic series and its partial sums are found as follows:

$$g_n(x)=f_n'(x)=\sum_{k=1}^{n}u_k'(x)=2xe^{-x^4}-4x^5e^{-x^4}+2\cdot 2xe^{-2^2x^4}-4\cdot 2^3x^5e^{-2^2x^4}-2xe^{-x^4}+4x^5e^{-x^4}$$

$$\ldots+2nxe^{-n^2x^4}-4n^3x^5e^{-n^2x^4}-2(n-1)xe^{-(n-1)^2x^4}+4(n-1)^3x^5e^{-(n-1)^2x^4}=2nxe^{-n^2x^4}-4n^3x^5e^{-n^2x^4}.$$

Again, at the point 0 we have $g(0) = \lim_{n\to\infty} g_n(0) = 0$, and at any other point $x \neq 0$ we calculate the limit by changing the variable $t = nx^2$ and using L'Hospital's rule:

$$g(x) = \lim_{n\to\infty} g_n(x) = \lim_{n\to\infty} \frac{2nx\left(1 - 2n^2 x^4\right)}{e^{n^2 x^4}} = \lim_{t\to+\infty} \frac{2}{x} \frac{t - 2t^3}{e^{t^2}} = \frac{2}{x} \lim_{t\to+\infty} \frac{1 - 6t^2}{2te^{t^2}}$$

$$= \frac{1}{x}\left(\lim_{t\to+\infty} \frac{1}{te^{t^2}} - \lim_{t\to+\infty} \frac{6t}{e^{t^2}}\right) = \frac{1}{x}\left(\lim_{t\to+\infty} \frac{1}{te^{t^2}} - \lim_{t\to+\infty} \frac{6}{2te^{t^2}}\right) = 0.$$

Therefore, the series of derivatives also converges to zero: $g(x) \equiv 0$, $\forall x \in \mathbb{R}$. From these results, we conclude that the original series can be differentiated term by term:

$$\left(\sum_{n=1}^{\infty} u_n(x)\right)' = f'(x) = 0 = g(x) = \sum_{n=1}^{\infty} u_n'(x), \forall x \in \mathbb{R},$$

that is, formula (4.11) is valid in this case.

Let us determine the character of convergence of the series $\sum_{n=1}^{\infty} u_n(x)$ and $\sum_{n=1}^{\infty} u_n'(x)$. For the first series, choosing $x_n = \frac{1}{\sqrt{n}}$, $\forall n \in \mathbb{N}$, we obtain

$$\left|\sum_{k=n+1}^{\infty} u_k(x_n)\right| = |f_n(x_n) - f(x_n)| = nx_n^2 e^{-n^2 x_n^4} = e^{-1} \underset{n\to\infty}{\not\to} 0.$$

This means that the series converges non-uniformly on $\mathbb{R}$ (and on any open interval containing 0). Analogously, using the same sequence of the points $x_n = \frac{1}{\sqrt{n}}$ for the second series, we get

$$\left|\sum_{k=n+1}^{\infty} u_k'(x_n)\right| = |g_n(x_n) - g(x_n)| = \left|2nx_n\left(1 - 2n^2 x_n^4\right)e^{-n^2 x_n^4}\right| = 2\sqrt{n}e^{-1} \underset{n\to\infty}{\not\to} 0,$$

which shows that the second series also converges non-uniformly on $\mathbb{R}$.

Thus, we have constructed the series which does not converge uniformly, and neither does the series of derivatives, but even so the rule of term-by-term differentiation is applicable.

Complement: Theorem of Differentiability with a Stronger Formulation

Let $\sum_{n=1}^{\infty} u_n(x)$ be a series of functions defined on an interval $[a, b]$. Suppose that each $u_n(x)$ is differentiable on $[a, b]$. If the series of derivatives $\sum_{n=1}^{\infty} u_n'(x)$ converges uniformly on $[a, b]$ to $g(x)$ and the original series $\sum_{n=1}^{\infty} u_n(x)$ converges at a point $x_0 \in [a, b]$, then the series $\sum_{n=1}^{\infty} u_n(x)$ converges uniformly on $[a, b]$, its sum $f(x)$ is *a differentiable function on* $[a, b]$ *and*

$$f'(x) = g(x), \forall x \in [a, b].$$

Proof This result follows immediately from the Complement to the property of differentiation of sequences (see Sect. 4.5 in Chap. 3).   □

Historical Remarks For mathematicians of the eighteenth century, including Euler and Lagrange, it went without saying that term-by-term operations (continuity, differentiation, integration) on series of functions are valid and equivalent to the corresponding operations on the sum of series. Cauchy in "Cours d'analyse" provided a "proof" that any infinite series of continuous functions is continuous. The first to pointed out the problem in the Cauchy statement was Abel in his paper on infinite series in 1826. Albeit Abel proposed some restrictions on the series to remedy this issue, he did not detected the principal problem—the kind of convergence used in the Cauchy formulation.

In the 1840s, three mathematicians discovered independently the necessity of a stronger notion of convergence to guarantee term-by-term operations: Weierstrass in 1841, Stokes in 1847 and Seidel in 1848. The most consistent and general concept of uniform convergence was proposed by Weierstrass. Cauchy returned to his theorem in a paper of 1853, but he did not show a progress in the solution of the problem, just observing that the theorem is applicable for a power series, while some restrictions are required for general series of functions.

The problem of term-by-term operations applied to series of functions was not solved until Weierstrass developed his concept of uniform convergence, popularized it and emphasized it importance in his lectures in the University of Berlin. Although he did not published the results on uniform convergence until 1894, his approach was widely disseminated by his students and pupils starting from the 1860s.

5 Complement: The Weierstrass Function—Everywhere Continuous and Nowhere Differentiable Function

> *Weierstrass gave examples of continuous functions of a real argument which, for any value of this argument, do not possess a finite derivative. A hundred years ago, such a function would have been regarded as an outrage to common sense. A continuous function, one would have said, is in essence susceptible of being represented by a curve, and a curve obviously always has a tangent.*
> *Jules Henri Poincaré, 1898*

Example of the Weierstrass Function A series $\sum u_n(x)$ of continuous functions converges uniformly on X, but nevertheless the sum of this series is not differentiable at any point of X.

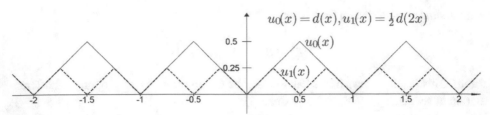

Fig. 4.1 The distance functions $u_0(x) = d(x)$ and $\tilde{u}_1(x) = \frac{1}{2}d(2x)$ (the scaled factor 2 is used instead of 10 for better visualization)

Let $d(x)$ be the distance from x to the nearest integer, which means that $d(x)$ is a continuous 1-periodic function with the range $\left[0, \frac{1}{2}\right]$ defined on the fundamental interval

$[0, 1]$ in the form: $d(x) = \begin{cases} x, & x \in \left[0, \frac{1}{2}\right] \\ 1 - x, & x \in \left[\frac{1}{2}, 1\right] \end{cases}$. Accordingly, $u_n(x) = \frac{1}{10^n}d(10^n x)$,

$n = 0, 1, \ldots$ is a continuous $\frac{1}{10^n}$-periodic function with the range $\left[0, \frac{1}{2 \cdot 10^n}\right]$ defined on

the fundamental interval $\left[0, \frac{1}{10^n}\right]$ in the form: $u_n(x) = \begin{cases} x, & x \in \left[0, \frac{1}{2 \cdot 10^n}\right] \\ \frac{1}{10^n} - x, & x \in \left[\frac{1}{2 \cdot 10^n}, \frac{1}{10^n}\right] \end{cases}$,

$n = 0, 1, \ldots$. The illustration of the first two distance functions is given in Fig. 4.1.

Let us consider the series $\sum_{n=0}^{\infty} u_n(x)$ on $X = [0, 1]$. Since $|u_n(x)| \leq \frac{1}{10^n}$ for all $x \in [0, 1]$ and the numerical series $\sum_{n=0}^{\infty} \frac{1}{10^n}$ converges, by the Weierstrass test, the series $\sum_{n=0}^{\infty} u_n(x)$ converges uniformly on $[0, 1]$ to its sum $f(x)$, and the continuity of $u_n(x)$ implies the continuity of $f(x)$. The partial sums of a slightly modified series, with $\tilde{u}_n(x) = \frac{1}{2^n}d(2^n x)$ instead of $u_n(x) = \frac{1}{10^n}d(10^n x)$ for better visualizations, are shown in Figs. 4.2 and 4.3.

Now let us show that $f(x)$ is differentiable nowhere in $[0, 1]$. Let us consider any fixed a in $[0, 1)$ with the decimal expansion $a = 0.a_1a_2\ldots a_na_{n+1}\ldots$, and choose a particular sequence h_m approaching 0, as m approaches infinity, in the form: $h_m = \begin{cases} -10^{-m}, & a_m = 4 \text{ or } 9 \\ 10^{-m}, & otherwise \end{cases}$. Then

$$\frac{f(a + h_m) - f(a)}{h_m} = \sum_{n=0}^{\infty} \frac{1}{10^n} \cdot \frac{d(10^n(a + h_m)) - d(10^n a)}{\pm 10^{-m}}$$

$$= \sum_{n=0}^{\infty} \pm 10^{m-n} \cdot \left[d\left(10^n(a + h_m)\right) - d\left(10^n a\right)\right].$$

Since $10^n h_m$ is an integer for $n \geq m$, we get $d(10^n(a + h_m)) - d(10^n a) = 0$. Therefore, the last series is actually a finite sum with m terms. For these terms (when $n < m$) we can represent $10^n a = a_1 \ldots a_n + 0. a_{n+1}a_{n+2}\ldots a_m \ldots$ and $10^n(a + h_m) = a_1 \ldots a_n +$

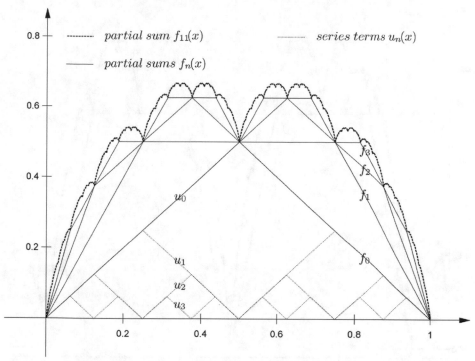

Fig. 4.2 Weierstrass function: partial sums of series $\sum_{n=0}^{\infty} \tilde{u}_n(x)$ with $\tilde{u}_n(x) = \frac{1}{2^n} d\left(2^n x\right)$ instead of $u_n(x) = \frac{1}{10^n} d\left(10^n x\right)$ for better visualization

$0.\,a_{n+1} a_{n+2} \ldots (a_m \pm 1) \ldots$, where $a_1 \ldots a_n$ is the integer part of both numbers. (Notice that except for m-th decimal digit, other digits are not changed, because we have chosen $h_m = -10^{-m}$ if $a_m = 9$). If it happens that $0.\,a_{n+1} a_{n+2} \ldots a_m \ldots < \frac{1}{2}$, then also $0.\,a_{n+1} a_{n+2} \ldots (a_m \pm 1) \ldots < \frac{1}{2}$ (in the special case $n = m - 1$, the last inequality is true because we have chosen $h_m = -10^{-m}$ when $a_m = 4$). Therefore,

$$d\left(10^n (a + h_m)\right) - d\left(10^n a\right) = d\left(0.a_{n+1} \ldots (a_m \pm 1) \ldots\right) - d\left(0.a_{n+1} \ldots a_m \ldots\right)$$

$$= 0.a_{n+1} \ldots (a_m \pm 1) \ldots - 0.a_{n+1} \ldots a_m \ldots = \pm 10^{n-m}.$$

Similarly, if $0.\,a_{n+1} a_{n+2} \ldots a_m \ldots \geq \frac{1}{2}$, then $0.\,a_{n+1} a_{n+2} \ldots (a_m \pm 1) \ldots \geq \frac{1}{2}$ (here, between two equivalent representations of the rational numbers $0.a_{n+1} \ldots a_k 00 \ldots = 0.a_{n+1} \ldots (a_k - 1) 99 \ldots$ we use the former), and consequently

$$d\left(10^n (a + h_m)\right) - d\left(10^n a\right) = d\left(0.a_{n+1} \ldots (a_m \pm 1) \ldots\right) - d\left(0.a_{n+1} \ldots a_m \ldots\right)$$

$$= (1 - 0.a_{n+1} \ldots (a_m \pm 1) \ldots) - (1 - 0.a_{n+1} \ldots a_m \ldots) = \mp 10^{n-m}.$$

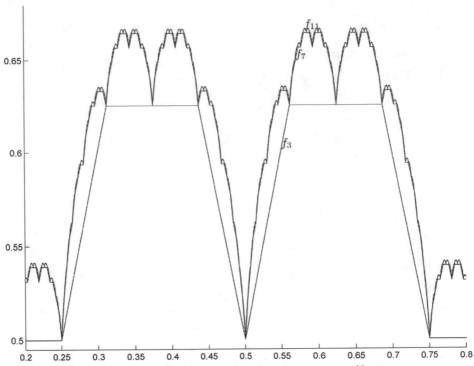

Fig. 4.3 Weierstrass function: amplified view of partial sums of series $\sum_{n=0}^{\infty} \tilde{u}_n(x)$

Thus, for $n < m$ we have

$$10^{m-n} \cdot \left[d \left(10^n \left(a + h_m \right) \right) - d \left(10^n a \right) \right] = \pm 1,$$

and, therefore,

$$\frac{f(a + h_m) - f(a)}{h_m} = \sum_{n=0}^{m-1} \pm 10^{m-n} \cdot \left[d \left(10^n \left(a + h_m \right) \right) - d \left(10^n a \right) \right] = \sum_{n=0}^{m-1} \pm 1.$$

The last sum is an even integer when m is even, and an odd integer when m is odd. Therefore, the sequence $\frac{f(a+h_m)-f(a)}{h_m} = \sum_{n=0}^{m-1} \pm 1$ does not converge as m approaches infinity, because this is a sequence of integers which are alternately odd and even. Hence, we have constructed a sequence $h_m = \pm 10^{-m}$ such that the limit of $\frac{f(a+h_m)-f(a)}{h_m}$ does not exist, which means that $f(x)$ is not differentiable at a.

Remark Of course, all the above considerations and conclusions are also true on $\mathbb{R}$, because all the functions $u_n(x) = \frac{1}{10^n} d(10^n x)$, $n = 0, 1, 2, \ldots$ have the common period 1, and so does $f(x)$. Therefore, their properties on $[0, 1]$ are automatically extended to $\mathbb{R}$.

Historical Remarks Examples of everywhere continuous and nowhere differentiable functions were discovered and rediscovered several times during the 19th and 20th centuries. Apparently, the first official register of such a function is the record of a talk given by Weierstrass at the Berlin Academy of Science in 1872, although it seems to be generally recognized that Weierstrass have shown an example of this kind as early as 1861 in the classroom lectures at the University of Berlin. However, examples of everywhere continuous and nowhere differentiable functions were created even before Weierstrass, but those findings of the researches were not published in their lifetimes. As far as we know now, the first construction of such a function was performed by Bolzano sometime in the 1830s, but his manuscript was not published until 1930. Bolzano succeeded in constructing a function everywhere continuous and nowhere differentiable, and proved that the function has these properties on a dense subset of argument values.

During the period between 1872 and 1930, a number of papers were published improving the construction of Weierstrass or simply rediscovering the same example. The main progress in simplification of the Weierstrass example was made by Takagi in 1903, whose construction was used in many subsequent presentations and references. This version is frequently presented in the modern books of Analysis, albeit with the reference to the work of van der Waerden, who rediscovered the Takagi example in 1930. In this text we follow closely the construction of the Weierstrass function proposed by Takagi.

Exercises

1. Find the sets of absolute and conditional convergence of the series of functions:

   (1) $\sum_{n=1}^{\infty} \frac{n}{x^n}$;

   (2) $\sum_{n=1}^{\infty} \frac{(-1)^n}{3n+2} \left(\frac{1-x}{1+x}\right)^n$;

   (3) $\sum_{n=1}^{\infty} \frac{n}{3n+1} \left(\frac{x}{3x+1}\right)^n$;

   (4) $\sum_{n=1}^{\infty} \frac{n3^{2n}}{2^n} x^n (1-x)^n$;

   (5) $\sum_{n=1}^{\infty} \frac{x^n}{1+x^{2n}}$;

   (6) $\sum_{n=1}^{\infty} \frac{x^n}{1-x^n}$;

   (7) $\sum_{n=1}^{\infty} \sin(nx)$;

   (8) $\sum_{n=1}^{\infty} \cos(nx)$;

   (9) $\sum_{n=1}^{\infty} n^2 e^{-nx}$;

   (10*) $\sum_{n=1}^{\infty} \frac{\ln(1+x^n)}{n^p}$, $x \geq 0$;

   (11) $\sum_{n=1}^{\infty} \frac{2^n \sin^n x}{n^2}$;

   (12) $\sum_{n=2}^{\infty} \frac{2^{n/2} \cos^n x}{n \ln n}$;

   (13) $\sum_{n=1}^{\infty} (a-x)(a-x^{1/2})(a-x^{1/3}) \cdot \ldots \cdot (a-x^{1/n})$, $x > 0, a > 1, a \neq 2$;

   (14*) $\sum_{n=1}^{\infty} (2-x)(2-x^{1/2})(2-x^{1/3}) \cdot \ldots \cdot (2-x^{1/n})$, $x > 0$.

2. Investigate the character of convergence (uniform or non-uniform) of the series of functions on the indicated set using the definition or the Cauchy criterion:

(1) $\sum_{n=0}^{\infty} x^{2n}$, (a) $X_1 = (-1, 1)$, (b) $X_2 = [-a, a]$, $0 < a < 1$;

(2) $\sum_{n=1}^{\infty} \frac{\sqrt{x}}{(1+x)^n}$, (a) $X_1 = [0, 1]$, (b) $X_2 = [1, +\infty)$;

(3) $\sum_{n=0}^{\infty} \frac{x^2}{(1+x^4)^n}$, $X = \mathbb{R}$;

(4) $\sum_{n=0}^{\infty} \frac{1}{(1+x^2)^n}$, (a) $X_1 = (0, +\infty)$, (b) $X_2 = [1, +\infty)$;

(5) $\sum_{n=1}^{\infty} \frac{x^4}{(1+x^2)^n}$, $X = \mathbb{R}$;

(6) $\sum_{n=1}^{\infty} \frac{x^n}{(1+x^2)^n}$, $X = \mathbb{R}$;

(7) $\sum_{n=1}^{\infty} \frac{x^{2n}}{(1+x^2)^n}$, $X = \mathbb{R}$;

(8) $\sum_{n=1}^{\infty} \frac{\cos nx - \cos(n+1)x}{n}$, $X = \mathbb{R}$;

(9) $\sum_{n=2}^{\infty} \frac{\sin n^2 x^2 - \sin(n+1)^2 x^2}{\ln n}$, $X = \mathbb{R}$;

(10) $\sum_{n=1}^{\infty} \left(\frac{x^n}{n} - \frac{x^{n+1}}{n+1} \right)$, $X = [-1, 1]$;

(11) $\sum_{n=1}^{\infty} \frac{x}{((n-1)x+1)(nx+1)}$, (a) $X_1 = [0, 1]$, (b) $X_2 = [1, +\infty)$;

(12) $\sum_{n=1}^{\infty} \frac{1}{((n-1)x+1)(nx+1)}$, $X = (0, +\infty)$;

(13) $\sum_{n=1}^{\infty} \frac{x^2}{((n-1)x+1)(nx+1)}$, (a) $X_1 = [0, 1]$, (b) $X_2 = [1, +\infty)$;

(14) $\sum_{n=1}^{\infty} \frac{1}{(x+n)(x+n+1)}$, $X = [0, +\infty)$;

(15) $\sum_{n=1}^{\infty} \frac{x}{(x+n)(x+n+1)}$, $X = [0, +\infty)$;

(16*) $\sum_{n=1}^{\infty} \frac{nx}{(1+x)(1+2x)\cdot...\cdot(1+nx)}$, (a) $X_1 = [0, 1]$, (b) $X_2 = [1, +\infty)$;

(17*) $\sum_{n=1}^{\infty} \sin \frac{x}{x^2+n^2}$, (a) $X_1 = \mathbb{R}$, (b) $X_2 = [-a, a]$, $a > 0$.

3. Investigate the character of convergence (uniform or non-uniform) of the series of functions on the indicated set applying the Weierstrass test, Dirichlet's or Abel's test:

(1) $\sum_{n=1}^{\infty} \frac{\cos nx}{n^2+x^2}$, $X = \mathbb{R}$;

(2) $\sum_{n=1}^{\infty} (-1)^n \frac{x^n}{\ln n + 2^n}$, (a) $X_1 = [-a, a]$, $0 < a < 2$, (b) $X_2 = (-2, 2)$;

(3) $\sum_{n=1}^{\infty} \frac{(-1)^n}{\sqrt[3]{n^4+x^2}}$, $X = \mathbb{R}$;

(4) $\sum_{n=1}^{\infty} \frac{(-1)^n}{\sqrt[3]{n^2+x^2}}$, $X = \mathbb{R}$;

(5) $\sum_{n=1}^{\infty} \frac{\sqrt{n} x^2}{5+n^2 x^2}$, $X = \mathbb{R}$;

(6) $\sum_{n=1}^{\infty} (-1)^n \sin \frac{x}{n^2+x^2}$, $X = \mathbb{R}$;

(7) $\sum_{n=2}^{\infty} (-1)^n \frac{x^n}{\ln n}$, $X = [0, 1]$;

(8) $\sum_{n=1}^{\infty} \frac{\sin 3nx}{\sqrt{n^3+x}}$, $X = [0, +\infty)$;

(9) $\sum_{n=2}^{\infty} \frac{\cos 5nx}{n \ln^2 n + x^4}$, $X = \mathbb{R}$;

(10) $\sum_{n=1}^{\infty} \sin \frac{x}{n^3+x^2}$, $X = \mathbb{R}$;

(11) $\sum_{n=2}^{\infty} \ln \left(1 + \frac{x^2}{n \ln^2 n} \right)$, (a) $X_1 = [-a, a]$, $a > 0$, (b) $X_2 = \mathbb{R}$;

(12) $\sum_{n=1}^{\infty} \arctan \frac{x^2}{x^4+n^4}$, $X = \mathbb{R}$;

(13) $\sum_{n=1}^{\infty} (-1)^n \arctan \frac{x^4}{1+nx^4}$, $X = \mathbb{R}$;

(14) $\sum_{n=1}^{\infty} x^2 e^{-nx}$, $X = [0, +\infty)$;

(15) $\sum_{n=2}^{\infty} \frac{\sin nx}{\ln n}$, (a) $X_1 = [a, 2\pi - a], 0 < a < \pi$, (b) $X_2 = [0, 2\pi]$, (c) $X_3 = \mathbb{R}$;

(16) $\sum_{n=1}^{\infty} \frac{\cos nx}{\sqrt[3]{n^4+x^4}}$, $X = \mathbb{R}$;

(17) $\sum_{n=1}^{\infty} \frac{\cos nx}{\sqrt[3]{n+x^4}}$, (a) $X_1 = [a, 2\pi - a], 0 < a < \pi$, (b) $X_2 = (0, 2\pi)$;

(18) $\sum_{n=1}^{\infty} \frac{\cos \frac{n\pi}{3}}{\sqrt[3]{n+x^2}}$, $X = \mathbb{R}$;

(19) $\sum_{n=2}^{\infty} \frac{\sin \frac{n\pi}{6}}{\sqrt[3]{\ln n+x}} \frac{x^n}{x^n+3}$, $X = [0, +\infty)$;

(20) $\sum_{n=2}^{\infty} \frac{(-1)^n}{n+\sin x}$, $X = \mathbb{R}$;

(21) $\sum_{n=2}^{\infty} \frac{\cos nx}{n+\cos x}$, (a) $X_1 = \left[\frac{\pi}{6}, \frac{11\pi}{6}\right]$, (b) $X_2 = (0, 2\pi)$;

(22) $\sum_{n=2}^{\infty} \frac{\sin x \cos nx}{n+\cos x}$, $X = \mathbb{R}$;

(23) $\sum_{n=1}^{\infty} \frac{\sin \frac{n\pi}{2}}{\sqrt[4]{n+x^2}} \frac{x^{4n}}{x^{4n}+5}$, $X = \mathbb{R}$;

(24) $\sum_{n=2}^{\infty} (-1)^n \frac{x^{2n}}{\ln n}$, $X = [-1, 1]$;

(25) $\sum_{n=2}^{\infty} \frac{\sin \frac{n\pi}{3}}{\sqrt{\ln(n^2+x^2)}}$, $X = \mathbb{R}$;

(26) $\sum_{n=1}^{\infty} \frac{\sin \frac{n\pi}{6}}{\sqrt[3]{n}} x^n$, $X = [0, 1]$;

(27) $\sum_{n=1}^{\infty} \frac{\cos \frac{n\pi}{4}}{\sqrt[3]{\ln(n+2)+x^2}} \frac{x^{2n}}{x^{2n}+1}$, $X = \mathbb{R}$;

(28) $\sum_{n=1}^{\infty} \frac{(-1)^n}{\sqrt{n}} \arctan \frac{n^2 x^2}{1+n^2 x^2}$, $X = \mathbb{R}$;

(29) $\sum_{n=1}^{\infty} \frac{\cos \frac{2}{3} n\pi}{\sqrt[5]{n+x}} \frac{x^n}{x^n+7}$, $X = [0, +\infty)$;

(30) $\sum_{n=2}^{\infty} \frac{(-1)^{n(n+1)/2}}{\sqrt{\ln n+x^2}} \sin \frac{n^2 x^2+2}{n^2 x^2+4}$, $X = \mathbb{R}$;

(31) $\sum_{n=2}^{\infty} \frac{\sin x \sin nx}{\sqrt{\ln(n^2+x^2)}} \frac{x^{2n}}{x^{2n}+1}$, $X = \mathbb{R}$.

4. Show that the series converges uniformly on X, but does not converge absolutely on this set:

(1) $\sum_{n=1}^{\infty} \frac{(-1)^n}{\sqrt[3]{n^2+x^2}}$, $X = \mathbb{R}$;

(2) $\sum_{n=1}^{\infty} \frac{(-1)^n}{x+n}$, $X = [0, +\infty)$;

(3) $\sum_{n=2}^{\infty} (-1)^n \frac{x^{2n}}{\ln n}$, $X = [-1, 1]$;

(4*) $\sum_{n=1}^{\infty} \frac{\cos \frac{n\pi}{6}}{\sqrt[3]{n+x^2}}$, $X = \mathbb{R}$;

(5) $\sum_{n=1}^{\infty} (-1)^n \arctan \frac{x}{1+nx^2}$, $X = \mathbb{R}$;

(6) $\sum_{n=1}^{\infty} \frac{(-1)^n}{\sqrt{n}} \cos \frac{nx}{2+nx}$, $X = [0, +\infty)$.

5. Show that the series converges absolutely on X, but does not converge uniformly on this set:

(1) $\sum_{n=1}^{\infty} (-1)^n x^n$, $X = (-1, 1)$;

(2) $\sum_{n=1}^{\infty} \frac{x^n}{n}$, $X = (-1, 1)$;

(3) $\sum_{n=1}^{\infty} \frac{(-1)^n}{(1+x)^n}$, $X = (0, +\infty)$;

(4) $\sum_{n=1}^{\infty} (-1)^n x^n (1 - x^n)$, $X = [0, 1]$.

6. Show that the series $\sum u_n(x)$ converges absolutely and uniformly on X, but the series $\sum |u_n(x)|$ does not converge uniformly on this set:
(1) $\sum_{n=1}^{\infty} (-1)^n x (1-x)^n$, $X = [0, 1]$;
(2) $\sum_{n=1}^{\infty} (-1)^n \frac{x}{(1+x)^n}$, $X = [0, +\infty)$;
(3) $\sum_{n=1}^{\infty} (-1)^n x (1-x^2)^n$, $X = [-1, 1]$;
(4) $\sum_{n=1}^{\infty} (-1)^n \sin \frac{x}{x^2+n^2}$, $X = \mathbb{R}$.

7*. Show that the series $\sum u_n(x)$ converges absolutely and uniformly on X, and the series $\sum |u_n(x)|$ converges uniformly on X, but this series does not have a convergent majorant series:

(1) $u_n(x) = \begin{cases} 0, & -2^{-(n+1)} \le x \le 2^{-(n+1)} \\ \frac{(-1)^n}{\sqrt{n}} \sin^2(2^{n+1}\pi x), & 2^{-(n+1)} < |x| < 2^{-n}, X = [-1, 1]; \\ 0, & 2^{-n} \le |x| \le 1 \end{cases}$

(2) $u_n(x) = \begin{cases} (-1)^n 2(n+1)\left(x - \frac{1}{n+1}\right), & \frac{1}{n+1} \le x \le \frac{2n+1}{2n(n+1)} \\ (-1)^{n+1} 2(n+1)\left(x - \frac{1}{n}\right), & \frac{2n+1}{2n(n+1)} \le x \le \frac{1}{n} \\ 0, & otherwise \end{cases}$, $X = [0, 1]$.

8. Let series $\sum u_n(x)$ and $\sum v_n(x)$ be defined on a set X. What can be stated about the series $\sum (u_n(x) + v_n(x))$, if:
(1) both original series converge uniformly on X;
(2) one of the original series converges uniformly, while another converges non-uniformly on X;
(3) both original series converge non-uniformly on X;
(4) both original series diverge on X.

9. Analyze the convergence/divergence of the series $\sum u_n(x)$ and $\sum v_n(x)$ on the given set. In the case of convergence, determine its type (uniform or non-uniform). Do the same for the series $\sum u_n(x) v_n(x)$. Use these examples to make a conclusion about the relationship between the behavior of the series $\sum u_n(x)$, $\sum v_n(x)$ and $\sum u_n(x) v_n(x)$:
(1) $u_n(x) = v_n(x) = \frac{(-1)^n}{\sqrt{n+x^2}}$, $X = \mathbb{R}$;

(2) $u_n(x) = (-1)^n \frac{x^{2n}}{\sqrt[3]{n+2}}$, $v_n(x) = (-1)^n \frac{x^{2n}}{\ln(n+2)}$, $X = [-1, 1]$;

(3) $u_n(x) = \frac{1}{\sqrt{n+x}}$, $v_n(x) = \frac{1}{n+x}$, $X = [0, +\infty)$;

(4) $u_n(x) = v_n(x) = \frac{\sin nx}{\sqrt[4]{n^3}}$, $X_1 = \mathbb{R}$, $X_2 = [a, 2\pi - a]$, $0 < a < \pi$;

(5) $u_n(x) = v_n(x) = \frac{\sin nx}{\sqrt[4]{n}}$, $X_1 = \mathbb{R}$, $X_2 = [a, 2\pi - a]$, $0 < a < \pi$;

(6) $u_n(x) = \sin nx$, $v_n(x) = \frac{\sin x}{\sqrt{n+x^2}}$, $X = \mathbb{R}$;

(7) $u_n(x) = (-1)^n \sin \frac{x}{\sqrt{n+x^2}}$, $v_n(x) = (-1)^n \frac{x}{\sqrt{n}+x^2}$, $X = \mathbb{R}$;

(8) $u_n(x) = (-1)^n \sin \frac{x}{n+x^2}$, $v_n(x) = (-1)^n \frac{x}{n+x^2}$, $X = \mathbb{R}$;

(9) $u_n(x) = (-1)^n \sin \frac{1}{n+x^2}$, $v_n(x) = (-1)^n \frac{1}{n+x^2}$, $X = \mathbb{R}$.

10. Check the conditions of Dini's theorem and investigate the character of convergence of the given series on the set X:

(1) $\sum_{n=1}^{\infty}(1-x)x^n$, $X = [0, 1]$;

(2) $\sum_{n=1}^{\infty}(1-x)x^n$, $X = [0, 1)$;

(3) $\sum_{n=1}^{\infty}(-1)^n(1-x)x^n$, $X = [0, 1]$;

(4) $\sum_{n=1}^{\infty}\left(\frac{x}{1+x}\right)^n$, $X = [0, +\infty)$;

(5) $\sum_{n=0}^{\infty}\frac{x^2}{(1+x^2)^n}$, $X = \mathbb{R}$;

(6) $\sum_{n=0}^{\infty}(-1)^n\frac{x}{(1+x^2)^n}$, $X = \mathbb{R}$;

(7) $\sum_{n=1}^{\infty}(1-x^2)x^{2n}$, $X = [-1, 1]$;

(8) $\sum_{n=1}^{\infty}(-1)^{n-1}D(x)x^n$, $X = (0, 1)$;

(9) $\sum_{n=1}^{\infty}(-1)^{n-1}D(x)\frac{x}{(1+x^2)^n}$, $X = (0, +\infty)$;

(10) $\sum_{n=1}^{\infty}\sin^2 x \cos^{2n} x$, $X = \left[-\frac{\pi}{2}, \frac{\pi}{2}\right]$;

(11) $\sum_{n=0}^{\infty}\sin^4 x \cos^{2n} x$, $X = \left[-\frac{\pi}{2}, \frac{\pi}{2}\right]$.

Recall that $D(x) = \begin{cases} 1, & x \in \mathbb{Q} \\ 0, & x \in \mathbb{R}\backslash\mathbb{Q} \end{cases}$.

11. Find the sum of the series, verify if the general term and the sum are bounded functions on the given set. Analyze the type of convergence of the series:

(1) $\sum_{n=0}^{\infty}\frac{1}{(1+x)^n}$, $X = (0, +\infty)$;

(2) $\sum_{n=0}^{\infty}\frac{(-1)^n}{(1+x)^n}$, $X = (0, +\infty)$;

(3) $\sum_{n=1}^{\infty}\frac{1}{n(n+1)x}$, $X = (0, +\infty)$;

(4) $\sum_{n=0}^{\infty}u_n(x)$, $u_0(x) = -\frac{1}{x}$, $u_n(x) = \frac{1}{n(n+1)x}$, $n \geq 1$, $X = (0, +\infty)$;

(5) $\sum_{n=0}^{\infty}\sin x \cos^{2n} x$, $X = \left[-\frac{\pi}{2}, \frac{\pi}{2}\right]$;

(6) $\sum_{n=1}^{\infty}(1-x^2)x^{2n}$, $X = [-1, 1]$;

(7) $\sum_{n=0}^{\infty}x^{2n}$, $X = (-1, 1)$;

(8) $\sum_{n=1}^{\infty}\frac{\sqrt{x}}{((n-1)x+1)(nx+1)}$, $X = [0, 1]$;

(9) $\sum_{n=1}^{\infty}\frac{x^2}{((n-1)x+1)(nx+1)}$, $X = [0, +\infty)$;

(10) $\sum_{n=0}^{\infty}\frac{1}{(x+n)(x+n+1)}$, $X = (0, +\infty)$;

(11) $\sum_{n=1}^{\infty}\frac{x}{(x+n)(x+n+1)}$, $X = [0, +\infty)$.

12. Find the sum $f(x)$ of the series and calculate the limit $\lim_{x \to a} f(x)$ if it exists. Investigate the character of convergence of the series on the given set and evaluate the possibility to calculate the limit term-by-term:

(1) $\sum_{n=0}^{\infty}xe^{-nx}$, $X = (0, +\infty)$, $a = 0$;

(2) $\sum_{n=0}^{\infty}\sqrt{x}e^{-nx}$, $X = (0, +\infty)$, $a = 0$;

(3) $\sum_{n=0}^{\infty}(-1)^n e^{-nx}$, $X = (0, +\infty)$, $a = 0$;

(4) $\sum_{n=0}^{\infty}(-1)^n xe^{-nx}$, $X = (0, +\infty)$, $a = 0$;

(5) $\sum_{n=1}^{\infty}(x^n - x^{2n})$, $X = (-1, 1)$, (a) $a = 1$, (b) $a = -1$;

(6) $\sum_{n=1}^{\infty}(-1)^{n-1}(x^n - x^{2n})$, $X = (-1, 1)$, (a) $a = 1$, (b) $a = -1$;

(7) $\sum_{n=0}^{\infty} \frac{(-1)^n}{(1+x)^n}$, $X = (0, +\infty)$, $a = 0$;

(8) $\sum_{n=0}^{\infty} \frac{x}{(1+x)^n}$, $X = (0, +\infty)$, $a = 0$;

(9) $\sum_{n=0}^{\infty} \frac{xD(x)}{(1+x)^n}$, $X = (0, +\infty)$, $a = 0$;

(10) $\sum_{n=0}^{\infty} \sin x \cos^{2n} x$, $X = \left(0, \frac{\pi}{2}\right]$, $a = 0$;

(11) $\sum_{n=0}^{\infty} (-1)^n x^{2n}$, $X = [0, 1)$, $a = 1$.

13. Find the sum of the series and verify if the general term and the sum of the series are continuous functions on the given set. Investigate the character of convergence of the series on this set:

(1) $\sum_{n=0}^{\infty} \sin x \cos^{2n} x$, (a) $X_1 = \left[-\frac{\pi}{2}, \frac{\pi}{2}\right]$, (b) $X_2 = \left(0, \frac{\pi}{2}\right]$;

(2) $\sum_{n=0}^{\infty} \frac{x}{(1+x)^n}$, $X = [0, +\infty)$;

(3) $\sum_{n=1}^{\infty} (x^n - x^{2n})$, $X = [0, 1]$;

(4) $\sum_{n=1}^{\infty} (-1)^{n-1}(x^n - x^{2n})$, $X = [0, 1]$;

(5) $\sum_{n=0}^{\infty} (-1)^n x^{2n}$, $X = (-1, 1)$;

(6) $\sum_{n=1}^{\infty} \frac{x^2}{(nx^2+1)((n+1)x^2+1)}$, (a) $X_1 = \mathbb{R}$, (b) $X_2 = (0, +\infty)$;

(7) $\sum_{n=0}^{\infty} u_n(x)$, $u_0(x) = -\frac{D(x)}{x^2+1}$, $u_n(x) = \frac{2D(x)}{(2n+x^2)^2-1}$, $n \geq 1$, $X = \mathbb{R}$;

(8) $\sum_{n=1}^{\infty} \frac{2D(x)}{(2n+x^2)^2-1}$, $X = \mathbb{R}$;

(9) $\sum_{n=0}^{\infty} xe^{-nx^2}$, $X = \mathbb{R}$;

(10) $\sum_{n=0}^{\infty} (-1)^n xe^{-nx^2}$, $X = \mathbb{R}$;

(11) $\sum_{n=1}^{\infty} (1 - x^2)x^{2n}$, (a) $X_1 = [-1, 1]$, (b) $X_2 = (-1, 1)$.

14. Find the sum of the series. Investigate the character of convergence of the series and the possibility of term-by-term integration on the given interval:

(1) $\sum_{n=0}^{\infty} \sin^3 x \cos^{2n} x$, $X = \left[0, \frac{\pi}{2}\right]$;

(2) $\sum_{n=1}^{\infty} x(1 - x)^n$, $X = [0, 1]$;

(3) $\sum_{n=1}^{\infty} \left(ne^{-nx} - (n - 1)e^{-(n-1)x}\right) \operatorname{sgn} x$, $X = [0, 1]$;

(4) $\sum_{n=1}^{\infty} \frac{x^2}{(nx^2+1)((n+1)x^2+1)}$, $X = [0, 1]$;

(5) $\sum_{n=1}^{\infty} \frac{x}{(nx+1)((n+1)x+1)}$, $X = [0, 1]$;

(6) $\sum_{n=1}^{\infty} \left(\sin \frac{x}{n} - \sin \frac{x}{n+1}\right)$, $X = \left[0, \frac{\pi}{2}\right]$;

(7) $\sum_{n=1}^{\infty} \left(x^{\frac{1}{n+1}} - x^{\frac{1}{n}}\right)$, $X = [0, 1]$;

(8) $\sum_{n=1}^{\infty} \left(x^{2n} - \frac{x^n}{2}\right) \operatorname{sgn}(1 - x)$, $X = [0, 1]$;

(9) $\sum_{n=1}^{\infty} \left(x^{2n+1} - \frac{x^n}{2}\right) \operatorname{sgn}(1 - x)$, $X = [0, 1]$;

(10*) $\sum_{n=1}^{\infty} (-1)^n 2nxe^{-nx^2}$, $X = [0, 1]$.

15. Construct an example of a series of non-integrable on $[a, b]$ by Riemann functions whose sum is Riemann integrable on $[a, b]$.

16. Given the series $\sum u_n(x)$ of differentiable on X functions, analyze the character of convergence of this series and the series $\sum u'_n(x)$ on the set X. Evaluate the possibility of term-by-term differentiation of the original series on X :

(1) $\sum_{n=0}^{\infty} \left(e^{-nx^2} - e^{-(n+1)x^2}\right)$, (a) $X_1 = \mathbb{R}$, (b) $X_2 = (0, +\infty)$;

(2) $\sum_{n=0}^{\infty} \left(xe^{-nx^2} - xe^{-(n+1)x^2} \right)$, $X = \mathbb{R}$;

(3*) $\sum_{n=0}^{\infty} xe^{-nx}$, (a) $X_1 = [0, +\infty)$, (b) $X_2 = (0, +\infty)$;

(4*) $\sum_{n=0}^{\infty} (-1)^n xe^{-nx}$, $X = [0, +\infty)$;

(5) $\sum_{n=1}^{\infty} \frac{x^2}{(nx^2+1)((n+1)x^2+1)}$, $X = \mathbb{R}$;

(6) $\sum_{n=1}^{\infty} \left(\frac{(n-1)x}{1+(n-1)^2 x^2} - \frac{nx}{1+n^2 x^2} \right)$, $X = \mathbb{R}$;

(7) $\sum_{n=1}^{\infty} \left(\frac{x^n}{n} - \frac{x^{n+1}}{n+1} \right)$, $X = [0, 1]$;

(8) $\sum_{n=1}^{\infty} (\arctan nx - \arctan(n-1)x)$, (a) $X_1 = \mathbb{R}$, (b)$X_2 = (0, +\infty)$;

(9*) $\sum_{n=1}^{\infty} \arctan \frac{x}{n^2}$, $X = \mathbb{R}$;

(10) $\sum_{n=1}^{\infty} \frac{x}{(1+(n-1)x)(1+nx)}$, (a) $X_1 = [0, +\infty)$, (b) $X_2 = (0, +\infty)$;

(11) $\sum_{n=1}^{\infty} \frac{x^2}{(1+(n-1)x)(1+nx)}$, (a) $X_1 = [0, +\infty)$, (b) $X_2 = (0, +\infty)$.

17*. Verify if a uniformly convergent on X series $\sum u_n(x)$ can have the series of derivatives $\sum u'_n(x)$ divergent at every point of X. Under the same condition on $\sum u_n(x)$, can the series $\sum u'_n(x)$ be divergent at infinitely many points of X and also convergent at infinitely many points of X?

Power Series

5

> *Power series are therefore especially convenient because one can*
> *compute with them almost as with polynomials.*
> *Constantin Carathéodory, 1950*

1   Introduction

> *In the beginning of the year 1665 I found the Method of*
> *approximating series & the Rule for reducing any dignity of any*
> *Binomial into such a series.*
> *Isaac Newton, 1718*

In this part of the text we consider a special type of the series of functions—power series.

▶ **Definition** (**Power Series**) A *power series* is a series of functions of the form

$$\sum_{n=0}^{\infty} c_n (x - a)^n, \tag{5.1}$$

where c_n are *coefficients* (constants) and a is a *central point* (also a constant).

Electronic Supplementary Material The online version of this article (https://doi.org/10.1007/978-3-030-79431-6_5) contains supplementary material, which is available to authorized users.

239
L. Bourchtein, A. Bourchtein, *Theory of Infinite Sequences and Series*,
https://doi.org/10.1007/978-3-030-79431-6_5

The domain of the general term $u_n(x) = c_n(x - a)^n$ is $\mathbb{R}$, but the set of convergence depends on the coefficients c_n. The only convergence property that does not depend on c_n is the fact that any power series converges at its central point $x = a$ to the coefficient c_0.

A particular case, found frequently in theory and applications, is a power series centered at $a = 0$:

$$\sum_{n=0}^{\infty} c_n x^n. \tag{5.2}$$

In the previous chapter we have already seen various power series (without using this terminology) and studied their convergence: the series $\sum_{n=0}^{\infty} x^n$, $\sum_{n=1}^{\infty} \frac{x^n}{n}$, $\sum_{n=0}^{\infty} \frac{x^n}{n!}$, $\sum_{n=1}^{\infty} (xn)^n$ are examples of power series. The first converges uniformly on $[-a, a]$, $0 < \forall a < 1$ and non-uniformly on $(-1, 1)$, the second converges uniformly on $[-a, a]$, $0 < \forall a < 1$ and non-uniformly on $[-1, 1)$, the third converges uniformly on $[-a, a]$, $\forall a > 0$ and non-uniformly on $\mathbb{R}$, and the last series converges only at the central point 0. In the following section we will prove that these are examples of all the kinds of the convergence that a power series may have: *a power series centered at a may converge only at the central point, or on an interval centered at a, or on the entire real axis.*

In this chapter we will analyze the properties of convergence of power series and properties of their sums. Partly we will transfer already obtained results of series of functions to this special case, and in other part we will derive specific results for this type of series. We will also study the methods of representation of functions in power series and find the power expansions for different functions. Finally, we will consider some applications of power series, which allow us to solve the problems of approximation, integration and differentiation.

One of the principal goals of this chapter is to find the sums of various power series and, on the other hand, represent elementary functions in power series. However, we will be prepared to introduce systematic approaches to solution of these problems only in the final parts of this chapter. Recall that until the moment, among different power series we have already met, there are very few whose exact sum we were able to found. Actually, we have already calculated only the sum of the geometric series

$$\sum_{n=0}^{\infty} x^n = \frac{1}{1-x}, \forall x \in (-1, 1)$$

and some of its immediate consequences. Therefore, to have a material for illustration of the properties of power series to be studied, in this introductory section we find the sums of two more rather simple power series.

First, consider the series $\sum_{n=0}^{\infty} nx^n$. The application of D'Alembert's test shows immediately that this series converges on $(-1, 1)$ and diverges outside $[-1, 1]$, since $|D_n| = \left|\frac{(n+1)x^{n+1}}{nx^n}\right| = \frac{n+1}{n}|x| \underset{n\to\infty}{\to} |x|$. At the endpoints $x = -1$ and $x = 1$ we get divergent series (the general terms $(-1)^n n$ and n do not approach 0). Finally, the general term nx^n converges to 0 on $(-1, 1)$, but this convergence is non-uniform on $(-1, 1)$:

$$\sup_{\forall x \in (-1,1)} |nx^n - 0| = n \cdot \sup_{\forall x \in (-1,1)} |x^n| = n \underset{n \to \infty}{\nrightarrow} 0,$$

which implies the non-uniform convergence of the series on $(-1, 1)$. However, the Weierstrass test reveals that the series converges uniformly on any interval $[-a, a], 0 < a < 1$: since $\frac{na^n}{b^n} \underset{n \to \infty}{\to} 0$ for any $0 < a < b < 1$, it follows that $|nx^n| \le na^n < b^n$ for all $x \in [-a, a]$ and for all sufficiently large n, and therefore, the convergence of the majorant geometric series $\sum_{n=0}^{\infty} b^n$ guarantees the uniform convergence of the original series on $[-a, a], 0 < a < 1$.

After the convergence properties are established, we find the sum of the series on the convergence interval $(-1, 1)$. To do this, we follow the procedure similar to the used for deriving the formula of geometric series. Consider the partial sum

$$S_n = \sum_{k=0}^{n} kx^k = x + 2x^2 + 3x^3 + \dots + nx^n,$$

multiply it by x:

$$x S_n = \sum_{k=0}^{n} kx^{k+1} = x^2 + 2x^3 + \dots + (n-1)x^n + nx^{n+1}$$

and calculate the difference

$$S_n - x S_n = x + x^2 + x^3 + \dots + x^n - nx^{n+1} = x \frac{1 - x^n}{1 - x} - nx^{n+1}.$$

Solving this relation for S_n, we have $S_n = x \frac{1-x^n}{(1-x)^2} - n \frac{x^{n+1}}{1-x}$, and consequently, the sum of the series is

$$f(x) = \sum_{n=0}^{\infty} nx^n = \lim_{n \to \infty} S_n = \lim_{n \to \infty} \left(x \frac{1 - x^n}{(1 - x)^2} - n \frac{x^{n+1}}{1 - x} \right) = \frac{x}{(1 - x)^2}, \forall x \in (-1, 1).$$

This is a very particular method of finding the sum of a series, which usually works only for geometric and closely related series.

Second, we consider the series $f(x) = \sum_{n=0}^{\infty} \frac{x^n}{n!}$. The series of derivatives $\sum_{n=1}^{\infty} \frac{x^{n-1}}{(n-1)!} = \sum_{m=0}^{\infty} \frac{x^m}{m!}$ represents the same original series. Since the original series satisfies the conditions of the property of term-by-term differentiation on any closed interval, we get

$$f'(x) = \sum_{n=0}^{\infty} \left(\frac{x^n}{n!} \right)' = \sum_{m=0}^{\infty} \frac{x^m}{m!} = f(x), \forall x \in \mathbb{R}.$$

Solving this differential equation for $f(x)$, we find $f(x) = Ce^x$, where C is an arbitrary constant. Substituting $x = 0$ in the series expression and in the exponential function, we specify $C = 1$. Then,

$$f(x) = \sum_{n=0}^{\infty} \frac{x^n}{n!} = e^x, \forall x \in \mathbb{R}.$$

This example is an elementary illustration of a general technique of finding the sums of power series via their differential properties, that will be studied later.

2   Set of Convergence of a Power Series

2.1   Convergence of a Power Series

We start the analysis of convergence of power series with the result of the extreme importance (specific for power series) that makes possible to establish the form of the sets of convergence of power series. To simplify the reasoning and notation, we provide demonstrations for a power series centered at the origin $\sum_{n=0}^{\infty} c_n x^n$ (see formula (5.2)). The general case is trivially recovered by translation of the coordinates $t = x - a$.

Abel's Lemma *If a power series (5.2) converges at a point $x_0 \neq 0$, then it converges absolutely and uniformly on the interval $[-b, b]$ for any $0 < b < |x_0|$. If a power series (5.2) diverges at a point x_0, then it diverges at all the points outside the interval $[-|x_0|, |x_0|]$.*

Proof Let us consider the first assumption: a power series $\sum_{n=0}^{\infty} c_n x^n$ converges at a point $x_0 \neq 0$. Then the general term of the series of numbers $\sum_{n=0}^{\infty} c_n x_0^n$ tends to 0: $c_n x_0^n \underset{n \to \infty}{\to} 0$. This in turn implies the boundedness of the sequence $c_n x_0^n$: there exists a constant M such that

$$\left| c_n x_0^n \right| \leq M, \forall n.$$

Consider now the interval $[-b, b]$, where $0 < b < |x_0|$. Picking an arbitrary x in $[-b, b]$, that is, $|x| \le b$, we have the following evaluation

$$|c_n x^n| = |c_n| \cdot |x^n| \le |c_n| \cdot b^n = |c_n| \cdot |x_0^n| \cdot \left| \frac{b}{x_0} \right|^n \le M \cdot \left| \frac{b}{x_0} \right|^n.$$

Since $\left| \frac{b}{x_0} \right| < 1$, the numerical geometric series $\sum_{n=0}^{\infty} M \cdot \left| \frac{b}{x_0} \right|^n$ converges. Now we define $M_n = M \cdot \left| \frac{b}{x_0} \right|^n$ and use the series $\sum_{n=0}^{\infty} M_n = \sum_{n=0}^{\infty} M \cdot \left| \frac{b}{x_0} \right|^n$ as the majorant series in the Weierstrass Test. Since the last series converges and $|c_n x^n| \le M_n$ for $\forall n$ and $\forall x$ such that $|x| \le b$, then, by the Weierstrass Test, the series $\sum_{n=0}^{\infty} |c_n x^n|$ converges uniformly on $[-b, b]$, which means that the series $\sum_{n=0}^{\infty} c_n x^n$ converges uniformly and absolutely on $[-b, b]$.

Now we turn to the second condition: a power series $\sum_{n=0}^{\infty} c_n x^n$ diverges at a point x_0. Suppose, on the contrary, that the series converges at a point x_1 such that $|x_1| > |x_0|$. Then, by the first part of the statement (already proved), the series converges on $[-b, b]$, where $0 < b < |x_1|$. In particular, choosing $b = |x_0|$, we arrive at the conclusion that the series $\sum_{n=0}^{\infty} c_n x^n$ converges at x_0 that contradicts the original condition of the second case. $\qquad\square$

Remark 1 From the result of the Lemma it follows immediately that the series $\sum_{n=0}^{\infty} c_n x^n$ converges absolutely on the entire interval $(-x_0, x_0)$ if it converges at $x_0 \ne 0$ (for any $x_1 \in (-x_0, x_0)$ just choose $|x_1| < b < |x_0|$ and apply the result of the Lemma).

Remark 2 The uniform convergence can not be extended to the interval $(-x_0, x_0)$. An example of this is the series $\sum_{n=1}^{\infty} \frac{x^n}{n}$ that converges at the point $x_0 = -1$, but does not converge uniformly on $(-1, 1)$ (although it converges uniformly on any interval $[-b, b] \subset (-1, 1)$).

Corollary *The corresponding result for a general power series $\sum_{n=0}^{\infty} c_n (x - a)^n$ follows directly from the result of Lemma by applying the change of the variable x by $x - a$ and can be formulated as follows. If a power series (5.1) converges at a point $x_0 \ne a$, it converges absolutely and uniformly on any interval $[a - b, a + b]$ where $0 < b < |a - x_0|$; if a power series (5.1) diverges at a point x_0, it diverges at all points outside the interval $[a - d, a + d]$, $d = |a - x_0|$.*

▶ **Definition** (**The Set of Convergence of a Power Series**) According to the general definition (see the definition of pointwise convergence on a set in Sect. 1 of Chap. 4), the *set of convergence of a power series* (5.1) is the set of all points where this series converges. Denote this set by E and define $R = \sup_{\forall x \in E} |x - a|$ (if E is not bounded, assume $R = +\infty$).

Theorem (The Set of Convergence of a Power Series) *If E is the set of convergence of a power series $\sum_{n=0}^{\infty} c_n x^n$ and $R = \sup_{\forall x \in E} |x|$, then there are only three options:*

(1) if $R = +\infty$, the series converges absolutely on $\mathbb{R}$ and uniformly on any interval $[-b, b]$, $b > 0$;

(2) if $R = 0$, the series converges only at the central point $x = 0$ and diverges at each other point;

(3) if $0 < R < +\infty$, the series converges absolutely on the interval $(-R, R)$ and uniformly on any interval $[-b, b]$, $0 < b < R$, and diverges outside the interval $[-R, R]$.

Proof In the first case, $R = +\infty$, the set E is unbounded. Therefore, for any $b > 0$ it can be found a point $x_0 \in E$ such that $|x_0| > b$ and, applying Abel's Lemma, we conclude that the series converges absolutely and uniformly on $[-b, b]$. The arbitrariness of b guarantees the absolute convergence on $\mathbb{R}$ and the uniform convergence on any interval $[-b, b]$, $b > 0$.

In the second case, $R = 0$, the set E can contain only the central point 0 (recall that any power series converges at its central point). This means that the series diverges at all other points.

Let us analyze the third case—$0 < R = \sup_{\forall x \in E} |x| < +\infty$. We start with the prove that series (5.2) converges absolutely on the interval $(-R, R)$ and uniformly on any interval $[-b, b]$, $0 < b < R$. Indeed, for any $0 < b < R$, by the definition of the supremum, there exists a point $x_0 \in E$ such that $b < |x_0| \leq R$. Then, applying Abel's Lemma with this x_0, we conclude that the series converges absolutely and uniformly on $[-b, b]$. Moreover, due to the arbitrariness of b, the convergence is absolute on $(-R, R)$.

Take now any $|x| > R$. Then x does not belong to E, and for this reason, the series diverges at this point. This means that the series diverges outside the interval $[-R, R]$. $\square$

Corollary *The corresponding result for a general power series $\sum_{n=0}^{\infty} c_n (x - a)^n$ follows directly from the statement of the Theorem substituting $x - a$ for x.*

More specifically, for a power series (5.1) there are only three options:

(1) if $R = +\infty$, the series (5.1) converges absolutely on $\mathbb{R}$ and uniformly on any interval $[a - b, a + b]$, $b > 0$;

(2) if $R = 0$, the series (5.1) converges only at the central point $x = a$ and diverges at all other points;

(continued)

(3) if $0 < R < +\infty$, the series (5.1) converges absolutely on the interval $(a - R, a + R)$ and uniformly on any interval $[a - b, a + b]$, $0 < b < R$, and diverges outside the interval $[a - R, a + R]$.

Remark The first two cases can be formally included in the third case if we set that $(a - \infty, a + \infty) = (-\infty, +\infty) = \mathbb{R}$ when $R = +\infty$, and that $(a - 0, a + 0) = \{a\}$, $[a - 0, a + 0] = \{a\}$ when $R = 0$.

▶ **Definition** The interval $(a - R, a + R)$, where $R = \sup_{\forall x \in E} |x - a|$ and E is the set of convergence of a series $\sum_{n=0}^{\infty} c_n (x - a)^n$, is called the *interval of convergence* of this series and R—the *radius of convergence*.

This terminology is very common and natural, because the Theorem states that the series converges on $(a - R, a + R)$ and diverges outside $[a - R, a + R]$. However, it is important to note that there is an ambiguity involved in this definition of the interval of convergence: the Theorem does not provide any result regarding convergence at the endpoints of the interval $(a - R, a + R)$, and we will see in the following examples that the behavior of power series can be different at the points $a - R$ and $a + R$, depending on a specific form of each series. Therefore, the *interval of convergence does not generally coincide with the complete set of convergence E* that can take any of the four forms: $(a - R, a + R)$, $[a - R, a + R)$, $(a - R, a + R]$ and $[a - R, a + R]$.

2.2   Determining the Radius of Convergence

The D'Alembert and Cauchy Formulas

The next natural question is how to determine the radius of convergence R. Let us start with the two simple situations when D'Alembert's and/or Cauchy's tests can be used.

Recall that the applicability of D'Alembert's Test (in the form with the limit) to the study of (absolute) convergence of a series of numbers $\sum_{n=0}^{\infty} a_n$ depends on the existence of the limit $D = \lim_{n \to \infty} \left| \frac{a_{n+1}}{a_n} \right|$. For a power series (5.1), at each fixed x we have $a_n = c_n (x - a)^n$, and therefore

$$D = \lim_{n \to \infty} \left| \frac{a_{n+1}}{a_n} \right| = \lim_{n \to \infty} \left| \frac{c_{n+1}}{c_n} \right| |x - a| = |x - a| \lim_{n \to \infty} \left| \frac{c_{n+1}}{c_n} \right|.$$

Let us suppose that the last limit exists. Then D'Alembert's test states that for $D < 1$ we have convergence, while for $D > 1$—divergence. This means that for $|x - a| < \dfrac{1}{\lim\limits_{n \to \infty} \left| \frac{c_{n+1}}{c_n} \right|}$

the series (5.1) converges, while for $|x - a| > \dfrac{1}{\lim\limits_{n \to \infty} \left| \frac{c_{n+1}}{c_n} \right|}$ it diverges. Therefore, the radius of convergence can be found by the D'Alembert formula

$$R = \frac{1}{\lim\limits_{n \to \infty} \left| \frac{c_{n+1}}{c_n} \right|} = \lim_{n \to \infty} \left| \frac{c_n}{c_{n+1}} \right| . \tag{5.3}$$

In the same manner, Cauchy's test involves the calculation of the limit

$$C = \lim_{n \to \infty} \sqrt[n]{|a_n|} = \lim_{n \to \infty} \sqrt[n]{|c_n (x - a)^n|} = |x - a| \cdot \lim_{n \to \infty} \sqrt[n]{|c_n|} .$$

If the last limit exists, then for $C < 1$ we have convergence, while for $C > 1$—divergence. Therefore, if $|x - a| < \dfrac{1}{\lim\limits_{n \to \infty} \sqrt[n]{|c_n|}}$, the series (5.1) converges, while if $|x - a| > \dfrac{1}{\lim\limits_{n \to \infty} \sqrt[n]{|c_n|}}$, it diverges. Hence, the radius of convergence can be determined by the Cauchy formula

$$R = \frac{1}{\lim\limits_{n \to \infty} \sqrt[n]{|c_n|}} . \tag{5.4}$$

Recall that the existence of the limit $\lim\limits_{n \to \infty} \left| \frac{c_{n+1}}{c_n} \right|$ guarantees the existence of the limit $\lim\limits_{n \to \infty} \sqrt[n]{|c_n|}$ equal to the former (for this reason, Cauchy's test is stronger than d'Alembert's), but in practice, technical calculation of the former (if it exist) is frequently easier. Nevertheless, it may happen that the latter limit (and consequently, the former) does not exist. In this case, we need to find an alternative (finer) option to determine the radius R.

Complement: The Cauchy-Hadamard Formula

The most useful formula for determining the radius of convergence in more problematic situations (when the simpler D'Alembert and Cauchy limits do not exist) is the *Cauchy-Hadamard formula* coming from the same Cauchy's test, albeit in a finer form. Recall that this test can be formulated with the use of the upper limit $\overline{C} = \limsup\limits_{n \to \infty} \sqrt[n]{|a_n|}$ in the following form: if $\overline{C} < 1$, the series $\sum_{n=0}^{\infty} a_n$ converges, while if $\overline{C} > 1$, the series diverges. Formally, this formulation resembles that with the general limit $C = \lim\limits_{n \to \infty} \sqrt[n]{|a_n|}$, but the important detail is that the upper limit always exists (if we allow infinite limits), although the general limit may not exist. In the case of power series, we have $a_n = c_n (x - a)^n$ at each fixed x, that leads to the following form of Cauchy's test with the upper limit: if $|x - a| < \dfrac{1}{\limsup\limits_{n \to \infty} \sqrt[n]{|c_n|}}$, the series (5.1) converges, while if $|x - a| > \dfrac{1}{\limsup\limits_{n \to \infty} \sqrt[n]{|c_n|}}$, it diverges. Therefore, the radius of convergence can be defined as follows

$$R = \frac{1}{\limsup\limits_{n \to \infty} \sqrt[n]{|c_n|}}. \tag{5.5}$$

This is the Cauchy-Hadamard formula and its advantage is that the upper limit $H = \limsup\limits_{n \to \infty} \sqrt[n]{|c_n|}$ always can be found (at least theoretically), while the limits in formulas (5.3) and (5.4) may not exist.

Remark There is no similar result for the upper limit of D'Alembert's test $\overline{D} = \limsup\limits_{n \to \infty} \left| \frac{a_{n+1}}{a_n} \right|$. The reason of this is that the complete formulation of this test requires the use of both upper and lower limits (the latter is defined in the form $\underline{D} = \liminf\limits_{n \to \infty} \left| \frac{a_{n+1}}{a_n} \right|$). More specifically: for $\overline{D} < 1$ we have the convergence, while for $\underline{D} > 1$—divergence. The upper limit $\overline{D}$ can not substitute the lower limit $\underline{D}$ in the second statement, neither $\underline{D}$ can substitute $\overline{D}$ in the first one. (Simple examples of the impossibility of such substitutions are the series $\sum_{n=1}^{\infty} a_n = \frac{1}{4} + \frac{1}{2} + \frac{1}{16} + \frac{1}{8} + \frac{1}{64} + \frac{1}{32} + \ldots$ and $\sum_{n=1}^{\infty} a_n = 4 + 2 + 16 + 8 + 64 + 32 + \ldots$, respectively, solved in detail in Sect. 3.6 of Chap. 2.) Due to this gap between $\overline{D}$ and $\underline{D}$ in the formulation of D'Alembert's test, none of the two limits $\limsup\limits_{n \to \infty} \left| \frac{c_{n+1}}{c_n} \right|$ and $\liminf\limits_{n \to \infty} \left| \frac{c_{n+1}}{c_n} \right|$ can be used to determine the radius of convergence.

In view of determining the radius of convergence by formula (5.5), the Theorem of the set of convergence can be reformulated in the following manner.

The Cauchy-Hadamard Theorem *A power series $\sum_{n=0}^{\infty} c_n(x-a)^n$ admits only one of the three options of convergence which can be formulated in terms of the limit $H = \limsup\limits_{n \to \infty} \sqrt[n]{|c_n|}$ as follows:*

(1) if $R = \dfrac{1}{\limsup\limits_{n \to \infty} \sqrt[n]{|c_n|}} = +\infty$, the series converges absolutely on $\mathbb{R}$ and uniformly on any interval $[a-b, a+b]$, $b > 0$;

(2) if $R = \dfrac{1}{\limsup\limits_{n \to \infty} \sqrt[n]{|c_n|}} = 0$, the series converges only at the central point $x = a$ and diverges at all other points;

(3) if $0 < R = \dfrac{1}{\limsup\limits_{n \to \infty} \sqrt[n]{|c_n|}} < +\infty$, the series converges absolutely on the interval $(a-R, a+R)$ and uniformly on any interval $[a-b, a+b]$, $0 < b < R$, and diverges outside the interval $[a-R, a+R]$. According to the definition in Sect. 2.1, the number $R = \frac{1}{H}$ is the radius of convergence, and $(a-R, a+R)$ is the interval of convergence.

Historical Remarks Formula (5.5) is generally attributed to Hadamard, who published it in 1888. It turns out that this result was first found and proven by Cauchy in 1821 in "Cours d'analyse", but was all but forgotten.

2.3   Convergence of the Series of Derivatives

The next problem to investigate is the behavior of the series of derivatives

$$\sum_{n=1}^{\infty} nc_n(x-a)^{n-1}, \tag{5.6}$$

or, at the central point $a = 0$,

$$\sum_{n=1}^{\infty} nc_n x^{n-1}. \tag{5.7}$$

Notice that this is one more power series with the coefficients $d_0 = 0$ and $d_n = nc_n$, $\forall n \neq 0$.

The sets of convergence of the series (5.1) and (5.6) generally do not coincide. For example, the series $\sum_{n=1}^{\infty} \frac{x^n}{n}$ converges at the point $x = -1$, but the series of derivatives $\sum_{n=1}^{\infty} x^{n-1}$ does not converge at this point. However, these two sets are intimately tied and we will reveal their connection in the next statements.

First, we prove an extension of Abel's Lemma, using the same technique of the proof as in the main Lemma. For simplicity, we use the power series centered at the origin.

Theorem (Extension of Abel's Lemma) *If a power series* (5.2) *converges at a point* $x_0 \neq 0$, *then the series of derivatives* (5.7) *converges absolutely and uniformly on the interval* $[-b, b]$ *for any* $0 < b < |x_0|$. *If a power series* (5.2) *diverges at a point* x_0, *then the series of derivatives* (5.7) *diverges at every point outside the interval* $[-x_0, x_0]$.

Proof Let us analyze the first situation: the power series $\sum_{n=0}^{\infty} c_n x^n$ converges at a point $x_0 \neq 0$. Then, the general term of the series of numbers $\sum_{n=0}^{\infty} c_n x_0^n$ tends to 0: $c_n x_0^n \underset{n \to \infty}{\to} 0$, that guarantees the boundedness of the sequence $c_n x_0^n$:

$$\left| c_n x_0^n \right| \leq M, \forall n.$$

Consider now the interval $[-b, b]$, where $0 < b < |x_0|$. Choose an arbitrary x in $[-b, b]$, that is, $|x| \leq b$, and get the following evaluation:

$$|nc_n x^{n-1}| = n|c_n| \cdot |x^{n-1}| \leq n|c_n| \cdot b^{n-1} = n\frac{|c_n|}{b} \cdot |x_0^n| \cdot \left|\frac{b}{x_0}\right|^n \leq \frac{M}{b} \cdot n \left|\frac{b}{x_0}\right|^n.$$

Since $\left|\frac{b}{x_0}\right| < 1$, the numerical series $\sum_{n=1}^{\infty} \frac{M}{b} \cdot n \left|\frac{b}{x_0}\right|^n$ converges (for example, apply

D'Alembert's test: $D = \lim\limits_{n\to\infty} \dfrac{(n+1)\left|\frac{b}{x_0}\right|^{n+1}}{n\left|\frac{b}{x_0}\right|^n} = \left|\frac{b}{x_0}\right| \lim\limits_{n\to\infty} \frac{n+1}{n} = \left|\frac{b}{x_0}\right| < 1$). If we define

$M_n = \frac{M}{b} \cdot n \left|\frac{b}{x_0}\right|^n$, then $|nc_n x^{n-1}| \le M_n$ for $\forall n$ and $\forall |x| \le b$, and consequently, the series $\sum_{n=1}^{\infty} M_n$ is a convergent majorant series for the power series of derivatives. Therefore, by the Weierstrass test, the series $\sum_{n=1}^{\infty} |nc_n x^{n-1}|$ converges uniformly on $[-b, b]$, which implies the uniform and absolute convergence of the series $\sum_{n=1}^{\infty} nc_n x^{n-1}$ on $[-b, b]$.

Consider now the second part: the power series $\sum_{n=0}^{\infty} c_n x^n$ diverges at a point x_0. Assume, by contradiction, that there exists a point $|x_1| > |x_0|$ at which the series of derivatives $\sum_{n=1}^{\infty} nc_n x^{n-1}$ converges. Since the last series is a power series, applying Abel's Lemma, we deduce that this series converges absolutely at x_0. Consequently, the original series $\sum_{n=0}^{\infty} c_n x^n$ also converges absolutely at x_0 according to the Comparison test (for series of numbers): $\lim\limits_{n\to\infty} \frac{|c_n x_0^n|}{|nc_n x_0^{n-1}|} = \lim\limits_{n\to\infty} \frac{|x_0|}{n} = 0$. However, this contradicts the hypothesis of the divergence of the series $\sum_{n=0}^{\infty} c_n x^n$ at x_0.   □

Remark As a by-product of this demonstration, we prove that the series of derivatives diverges at each point where the original power series diverges. This result can be useful in practice when we investigate the convergence of the two series—original and of the derivatives—at the endpoints of the interval of convergence.

The next Theorem about the convergence of the series of derivatives follows directly from this extension of Abel's Lemma.

Theorem (Convergence of the Series of Derivatives) *Let R be the radius of convergence of a series $\sum_{n=0}^{\infty} c_n x^n$. The series of derivatives $\sum_{n=1}^{\infty} c_n n x^{n-1}$ converges absolutely on the interval $(-R, R)$ and uniformly on any interval $[-b, +b]$, $0 < b < R$, and diverges outside the interval $[-R, R]$.*

Proof As was shown before, the series $\sum_{n=0}^{\infty} c_n x^n$ converges on the interval $(-R, R)$ and diverges outside the interval $[-R, R]$, where R is the radius of convergence. From the extension of Abel's Lemma, it follows that the series of derivatives $\sum_{n=1}^{\infty} c_n n x^{n-1}$ converges on the same interval $(-R, R)$ and diverges outside $[-R, R]$.   □

Corollary 1 *The same result is true for the series centered at a: if R is the radius of convergence of a series $\sum_{n=0}^{\infty} c_n (x - a)^n$, then the series of derivatives $\sum_{n=1}^{\infty} c_n n (x - a)^{n-1}$ converges absolutely on the interval $(a - R, a + R)$ and uniformly on any interval $[a - b, a + b]$, $0 < b < R$, and diverges outside the interval $[a - R, a + R]$.*

Corollary 2 *Since the series of derivatives is a power series, it can be considered in turn as the original series and the same result can be applied to the series of its derivatives,* that is, the series of second derivatives of the series $\sum_{n=0}^{\infty} c_n (x - a)^n$, and so on. Thus, we arrive at the following generalization of the last Theorem: if R is the radius of convergence of a series $\sum_{n=0}^{\infty} c_n (x - a)^n$, then the series of derivatives of any order

$$\sum_{n=0}^{\infty} c_n ((x - a)^n)^{(k)} = \sum_{n=k}^{\infty} c_n n(n - 1) \cdot \ldots \cdot (n - k + 1)(x - a)^{n-k}$$

converges absolutely on the interval $(a - R, a + R)$ and uniformly on any interval $[a - b, a + b]$, $0 < b < R$, and diverges outside the interval $[a - R, a + R]$.

2.4   Behavior at the Endpoints of the Interval of Convergence

Until the moment, we do not pay attention to the behavior of power series at the endpoints of the interval of convergence. The next Theorem fills this gap. As usual, we prove the statement for a power series centered at the origin.

Abel's Theorem *Let $R > 0$ be the radius of convergence of a power series (5.2)—$\sum_{n=0}^{\infty} c_n x^n$. If series (5.2) converges (at least conditionally) at the endpoint $x = R$, then it converges uniformly on $[0, R]$. A similar statement is true for the endpoint $x = -R$.*

Proof We demonstrate this statement for the point $x = R$, since the proof for $x = -R$ is analogous. To analyze the behavior of series (5.2) on the interval $[0, R]$, we use the following representation:

$$\sum_{n=0}^{\infty} c_n x^n = \sum_{n=0}^{\infty} c_n R^n \cdot \left(\frac{x}{R}\right)^n.$$

By the condition of the Theorem, the series of numbers $\sum_{n=0}^{\infty} c_n R^n$ converges (and consequently, it converges uniformly on $[0, R]$ because it does not depend on x). The functions $\left(\frac{x}{R}\right)^n$ form a sequence, which is monotone (decreasing) for every fixed point $x \in [0, R]$ and uniformly bounded on $[0, R]$: $0 \leq \left(\frac{x}{R}\right)^n \leq 1$, $\forall x \in [0, R]$, $\forall n \in \mathbb{N}$. Then, according to Abel's test for uniform convergence, series (5.2) converges uniformly on the interval $[0, R]$.   □

Remark In the case of series (5.1)—$\sum_{n=0}^{\infty} c_n (x - a)^n$—the statement takes the following form: if a power series (5.1) converges at the endpoint $x = a + R$ of the interval of convergence, then it converges uniformly on $[a, a + R]$. A similar statement is true for the endpoint $x = a - R$.

3   Properties of Power Series and Their Sums

> *Newton, Leibniz, Euler, and even Lagrange regarded series as an*
> *extension of the algebra of polynomials and hardly realized that*
> *they were introducing new problems by extending sums to an*
> *infinite number of terms.*
> *Morris Kline, 1972*

3.1   Arithmetic Properties

The properties of a linear combination of uniformly convergent series (Sect. 2.2 of Chap. 4)
are directly transferred to the power series (5.1) due to their uniform convergence on any
closed interval inside the interval of convergence $(a - R, a + R)$.

Property 1a *If power series* $\sum_{n=0}^{\infty} c_n(x - a)^n$ *and* $\sum_{n=0}^{\infty} d_n(x - a)^n$ *have the intervals
of convergence* $(a - R_c, a + R_c)$ *and* $(a - R_d, a + R_d)$, *respectively, then the series*
$\sum_{n=0}^{\infty}(c_n + d_n)(x - a)^n$ *converges on the interval* $(a - R, a + R)$, $R = \min\{R_c, R_d\}$ *and
converges uniformly on any closed subinterval of this interval.*

Proof The two power series converge on the smaller interval between $(a - R_c, a + R_c)$
and $(a - R_d, a + R_d)$, that is, on $(a - R, a + R)$, $R = \min\{R_c, R_d\}$, and consequently,
they converge uniformly on any closed subinterval of this interval. Then, according to the
arithmetic properties of series of functions (Property 1 in Sect. 2.2, Chap. 4), the series
$\sum_{n=0}^{\infty}(c_n + d_n)(x - a)^n$ converges on $(a - R, a + R)$ and converges uniformly on any
closed subinterval of this interval. $\square$

Remark The interval of convergence of the series $\sum_{n=0}^{\infty}(c_n + d_n)(x - a)^n$ contains the
interval $(a - R, a + R)$, $R = \min\{R_c, R_d\}$, but, in general, we cannot state that $(a -
R, a + R)$ is the interval of convergence of this series. A trivial example of this case is
$\sum_{n=0}^{\infty} c_n(x - a)^n = \sum_{n=0}^{\infty} x^n$ and $\sum_{n=0}^{\infty} d_n(x - a)^n = \sum_{n=0}^{\infty} -x^n$, where both series
have the interval of convergence $(a - R_c, a + R_c) = (a - R_d, a + R_d) = (-1, 1)$, but
their sum is the series of zeros that converges (uniformly) on $\mathbb{R}$.

Property 2a *If a power series* $\sum_{n=0}^{\infty} c_n(x - a)^n$ *has the interval of convergence* $(a -
R, a + R)$ *and a function* $g(x)$ *is bounded on* $(a - R, a + R)$, *then the series*
$\sum_{n=0}^{\infty} g(x)c_n(x - a)^n$ *converges uniformly on any closed subinterval of* $(a - R, a + R)$.

Proof The series $\sum_{n=0}^{\infty} c_n(x - a)^n$ converges uniformly on any closed subinterval of the
interval of convergence. Then, according to the arithmetic properties of series of functions
(Property 2 in Sect. 2.2, Chap. 4), the boundedness of $g(x)$ on $(a-R, a+R)$ guarantees that

the series $\sum_{n=0}^{\infty} g(x)c_n(x-a)^n$ also converges uniformly on the same closed subintervals of $(a-R, a+R)$. $\square$

Remark The condition of boundedness of $g(x)$ on $(a-R, a+R)$ can be relaxed to the boundedness of $g(x)$ on any closed subinterval of $(a-R, a+R)$. The function $g(x) = \frac{1}{x}$ is an example of an unbounded on $(0, 1)$ function, which is bounded on any closed subinterval of $(0, 1)$.

Product of Power Series

The third property deals with the *Cauchy product of two power series*. This operation between series was introduced in Chap. 2 (Sects. 2 and 4). For two series $\sum_{n=0}^{\infty} c_n x^n$ and $\sum_{n=0}^{\infty} d_n x^n$ centered at 0, their Cauchy product is a new power series (centered at the same point) defined by the formula

$$\sum_{n=0}^{\infty} c_n x^n \cdot \sum_{n=0}^{\infty} d_n x^n = (c_0 + c_1 x + c_2 x^2 + \ldots)(d_0 + d_1 x + d_2 x^2 + \ldots)$$

$$= c_0 d_0 + (c_0 d_1 + c_1 d_0)x + (c_0 d_2 + c_1 d_1 + c_2 d_0)x^2 + \ldots = \sum_{n=0}^{\infty} e_n x^n, \ e_n = \sum_{k=0}^{n} c_k d_{n-k}.$$

Notice that this type of the product between two series is specially suitable for power series since the resulting series is obtained by grouping the terms of the original series according to the powers of x and represents a generalization of the product of two polynomials in the form of distribution by powers.

Property 3a *If series $f(x) = \sum_{n=0}^{\infty} c_n(x-a)^n$ and $g(x) = \sum_{n=0}^{\infty} d_n(x-a)^n$ have the intervals of convergence $(a-R_c, a+R_c)$ and $(a-R_d, a+R_d)$, respectively, then their Cauchy product $\sum_{n=0}^{\infty} e_n(x-a)^n$, $e_n = \sum_{k=0}^{n} c_k d_{n-k}$ converges to the function $f(x)g(x)$ on the interval $(a-R, a+R)$, $R = \min\{R_c, R_d\}$ and converges uniformly on any closed subinterval of this interval.*

Proof Pick an arbitrary point $x_1 \neq a$ in the interval $(a-R, a+R)$ and fix it. The two series of numbers $\sum_{n=0}^{\infty} c_n(x_1-a)^n$ and $\sum_{n=0}^{\infty} d_n(x_1-a)^n$ converge absolutely, and according to the property of product (Sect. 2.1 in Chap. 2) their Cauchy product also converges and the sum of this series is $f(x_1)g(x_1)$, which means that the series $\sum_{n=0}^{\infty} e_n(x-a)^n$ converges at x_1 to $f(x_1)g(x_1)$. Then, by Abel's Lemma, the last power series converges absolutely and uniformly on the interval $[a-R_b, a+R_b]$ for any $0 < R_b < |x_1-a|$ and, besides, its sum is equal to $f(x)g(x)$. Due to the arbitrariness of x_1 (such that $0 < |x_1-a| < R$), it follows that the series $\sum_{n=0}^{\infty} e_n(x-a)^n$ converges absolutely and uniformly to $f(x)g(x)$ on any interval $[a-R_b, a+R_b]$ where $0 < R_b < R = \min\{R_c, R_d\}$. Hence, the series $\sum_{n=0}^{\infty} e_n(x-a)^n$ has the interval of convergence $(a-R, a+R)$, $R \geq \min\{R_c, R_d\}$. $\square$

Remark As it was in the case of the sum of two power series, the Cauchy product can converge on an interval with the radius larger than $R = \min\{R_c, R_d\}$. Indeed, the series $\sum_{n=0}^{\infty} x^n = 1 + x + x^2 + \ldots = \frac{1}{1-x}$ $(c_n = 1, \forall n)$ converges on $(-1, 1)$, that is, $R_c = 1$, and the series $1 - x$ $(d_0 = 1, d_1 = -1, d_n = 0, \forall n > 1)$ converges on $\mathbb{R}$, that is, $R_d = +\infty$. By definition, their Cauchy product has the form $1 + (1 - 1)x + (1 - 1)x^2 + \ldots = 1 = \sum_{n=0}^{\infty} e_n x^n$, where $e_n = 0, \forall n > 0$. Consequently, this product is a power series convergent on $\mathbb{R}$, although $\min\{R_c, R_d\} = 1$.

Example 1p Consider the two geometric power series with known sums: $\sum_{n=0}^{\infty} x^n = \frac{1}{1-x}$ and $\sum_{n=1}^{\infty} x^n = \frac{x}{1-x}$ and verify that their Cauchy product results in the product of the corresponding sums. Indeed,

$$\sum_{n=0}^{\infty} x^n \cdot \sum_{n=1}^{\infty} x^n = (1 + x + x^2 + \ldots)(x + x^2 + x^3 + \ldots) = x + 2x^2 + 3x^3 + \ldots = \sum_{n=1}^{\infty} n x^n.$$

The sum of the last power series was found in the introductory section of this chapter and is equal to the product of the two functions:

$$\sum_{n=1}^{\infty} n x^n = \frac{x}{(1-x)^2} = \frac{1}{1-x} \cdot \frac{x}{1-x}.$$

According to Property 3a, the obtained series converges, at least, on the interval $(-1, 1)$, and the independent study of its convergence (see the introductory section) confirms that this is the interval of convergence of this series.

Example 2p Find the product of the series $\frac{1}{1-x} = \sum_{n=0}^{\infty} x^n$ with the series $e^x = \sum_{n=0}^{\infty} \frac{x^n}{n!}$, that is, determine the expansion of the function $\frac{1}{1-x} e^x$ in power series. According to the definition, grouping the terms with the same powers of x, we obtain:

$$\frac{1}{1-x} e^x = \sum_{n=0}^{\infty} x^n \cdot \sum_{n=0}^{\infty} \frac{x^n}{n!} = \left(1 + x + x^2 + x^3 + \ldots\right)\left(1 + \frac{x}{1!} + \frac{x^2}{2!} + \frac{x^3}{3!} + \ldots\right)$$

$$= 1 + x\left(1 + \frac{1}{1!}\right) + x^2\left(1 + \frac{1}{1!} + \frac{1}{2!}\right) + x^3\left(1 + \frac{1}{1!} + \frac{1}{2!} + \frac{1}{3!}\right) + \ldots$$

$$+ x^n\left(1 + \frac{1}{1!} + \frac{1}{2!} + \ldots + \frac{1}{n!}\right) + \ldots$$

$$= \sum_{n=0}^{\infty} x^n\left(1 + \frac{1}{1!} + \frac{1}{2!} + \ldots + \frac{1}{n!}\right).$$

Since the series $\sum_{n=0}^{\infty} x^n$ converges on $(-1, 1)$ and $\sum_{n=0}^{\infty} \frac{x^n}{n!}$—on $\mathbb{R}$, the series of the product converges (at least) on $(-1, 1)$.

3.2   Functional Properties

Composition of Power Series

Consider now a composition of functions and the corresponding composition of their series. General rule and formula for *composition of two power series* are rather complicated, but one very particular form is frequently used and has a simple formulation. We start with this special form of composition of two functions, which we formulate separately from a general result (presented later).

Property 1f—Change of Variable *Suppose that a series $f(x) = \sum_{n=0}^{\infty} c_n (x-a)^n$ converges on $(a-R, a+R)$, $R > 0$ and that $x = g(t) = a + \alpha(t-b)^k$, $k \in \mathbb{N}$, $\alpha \neq 0$. In this case, the power series $h(t) = \sum_{n=0}^{\infty} c_n \alpha^n (t-b)^{kn}$ converges to the function $f(g(t))$ on $(b - R_1, b + R_1)$, $R_1 = \left(\frac{R}{|\alpha|} \right)^{1/k}$.*

Proof First, notice that

$$
h(t) = \sum_{n=0}^{\infty} c_n \alpha^n (t-b)^{kn} = c_0 + c_1 \alpha (t-b)^k + c_2 \alpha^2 (t-b)^{2k} + \dots
$$

is a power series of the variable t (with various zero coefficients if $k > 1$). Second, we analyze the convergence of this series. The condition $0 < |t - b| < R_1$, $R_1 = \left(\frac{R}{|\alpha|} \right)^{1/k}$ is equivalent to $0 < |x - a| < R$ due to the relation $x - a = \alpha(t - b)^k$ between x and t. Therefore, taking any point $t_1 \neq b$ in the interval $(b - R_1, b + R_1)$, that is, $0 < |t_1 - b| < R_1$, we have the corresponding point $x_1 = a + \alpha(t_1 - b)^k$ that satisfies the inequality $0 < |x_1 - a| < R$, that is, the point x_1 lies in $(a - R, a + R)$. Then, the series of numbers $\sum_{n=0}^{\infty} c_n (x_1 - a)^n$ converges absolutely to $f(x_1)$, and since $x_1 - a = \alpha(t_1 - b)^k$, the same series of numbers can be expressed in the form $\sum_{n=0}^{\infty} c_n \alpha^n (t_1 - b)^{kn}$ and its sum—in the form $f(a + \alpha(t_1 - b)^k) = f(g(t_1))$. Since the last series converges, by Abel's Lemma, the power series $\sum_{n=0}^{\infty} c_n \alpha^n (t - b)^{kn}$ converges absolutely and uniformly on the interval $[b - r, b + r]$ for any $0 < r < |t_1 - b|$, and besides, its sum is equal to $f(g(t))$. Then, due to the arbitrariness of t_1 (such that $0 < |t_1 - b| < R_1$), it follows that the series $h(t) = \sum_{n=0}^{\infty} c_n \alpha^n (t - b)^{kn}$ converges absolutely and uniformly to $f(g(t))$ on any interval $[b - r, b + r]$ where $0 < r < R_1$. Hence, the series $\sum_{n=0}^{\infty} c_n \alpha^n (t - b)^{kn}$ has the interval of convergence $(b - R_1, b + R_1)$, $R_1 = \left(\frac{R}{|\alpha|} \right)^{1/k}$. $\qquad \square$

Complement: Composite Series

Now we consider a general formulation of the rule of composite power series.

Property 2f—Theorem of Composite Power Series *Suppose that a power series*

$$y = f(x) = \sum_{n=1}^{\infty} c_n x^n = c_1 x + c_2 x^2 + c_3 x^3 + \dots \tag{5.8}$$

has the radius of convergence $R_x > 0$, and a power series

$$u = g(y) = \sum_{n=0}^{\infty} d_n y^n = d_0 + d_1 y + d_2 y^2 + d_3 y^3 + \dots \tag{5.9}$$

has the radius of convergence $R_y > 0$. Then, in a neighborhood of the origin, the composite function $h(x) = g(f(x))$ can be expressed in a power series of the variable x, obtained by the substitution of series (5.8) for y in (5.9) with corresponding reordering and regrouping of the terms according to the powers of x:

$$\begin{aligned}
h(x) = g(f(x)) &= \sum_{n=0}^{\infty} d_n \left(\sum_{m=1}^{\infty} c_m x^m \right)^n \\
&= d_0 + d_1(c_1 x + c_2 x^2 + c_3 x^3 + \dots) + d_2(c_1 x + c_2 x^2 + c_3 x^3 + \dots)^2 \\
&\quad + d_3(c_1 x + c_2 x^2 + c_3 x^3 + \dots)^3 + \dots \\
&= d_0 + d_1 c_1 x + (d_1 c_2 + d_2 c_1^2)x^2 + (d_1 c_3 + 2d_2 c_1 c_2 + d_3 c_1^3)x^3 \\
&\quad + (d_1 c_4 + 2d_2 c_1 c_3 + d_2 c_2^2 + 3d_3 c_1^2 c_2 + d_4 c_1^4)x^4 + \dots.
\end{aligned} \tag{5.10}$$

Proof First, the property of continuity of the sum of uniformly convergent series of continuous functions (Sect. 4.3 in Chap. 4) guarantees that $f(x)$ is continuous in a neighborhood of the origin. Using this fact and also that $f(0) = 0$, we can conclude that for sufficiently small values of $|x|$, the corresponding values of $y = f(x)$ (defined by series (5.8)) lie in a small neighborhood of 0, and consequently, belong to the interval of convergence of series (5.9). Therefore, if we choose an arbitrary x_1 in a small neighborhood of 0, then the corresponding $y_1 = f(x_1)$ stays inside the interval of convergence of (5.9) and, consequently, the series

$$g(y_1) = \sum_{n=0}^{\infty} d_n y_1^n$$

converges absolutely. Since $y_1 = f(x_1) = \sum_{n=1}^{\infty} c_n x_1^n$, the numerical series

$$h(x_1) = \sum_{n=0}^{\infty} d_n \left(\sum_{m=1}^{\infty} c_m x_1^m \right)^n$$

$$= d_0 + d_1(c_1 x_1 + c_2 x_1^2 + c_3 x_1^3 + \ldots) + d_2(c_1 x_1 + c_2 x_1^2 + c_3 x_1^3 + \ldots)^2$$
$$+ d_3(c_1 x_1 + c_2 x_1^2 + c_3 x_1^3 + \ldots)^3 + \ldots \tag{5.11}$$

converges absolutely.

Now, to arrive at the desired result, we need to justify the omission of parentheses in the last formula and regrouping the terms of the series according to the powers of x_1. To do this, we apply the Theorem of double series (Sect. 6 in Chap. 2) with the following infinite matrix of coefficients:

$$
\begin{array}{llll}
d_0 & 0 & 0 & 0 \quad \ldots \\
0 \; d_1 c_1 x_1 & 0 & 0 \quad \ldots \\
0 \; d_1 c_2 x_1^2 & d_2 c_1^2 x_1^2 & 0 \quad \ldots \\
0 \; d_1 c_3 x_1^3 & 2 d_2 c_1 c_2 x_1^3 & d_3 c_1^3 x_1^3 \; \ldots
\end{array}
\tag{5.12}
$$

$$\ldots$$

For this matrix, the repeated row series $\sum_{i=1}^{\infty} \left(\sum_{j=1}^{\infty} a_{ij} \right)$ (see formulas (2.23) and (2.27) in Chap. 2) is exactly the series we intend to obtain (the constants x_1^k can always be placed inside the parentheses):

$$d_0 + d_1 c_1 x_1 + (d_1 c_2 + d_2 c_1^2)x_1^2 + (d_1 c_3 + 2 d_2 c_1 c_2 + d_3 c_1^3)x_1^3$$
$$+ (d_1 c_4 + 2 d_2 c_1 c_3 + d_2 c_2^2 + 3 d_3 c_1^2 c_2 + d_4 c_1^4)x_1^4 + \ldots \tag{5.13}$$

On the other hand, series (5.11) is the repeated column series $\sum_{j=1}^{\infty} \left(\sum_{i=1}^{\infty} a_{ij} \right)$ (see (2.28) in Chap. 2) (the constants d_k can be placed inside the parentheses). Then, by the Corollary to the Theorems about double series (Sect. 6 in Chap. 2), the absolute convergence of series (5.11) guarantees the absolute convergence of series (5.13), and both repeated series converge to the same sum $h(x_1)$. Since this is true for any x_1 in a neighborhood of 0, the composite rule (5.10) is proved. Moreover, the absolute convergence of series (5.13) implies (according to Abel's Lemma) that the last series in (5.10) converges uniformly in a neighborhood of 0.   □

Remark 1 The condition $c_0 = 0$ simplifies the technical part of the demonstration, but the statement continues to be valid even in the case $c_0 \neq 0$, under the additional condition that

$|c_0| = |f(0)| < R_y$. The proof in this case, albeit a bit more involved, follows the same lines of the presented demonstration.

Remark 2 In practice, the restriction $c_0 = 0$ usually does not cause problems, for if $c_0 \neq 0$, then we can consider the function $f(x) - c_0$ instead of $f(x)$.

Example 1f Let us find the power series expansion of e^{2x} starting from the series $e^x = \sum_{n=0}^{\infty} \frac{x^n}{n!}$ and using the two methods: first, we apply the change of the variable $t = 2x$, and second, we calculate the Cauchy product of the series of e^x with itself. Using the first approach, according to Property 1f, we obtain

$$e^{2x} = e^t = \sum_{n=0}^{\infty} \frac{t^n}{n!} = \sum_{n=0}^{\infty} \frac{(2x)^n}{n!}.$$

Since the series of e^t converges on $\mathbb{R}$, the same is true for the last series.

Employing the second technique, we obtain the following product:

$$\sum_{n=0}^{\infty} \frac{x^n}{n!} \cdot \sum_{n=0}^{\infty} \frac{x^n}{n!} = \left(1 + \frac{x}{1!} + \frac{x^2}{2!} + \cdots\right)\left(1 + \frac{x}{1!} + \frac{x^2}{2!} + \cdots\right)$$

$$= 1 + x\left(\frac{1}{1!} + \frac{1}{1!}\right) + x^2\left(\frac{1}{0!}\frac{1}{2!} + \frac{1}{1!}\frac{1}{1!} + \frac{1}{2!}\frac{1}{0!}\right) + x^3\left(\frac{1}{0!}\frac{1}{3!} + \frac{1}{1!}\frac{1}{2!} + \frac{1}{2!}\frac{1}{1!} + \frac{1}{3!}\frac{1}{0!}\right) + \cdots$$

$$+ x^n\left(\frac{1}{0!}\frac{1}{n!} + \frac{1}{1!}\frac{1}{(n-1)!} + \frac{1}{2!}\frac{1}{(n-2)!} + \cdots + \frac{1}{(n-1)!}\frac{1}{1!} + \frac{1}{n!}\frac{1}{0!}\right) + \cdots$$

$$= \sum_{n=0}^{\infty} \frac{x^n}{n!}\left(1 + \binom{n-1}{n} + \binom{n-2}{n} + \cdots + \binom{1}{n} + 1\right) = \sum_{n=0}^{\infty} \frac{x^n}{n!}(1+1)^n = \sum_{n=0}^{\infty} \frac{(2x)^n}{n!} = e^{2x}.$$

Since the original series $\sum_{n=0}^{\infty} \frac{x^n}{n!}$ converges on $\mathbb{R}$, it implies (by Property 3a) that its square also converges on $\mathbb{R}$.

Notice that although the two procedures provide the same result, the first is much simpler. This frequently happens with developments in power series: the validity (theoretical) of two (or more) methods does not mean that they have a similar simplicity/complexity of application in a specific situation. The next example shows that it may happen that one of the techniques gives a simple expansion in power series, while another does not allow us to find a general form of the coefficients of the series.

Example 2f Find the power series expansion of e^{-x} starting with the series $e^x = \sum_{n=0}^{\infty} \frac{x^n}{n!}$ and using the two methods: first, change the variable $t = -x$, and second, rewrite e^{-x} in the form $\frac{1}{e^x}$ and consider the composite function $h(x) = g(f(x)) = \frac{1}{1-(1-e^x)}$, $f(x) = 1 - e^x$, $g(y) = \frac{1}{1-y}$.

Following Property 1f, the change of the variable $t = -x$ gives

$$e^{-x} = e^t = \sum_{n=0}^{\infty} \frac{t^n}{n!} = \sum_{n=0}^{\infty} \frac{(-x)^n}{n!} = \sum_{n=0}^{\infty} \frac{(-1)^n}{n!} x^n.$$

Since the series of e^t converges on $\mathbb{R}$, the same is true for the obtained power series.

To apply the second technique, notice that the power series $e^x = \sum_{n=0}^{\infty} \frac{x^n}{n!}$ has the first coefficient different from zero ($c_0 = 1$), and for this reason we use the representation $h(x) = g(f(x)) = \frac{1}{e^x} = \frac{1}{1-f(x)}$, where $y = f(x) = 1 - e^x$, $g(y) = \frac{1}{1-y}$. The series

$$y = f(x) = 1 - e^x = -\sum_{n=1}^{\infty} \frac{x^n}{n!} = -\frac{x}{1!} - \frac{x^2}{2!} - \frac{x^3}{3!} - \frac{x^4}{4!} + \dots, \quad \forall x \in \mathbb{R}$$

and

$$g(y) = \frac{1}{1-y} = \sum_{n=0}^{\infty} y^n = 1 + y + y^2 + y^3 + \dots, \quad \forall y \in (-1, 1)$$

satisfy the conditions of the Theorem of composite power series. The first series has the radius of convergence $R_x = \infty$ and the second—$R_y = 1$. Therefore, in a neighborhood of $x = 0$, we can obtain the power series expansion of $h(x) = g(f(x))$ by substituting the series of $f(x) = 1 - e^x$ for y in the series of $\frac{1}{1-y}$ and joining the terms with the same powers of x:

$$h(x) = e^{-x} = 1 - \left(\frac{x}{1!} + \frac{x^2}{2!} + \frac{x^3}{3!} + \frac{x^4}{4!} + \dots \right) + \left(\frac{x}{1!} + \frac{x^2}{2!} + \frac{x^3}{3!} + \frac{x^4}{4!} + \dots \right)^2$$

$$- \left(\frac{x}{1!} + \frac{x^2}{2!} + \frac{x^3}{3!} + \frac{x^4}{4!} + \dots \right)^3 + \left(\frac{x}{1!} + \frac{x^2}{2!} + \frac{x^3}{3!} + \frac{x^4}{4!} + \dots \right)^4 + \dots$$

Unfortunately, this form make impossible to find a general form of the coefficients of the resulting series. Instead, we can calculate the first terms upto the fourth power of x and confirm that we arrive at the same terms as in the first method:

$$h(x) = e^{-x} = 1 - x + x^2 \left(-\frac{1}{2!} + 1 \right) - x^3 \left(\frac{1}{3!} - 2\frac{1}{1!\,2!} + 1 \right)$$

$$+ x^4 \left(-\frac{1}{4!} + 2\frac{1}{1!\,3!} + \frac{1}{(2!)^2} - 3\frac{1}{2!} + 1 \right) + \dots$$

$$= 1 - \frac{1}{1!} x + \frac{1}{2!} x^2 - \frac{1}{3!} x^3 + \frac{1}{4!} x^4 + \dots.$$

The convergence of the power series of $g(y)$ is guaranteed when $|y| = |1 - e^x| < 1$, which amounts to the inequality $x < \ln 2$. Therefore, in the second method the interval of convergence guaranteed by this procedure is less than obtained in the first method.

Thus, although the use of both methods is theoretically justified, the results of their application are different: the first allows us to find immediately the desired power series and determine the largest possible interval of convergence—$\mathbb{R}$, while the second gives only some first terms of the composite series and shows the reduced interval of convergence.

Example 3f Find the power series expansion of $\frac{1}{1+x^2}$ starting with the series $\frac{1}{1-x} = \sum_{n=0}^{\infty} x^n$ and using the change of variable of Property 1f. Denoting $t = -x^2$, we have the desired series

$$\frac{1}{1+x^2} = \frac{1}{1-t} = \sum_{n=0}^{\infty} t^n = \sum_{n=0}^{\infty} (-x^2)^n = \sum_{n=0}^{\infty} (-1)^n x^{2n}.$$

Since the original series for $\frac{1}{1-x}$ converges on $(-1, 1)$, according to Property 1f, the same is true for the derived series.

Example 4f Find the power series expansion of a^x $(a > 0, a \neq 1)$ starting with the series $e^x = \sum_{n=0}^{\infty} \frac{x^n}{n!}$ and using the change of variable of Property 1f. To solve this task, we represent $a^x = e^{\ln a^x} = e^{x \ln a}$ and change the variable by the formula $t = x \ln a$:

$$a^x = e^{x \ln a} = e^t = \sum_{n=0}^{\infty} \frac{t^n}{n!} = \sum_{n=0}^{\infty} \frac{(x \ln a)^n}{n!} = \sum_{n=0}^{\infty} \frac{\ln^n a}{n!} x^n.$$

Since the series of e^t converges on $\mathbb{R}$, by Property 1f, the same is true for the obtained series.

Change of the Central Point

Property 3f—Change of the Central Point *Let $f(x) = \sum_{n=0}^{\infty} c_n x^n$ be a series convergent on $|x| < R$. If $a \in (-R, R)$, then $f(x)$ can be expanded in a power series centered at a that converges on $|x - a| < R - |a|$.*

Proof Since $\sum_{n=0}^{\infty} c_n x^n$ converges on $(-R, +R)$, it converges absolutely at every point of this interval. Choose $a \in (-R, R)$ and represent the original series in the form

$$f(x) = \sum_{n=0}^{\infty} c_n x^n = \sum_{n=0}^{\infty} c_n ((x - a) + a)^n = \sum_{n=0}^{\infty} c_n \left(\sum_{m=0}^{n} \binom{n}{m} a^{n-m} (x - a)^m \right).$$

For any fixed x the last series is a repeated series of numbers (with various zero coefficients) that converges absolutely on $|x - a| < R - |a|$ (this inequality is the condition for a neighborhood of a be located inside the interval $(-R, +R)$). Then, for these x, we can use the result on equality of the repeated series (see the Corollary in Sect. 6 of Chap. 2) to change the order of summation:

$$f(x) = \sum_{n=0}^{\infty} c_n \left(\sum_{m=0}^{n} \binom{n}{m} a^{n-m} (x-a)^m \right) = \sum_{m=0}^{\infty} \left(\sum_{n=m}^{\infty} c_n \binom{n}{m} a^{n-m} \right) (x-a)^m.$$

The last series is a power series centered at a with the coefficients $d_m = \sum_{n=m}^{\infty} c_n \binom{n}{m} a^{n-m}$, which converges on the interval $|x-a| < R - |a|$.   □

Remark As usual, this result can be extended to the case of a general central point. In this case, if a power series $f(x) = \sum_{n=0}^{\infty} c_n (x-a)^n$ converges on the interval $|x-a| < R$, then for any $b \in (a-R, a+R)$, the same function can be developed in a power series centered at b: $f(x) = \sum_{n=0}^{\infty} d_n (x-b)^n$, which converges on the interval $|x-b| < R - |b-a|$.

3.3   Analytic Properties

The Theorems of uniform convergence of power series and series of its derivatives (proved in Sect. 2) together with the properties of the sums of series convergent uniformly (derived in Chap. 4), allow us to easily obtain results about *analytic properties of the sums of power series*. In what follows, we use the introduced notations of the radius of convergence R and the interval of convergence $(a - R, a + R)$ of a power series $\sum_{n=0}^{\infty} c_n (x-a)^n$, and in the formulation of properties we suppose that $R > 0$. In the demonstrations of the properties we use (without mentioning it again) the results of the Theorem of convergence of power series (Sect. 2.1) and the Theorem of convergence of the series of derivatives (Sect. 2.3) which guarantee the uniform convergence of both series ((5.1) and (5.6)) on any interval $[a - b, a + b]$, $0 < b < R$.

Property 1—Boundedness of Sum *The sum of a power series* (5.1) *is bounded on any closed subinterval of the interval of convergence* $(a - R, a + R)$.

Proof All the functions $c_n (x-a)^n$ are bounded on any finite interval, in particular, on any closed finite interval. Then, according to the property of boundedness of series of functions (Sect. 4.1 of Chap. 4), the uniform convergence of series (5.1) on any interval $[a-b, a+b]$, $0 < b < R$ guarantees that the sum of the series (5.1) is bounded on $[a - b, a + b]$.   □

Remark This property cannot be extended to the entire interval of convergence, since a power series may not converge uniformly on the interval $[a - R, a + R]$. One example of this is the series $\sum_{n=0}^{\infty} x^n$ that has the interval of convergence $(-1, 1)$, but does not converge at the endpoints -1 and 1. For this reason, the sum $f(x) = \frac{1}{1-x}$ is unbounded on $(-1, 1)$.

Property 2—Limit of Sum *The sum of a power series* (5.1) *has a limit at any point* x_0 *of its interval of convergence* $(a - R, a + R)$ *and the following formula holds:*

$$\lim_{x \to x_0} \sum_{n=0}^{\infty} c_n (x - a)^n = \sum_{n=0}^{\infty} c_n (x_0 - a)^n = \sum_{n=0}^{\infty} \lim_{x \to x_0} c_n (x - a)^n, \qquad (5.14)$$

that is, the order of the limit and sum can be interchanged.

Proof For any point $x_0 \in (a - R, a + R)$ there exists a closed interval $[a - b, a + b] \subset (a - R, a + R)$ with $|a - x_0| < b < R$ that contains x_0, which implies that x_0 is an accumulation point of $[a - b, a + b]$. Since the functions $c_n (x - a)^n$ have limit at every real point, in particular, at x_0, by the property of the limit of series of functions (Sect. 4.2 of Chap. 4), the uniform convergence of series (5.1) on the interval $[a - b, a + b]$ guarantees the validity of formula (5.14). □

Remark This property cannot be extended to any accumulation point of the interval of convergence, since at the endpoints $a - R$ and $a + R$ the conditions of the property of the limit of series may not be satisfied: the interval $(a - R, a + R)$ is generally not the set of uniform convergence and its endpoints are not accumulation points of the sets of uniform convergence. One example of this is the series $\sum_{n=0}^{\infty} \frac{x^n}{n}$ whose interval of convergence is $[-1, 1)$ and that diverges at the endpoint 1.

Property 3: Continuity

Theorem. Continuity on the Interval of Convergence *The sum of a power series* (5.1) *is continuous on the entire interval of convergence* $(a - R, a + R)$.

Proof Any point x_0 of the interval $(a - R, a + R)$ can be inserted in the closed interval $[a - b, a + b]$ with $|a - x_0| < b < R$. Since the functions $c_n (x - a)^n$ are continuous on $\mathbb{R}$, in particular, on any closed interval, by the property of continuity of series of functions (Sect. 4.3 of Chap. 4), the uniform convergence of series (5.1) on the interval $[a - b, a + b]$ guarantees that the sum of series (5.1) is continuous on $[a - b, a + b]$ and, in particular, at x_0. Since x_0 is an arbitrary point of $(a - R, a + R)$, the sum of the series is continuous on $(a - R, a + R)$. □

Remark Unlike the boundedness property, the continuity of the sum is applied to the entire interval of convergence $(a - R, a + R)$. This is a consequence of the fact that boundedness is a global property, considered on a set, while continuity is a local concept: continuity on a set is simply continuity at each point of this set.

Theorem. Continuity at the Endpoints of the Interval of Convergence *If a power series (5.1) converges (at least conditionally) at the point $x = a + R$, then it is continuous on $[a, a + R]$. A similar statement is true for the endpoint $x = a - R$.*

Proof Abel's Theorem states that series (5.1) converges uniformly on $[a, a + R]$ if it is convergent at $a + R$. Then the property of continuity of a series of functions (Sect. 4.3 of Chap. 4) ensures the continuity of the sum on $[a, a + R]$.   □

Property 4: Integrability

Theorem About Definite (Riemann) Integral *The sum $f(x)$ of a power series (5.1) is Riemann integrable on any closed subinterval $[\alpha, \beta]$ of the interval of convergence and the following formula is true:*

$$\int_\alpha^\beta f(x)dx = \int_\alpha^\beta \sum_{n=0}^\infty c_n(x-a)^n dx = \sum_{n=0}^\infty \int_\alpha^\beta c_n(x-a)^n dx$$

$$= \sum_{n=0}^\infty c_n \frac{(\beta-a)^{n+1} - (\alpha-a)^{n+1}}{n+1}. \tag{5.15}$$

Thus, a power series can be integrated term by term inside its interval of convergence.

Proof The functions $c_n(x-a)^n$ are continuous on $\mathbb{R}$, in particular, on any closed interval. Then, according to the Corollary to the Theorem of integrability of series of functions (Sect. 4.4 of Chap. 4), the uniform convergence of series (5.1) on $[a - b, a + b]$, $\forall b < R$ guarantees the integrability of $f(x)$ on any closed interval $[\alpha, \beta] \subset (a - R, a + R)$ and the validity of formula (5.15).   □

Theorem About Indefinite Integral *The indefinite integral of the sum $f(x)$ of a power series (5.1) exists on $(a - R, a + R)$ and can be found by the formula*

$$F(x) = \int f(x)dx = C + \sum_{n=0}^\infty \frac{c_n}{n+1}(x-a)^{n+1}, \ \forall x \in (a - R, a + R). \tag{5.16}$$

Proof If we choose $\alpha = a$ and $\beta = x \in (a - R, a + R)$ in formula (5.15), then we obtain the integral with variable upper limit (one of the antiderivatives of $f(x)$) which is defined by the formula

$$h(x) - h(a) = \int_a^x f(t)dt = \int_a^x \sum_{n=0}^{\infty} c_n(t-a)^n dt = \sum_{n=0}^{\infty} \int_a^x c_n(t-a)^n dt$$

$$= \sum_{n=0}^{\infty} \frac{c_n}{n+1}(x-a)^{n+1}. \tag{5.17}$$

Therefore, the indefinite integral of the power series (5.1) takes form (5.16).

In particular, one of the antiderivatives of series (5.2) centered at $a = 0$ has the representation

$$h(x) - h(0) = \int_0^x f(t)dt = \sum_{n=0}^{\infty} \frac{c_n}{n+1}x^{n+1} \tag{5.18}$$

and its indefinite integral is

$$F(x) = \int f(x)dx = C + \sum_{n=0}^{\infty} \frac{c_n}{n+1}x^{n+1}, \ \forall x \in (-R, R). \tag{5.19}$$

$\square$

Property 5: Differentiability

Theorem. Differentiability on the Interval of Convergence *The sum $f(x)$ of a power series (5.1) is differentiable on the entire interval of convergence $(a - R, a + R)$ and its derivative is equal to the sum $g(x)$ of the series of derivatives (5.6):*

$$f'(x) = \left(\sum_{n=0}^{\infty} c_n(x-a)^n \right)' = \sum_{n=0}^{\infty} \left(c_n(x-a)^n \right)'$$

$$= \sum_{n=1}^{\infty} c_n n(x-a)^{n-1}$$

$$= g(x), \ \forall x \in (a - R, a + R). \tag{5.20}$$

Thus, a power series can be differentiated term by term on its interval of convergence.

Proof Any point x_0 of the interval $(a - R, a + R)$ can be inserted in the closed interval $[a - b, a + b]$ with $|a - x_0| < b < R$. The original power series (5.1) and the series of derivatives (5.6) converge uniformly on $[a - b, a + b]$, and the functions $c_n(x - a)^n$ are continuously differentiable on $\mathbb{R}$, in particular, on $[a-b, a+b]$. Therefore, according to the Corollary to the Theorem of differentiability of series of functions (Sect. 4.5 of Chap. 4), the function $f(x)$ is differentiable on $[a-b, a+b]$ and $f'(x) = g(x), \forall x \in [a-b, a+b]$. In particular, this property holds at the point x_0. Since x_0 is an arbitrary point in $(a-R, a+R)$, we arrive at the statement of this Property. $\square$

Corollary. Infinite Differentiability on the Interval of Convergence *Since the series of derivatives in turn is a power series, the same Property is true for this series. Next, this Property can be applied to the series of second derivatives, and so on. Hence, we have the following generalization of the Property of differentiation:*

The sum $f(x)$ of a power series (5.1) is infinitely differentiable on the entire interval of convergence $(a - R, a + R)$ and its m-th derivative is equal to the power series of m-th derivatives:

$$\left(\sum_{n=0}^{\infty} c_n(x-a)^n \right)^{(m)} = \sum_{n=0}^{\infty} \left(c_n(x-a)^n \right)^{(m)}$$

$$= \sum_{n=m}^{\infty} c_n n(n-1) \cdot \ldots \cdot (n-m+1)(x-a)^{n-m}, \forall x \in (a - R, a + R).$$

$$(5.21)$$

Thus, a power series is infinitely differentiable term by term.

Remark The properties of differentiability and integrability in the sense of indefinite integral are valid for the entire interval $(a - R, a + R)$, while the Riemann integrability holds only on the closed subintervals of $(a - R, a + R)$ where power series (5.1) converges uniformly.

Example 1a The series $\sum_{n=0}^{\infty} x^n$ converges on $|x| < 1$, converges uniformly on $|x| \leq a$, $0 < \forall a < 1$ and diverges on $|x| \geq 1$. Its sum $\frac{1}{1-x}$ is bounded on any closed subinterval of $(-1, 1)$, but is unbounded on the entire interval of convergence $(-1, 1)$.

The series $\sum_{n=0}^{\infty} x^n$ is differentiable term by term on its interval of convergence $(-1, 1)$, that is, the derivative of the sum is equal to the sum of the series of derivatives:

$$\frac{1}{(1-x)^2} = \left(\frac{1}{1-x} \right)' = \left(\sum_{n=0}^{\infty} x^n \right)' = \sum_{n=0}^{\infty} (x^n)' = \sum_{n=1}^{\infty} n x^{n-1}.$$

Notice that multiplication of the last equality by x results in the series whose sum was found in the introductory section by applying another (rudimentary) technique: $\frac{x}{(1-x)^2} = \sum_{n=1}^{\infty} n x^n$.

The series $\sum_{n=0}^{\infty} x^n$ is term-by-term integrable on its interval of convergence $(-1, 1)$, that is, the integral of the sum is equal to the sum of the series of integrals:

$$- \ln(1 - x) = \int_0^x \frac{1}{1 - t} dt = \int_0^x \left(\sum_{n=0}^{\infty} t^n \right) dt = \sum_{n=0}^{\infty} \int_0^x t^n dt = \sum_{n=0}^{\infty} \frac{x^{n+1}}{n + 1} = \sum_{n=1}^{\infty} \frac{x^n}{n}.$$

The last series converges on $[-1, 1)$ (at the left endpoint we have the convergent alternating harmonic series, and at the right—the divergent harmonic series), and consequently, it converges uniformly on $[-1, b), -1 < \forall b < 1$.

Example 2a As was shown before (Example 3f in Sect. 3.2), the series $\sum_{n=0}^{\infty}(-1)^n x^{2n}$ converges on $|x| < 1$, converges uniformly on $|x| \le a, 0 < \forall a < 1$ and diverges on $|x| \ge 1$. Its sum $\frac{1}{1+x^2}$ is bounded on the entire interval of convergence $(-1, 1)$. Notice that the function $\frac{1}{1+x^2}$ is defined, bounded and infinitely differentiable on $\mathbb{R}$, but outside the interval $(-1, 1)$ it does not represent the sum of the series $\sum_{n=0}^{\infty}(-1)^n x^{2n}$.

The series $\sum_{n=0}^{\infty}(-1)^n x^{2n}$ is term-by-term differentiable on its interval of convergence, that is, the derivative of the sum is equal to the sum of the series of derivatives:

$$-\frac{2x}{(1 + x^2)^2} = \left(\frac{1}{1 + x^2} \right)' = \left(\sum_{n=0}^{\infty}(-1)^n x^{2n} \right)' = \sum_{n=0}^{\infty}(-1)^n (x^{2n})' = \sum_{n=1}^{\infty}(-1)^n 2n x^{2n-1}.$$

The last series converges on $|x| < 1$ and diverges on $|x| \ge 1$.

The series $\sum_{n=0}^{\infty}(-1)^n x^{2n}$ is term-by-term integrable on its interval of convergence $(-1, 1)$, that is, the integral of the sum is equal to the sum of the series of integrals:

$$\arctan x = \int_0^x \frac{1}{1 + t^2} dt = \int_0^x \left(\sum_{n=0}^{\infty}(-1)^n t^{2n} \right) dt = \sum_{n=0}^{\infty}(-1)^n \int_0^x t^{2n} dt = \sum_{n=0}^{\infty}(-1)^n \frac{x^{2n+1}}{2n + 1}.$$

The last series converges uniformly on $[-1, 1]$ (at both endpoints we have a convergent alternating series). The function $\arctan x$ is defined, bounded and infinitely differentiable on $\mathbb{R}$, but it represents the sum of the obtained series only on the interval $[-1, 1]$.

Example 3a The series $\sum_{n=0}^{\infty} \frac{x^n}{n!}$ converges on $\mathbb{R}$ and converges uniformly on $[-b, b]$, $\forall b > 0$. Its sum e^x is bounded on $[-b, b], \forall b > 0$, but it is not bounded on $\mathbb{R}$.

The series $\sum_{n=0}^{\infty} \frac{x^n}{n!}$ is term-by-term differentiable on $\mathbb{R}$, that is, the derivative of the sum is equal to the sum of the series of derivatives:

$$e^x = (e^x)' = \left(\sum_{n=0}^{\infty} \frac{x^n}{n!} \right)' = \sum_{n=0}^{\infty} \frac{1}{n!}(x^n)' = \sum_{n=1}^{\infty} \frac{x^{n-1}}{(n - 1)!} = \sum_{n=0}^{\infty} \frac{x^n}{n!}.$$

The last series coincides with the original and satisfies the same properties.

The series $\sum_{n=0}^{\infty} \frac{x^n}{n!}$ is term-by-term integrable on $\mathbb{R}$, that is, the integral of the sum is equal to the sum of the series of integrals:

$$e^x + C = \int e^t dt = \int \left(\sum_{n=0}^{\infty} \frac{t^n}{n!} \right) dt = \sum_{n=0}^{\infty} \frac{1}{n!} \int t^n dt = \tilde{C} + \sum_{n=0}^{\infty} \frac{x^{n+1}}{(n+1)!} = C + \sum_{n=0}^{\infty} \frac{x^n}{n!}$$

(where $\tilde{C} = C + 1$). Except for an arbitrary constant of integration C, the last series coincides with the original and satisfies the same properties.

3.4   Uniqueness of Power Series Expansion, Analytic Functions

Theorem of Uniqueness 1 *The power series expansion $\sum_{n=0}^{\infty} c_n (x - a)^n$ of a function $f(x)$ is unique on any interval $(a - c, a + c), c > 0$ where the series converges to $f(x)$.*

Proof For contradiction, let us assume that there are two different series $\sum_{n=0}^{\infty} c_n (x-a)^n$ and $\sum_{n=0}^{\infty} d_n (x - a)^n$ converging to $f(x)$ on the intervals $(a - c, a + c), c > 0$ and $(a - d, a + d), d > 0$, respectively. Then, on the common interval of convergence to $f(x)$ we have

$$\sum_{n=0}^{\infty} c_n (x - a)^n = f(x) = \sum_{n=0}^{\infty} d_n (x - a)^n$$

and setting $x = a$ in the last equality we obtain $c_0 = d_0$. Next, deriving both series term by term

$$\sum_{n=1}^{\infty} n c_n (x - a)^{n-1} = f'(x) = \sum_{n=1}^{\infty} n d_n (x - a)^{n-1}$$

and setting again $x = a$, we see that $c_1 = d_1$; and so on. After k-th differentiation term by term we obtain

$$\sum_{n=k}^{\infty} n(n-1) \ldots (n-k+1) c_n (x-a)^{n-k} = f^{(k)}(x) = \sum_{n=k}^{\infty} n(n-1) \ldots (n-k+1) d_n (x-a)^{n-k}$$

and substitution $x = a$ shows that $c_k = d_k$. Hence, all the coefficients of the two series coincide, that is, these two series are the same. $\square$

There is a stronger version of this Theorem: a power series $\sum_{n=0}^{\infty} c_n (x-a)^n$ convergent on $(a - r, a + r)$ is uniquely determined by its values on a subset E of this interval such that a is an accumulation point of E. This result is specified and proved in the following theorem.

Theorem of Uniqueness 2 *If two power series $\sum_{n=0}^{\infty} c_n (x - a)^n$ and $\sum_{n=0}^{\infty} d_n (x - a)^n$ converge on $(a - r, a + r)$ and coincide on a set $E \subset (a - r, a + r)$ that has the accumulation point a, then these two series coincide on $(a - r, a + r)$.*

Proof Since a is an accumulation point of E, there exists a sequence of points $x_k \in E$ such that $x_k \underset{k \to \infty}{\to} a$ and $x_k \neq a$, $\forall k$. At these points

$$\sum_{n=0}^{\infty} c_n (x_k - a)^n = \sum_{n=0}^{\infty} d_n (x_k - a)^n, \ \forall x_k \in E,$$

and applying the limit in this equality as $x_k \underset{k \to \infty}{\to} a$, by Property 2 (Sect. 3.3), we obtain:

$$\lim_{x_k \to a} \sum_{n=0}^{\infty} c_n (x_k - a)^n = \sum_{n=0}^{\infty} \lim_{x_k \to a} c_n (x_k - a)^n = c_0$$

$$= d_0 = \sum_{n=0}^{\infty} \lim_{x_k \to a} d_n (x_k - a)^n = \lim_{x_k \to a} \sum_{n=0}^{\infty} d_n (x_k - a)^n.$$

Knowing now that $c_0 = d_0$, the equality between the two power series (at the points $x_k \in E$) takes the form

$$\sum_{n=1}^{\infty} c_n (x_k - a)^n = \sum_{n=1}^{\infty} d_n (x_k - a)^n,$$

or, dividing by $(x_k - a)$, we have

$$\sum_{n=1}^{\infty} c_n (x_k - a)^{n-1} = \sum_{n=1}^{\infty} d_n (x_k - a)^{n-1}.$$

Again, applying the limit as $x_k \underset{k\to\infty}{\to} a$ to the last equality, according to Property 2, we obtain:

$$\lim_{x_k\to a}\sum_{n=1}^{\infty} c_n(x_k-a)^{n-1} = \sum_{n=1}^{\infty}\lim_{x_k\to a} c_n(x_k-a)^{n-1} = c_1$$

$$= d_1 = \sum_{n=1}^{\infty}\lim_{x_k\to a} d_n(x_k-a)^{n-1} = \lim_{x_k\to a}\sum_{n=1}^{\infty} d_n(x_k-a)^{n-1}.$$

This shows that $c_1 = d_1$. And so on. At the m-th step of this procedure, we obtain the equality between the series

$$\sum_{n=m}^{\infty} c_n(x_k-a)^{n-m} = \sum_{n=m}^{\infty} d_n(x_k-a)^{n-m},$$

and calculating the limits as $x_k \underset{k\to\infty}{\to} a$ on the both sides, we conclude that $c_m = d_m$:

$$\lim_{x_k\to a}\sum_{n=m}^{\infty} c_n(x_k-a)^{n-m} = \sum_{n=m}^{\infty}\lim_{x_k\to a} c_n(x_k-a)^{n-m} = c_m$$

$$= d_m = \sum_{n=m}^{\infty}\lim_{x_k\to a} d_n(x_k-a)^{n-m} = \lim_{x_k\to a}\sum_{n=m}^{\infty} d_n(x_k-a)^{n-m}.$$

In this manner, we show that $c_n = d_n$, $\forall n$, that is, the two series are the same.   □

Even and Odd Functions

Theorem *If a series $\sum_{n=0}^{\infty} c_n x^n$ is a representation of an even function $f(x)$ in a neighborhood of the origin, then this series contains only even powers of x. Analogously, if a series $\sum_{n=0}^{\infty} c_n x^n$ is a development of an odd function $f(x)$ in a neighborhood of the origin, then this series contains only odd powers of x.*

Proof We demonstrate only the case of even function, since the second case is handled similarly.

In a neighborhood of the origin we have

$$f(x) = c_0 + c_1 x + c_2 x^2 + c_3 x^3 + \dots$$

and

$$f(-x) = c_0 - c_1 x + c_2 x^2 - c_3 x^3 + \dots .$$

Then, subtracting one result from another and using the condition of even function $f(x) = f(-x)$, we obtain

$$f(x) - f(-x) = 2c_1 x + 2c_3 x^3 + \ldots = 0 .$$

Due to the uniqueness of the power series, the last equality implies that $c_1 = c_3 = \ldots = 0$, that is, all the odd coefficients of the original series equal zero. □

Analytic Functions

▶ **Definition** (**Analytic Function**) A function $f(x)$ represented by a power series $f(x) = \sum_{n=0}^{\infty} c_n (x-a)^n$ convergent on the interval $(a - R, a + R)$, $R > 0$ is called an *analytic function in this interval*. If $R = +\infty$, a function $f(x)$ is analytic in $\mathbb{R}$. A *function $f(x)$ is analytic at a point a* if it is analytic in a neighborhood of this point.

Notice that Property 3f about the change of the central point (Sect. 3.2) can be reformulated in the following way: if a function $f(x)$ is analytic in the interval $(a - R, a + R)$, then it is analytic in any subinterval of $(a - R, a + R)$, including $f(x)$ is analytic at every point of $(a - R, a + R)$.

4   Taylor Series

> *The Gregory-Newton interpolation formula was used by Brook Taylor to develop the most powerful single method for expanding a function into an infinite series.*
> *Morris Kline, 1972*

4.1   Taylor Coefficients and Taylor Series

In the preceding section we have proved that a power series (5.1) with the radius of convergence $R > 0$ is infinitely differentiable term by tem on the interval $(a - R, a + R)$. Using this property, we express now the coefficients of the series (5.1) in terms of derivatives of its sum $f(x)$.

Theorem 1 (Taylor Coefficients) *The coefficients of a power series* $f(x) = \sum_{n=0}^{\infty} c_n (x - a)^n$ *(formula (5.1)) can be found by the formula*

$$c_n = \frac{f^{(n)}(a)}{n!}, \ \forall n. \tag{5.22}$$

(As usual, the derivative of order 0 is the function itself.)

Proof Let us differentiate the power series term by term m times successively:

$$f(x) = \sum_{n=0}^{\infty} c_n (x - a)^n = c_0 + c_1(x - a) + c_2(x - a)^2 + \ldots + c_n(x - a)^n + \ldots ,$$

$$f'(x) = \sum_{n=1}^{\infty} c_n n(x - a)^{n-1} = c_1 + 2c_2(x - a) + \ldots + nc_n(x - a)^{n-1} + \ldots ,$$

$$f''(x) = \sum_{n=2}^{\infty} c_n n(n-1)(x-a)^{n-2} = 2c_2 + 3 \cdot 2c_3(x-a) + \ldots + n(n-1)c_n(x-a)^{n-2} + \ldots ,$$

$$\ldots$$

$$f^{(m)}(x) = \sum_{n=m}^{\infty} c_n n(n-1) \cdot \ldots \cdot (n-m+1)(x-a)^{n-m} = m!c_m + (m+1)m \cdot \ldots \cdot 2c_{m+1}(x-a) + \ldots .$$

Evaluating the obtained expressions for the derivatives at the central point $x = a$, we notice that in each series the only term which does not vanish is the first one:

$$f(a) = c_0, \ f'(a) = c_1, \ f''(a) = 2c_2, \ \ldots , \ f^{(m)}(a) = m!c_m,$$

or, solving for c_n:

$$c_0 = f(a), \ c_1 = \frac{f'(a)}{1!}, \ c_2 = \frac{f''(a)}{2!}, \ \ldots , \ c_m = \frac{f^{(m)}(a)}{m!}.$$

In this manner, we arrive at formula (5.22). $\square$

> ▶ **Definition** The coefficients $\frac{f^{(n)}(a)}{n!}$ are called the *Taylor coefficients* and the power
> series (5.1) written in the form
>
> $$f(x) = \sum_{n=0}^{\infty} \frac{f^{(n)}(a)}{n!}(x-a)^n \qquad (5.23)$$
>
> is called the *Taylor series of the function* $f(x)$ *at the point* a. The partial sums of
> the series (5.23) are denoted by $T_n(x) = \sum_{k=0}^{n} \frac{f^{(k)}(a)}{k!}(x-a)^k$ and called the *Taylor*
> *polynomials*.

Correspondingly, the power series (5.2) (centered at $a = 0$) written in the form

$$f(x) = \sum_{n=0}^{\infty} \frac{f^{(n)}(0)}{n!}x^n \qquad (5.24)$$

is the *Taylor series of* $f(x)$ *at the origin* (also called the *Maclaurin series*) and $T_n(x) = \sum_{k=0}^{n} \frac{f^{(k)}(0)}{k!}x^k$ are the *Taylor polynomials at* $a = 0$.

4.2   Relation Between the Taylor Series and Formula

The expression of the Taylor polynomial is the same as is given in the famous *Taylor's*
theorem. Let us recall this important result of Calculus and Analysis, and compare it with
the Taylor series expansion.

> **Taylor's Theorem** *If a function* $f(x)$ *is differentiable* $(n+1)$ *times in a neighbor-*
> *hood of a point* a, *then the following formula, called the Taylor formula for* $f(x)$ *at*
> a, *is true for any* x *of this neighborhood:*
>
> $$f(x) = T_n(x) + R_n(x), \quad T_n(x) = \sum_{k=0}^{n} \frac{f^{(k)}(a)}{k!}(x-a)^k,$$
>
> $$R_n(x) = \frac{f^{(n+1)}(c)}{n!\,p}\left(\frac{x-a}{x-c}\right)^p (x-c)^{n+1}, \qquad (5.25)$$
>
> *where* $T_n(x)$ *is the Taylor polynomial of degree* n, $R_n(x)$ *is the remainder,* $p > 0$
> *and* c *is a point between* a *and* x.

In particular, if $p = n + 1$, the remainder takes the Lagrange form, the most known in Calculus:

$$R_n(x) = \frac{f^{(n+1)}(c)}{(n+1)!}(x-a)^{n+1}.$$

Another frequently used representation of the remainder is the Cauchy form ($p = 1$):

$$R_n(x) = \frac{f^{(n+1)}(c)}{n!}(x-c)^n \cdot (x-a).$$

Since c is a point in one of the intervals (a, x) or (x, a), we can represent it in the form $c = a + \theta(x - a)$, where $0 < \theta < 1$. Then, $x - c = x - a - \theta(x - a) = (x - a)(1 - \theta)$. Substituting this representation of c in the Cauchy remainder, we get the following formula:

$$R_n(x) = \frac{f^{(n+1)}(a + \theta(x-a))}{n!}(x-a)^{n+1} \cdot (1-\theta)^n, \ 0 < \theta < 1.$$

Complement—Proof of Taylor's Theorem Due to the importance of this result and because Taylor's Theorem is usually proved in Calculus only in the form with the Lagrange remainder, we offer below the demonstration of Taylor's Theorem with the remainder in a more general form.

Let us denote the difference between the function $f(x)$ and its Taylor polynomial $T_n(x) = \sum_{k=0}^{n} \frac{f^{(k)}(a)}{k!}(x-a)^k$ by $R_n(x)$: $R_n(x) = f(x) - T_n(x)$ and show that $R_n(x)$ has the form specified in Taylor's Theorem. To do this, we take any point $x > a$ in a neighborhood of a where the conditions of the Theorem are satisfied and fix this point (for $x < a$ the proof is exactly the same). Consider now the auxiliary function

$$\varphi(t) = f(x) - g(x,t) - \frac{(x-t)^p}{(x-a)^p}R_n(x),$$

where

$$g(x,t) = \sum_{k=0}^{n} \frac{f^{(k)}(t)}{k!}(x-t)^k = f(t) + \frac{f'(t)}{1!}(x-t) + \ldots + \frac{f^{(n)}(t)}{n!}(x-t)^n.$$

Notice that $g(x, a) = T_n(x)$. Since all the functions involved in the definition of $\varphi(t)$ are, at least, differentiable in a neighborhood of a, the function $\varphi(t)$ is differentiable on $[a, x]$. At the endpoints of this interval, we have:

$$\varphi(a) = f(x) - g(x,a) - \frac{(x-a)^p}{(x-a)^p}R_n(x) = f(x) - T_n(x) - R_n(x) = 0$$

and

$$\varphi(x) = f(x) - g(x, x) - \frac{(x-x)^p}{(x-a)^p} R_n(x) = f(x) - \sum_{k=0}^{n} \frac{f^{(k)}(x)}{k!}(x-x)^k = 0.$$

Therefore, $\varphi(t)$ satisfies the conditions of the Lagrange theorem (the mean-value theorem) on the interval $[a, x]$, and consequently, there exists a point $c \in (a, x)$ such that $\varphi'(c) = \frac{\varphi(x) - \varphi(a)}{x-a} = 0$.

Next, we find the expression for the derivative of $\varphi(t)$:

$$\varphi'(t) = -g'_t(x, t) - ((x-t)^p)'_t \frac{R_n(x)}{(x-a)^p}$$

$$= -f'(t) + \frac{f'(t)}{1!} - \frac{f''(t)}{1!}(x-t) + \frac{f''(t)}{2!}2(x-t) - \frac{f'''(t)}{2!}(x-t)^2 +$$

$$\dots + \frac{f^{(n)}(t)}{n!} n(x-t)^{n-1} - \frac{f^{(n+1)}(t)}{n!}(x-t)^n + p(x-t)^{p-1} \frac{R_n(x)}{(x-a)^p}.$$

Evidently, all the terms on the right-hand side cancel each other, except for the last two terms, and therefore,

$$\varphi'(t) = -\frac{f^{(n+1)}(t)}{n!}(x-t)^n + p(x-t)^{p-1} \frac{R_n(x)}{(x-a)^p}.$$

Since $\varphi'(c) = 0$, it follows that

$$\frac{f^{(n+1)}(c)}{n!}(x-c)^n = p(x-c)^{p-1} \frac{R_n(x)}{(x-a)^p},$$

or, expressing $R_n(x)$,

$$R_n(x) = \frac{f^{(n+1)}(c)}{n!}(x-c)^n \frac{(x-a)^p}{p(x-c)^{p-1}} = \frac{f^{(n+1)}(c)}{n!p}(x-a)^p(x-c)^{n-p+1}.$$

This completes the proof.   □

Suppose now that a function $f(x)$ has the Taylor series expansion (and, in particular, $f(x)$ is infinitely differentiable) in a neighborhood of a. Then, the two representations of $f(x)$ are valid—by formula (5.23) and by formula (5.25). Comparing these two formulas, we conclude that, under these conditions, the remainder $r_n(x)$ of series (5.23) is equal to the remainder $R_n(x)$ in formula (5.25): $r_n(x) = f(x) - T_n(x) = R_n(x)$. If $f(x)$ is represented in the Taylor series (5.23) with the interval of convergence $(a - R, a + R)$ (which is R-neighborhood of a), then in the interval $(a - R, a + R)$ the conditions of

Taylor's Theorem hold, and consequently, $r_n(x) = R_n(x)$ on $(a - R, a + R)$. Thus, we arrive at the following result.

Theorem 2 (Remainder in the Taylor Series) *If $f(x)$ is represented in the Taylor series (5.23) with the interval of convergence $(a - R, a + R)$, $R > 0$, then its partial sums $f_n(x) = \sum_{k=0}^{n} \frac{f^{(k)}(a)}{k!}(x - a)^k$ are the Taylor polynomials: $f_n(x) = T_n(x)$, and the remainders $r_n(x)$ coincide with the remainders $R_n(x)$ of the Taylor formula (5.25): $r_n(x) = f(x) - T_n(x) = R_n(x)$ on $(a - R, a + R)$.*

   This representation of the remainder provides one more option for the investigation of convergence of a power series and the evaluation of remainders in the problems of approximation of the sum of power series by the Taylor polynomials.

4.3   Conditions of Expansion in the Taylor Series

Theorem 1 of Sect. 4.1 says that a power series (5.1) with the radius of convergence $R > 0$ is always a Taylor series for the function $f(x)$ which is the sum of this power series. In other words, joining the results of Property 5 in Sect. 3.2 and Theorem 1 in Sect. 4.1, we can state that an analytic in interval $(a - R, a + R)$, $R > 0$ function $f(x)$ is infinitely differentiable in this interval and its coefficients in the power series (5.1) are the Taylor coefficients (5.22).

   Now we consider the inverse problem: if a function $f(x)$ is infinitely differentiable in an interval, is it possible to represent this function in a Taylor series that converges on this interval to $f(x)$? Or rewording: if a function $f(x)$ is infinitely differentiable in an interval, can we state that it is analytic in the same interval or, at least, in a neighborhood of the central point of this interval? The answer, in general, is negative and there are two reasons for this. First, a Taylor series may diverge. And second, even if a series converges, it may converge to another function, different from the original.

   Next, we consider examples of these two situations.

Example 1 This example shows that Taylor series can diverge on sets where the original function is infinitely differentiable. Actually, we have already seen different functions of this type, without paying attention to this aspect of the behavior of their power series. Based on the geometric series $\sum_{n=0}^{\infty} x^n = \frac{1}{1-x}$ convergent on $(-1, 1)$, we can consider similar power series, such that $\sum_{n=0}^{\infty}(-1)^n x^n$ and $\sum_{n=0}^{\infty}(-1)^n x^{2n}$, and rewriting them in the form $\sum_{n=0}^{\infty}(-x)^n$ and $\sum_{n=0}^{\infty}(-x^2)^n$ we immediately find their sums: $\sum_{n=0}^{\infty}(-1)^n x^n = \frac{1}{1+x}$ and $\sum_{n=0}^{\infty}(-1)^n x^{2n} = \frac{1}{1+x^2}$. Just like the original geometric series, both series converge on $(-1, 1)$ and diverge outside this interval. Notice that the first two functions—$\frac{1}{1-x}$ and $\frac{1}{1+x}$—are not defined at one of the endpoint of the interval of convergence, that may, until a certain degree, justify the impossibility to use the same series for representation of these functions outside the interval $(-1, 1)$. However, the third function—$\frac{1}{1+x^2}$—is defined and infinitely differentiable on the entire real axis, and even so, its power series diverges outside $(-1, 1)$.

Example 2 This example is a more radical version of Example 1. It shows that there are functions whose Taylor series diverge at every point, except the central one.

Consider the function $f(x)$ defined via series: $f(x) = \sum_{n=0}^{+\infty} e^{-n} \cos(n^2 x)$. Notice first that this series converges uniformly on $\mathbb{R}$ according to the Weierstrass test: the evaluation $|e^{-n} \cos(n^2 x)| \le e^{-n}$ holds for $\forall x \in \mathbb{R}$ and the majorant numerical series $\sum_{n=0}^{+\infty} e^{-n}$ is a convergent geometric series. The series of derivatives (with respect to x) of the general term $e^{-n} \cos(n^2 x)$ has the form $\pm \sum_{n=0}^{+\infty} n^{2k} e^{-n} \cos(n^2 x)$ if the order k of derivative is even and $\pm \sum_{n=0}^{+\infty} n^{2k} e^{-n} \sin(n^2 x)$ if k is odd. Since $\lim_{n \to \infty} \frac{n^{2k} e^{-n}}{e^{-\frac{n}{2}}} = \lim_{n \to \infty} \frac{n^{2k}}{e^{\frac{n}{2}}} = 0$ (the indeterminate form is easily calculated applying L'Hospital's rule $2k$ times), for any fixed k there exists sufficiently large n such that $|n^{2k} e^{-n} \cos(n^2 x)| \le e^{-\frac{n}{2}}$ and $|n^{2k} e^{-n} \sin(n^2 x)| \le e^{-\frac{n}{2}}$. Therefore, the convergence of the majorant series $\sum_{n=0}^{+\infty} e^{-\frac{n}{2}}$ implies the uniform convergence on $\mathbb{R}$ of the series $\pm \sum_{n=0}^{+\infty} n^{2k} e^{-n} \cos(n^2 x)$ and $\pm \sum_{n=0}^{+\infty} n^{2k} e^{-n} \sin(n^2 x)$, according to the Weierstrass test. Then, the function $f(x)$ has derivatives of any order on $\mathbb{R}$ and they can be calculated applying term-by-term differentiation of the original series, that is, $f^{(k)}(x) = \pm \sum_{n=0}^{+\infty} n^{2k} e^{-n} \cos(n^2 x)$ if k is even and $f^{(k)}(x) = \pm \sum_{n=0}^{+\infty} n^{2k} e^{-n} \sin(n^2 x)$ if k is odd. In particular, at the point $x = 0$, we have $f^{(k)}(0) = \pm \sum_{n=0}^{+\infty} n^{2k} e^{-n}$ if k is even and $f^{(k)}(0) = 0$ if k is odd. Consequently, the formal Taylor series at the central point $a = 0$ can be formed, and the absolute value of its general term with the even index k has the form $|c_k x^k| = \frac{|f^{(k)}(0)|}{k!} x^k = \sum_{n=0}^{+\infty} n^{2k} e^{-n} \frac{x^k}{k!}$. Fixing now any $x \ne 0$, we can see that $|c_k x^k| > 1$ for all k large enough: $|c_k x^k| > k^{2k} e^{-k} \frac{x^k}{k!} > \frac{k^{2k} x^k}{k^k} e^{-k} = \left(\frac{kx}{e} \right)^k > 1$ for $\forall x \ne 0$ and $k > \left[\frac{e}{x} \right]$. Since the general term does not tend to 0 at any $x \ne 0$, the Taylor series diverges at every point $x \ne 0$, that is, it diverges everywhere (except for the central point).

Example 3 (Cauchy's Example) This is an example of the second type, when the Taylor series for $f(x)$ converges, but its sum differs from $f(x)$ at every point, except the central one.

Consider the function $f(x) = \begin{cases} e^{-\frac{1}{x^2}}, x \ne 0, \\ 0, x = 0. \end{cases}$. Calculating the limit $\lim_{x \to 0} f(x) = \lim_{x \to 0} e^{-\frac{1}{x^2}} = 0 = f(0)$, we see that the function $f(x)$ is continuous at 0, and consequently, on the entire real axis. Let us find the derivatives of this function. If $x \ne 0$ then

$$f'(x) = e^{-\frac{1}{x^2}} \cdot \frac{2}{x^3} = \frac{2}{x^3} e^{-\frac{1}{x^2}}, \quad f''(x) = \left(\frac{2}{x^3} \right)^2 e^{-\frac{1}{x^2}} - \frac{2 \cdot 3}{x^4} e^{-\frac{1}{x^2}}$$

$$= \left(\left(\frac{2}{x^3} \right)^2 - \frac{6}{x^4} \right) e^{-\frac{1}{x^2}},$$

$$f'''(x) = \left(\frac{2}{x^3}\right)^3 e^{-\frac{1}{x^2}} + \frac{4 \cdot (-6)}{x^7} e^{-\frac{1}{x^2}} - \frac{6}{x^4} \cdot \frac{2}{x^3} \cdot e^{-\frac{1}{x^2}} + \frac{2 \cdot 3 \cdot 4}{x^5} e^{-\frac{1}{x^2}}$$

$$= \left(\left(\frac{2}{x^3}\right)^3 - \frac{36}{x^7} + \frac{4!}{x^5}\right) e^{-\frac{1}{x^2}},$$

$$\ldots$$

$$f^{(n)}(x) = \left(\left(\frac{2}{x^3}\right)^n + \frac{a_{3n-2}}{x^{3n-2}} + \frac{a_{3n-4}}{x^{3n-4}} + \ldots \pm \frac{(n+1)!}{x^{n+2}}\right) e^{-\frac{1}{x^2}} = P_{3n}\left(\frac{1}{x}\right) \cdot e^{-\frac{1}{x^2}}.$$

Using these results, we find the derivatives at $x = 0$ by definition. To solve the indeterminate forms that appear in definitions we use the change of variable $t = \frac{1}{x}$ and L'Hospital's rule:

$$f'(0) = \lim_{x \to 0} \frac{f(x) - f(0)}{x - 0} = \lim_{x \to 0} \frac{e^{-\frac{1}{x^2}}}{x} = \lim_{t \to \infty} \frac{e^{-t^2}}{\frac{1}{t}} = \lim_{t \to \infty} \frac{t}{e^{t^2}} = \lim_{t \to \infty} \frac{1}{2t \cdot e^{t^2}} = 0,$$

$$f''(0) = \lim_{x \to 0} \frac{f'(x) - f'(0)}{x - 0} = \lim_{x \to 0} \frac{\frac{2}{x^3} e^{-\frac{1}{x^2}}}{x} = \lim_{x \to 0} \frac{2e^{-\frac{1}{x^2}}}{x^4} = \lim_{t \to \infty} \frac{2t^4}{e^{t^2}}$$

$$= \lim_{t \to \infty} \frac{2 \cdot 4t^3}{2t \cdot e^{t^2}} = \lim_{t \to \infty} \frac{4 \cdot 2t}{2t \cdot e^{t^2}} = 0,$$

$$f'''(0) = \lim_{x \to 0} \frac{f''(x) - f''(0)}{x - 0} = \lim_{x \to 0} \left(\frac{4}{x^7} - \frac{6}{x^5}\right) e^{-\frac{1}{x^2}} = \lim_{t \to \infty} \frac{4t^7 - 6t^5}{e^{t^2}}$$

$$= \lim_{t \to \infty} \frac{28t^6 - 30t^4}{2t \cdot e^{t^2}} = \ldots = 0,$$

$$\ldots$$

$$f^{(n+1)}(0) = \lim_{x \to 0} \frac{f^{(n)}(x) - f^{(n)}(0)}{x - 0} = \lim_{x \to 0} \frac{1}{x} \cdot P_{3n}\left(\frac{1}{x}\right) \cdot e^{-\frac{1}{x^2}} = \lim_{x \to 0} P_{3n+1}\left(\frac{1}{x}\right) \cdot e^{-\frac{1}{x^2}}$$

$$= \lim_{t \to \infty} \frac{P_{3n+1}(t)}{e^{t^2}} = \lim_{t \to \infty} \frac{P'_{3n+1}(t)}{2t \cdot e^{t^2}} = \ldots = 0.$$

Considering the central point a at the origin, we obtain the following Taylor series of $f(x)$:

$$\sum_{n=0}^{\infty} c_n x^n = \sum_{n=0}^{\infty} \frac{f^{(n)}(0)}{n!} x^n = \sum_{n=0}^{\infty} 0 \cdot x^n = 0 + 0 + 0 + \ldots = 0.$$

This series converges at every real point and the sum of the series is zero at $\forall x \in (-\infty, +\infty)$, that is, the Taylor series constructed for the function $f(x)$ converges to the zero function that has nothing to do with $f(x)$.

From the Theorem about the Taylor coefficients and considered examples it follows that infinite differentiability of a function is necessary but not sufficient condition for the expansion of function in the Taylor series, that is, for the analyticity of function. Let us consider now some sufficient conditions which ensure the representation of a function in the power series. As before, in deriving the results we use the central point $a = 0$.

Theorem 1 (Criterion for Expansion in a Power Series) *A function $f(x)$ is analytic on an interval $(-R, R)$, $R > 0$ if and only if $f(x)$ is infinitely differentiable on $(-R, R)$ and its remainder in the Taylor formula tends to 0 as $n \to \infty$ for $\forall x \in (-R, R)$.*

Proof *Necessity* Suppose that $f(x)$ is analytic, that is, it allows expansion in a power series (5.2) that converges to $f(x)$ on the interval $(-R, R)$, $R > 0$: $f(x) = \sum_{n=0}^{\infty} c_n x^n$. As was shown in previous sections, the function $f(x)$ is infinitely differentiable on $(-R, R)$ and its power series takes the form of the Taylor series (5.24). The convergence of a series is equivalent to the tendency of its remainder to 0. From Theorem 2 of the preceding section we know that the remainder r_n is equal to the remainder R_n in the Taylor formula. Therefore, R_n approaches 0 as $n \to \infty$ at every $\forall x \in (-R, R)$.

Sufficiency If $f(x)$ is infinitely differentiable on $(-R, R)$, we can construct its formal Taylor series (5.24) using formulas (5.22) (with $a = 0$) to determine all the coefficients. To show that this series converges to $f(x)$ on $(-R, R)$, we evaluate the difference between $f(x)$ and the partial sums $f_n(x) = \sum_{k=0}^{n} c_k x^k$ of the constructed series. Recalling that $f_n(x) = T_n(x)$ and noting that the conditions of Taylor's Theorem are satisfied on $(-R, R)$, we obtain the Taylor formula: $f(x) - f_n(x) = f(x) - T_n(x) = R_n(x)$, $\forall x \in (-R, R)$. Since the remainder $R_n(x)$ tends to 0 on $(-R, R)$, we conclude that $f(x) - f_n(x) = R_n(x) \underset{n \to \infty}{\to} 0$, that is, $f_n(x) \underset{n \to \infty}{\to} f(x)$, which means that the series (5.24) converges to $f(x)$ on the interval $(-R, R)$, or equivalently, $f(x)$ is analytic on $(-R, R)$.

□

Theorem 2 (Test for a Power Series Expansion) *If $f(x)$ is infinitely differentiable on an interval $(-R, R)$, $R > 0$ and its derivatives satisfy the evaluations*

$$\left| f^{(n)}(x) \right| \le M \frac{n!}{R^n}, \forall x \in (-R, R), \forall n \in \mathbb{N}, \tag{5.26}$$

where M is a constant, then $f(x)$ is analytic on $(-R, R)$.

Proof Let us check that $f(x)$ satisfies the hypotheses of the preceding Theorem 1. The condition of infinite differentiability is found among the hypotheses of this Theorem, then it remains to show that the remainder in the Taylor formula (5.25) for the function $f(x)$ tends to 0 as $n \to \infty$ at every $\forall x \in (-R, R)$. Using the remainder (at $a = 0$) in the Lagrange form $R_n(x) = \frac{f^{(n+1)}(c)}{(n+1)!} x^{n+1}$ and employing the evaluations of derivatives (5.26), we obtain for every fixed x in $(-R, R)$:

$$|R_n(x)| = \left| \frac{f^{(n+1)}(c)}{(n+1)!} x^{n+1} \right| \le M \left(\frac{|x|}{R} \right)^{n+1}.$$

Since $\frac{|x|}{R} < 1, \forall x \in (-R, R)$, it follows that $\left(\frac{|x|}{R} \right)^{n+1} \underset{n \to \infty}{\to} 0$, which implies that $R_n(x) \underset{n \to \infty}{\to} 0, \forall x \in (-R, R)$. Therefore, the conditions of Theorem 1 are satisfied and $f(x)$ can be represented in a power series on the interval $(-R, R)$. $\square$

Remark 1 As usual, inequalities (5.26) can be imposed starting from some index n—the violation of these inequalities for a finite number of derivatives does not affect the resulting convergence of R_n.

Remark 2 In practice, weaker versions of Theorem 2 are frequently used, which impose stronger restrictions on derivative. One of these formulations uses the condition

$$\left| f^{(n)}(x) \right| \le Mq^n, \forall x \in (-R, R), \forall n \in \mathbb{N}, \tag{5.27}$$

where q is a constant, and another employs the uniform boundedness of derivatives

$$\left| f^{(n)}(x) \right| \le M, \forall x \in (-R, R), \forall n \in \mathbb{N}. \tag{5.28}$$

It is easy to show that the last restriction guarantees inequality (5.27), which in turn guarantees condition (5.26) of Theorem 2. Indeed, without a loss of generality, we can consider the constant $q > 1$ in (5.27) (if (5.27) holds for $q \le 1$, it is also valid for $q > 1$). Then $M < Mq^n$, that is, (5.28) is more restrictive than (5.27). At the next step, we compare q^n and $\frac{n!}{R^n}$. Notice that for any two constants q and R the ratio $\frac{q^n}{\frac{n!}{R^n}} = \frac{(qR)^n}{n!}$ approaches 0. To show this, we choose a fixed index $N = [qR]$ and use the representation $n = N + k$ $(n > N)$ to obtain

$$\frac{(qR)^n}{n!} = \frac{(qR)^{N+k}}{(N+k)!} = \frac{(qR)^N}{N!} \cdot \frac{(qR)^k}{(N+1) \cdot \ldots \cdot (N+k)} < \frac{(qR)^N}{N!} \cdot \left(\frac{qR}{N+1} \right)^k \underset{k \to \infty}{\to} 0$$

(notice that here $\frac{(qR)^N}{N!}$ is a constant and $\frac{qR}{N+1} < 1$). Therefore, $q^n < \frac{n!}{R^n}$ at least for sufficiently large values of n, which means that inequality (5.27) is more restrictive than (5.26). Hence, condition (5.28) guarantees the validity of (5.27) and the latter in turn ensures the validity of (5.26). Then we conclude that the formulation of Theorem 2 with inequality (5.28) is the weakest version (among considered here), the formulation with inequality (5.27) is the intermediate version, and the original formulation is the strongest result. However, due to its simplicity, inequality (5.28) is frequently used in practical investigations of functions and series.

Remark 3 The obtained results show that the analyticity of a function is a quite restrictive condition, even stronger than infinite differentiability. However, if we give up representing a function in the very special form of series of functions—the power series, then extending the choice of the terms of series onto the class of polynomials, we can guarantee the uniform convergence of this type of series to any continuous function. In fact, the Weierstrass approximation Theorem states that any continuous on an interval $[a, b]$ function is the limit of the uniformly convergent on $[a, b]$ sequence of polynomials. Choosing, for instance, a function $f(x)$ continuous on the interval $[0, 1]$, we can define the terms $u_n(x)$ via the Bernstein polynomials $B_n(x)$ by the formulas $u_1(x) = B_1(x)$, $u_n(x) = B_n(x) - B_{n-1}(x)$, $n = 2, 3, \ldots$ and ensure therefore (by the Weierstrass approximation Theorem) that the series $\sum_{n=1}^{\infty} u_n(x)$ converges uniformly on $[0, 1]$ to $f(x)$, because its partial sums are the Bernstein polynomials $B_n(x)$.

Historical Remarks 1 Taylor published his work on the Taylor series in 1715, thirty years after Newton discovered, but did not publish, the same series expansions. Even earlier, in 1665–1668, Newton and Mercator first found the expansion of the logarithmic function in an infinite series.

Newton extensively used infinite series to represent functions in 1665–1670, but it appears that initially he did not note the relation between the derivatives of functions and the coefficients of their series expansions, which is an important feature of the Taylor series. However, some explanations on series expansions in 1687 "Principia" show he already used these relations to find the Taylor series. So it is almost certain that Newton already discovered the Taylor series expansions at that time. Moreover, in 1692 paper Newton gave an explicit statement of the Taylor series and formula of the Taylor coefficients and presented its particular case named for Maclaurin.

Even before Newton, in 1671 Gregory invented and used the Taylor series to obtain expansions of a wide range of functions. A few comments in his letters evidence that he derived some series expansions by using the derivatives of the original functions. Being informed about Newton work on representation of functions by series, Gregory seems to have mistakenly assumed that his method of successive differentiation for deriving the series was anticipated by Newton. In consequence of this, he felt that Newton deserved the privilege of first publication. As a result of this confusion, Newton's unwillingness to publish his works leaded to the oblivion of the Gregory findings, which were discovered and recognized only in the twentieth century.

Historical Remarks 2 The expression for the remainder in the Taylor formula was first derived by Lagrange in 1797 "Théorie des fonctions analytiques". The integral form of the remainder follows immediately from the results of this book, but it was not stated explicitly by Lagrange. This gap was filled by Prony in 1805. The commonly used in modern textbooks proof of Taylor's theorem by employing Rolle's theorem is attributed to Bonnet and it appeared first in a calculus textbook in 1868.

Historical Remarks 3 Up to the beginning of the nineteenth century it was a common belief that any Taylor series converges to the function by which it was originated. This general opinion was clearly expressed by Lagrange in 1797 in his famous book on differential calculus "Théorie des fonctions analytiques". The critique of this view by Abel and the counterarguments of Cauchy, who published his example of non-analytic function in 1823 (see Example 3 in this section), were the first steps in changing this belief.

About fifty years later, in 1876, du Bois-Raymond gave the first example of infinitely differentiable function whose Taylor series diverges everywhere except the central point. Since then, numerous examples were constructed illustrating the same ill-behavior of Taylor series (Example 2 in this section is one of them).

5   Power Series Expansion of Elementary Functions

In this section we find power series expansions of many familiar functions. This means that we systematically solve the following problem: for a given function $f(x)$ find a power series that converges to $f(x)$ on its interval of convergence (but generally not on the entire domain of $f(x)$). A few of these series were derived before, but for the sake of completeness we present here all the relevant series.

To construct a power series expansion of a given function we use the *three main methods*. The first technique bases the construction of the desired series on the use of elementary power series with known sums (such as geometric series) by reducing a given function to the form of the sum of elementary series via application of analytic properties. The second approach consists in discovering the differential relations for the analyzed power series and finding a function that satisfies the same differential equations. Finally, the third method uses the formula for the Taylor coefficients, which can be found in general form for some (but not all) elementary functions.

The advantage of the first approach is its simplicity: the Taylor coefficients are obtained automatically and, due to strong connection with the series whose properties were already analyzed, there is no need to search for the interval of convergence and show that the constructed series converges to a given function. The unique point that remains to study is the behavior of the series at the endpoints of the interval of convergence. However, a family of functions to which this technique is applicable is very restrictive.

The second method requires a search for differential relations that the chosen series can satisfy, which can be a complicated problem. Besides, even if a (ordinary) differential equation is found, it should have relatively simple solution in terms of elementary functions. The advantage of this technique is that the convergence of the obtained power series to the corresponding solution of a differential equation follows automatically from the used construction.

The third procedure can be more technically involved in the part related to finding a general form of higher order derivatives, and sometimes it cannot be solved in a suitable form. Besides, after a formal Taylor series is constructed, we have to determine its interval of convergence, investigate the behavior at the endpoints of this interval, and show that the derived power series converges to the original function. However, the range of application of the third technique is sufficiently large. Finally, we can use a combination of these three methods to achieve the desired power series expansion in the most efficient way.

5.1 Using Analytic Properties of Power Series

The formulas of these group are related in one way or another to the geometric series convergent on $(-1, 1)$:

$$\frac{1}{1-x} = \sum_{n=0}^{\infty} x^n = 1 + x + x^2 + \ldots + x^n + \ldots. \tag{5.29}$$

1. Function $\frac{1}{1-x}$ and Its Derivatives and Integrals

Recalling that a power series is differentiable term by term on the entire interval of convergence, we differentiate the series (5.29) and find

$$\frac{1}{(1-x)^2} = \left(\frac{1}{1-x}\right)' = \sum_{n=0}^{\infty} (x^n)' = \sum_{n=1}^{\infty} n x^{n-1}.$$

Differentiating one more time, we obtain the expansion of the next function

$$\frac{2}{(1-x)^3} = \left(\frac{1}{(1-x)^2}\right)' = \sum_{n=1}^{\infty} (n x^{n-1})' = \sum_{n=2}^{\infty} n(n-1) x^{n-2}.$$

Continuing this procedure, we arrive at the general formula for the functions $\frac{1}{(1-x)^p}$, $\forall p \in$ $\mathbb{N}$:

$$\frac{(p-1)!}{(1-x)^p} = \left(\frac{(p-2)!}{(1-x)^{p-1}}\right)' = \sum_{n=p-2}^{\infty} (n(n-1)\cdot\ldots\cdot(n-p+3)x^{n-p+2})' = \sum_{n=p-1}^{\infty} n(n-1)\cdot\ldots\cdot(n-p+2)x^{n-p+1}$$

or

$$\frac{1}{(1-x)^p} = \sum_{n=p-1}^{\infty} \frac{n(n-1)\cdot\ldots\cdot(n-p+2)}{(p-1)!} x^{n-p+1}$$

$$= \sum_{m=0}^{\infty} \frac{(m+p-1)(m+p-2)\cdot\ldots\cdot(m+1)}{(p-1)!} x^m.$$

(In the case $p = 1$ we understand that $n(n-1)\cdot\ldots\cdot(n-p+2) = (m+p-1)(m+p-2)\cdot\ldots\cdot(m+1) = 1$ since the last term of the product is larger than the first one.) The interval of convergence of the last series is $(-1, 1)$, the same as for the geometric series from which it was derived.

Let us investigate the behavior of these series with $p > 1$ at the endpoints of the interval $(-1, 1)$. For both points $x = \pm 1$ we have the same evaluation of the general term:

$$\left| \frac{(m+p-1)(m+p-2)\cdot\ldots\cdot(m+1)}{(p-1)!} x^m \right|_{x=\pm 1}$$

$$= \frac{(m+p-1)(m+p-2)\cdot\ldots\cdot(m+1)}{(p-1)!} \geq 1,$$

which shows that the general term does not approaches 0, and consequently, the series diverges at the endpoints of the interval of convergence. (The same result follows directly from the extension of Abel's Lemma in Sect. 2.3, since the original geometric series diverges at both endpoints.)

Integrating the geometric series term by term, we obtain the following representation of the logarithmic function valid in $(-1, 1)$:

$$\ln(1-x) = -\int_0^x \frac{1}{1-t} dt = -\int_0^x \sum_{n=0}^{\infty} t^n dt = -\sum_{n=0}^{\infty} \int_0^x t^n dt = -\sum_{n=0}^{\infty} \frac{x^{n+1}}{n+1} = -\sum_{m=1}^{\infty} \frac{x^m}{m},$$

that is,

$$\ln(1-x) = -\sum_{n=1}^{\infty} \frac{x^n}{n} = -x - \frac{x^2}{2} - \frac{x^3}{3} - \ldots - \frac{x^n}{n} - \ldots.$$

Let us check the behavior of this series at the endpoints of the interval $(-1, 1)$. At the point $x = 1$ the function $\ln(1 - x)$ is not defined and the series diverges because it takes the form of the harmonic series $- \sum_{n=1}^{\infty} \frac{1}{n}$. At the point $x = -1$ the function $\ln(1-x)$ is defined and the series converges since it becomes the alternating harmonic series $\sum_{n=1}^{\infty} \frac{(-1)^{n-1}}{n}$. Then, by Abel's Theorem (Sect. 2.4), the series converges on $[-1, 1)$ and converges uniformly on $[-1, b]$, $-1 < \forall b < 1$, and the formula $\ln(1 - x) = - \sum_{n=1}^{\infty} \frac{x^n}{n}$ is true on the interval $[-1, 1)$. In particular, at $x = -1$ we get the exact sum of the alternating harmonic series:

$$\sum_{n=1}^{\infty} \frac{(-1)^{n-1}}{n} = \ln 2.$$

2. Function $\frac{1}{1+x}$ and Its Derivatives and Integrals

Just as in the anterior case, changing x by $-x$ in the above formulas, we obtain the following power series expansions valid on the interval $(-1, 1)$

$$\frac{1}{1+x} = \sum_{n=0}^{\infty} (-1)^n x^n = 1 - x + x^2 - x^3 + \ldots + (-1)^n x^n + \ldots,$$

$$\frac{1}{(1+x)^p} = \sum_{n=0}^{\infty} (-1)^n \frac{(n+p-1)(n+p-2) \cdot \ldots \cdot (n+1)}{(p-1)!} x^n, \ \forall p \in \mathbb{N},$$

$$\ln(1 + x) = \sum_{n=1}^{\infty} (-1)^{n-1} \frac{x^n}{n} = x - \frac{x^2}{2} + \frac{x^3}{3} - \frac{x^4}{4} + \ldots + (-1)^{n-1} \frac{x^n}{n} + \ldots.$$

From the properties of the Taylor series of $\frac{1}{(1-x)^p}$, $p \in \mathbb{N}$ and $\ln(1 - x)$ it follows that the first two series diverge at the endpoints $x = \pm 1$ and the third series diverges at $x = -1$ and converges at $x = 1$ to $\ln 2$.

3. Function $\frac{1}{1+x^2}$ and Its Derivatives and Integrals

Substituting x^2 for x in the second group of the formulas, we obtain the following power series with the interval of convergence $(-1, 1)$:

$$\frac{1}{1+x^2} = \sum_{n=0}^{\infty} (-1)^n x^{2n} = 1 - x^2 + x^4 - x^6 + \ldots + (-1)^n x^{2n} + \ldots,$$

$$\frac{1}{(1+x^2)^p} = \sum_{n=0}^{\infty} (-1)^n \frac{(n+p-1)(n+p-2) \cdot \ldots \cdot (n+1)}{(p-1)!} x^{2n}, \ \forall p \in \mathbb{N},$$

$$\ln(1 + x^2) = \sum_{n=1}^{\infty} (-1)^{n-1} \frac{x^{2n}}{n} = x^2 - \frac{x^4}{2} + \frac{x^6}{3} - \frac{x^8}{4} + \ldots + (-1)^{n-1} \frac{x^{2n}}{n} + \ldots.$$

Integrating the first formula, we obtain also the expansion of $\arctan x$:

$$\arctan x = \int_0^x \frac{1}{1+t^2} dt = \int_0^x \sum_{n=0}^\infty (-1)^n t^{2n} dt = \sum_{n=0}^\infty \int_0^x (-1)^n t^{2n} dt = \sum_{n=0}^\infty (-1)^n \frac{x^{2n+1}}{2n+1},$$

with the same interval of convergence $(-1, 1)$. Hence,

$$\arctan x = \sum_{n=0}^\infty (-1)^n \frac{x^{2n+1}}{2n+1} = x - \frac{x^3}{3} + \frac{x^5}{5} - \frac{x^7}{7} + \ldots + (-1)^n \frac{x^{2n+1}}{2n+1} + \ldots.$$

At the endpoints of the interval $(-1, 1)$ the first two series (for $\frac{1}{1+x^2}$ and $\frac{1}{(1+x^2)^p}$) diverge because the general term does not approach 0 when $x = \pm 1$. The third series (for $\ln(1 + x^2)$) converges at both endpoints $x = \pm 1$ because it becomes the alternating harmonic series. Therefore, by Abel's Theorem (Sect. 2.4), this series converges uniformly on $[-1, 1]$ and the formula $\ln(1 + x^2) = \sum_{n=1}^\infty (-1)^{n-1} \frac{x^{2n}}{n}$ is true on the interval $[-1, 1]$. The fourth series (for $\arctan x$) also converges at both endpoints, because at $x = 1$ it is the convergent alternating series $\sum_{n=0}^\infty \frac{(-1)^n}{2n+1}$ and at $x = -1$ it is a similar alternating series $\sum_{n=0}^\infty \frac{(-1)^{n-1}}{2n+1}$. Therefore, by Abel's Theorem (Sect. 2.4), this series converges uniformly on $[-1, 1]$ and the formula $\arctan x = \sum_{n=0}^\infty (-1)^n \frac{x^{2n+1}}{2n+1}$ is true on the interval $[-1, 1]$. In particular,

$$\sum_{n=0}^\infty \frac{(-1)^n}{2n+1} = \arctan 1 = \frac{\pi}{4}.$$

5.2   Finding the Sum of Power Series via Differential Relations

In the *method of differential relations*, we start with a chosen power series and determine its interval of convergence. Next, we find derivatives of the series on the interval of convergence and try to find relations between the original series and its derivatives. If we succeed in this, we use the theory of ordinary differential equations to solve the differential equation in the terms of elementary functions. If we find such a solution, it represents the sum of the original series with precision up to a few constants (depending on the order of equation), that can be specified by choosing some values of variable x inside the interval of convergence.

1. Function $\frac{1}{1-x}$

For the sake of illustration and comparison with the previous method, we derive the representation of $\frac{1}{1-x}$ using the new technique. Let us start with the series $\sum_{n=0}^\infty x^n$, supposing at the moment that its sum is unknown. First, we determine that the series converges on the interval $(-1, 1)$, denote its sum by $f(x)$ and differentiate it in this interval:

$$f'(x) = \sum_{n=0}^{\infty} (x^n)' = \sum_{n=1}^{\infty} nx^{n-1}, \forall x \in (-1, 1).$$

Next, we transform this result to the form that allows us to connect $f'(x)$ with $f(x)$:

$$f'(x) = \sum_{n=0}^{\infty} (n+1)x^n = \sum_{n=0}^{\infty} nx^n + \sum_{n=0}^{\infty} x^n = x \sum_{n=1}^{\infty} nx^{n-1} + \sum_{n=0}^{\infty} x^n = xf'(x) + f(x), \forall x \in (-1, 1).$$

The obtained differential relation can be written in the form $(1 - x)f'(x) = f(x)$ which evidences that we have a separable ordinary differential equation whose general solution is found by integrating independently the parts involving unknown function and independent variable: $\int \frac{df}{f} = \int \frac{dx}{1-x}$. The result of integration we can represent in the form $f(x) = \frac{C}{1-x}$, where C is an arbitrary constant. To specify C, we substitute $x = 0$ in the original series: $f(0) = \sum_{n=0}^{\infty} 0^n = 1$, that implies $C = 1$. Hence, $\sum_{n=0}^{\infty} x^n = \frac{1}{1-x}$ on the interval $(-1, 1)$.

Of course, this procedure is more technically complicated than a direct calculation of the sum in the case of the geometric series $\sum_{n=0}^{\infty} x^n$. However, this is a useful example to illustrate the new technique and its principal steps, and this method can be applied to a broader class of functions. In this approach, the sum of the series is denoted by $f(x)$ from the beginning, which dismisses the necessity to show that the power series converges to a specific function $f(x)$, which is certainly an advantage of this method. On the negative side, this method is somewhat unnatural since it requires to have a preliminary idea what kind of power series can be handled in this way. It may happen that finding an appropriate differential relation will be quite involved or even impossible task, and also there is no guarantee that the obtained differential equation will have relatively simple solution (if any) in elementary functions.

In what follows, we consider a family of power series that lead to representation of new functions, but later, in the next section, we will "rediscover" these power series expansions by applying more direct and common approach of the Taylor coefficients.

2. Functions $f(x) = (1 + x)^p, \forall p \notin \mathbb{N} \cup \{0\}$
Let us consider the power series defined in the form

$$\sum_{n=0}^{\infty} \frac{p(p-1) \cdot \ldots \cdot (p-n+1)}{n!} x^n$$

$$= 1 + \frac{p}{1!}x + \frac{p(p-1)}{2!}x^2 + \ldots + \frac{p(p-1)\ldots(p-n+1)}{n!}x^n + \ldots,$$

where p is a real parameter different from natural numbers and zero (if $p \in \mathbb{N}$ or $p = 0$, we have a polynomial of degree p, which is a trivial case). These series are called binomial. Let us start with finding its interval of convergence, by applying D'Alembert's test:

$$R = \lim_{n \to \infty} \left| \frac{c_n}{c_{n+1}} \right| = \lim_{n \to \infty} \left| \frac{p\,(p-1)\ldots(p-n+1) \cdot (n+1)!}{n! \cdot p\,(p-1)\ldots(p-n+1)\,(p-n)} \right| = \lim_{n \to \infty} \frac{n+1}{|p-n|} = 1.$$

This means that the series converges on the interval $(-1, 1)$ to its sum $f(x)$, and consequently, it converges uniformly on any closed subinterval in $(-1, 1)$. Therefore, the series is infinitely differentiable in $(-1, 1)$ and the derivatives can be calculated term by term, in particular, the first derivative is

$$f'(x) = \sum_{n=1}^{\infty} \frac{p\,(p-1) \cdot \ldots \cdot (p-n+1)}{(n-1)!} x^{n-1}, \forall x \in (-1, 1).$$

Now let us search for differential relations involving $f(x)$ and $f'(x)$. Using the arithmetic properties of power series, we perform some algebraic manipulations with the series:

$$(1+x)f'(x) = \sum_{n=1}^{\infty} \frac{p\,(p-1) \cdot \ldots \cdot (p-n+1)}{(n-1)!} x^{n-1}$$

$$+ x \sum_{n=1}^{\infty} \frac{p\,(p-1) \cdot \ldots \cdot (p-n+1)}{(n-1)!} x^{n-1}$$

$$= \sum_{n=0}^{\infty} \frac{p\,(p-1) \cdot \ldots \cdot (p-n)}{n!} x^n + \sum_{n=1}^{\infty} \frac{p\,(p-1) \cdot \ldots \cdot (p-n+1)}{(n-1)!} x^n$$

$$= p + \sum_{n=1}^{\infty} \left(\frac{p\,(p-1) \cdot \ldots \cdot (p-n)}{n!} + \frac{p\,(p-1) \cdot \ldots \cdot (p-n+1)}{(n-1)!} \right) x^n$$

$$= p + \sum_{n=1}^{\infty} \frac{p\,(p-1) \cdot \ldots \cdot (p-n)}{(n-1)!} \left(\frac{1}{n} + \frac{1}{p-n} \right) x^n$$

$$= p + \sum_{n=1}^{\infty} \frac{p\,(p-1) \cdot \ldots \cdot (p-n)}{(n-1)!} \cdot \frac{p}{n(p-n)} x^n$$

$$= p \left(1 + \sum_{n=1}^{\infty} \frac{p\,(p-1) \cdot \ldots \cdot (p-n+1)}{n!} x^n \right) = pf(x), \forall x \in (-1, 1).$$

Hence, we find the ordinary differential equation for $f(x)$ in the form $(1+x)f'(x) = pf(x)$, $\forall x \in (-1, 1)$. This is an elementary separable equation that allows for a simple solution. The equation can be rewritten in the form $\left(\frac{f(x)}{(1+x)^p}\right)' = 0$ which shows that $\frac{f(x)}{(1+x)^p} = C$ or $f(x) = C(1+x)^p$ in $(-1, 1)$, where C is an arbitrary constant. Taking into account that $f(0) = \sum_{n=0}^{\infty} \frac{p(p-1)\cdots(p-n+1)}{n!} 0^n = 1$, we specify $C = 1$ and conclude that $f(x) = (1+x)^p$ in $(-1, 1)$.

In the next section, we derive and investigate the binomial series using the formula of the Taylor coefficients.

3. Function $f(x) = e^x$

Consider the series $\sum_{n=0}^{\infty} \frac{x^n}{n!}$. By D'Alembert's test, its radius of convergence is

$$R = \lim_{n \to \infty} \left| \frac{c_n}{c_{n+1}} \right| = \lim_{n \to \infty} \frac{(n+1)!}{n!} = \lim_{n \to \infty} (n+1) = +\infty,$$

that is, the series converges on the entire real axis, and consequently, it converges uniformly on any finite interval. Then the series is infinitely differentiable on $\mathbb{R}$ and the derivatives can be found using term-by-term differentiation. Denote the sum of the series by $f(x)$ and calculate its first derivative:

$$f'(x) = \sum_{n=1}^{\infty} \frac{x^{n-1}}{(n-1)!} = \sum_{n=0}^{\infty} \frac{x^n}{n!} = f(x), \forall x \in \mathbb{R}.$$

The ordinary differential equation $f'(x) = f(x)$ is a trivial separable equation whose general solution is $f(x) = Ce^x$, where C is an arbitrary constant. Choosing $x = 0$, we have $f(0) = \sum_{n=0}^{\infty} \frac{0^n}{n!} = 1$ and then $C = 1$. Hence, the sum of the series $\sum_{n=0}^{\infty} \frac{x^n}{n!}$ is $f(x) = e^x, \forall x \in \mathbb{R}$.

In the next section we will derive the same power series expansion of e^x using the method of the Taylor coefficients.

5.3   Method of the Taylor Coefficients

In this section we apply the *method of the Taylor coefficients* both to obtain already known expansions with intention to compare different methods and to derive power series representations of some new functions.

1. Functions $f(x) = (1+x)^p, \forall p \in \mathbb{R}$

We start this section by applying the method of the Taylor coefficients to the function that has the binomial series expansion, already found in Sect. 5.2 using the technique of differential relations.

If p is a natural number or zero, then the given function is a polynomial of degree p, which is a particular (trivial and singular) case of a power series (all the terms of this series with powers greater than p are not present). For this reason, in what follows we consider only the cases when p is not a natural number or 0. In this case, the domain of $f(x)$ is $(-1, +\infty)$. Let us calculate the derivatives of this function:

$$f(x) = (1+x)^p, \; f'(x) = p(1+x)^{p-1}, \; f''(x) = p(p-1)(1+x)^{p-2}, \; \dots \, ,$$

$$f^{(n)}(x) = p(p-1)(p-2) \cdot \dots \cdot (p-n+1)(1+x)^{p-n}, \; \dots \, .$$

Notice that all the derivatives exist on the interval $(-1, +\infty)$ and since $p \notin \mathbb{N} \cup \{0\}$, there is an infinite number of non-zero derivatives. To construct the power series centered at $a = 0$, we evaluate the derivatives at $a = 0$:

$$f(0) = 1, \; f'(0) = p, \; f''(0) = p(p-1), \; \dots \, , \; f^{(n)}(0) = p(p-1)(p-2) \cdot \dots \cdot (p-n+1), \; \dots$$

and substitute these results in the formal Taylor series (5.24):

$$\sum_{n=0}^{\infty} \frac{f^{(n)}(0)}{n!} x^n = \sum_{n=0}^{\infty} \frac{p(p-1) \cdot \dots \cdot (p-n+1)}{n!} x^n$$

$$= 1 + \frac{p}{1!}x + \frac{p(p-1)}{2!}x^2 + \dots + \frac{p(p-1)\dots(p-n+1)}{n!}x^n + \dots .$$

Now we find the radius of convergence of the obtained series using, for example, D'Alembert's formula:

$$R = \lim_{n \to \infty} \left| \frac{c_n}{c_{n+1}} \right| = \lim_{n \to \infty} \left| \frac{p(p-1)\dots(p-n+1) \cdot (n+1)!}{n! \cdot p(p-1)\dots(p-n+1)(p-n)} \right| = \lim_{n \to \infty} \frac{n+1}{|p-n|} = 1.$$

This means that the series converges on the interval $(-1, 1)$, and consequently, it is sufficient to analyze its behavior only on this interval. Next we verify if this formal series converges to the original function $f(x)$. To do this, we analyze the behavior of the remainder in the Taylor formula of $f(x)$ using the Cauchy form:

$$R_n(x) = \frac{f^{(n+1)}(\theta x)}{n!} x^{n+1} (1-\theta)^n, \; 0 < \theta < 1.$$

Substituting in this formula the expression for $(n+1)$-th derivative of the given function, we obtain

$$R_n(x) = \frac{p(p-1)\dots(p-n)}{n!}(1+\theta x)^{p-n-1} x^{n+1} (1-\theta)^n$$

$$= \left(\frac{(p-1)(p-2)\dots(p-n)}{n!} x^n \right) \cdot \left(\frac{1-\theta}{1+\theta x} \right)^n \cdot \left(px \cdot (1+\theta x)^{p-1} \right).$$

Now we evaluate separately each of the three factors on the right-hand side. Since $-1 < x < 1$ and $0 < \theta < 1$, we get

$$0 < \frac{1-\theta}{1+\theta x} < 1 \text{ and } 0 < \left(\frac{1-\theta}{1+\theta x}\right)^n < 1, \forall n \in N, \forall x \in (-1, 1),$$

that is, the second term is bounded. The third factor satisfies the estimate

$$\left| px (1+\theta x)^{p-1} \right| \leq \begin{cases} |px| \cdot (1+|x|)^{p-1}, p > 1 \\ |px| \cdot (1-\theta)^{p-1}, p < 1 \end{cases} \leq \begin{cases} p \cdot 2^{p-1}, p > 1 \\ |p| \cdot (1-\theta)^{p-1}, p < 1, \end{cases}$$

that is, it is also bounded for any fixed x in $(-1, 1)$. The first factor can be considered as the general term of the power series for the function $(1+x)^{p-1}$. We have already verified that the power series for $(1+x)^p$, and consequently, for $(1+x)^{p-1}$ converges on the interval $(-1, 1)$ for any p, which implies the convergence of its general term to 0 as $n \to \infty$:

$$\frac{(p-1)(p-2)\ldots(p-n)}{n!} x^n \underset{n\to\infty}{\to} 0, \forall x \in (-1, 1).$$

Hence, the second and third factors in the expression for the remainder are bounded and the first factor approaches 0. This implies that the remainder approaches 0 as $n \to \infty$: $R_n(x) \underset{n\to\infty}{\to} 0$. Therefore, by the criterion for expansion of a function in a power series (Sect. 4.3), we conclude that the constructed series converges to the function $(1+x)^p$ on the interval $(-1, 1)$, that is, the given function has the following representation:

$$(1+x)^p = \sum_{n=0}^{\infty} \frac{p(p-1)\ldots(p-n+1)}{n!} x^n = 1 + \frac{p}{1!} x + \frac{p(p-1)}{2!} x^2 + \ldots$$
$$+ \frac{p(p-1)\ldots(p-n+1)}{n!} x^n + \ldots .$$

This series is called the binomial series or binomial representation.

In the preceding section the same binomial series was obtained using the technique of differential relations. Comparing the two approaches, we see that the method of the Taylor coefficients is more technically demanding due to the relative complexity of the evaluation of its remainder required to show the convergence of the power series to the specific function.

Let us consider a special case of the binomial series with $p = -\frac{1}{2}$:

$$\frac{1}{\sqrt{1+x}} = (1+x)^{-\frac{1}{2}} = \sum_{n=0}^{\infty} \frac{\left(-\frac{1}{2}\right)\left(-\frac{1}{2}-1\right)\cdots\left(-\frac{1}{2}-n+1\right)}{n!} x^n$$

$$= \sum_{n=0}^{\infty} (-1)^n \frac{\frac{1}{2}\cdot\frac{3}{2}\cdot\ldots\cdot\left(n-\frac{1}{2}\right)}{n!} x^n$$

$$= \sum_{n=0}^{\infty} (-1)^n \frac{1\cdot 3\cdot\ldots\cdot(2n-1)}{n!}\left(\frac{x}{2}\right)^n = \sum_{n=0}^{\infty} (-1)^n \frac{(2n-1)!!}{(2n)!!} x^n.$$

As was shown in the general case of the binomial representation, this series converges to $f(x) = \frac{1}{\sqrt{1+x}}$ on the interval $(-1, 1)$. At the endpoints $x = -1$ and $x = 1$ we have the series $\sum_{n=1}^{\infty} \frac{(2n-1)!!}{(2n)!!}$ and $\sum_{n=1}^{\infty} (-1)^n \frac{(2n-1)!!}{(2n)!!}$, respectively. To analyze the convergence/divergence of these two series let us prove the following double inequality: $\frac{1}{2\sqrt{n}} < \frac{(2n-1)!!}{(2n)!!} < \frac{1}{\sqrt{2n+1}}$, $\forall n \geq 2$. For both parts of the inequality we use mathematical induction. If $n = 2$, both inequalities hold: $\frac{1}{2\sqrt{2}} < \frac{3}{8} < \frac{1}{\sqrt{5}}$. Suppose that the left-hand inequality is true for $k = n$ and demonstrate that it holds for $k = n + 1$:

$$\frac{(2n+1)!!}{(2n+2)!!} = \frac{(2n-1)!!}{(2n)!!}\frac{2n+1}{2n+2} > \frac{1}{2\sqrt{n}}\frac{2n+1}{2n+2} = \frac{1}{2\sqrt{n+1}}\frac{\sqrt{n+1}}{\sqrt{n}}\frac{2n+1}{2n+2}$$

$$= \frac{1}{2\sqrt{n+1}}\frac{2n+1}{2\sqrt{n}\sqrt{n+1}}$$

$$= \frac{1}{2\sqrt{n+1}}\sqrt{\frac{(2n+1)^2}{4n(n+1)}} = \frac{1}{2\sqrt{n+1}}\sqrt{\frac{4n^2+4n+1}{4n^2+4n}} > \frac{1}{2\sqrt{n+1}}.$$

Similarly, suppose that the right-hand inequality is true for $k = n$ and show that it holds for $k = n + 1$:

$$\frac{(2n+1)!!}{(2n+2)!!} = \frac{(2n-1)!!}{(2n)!!}\frac{2n+1}{2n+2} < \frac{1}{\sqrt{2n+1}}\frac{2n+1}{2n+2} = \frac{1}{\sqrt{2n+3}}\frac{\sqrt{2n+3}}{\sqrt{2n+1}}\frac{2n+1}{2n+2}$$

$$= \frac{1}{\sqrt{2n+3}}\sqrt{\frac{(2n+1)(2n+3)}{(2n+2)^2}} = \frac{1}{\sqrt{2n+3}}\sqrt{\frac{4n^2+8n+3}{4n^2+8n+4}} < \frac{1}{\sqrt{2n+3}}.$$

Using the left-hand inequality, we immediately conclude that the series at $x = -1$ diverges since the corresponding $p = \frac{1}{2}$-series $\sum_{n=1}^{\infty} \frac{1}{2\sqrt{n}}$ diverges. The series at $x = 1$ is an alternating series convergent according to Leibniz's test, since $b_n = \frac{(2n-1)!!}{(2n)!!}$ converges

monotonically to 0: $b_{n+1} = \frac{(2n+1)!!}{(2n+2)!!} = \frac{(2n-1)!!}{(2n)!!}\frac{2n+1}{2n+2} < b_n$ and $b_n < \frac{1}{\sqrt{2n+1}} \underset{n\to\infty}{\to} 0$.
However, this convergence is conditional, since the series of absolute values coincides with the divergent series at $x = -1$. Thus, the series for $\frac{1}{\sqrt{1+x}}$ converges on $(-1, 1]$ and, according to Abel's Theorem, converges uniformly on $(a, 1], -1 < \forall a < 1$.

2. Function $f(x) = \frac{1}{\sqrt{1-x}}$ and Related Functions

The power series expansion of $f(x) = \frac{1}{\sqrt{1-x}}$ can be obtained from the binomial series for $\frac{1}{\sqrt{1+x}}$ by substituting $-x$ for x:

$$\frac{1}{\sqrt{1-x}} = (1+(-x))^{-\frac{1}{2}} = \sum_{n=0}^{\infty} \frac{\left(-\frac{1}{2}\right)\left(-\frac{1}{2}-1\right)\cdots\left(-\frac{1}{2}-n+1\right)}{n!}(-x)^n$$

$$= \sum_{n=0}^{\infty} \frac{\frac{1}{2}\cdot\frac{3}{2}\cdot\ldots\cdot\left(n-\frac{1}{2}\right)}{n!}x^n$$

$$= \sum_{n=0}^{\infty} \frac{1\cdot 3\cdot\ldots\cdot(2n-1)}{n!}\left(\frac{x}{2}\right)^n = \sum_{n=0}^{\infty} \frac{(2n-1)!!}{(2n)!!}x^n.$$

From the result for $\frac{1}{\sqrt{1+x}}$ shown in the previous subsection, it follows directly that this series converges to $f(x) = \frac{1}{\sqrt{1-x}}$ on the interval $[-1, 1)$ and converges uniformly on $[-1, a), -1 < \forall a < 1$.

Using the series for $\frac{1}{\sqrt{1-x}}$ with the argument x^2, we have

$$\frac{1}{\sqrt{1-x^2}} = \sum_{n=0}^{\infty} \frac{1\cdot 3\cdot\ldots\cdot(2n-1)}{n!}\left(\frac{x^2}{2}\right)^n = \sum_{n=0}^{\infty} \frac{(2n-1)!!}{(2n)!!}x^{2n}$$

with the same interval of convergence $(-1, 1)$, since the condition $|x| < 1$ is equivalent to $x^2 < 1$. However, at both endpoints $x = \pm 1$ the series $\sum_{n=1}^{\infty} \frac{(2n-1)!!}{(2n)!!}$ diverges (see the previous subsection).

Finally, integrating the last formula on $(-1, 1)$, we arrive at the series for the function $\arcsin x$ convergent on $(-1, 1)$:

$$\arcsin x = \sum_{n=0}^{\infty} \frac{1\cdot 3\cdot\ldots\cdot(2n-1)}{n!}\frac{x^{2n+1}}{2^n(2n+1)} = \sum_{n=0}^{\infty} \frac{(2n-1)!!}{(2n)!!}\frac{x^{2n+1}}{2n+1}.$$

At the endpoint $x = 1$, we get the positive series of numbers

$$\sum_{n=0}^{\infty} \frac{(2n-1)!!}{(2n)!!}\frac{1}{2n+1}.$$

Using the inequality $\frac{(2n-1)!!}{(2n)!!} < \frac{1}{\sqrt{2n+1}}$, $\forall n \geq 2$ shown in the previous subsection, the general term can be evaluate in the form $\frac{(2n-1)!!}{(2n)!!} \frac{1}{2n+1} < \frac{1}{(2n+1)^{3/2}} < \frac{1}{n^{3/2}}$, which implies the convergence of the series at $x = 1$ since the $p = \frac{3}{2}$-series $\sum_{n=1}^{\infty} \frac{1}{n^{3/2}}$ converges. Then, the series also converges at $x = -1$ (the series of absolute values at $x = -1$ coincides with the series at $x = 1$). Therefore, according to Abel's Theorem, the series converges uniformly on the closed interval $[-1, 1]$ to the function $\arcsin x$.

Due to the relation $\arccos x = \frac{\pi}{2} - \arcsin x$, the power series expansion for $\arccos x$ follows immediately from that for $\arcsin x$:

$$\arccos x = \frac{\pi}{2} - \sum_{n=0}^{\infty} \frac{1 \cdot 3 \cdot \ldots \cdot (2n-1)}{n!} \frac{x^{2n+1}}{2^n(2n+1)} = \frac{\pi}{2} - \sum_{n=0}^{\infty} \frac{(2n-1)!!}{(2n)!!} \frac{x^{2n+1}}{2n+1}.$$

3. Function $f(x) = e^x$

Let us analyze this function on the entire real axis and choose the central point $a = 0$ for the construction of power series. It is easy to find a general formula for n-th derivative:

$$f(x) = e^x, \ f'(x) = e^x, \ldots, f^{(n)}(x) = e^x.$$

Then,

$$f(0) = 1, \ f'(0) = 1, \ldots, f^{(n)}(0) = 1$$

and we find the formal Taylor series in the form

$$\sum_{n=0}^{\infty} \frac{f^{(n)}(0)}{n!} x^n = \sum_{n=0}^{\infty} \frac{x^n}{n!} = 1 + \frac{x}{1!} + \frac{x^2}{2!} + \ldots + \frac{x^n}{n!} + \ldots.$$

Calculating the radius of the convergence, we get

$$R = \lim_{n \to \infty} \left| \frac{c_n}{c_{n+1}} \right| = \lim_{n \to \infty} \frac{(n+1)!}{n!} = \lim_{n \to \infty} (n+1) = +\infty,$$

that is, the series converges on the entire real axis. To verify whether this series converges to the given function, we take any $\tilde{R} > 0$ and evaluate the derivatives of the function $f(x) = e^x$ on the interval $\left[-\tilde{R}, \tilde{R}\right]$:

$$\left| f^{(n)}(x) \right| = e^x \leq e^{\tilde{R}}, \ \forall n \in N, \ \forall x \in \left[-\tilde{R}, \tilde{R}\right].$$

Since the derivatives are uniformly bounded, the conditions of Remark 2 to Theorem 2 in Sect. 4.3 are satisfied for the function $f(x) = e^x$ on the interval $\left[-\tilde{R}, \tilde{R}\right]$. Since $\tilde{R}$ is an arbitrary positive number, we conclude that the series converges to the given function e^x absolutely on the entire real axis and uniformly on any interval $\left[-\tilde{R}, \tilde{R}\right]$. Therefore,

$$e^x = \sum_{n=0}^{\infty} \frac{x^n}{n!} = 1 + \frac{x}{1!} + \frac{x^2}{2!} + \ldots + \frac{x^n}{n!} + \ldots, \quad \forall x \in \mathbb{R}.$$

In particular, at $x = 1$ we find the representation of the number e in the series

$$e = \sum_{n=0}^{\infty} \frac{1}{n!}.$$

Notice that the convergence of the series $\sum_{n=0}^{\infty} \frac{x^n}{n!}$ is not uniform on $\mathbb{R}$, that can be seen from the evaluation of its general term. Indeed, using the sequence of points $x_n = n$, we have $u_n(x_n) = \frac{n^n}{n!} > n \underset{n \to \infty}{\to} \infty \neq 0$, that is, the general term does not converge uniformly on $\mathbb{R}$ to 0, and consequently, according to the Test for non-uniform convergence, the power series for e^x does not converge uniformly on $\mathbb{R}$.

In the preceding section we have obtained the same power series expansion of e^x using the method of differential relations. Comparing the two methods, we can conclude that both have a similar simplicity in this case. Even so, the method of the Taylor coefficients seems to be more direct and natural.

Substituting $-x$ for x in the series of e^x, we obtain the representation of e^{-x} in the Taylor series:

$$e^{-x} = \sum_{n=0}^{\infty} (-1)^n \frac{x^n}{n!} = 1 - \frac{x}{1!} + \frac{x^2}{2!} - \ldots + (-1)^n \frac{x^n}{n!} + \ldots, \quad \forall x \in \mathbb{R}.$$

Using the arithmetic properties of uniformly convergent series, we immediately deduce that the functions $\cosh x = \frac{e^x + e^{-x}}{2}$ and $\sinh x = \frac{e^x - e^{-x}}{2}$ have the following power series expansions on $\mathbb{R}$:

$$\cosh x = \sum_{n=0}^{\infty} \frac{x^{2n}}{(2n)!} = 1 + \frac{x^2}{2!} + \frac{x^4}{4!} + \ldots + \frac{x^{2n}}{(2n)!} + \ldots, \quad \forall x \in \mathbb{R}$$

and

$$\sinh x = \sum_{n=0}^{\infty} \frac{x^{2n+1}}{(2n+1)!} = x + \frac{x^3}{3!} + \frac{x^5}{5!} + \ldots + \frac{x^{2n+1}}{(2n+1)!} + \ldots, \quad \forall x \in \mathbb{R}.$$

4. Functions $f(x) = \sin x$ and $f(x) = \cos x$

The function $f(x) = \sin x$ is infinitely differentiable on $\mathbb{R}$ and its n-th derivative can be found in the following form:

$$f'(x) = \cos x = \sin\left(x + \frac{\pi}{2}\right), \ f''(x) = \cos\left(x + \frac{\pi}{2}\right)$$

$$= \sin\left(x + \frac{\pi}{2} + \frac{\pi}{2}\right) = \sin\left(x + 2 \cdot \frac{\pi}{2}\right), \dots,$$

$$f^{(n)}(x) = \sin\left(x + n \cdot \frac{\pi}{2}\right).$$

Since derivatives are uniformly bounded on $\mathbb{R} - \left|f^{(n)}(x)\right| = \left|\sin\left(x + n \cdot \frac{\pi}{2}\right)\right| \le 1, \forall n \in \mathbb{N}, \forall x \in \mathbb{R}-$, the function $f(x) = \sin x$ satisfies the conditions of Remark 2 to Theorem 2 (Sect. 4.3) on the entire real axis. Choosing the central point $a = 0$, we specify the values of derivatives

$$f(0) = 0, \ f'(0) = 1, \ f''(0) = 0, \ f'''(0) = -1, \dots, f^{(n)}(0) = \sin n \cdot \frac{\pi}{2} = \begin{cases} 0, n = 2k, \\ (-1)^k, n = 2k + 1, \end{cases}$$

and substituting these values in the Taylor series (5.24), we obtain:

$$\sin x = x - \frac{x^3}{3!} + \frac{x^5}{5!} - \frac{x^7}{7!} + \dots = \sum_{n=0}^{\infty} (-1)^n \frac{x^{2n+1}}{(2n + 1)!}, \ \forall x \in \mathbb{R}.$$

Notice that the obtained series contains only odd powers, in accordance with the Theorem on representation of even and odd functions (Sect. 3.4).

Although the constructed series converges uniformly on any interval $\left[-\tilde{R}, \tilde{R}\right]$, it does not converge uniformly on $\mathbb{R}$, that can be seen from the evaluation of its general term: at the points $x_n = 2n + 1$ we have $|u_n(x_n)| = \frac{(2n+1)^{2n+1}}{(2n+1)!} > 2n + 1 \underset{n \to \infty}{\to} \infty \ne 0$, that is, the general term does not converge uniformly to 0 on $\mathbb{R}$, and consequently, according to the Test for non-uniform convergence, the series does not converge uniformly on $\mathbb{R}$.

In the same way as for $\sin x$, we find the power series expansion centered at $a = 0$ for $\cos x$. The function is infinitely differentiable on $\mathbb{R}$ and its derivatives can be expressed in the form

$$f'(x) = -\sin x = \cos\left(x + \frac{\pi}{2}\right), \ f''(x) = -\sin\left(x + \frac{\pi}{2}\right)$$

$$= \cos\left(x + \frac{\pi}{2} + \frac{\pi}{2}\right) = \cos\left(x + 2 \cdot \frac{\pi}{2}\right), \dots,$$

$$f^{(n)}(x) = \cos\left(x + n \cdot \frac{\pi}{2}\right).$$

Then, at the point 0 we get:

$$f(0) = 1, \; f'(0) = 0, \; f''(0) = -1, \; f'''(0) = 0, \ldots, f^{(n)}(0) = \cos n \cdot \frac{\pi}{2} = \begin{cases} 0, n = 2k+1, \\ (-1)^k, n = 2k. \end{cases}$$

Since derivatives are uniformly bounded on $\mathbb{R} - \left| f^{(n)}(x) \right| = \left| \cos \left(x + n \cdot \frac{\pi}{2} \right) \right| \le 1, \forall n \in \mathbb{N}, \forall x \in \mathbb{R}-$, the function $f(x) = \cos x$ satisfies the conditions of Remark 2 to Theorem 2 (Sect. 4.3) on the entire real axis and its Taylor series is

$$\cos x = 1 - \frac{x^2}{2!} + \frac{x^4}{4!} - \frac{x^6}{6!} + \ldots = \sum_{n=0}^{\infty} (-1)^n \frac{x^{2n}}{(2n)!}, \quad \forall x \in \mathbb{R}.$$

According to the Theorem on representation of even and odd functions (Sect. 3.4), the obtained series contains only even powers. Using the same considerations as for the series of $\sin x$, it can be shown that the series for $\cos x$ does not converge uniformly on $\mathbb{R}$.

Notice that the same series of $\cos x$ can be obtained by differentiating term by term the series of $\sin x$, which can be made on the entire real axis, since the series of $\sin x$ is uniformly convergent on any closed interval. This task is left to the reader.

5. Functions $f(x) = \tan x$ and $f(x) = \cot x$

For many infinitely differentiable functions it is impossible to find a general formula of n-th derivative. One example of this is the function $f(x) = \tan x$. However, theoretically we can calculate successively all derivatives until the required order and evaluate them at the central point to specify the first terms of the Taylor series. For $f(x) = \tan x$ the first seven derivatives can be expressed in the following manner:

$$f'(x) = \frac{1}{\cos^2 x}, \; f''(x) = \frac{2\sin x}{\cos^3 x}, \; f'''(x) = \frac{2\cos^2 x + 6\sin^2 x}{\cos^4 x} = -4f'(x) + 6\frac{1}{\cos^4 x},$$

$$f^{(4)}(x) = -4f''(x) + 6\left(\frac{1}{\cos^4 x} \right)' = -4f''(x) + 6\frac{4\sin x}{\cos^5 x},$$

$$f^{(5)}(x) = -4f'''(x) + 6\left(\frac{1}{\cos^4 x} \right)'' = -4f'''(x) + 24\frac{\cos^2 x + 5\sin^2 x}{\cos^6 x}$$

$$= -4\left(-4f'(x) + \frac{6}{\cos^4 x} \right) + 24\left(-\frac{4}{\cos^4 x} + \frac{5}{\cos^6 x} \right) = 16f'(x) - 120\frac{1}{\cos^4 x} + 120\frac{1}{\cos^6 x},$$

$$f^{(6)}(x) = 16f''(x) - 120\left(\frac{1}{\cos^4 x} \right)' + 120\frac{6\sin x}{\cos^7 x},$$

$$f^{(7)}(x) = 16f'''(x) - 120\left(\frac{1}{\cos^4 x} \right)'' + 720\frac{\cos^2 x + 7\sin^2 x}{\cos^8 x}.$$

As is seen, some results for derivatives of lower order can be used to find the derivatives of higher order. Specifying the values at the central point 0, we have

$$f(0) = 0,\ f'(0) = 1,\ f''(0) = 0,\ f'''(0) = -4f'(0) + 6 = 2,\ f^{(4)}(0) = -4f''(0) + 0 = 0,$$

$$f^{(5)}(0) = 16f'(0) - 120 + 120 = 16,\ f^{(6)}(0) = 16f''(0) - 0 + 0 = 0,$$

$$f^{(7)}(0) = 16f'''(0) - 120 \cdot 4 + 720 \cdot 1 = 272.$$

Then, the first terms of the series for $\tan x$ have the form:

$$\tan x = x + \frac{2}{3!}x^3 + \frac{16}{5!}x^5 + \frac{272}{7!}x^7 + \ldots = x + \frac{x^3}{3} + \frac{2x^5}{15} + \frac{17x^7}{315} + \ldots.$$

As was expected (according to the theory), all even powers disappear because the function $f(x) = \tan x$ is odd.

To find the first terms of the power series expansion of $\cot x$ centered at $\frac{\pi}{2}$, we can employ the trigonometric relation and make use of the found series of $\tan x$:

$$\cot x = \tan\left(\frac{\pi}{2} - x\right) = \left(\frac{\pi}{2} - x\right) + \frac{1}{3}\left(\frac{\pi}{2} - x\right)^3 + \frac{2}{15}\left(\frac{\pi}{2} - x\right)^5 + \frac{17}{315}\left(\frac{\pi}{2} - x\right)^7 + \ldots.$$

The reader can confirm this result using direct calculation of derivatives of $\cot x$ at the point $\frac{\pi}{2}$.

Since the general form of n-th derivative, and consequently, the form of general term of the series for $\tan x$ and $\cot x$ is unknown, using this method of construction of the Taylor series it is impossible to determine the interval of convergence of the obtained series. In one of the following examples we will apply another approach to find the power series of $\tan x$, which will allow us to determine the interval of convergence of the series of $\tan x$, and consequently, of $\cot x$.

5.4   Taylor Series for Various Functions

Example 1t Find the power series expansion of e^{-x^2} at the central point 0.

Applying the properties of the power series, the required series can be found by substituting $-x^2$ for x in the known series of e^x. Since the series of e^x converges on $\mathbb{R}$ and converges uniformly on any closed interval, the same is true for the series of e^{-x^2}:

$$e^{-x^2} = \sum_{n=0}^{\infty} (-1)^n \frac{x^{2n}}{n!},\ \forall x \in \mathbb{R}.$$

Example 2t Find the Taylor series centered at 0 for $x \cos x^2$.
The power series of $\cos x^2$ can be obtained by substituting x^2 for x in the series of $\cos x$. After this, it only remains to multiply the obtained series by x:

$$x \cos x^2 = x \sum_{n=0}^{\infty} \frac{(-1)^n x^{4n}}{(2n)!} = \sum_{n=0}^{\infty} \frac{(-1)^n x^{4n+1}}{(2n)!}.$$

This series inherits the properties of convergence of the series for $\cos x$: it converges (non-uniformly) on $\mathbb{R}$ and converges uniformly on any closed interval.

Example 3t Find the Taylor series centered at 0 for $\cos^2 x$.
Using the trigonometric formula $\cos^2 x = \frac{1+\cos 2x}{2}$, we reduce this problem to expansion of $\cos x$:

$$\cos^2 x = \frac{1}{2} + \frac{1}{2} \cos 2x = \frac{1}{2} + \frac{1}{2} \sum_{n=0}^{\infty} \frac{(-1)^n (2x)^{2n}}{(2n)!} = 1 + \frac{1}{2} \sum_{n=1}^{\infty} \frac{(-1)^n (2x)^{2n}}{(2n)!}.$$

This series converges (non-uniformly) on $\mathbb{R}$ and converges uniformly on any closed interval.

Example 4t Find the power series for a^x, $a > 0$, $a \neq 1$ with the central point 0.
The first method: find the general formula for n-th derivative—$f^{(n)}(x) = \ln^n a \cdot a^x$ and $f^{(n)}(0) = \ln^n a$, and next compose the formal Taylor series

$$a^x = \sum_{n=0}^{\infty} \frac{f^{(n)}(0)}{n!} x^n = \sum_{n=0}^{\infty} \frac{x^n \ln^n a}{n!}.$$

The investigation of the properties of convergence can be made in the same way as for e^x and it reveals that the series converges (non-uniformly) to a^x on $\mathbb{R}$ and converges uniformly on any closed interval.

Another manner is to make use of the known series for e^x via representation $a^x = e^{\ln a^x} = e^{x \ln a}$:

$$a^x = e^{x \ln a} = \sum_{n=0}^{\infty} \frac{(x \ln a)^n}{n!}.$$

Example 5t Find the Taylor series of $\sin x$ using the central point $\frac{\pi}{4}$.
The general formula for derivatives was already found:

$$f^{(n)}(x) = \sin\left(x + n \cdot \frac{\pi}{2}\right), \forall x \in \mathbb{R}.$$

In particular, at the point $\frac{\pi}{4}$ we have

$$f^{(n)}(x) = \sin\left(\frac{\pi}{4} + n \cdot \frac{\pi}{2}\right) = \frac{\sqrt{2}}{2}\begin{cases} 1, n = 4k, 4k+1 \\ -1, n = 4k+2, 4k+3 \end{cases}, k \in \mathbb{N} \end{cases} = (-1)^{\frac{n(n-1)}{2}} \cdot \frac{\sqrt{2}}{2}.$$

Therefore, the Taylor series takes the form

$$\sin x = \frac{\sqrt{2}}{2}\sum_{n=0}^{\infty}(-1)^{\frac{n(n-1)}{2}}\frac{(x-\frac{\pi}{4})^n}{n!}, \quad \forall x \in \mathbb{R}.$$

The investigation of convergence can be made in the same way as for the central point 0, and it shows that the series converges (non-uniformly) to $\sin x$ on $\mathbb{R}$ and converges uniformly on any closed interval.

Another method of solution consists of the substitution of the central point $\frac{\pi}{4}$ by the point 0 via the change of variable $t = x - \frac{\pi}{4}$ and the use of the already derived Taylor series of $\sin x$ and $\cos x$ at the point 0. Using these transformations we arrive at the same final series:

$$\sin x = \sin\left(t + \frac{\pi}{4}\right) = \sin t \cdot \cos\frac{\pi}{4} + \cos t \cdot \sin\frac{\pi}{4}$$

$$= \frac{\sqrt{2}}{2}\sum_{n=0}^{\infty}(-1)^n\frac{t^{2n+1}}{(2n+1)!} + \frac{\sqrt{2}}{2}\sum_{n=0}^{\infty}(-1)^n\frac{t^{2n}}{(2n)!}$$

$$= \frac{\sqrt{2}}{2}\left(\sum_{n=0}^{\infty}(-1)^n\frac{(x-\frac{\pi}{4})^{2n+1}}{(2n+1)!} + \sum_{n=0}^{\infty}(-1)^n\frac{(x-\frac{\pi}{4})^{2n}}{(2n)!}\right)$$

$$= \frac{\sqrt{2}}{2}\sum_{n=0}^{\infty}(-1)^{\frac{n(n-1)}{2}}\frac{(x-\frac{\pi}{4})^n}{n!}, \quad \forall x \in \mathbb{R}.$$

Example 6t Find a representation of the function $e^{\sin x}$ in the Taylor series at 0 until the terms of the fifth order.

Recall the Taylor formulas for $f(x) = \sin x$ and $g(y) = e^y$ at the points $x_0 = 0$ and $y_0 = \sin x_0 = 0$:

$$y = f(x) = \sin x = \sum_{n=0}^{\infty}(-1)^n\frac{x^{2n+1}}{(2n+1)!} = x - \frac{x^3}{3!} + \frac{x^5}{5!} - \dots, \forall x \in \mathbb{R}$$

and

$$u = g(y) = e^y = \sum_{n=0}^{\infty}\frac{y^n}{n!} = 1 + \frac{y}{1!} + \frac{y^2}{2!} + \frac{y^3}{3!} + \frac{y^4}{4!} + \frac{y^5}{5!} + \dots, \forall y \in \mathbb{R}.$$

Both series have the radius of convergence ∞.

According to the Theorem of composite power series (Sect. 3.2), we substitute the first series (series of $\sin x$) for y in the second series (series of e^y) and group the terms with the same power of x, keeping the terms until the fifth power:

$$h(x) = g(f(x)) = e^{\sin x} = 1 + \frac{1}{1!}\left(x - \frac{x^3}{3!} + \frac{x^5}{5!} - \cdots\right) + \frac{1}{2!}\left(x - \frac{x^3}{3!} + \frac{x^5}{5!} - \cdots\right)^2$$

$$+ \frac{1}{3!}\left(x - \frac{x^3}{3!} + \frac{x^5}{5!} - \cdots\right)^3 + \frac{1}{4!}\left(x - \frac{x^3}{3!} + \frac{x^5}{5!} - \cdots\right)^4 + \frac{1}{5!}\left(x - \frac{x^3}{3!} + \frac{x^5}{5!} - \cdots\right)^5 + \cdots$$

$$= 1 + \frac{1}{1!}x + \frac{1}{2!}x^2 + \left(-\frac{1}{1!}\frac{1}{3!} + \frac{1}{3!}\right)x^3 + \left(-2\frac{1}{2!}\frac{1}{3!} + \frac{1}{4!}\right)x^4 + \left(\frac{1}{1!}\frac{1}{5!} - 3\frac{1}{3!}\frac{1}{3!} + \frac{1}{5!}\right)x^5 + \cdots$$

$$= 1 + x + \frac{1}{2}x^2 - \frac{1}{8}x^4 - \frac{1}{15}x^5 + \cdots .$$

Example 7t Find the power series expansion of the function $e^{\cos x}$ at the point 0 until the terms of the sixth order.

Recall that the Taylor series of $f(x) = \cos x$ at $x_0 = 0$ has the non-zero first coefficient:

$$f(x) = \cos x = \sum_{n=0}^{\infty} (-1)^n \frac{x^{2n}}{(2n)!} = 1 - \frac{x^2}{2!} + \frac{x^4}{4!} - \frac{x^6}{6!} + \cdots, \quad \forall x \in \mathbb{R}.$$

To satisfy the conditions of the Theorem of composite power series (Sect. 3.2) we introduce an auxiliary function $\tilde{f}(x) = f(x) - 1 = \cos x - 1$, whose Taylor series has zero constant term, and use the representation $h(x) = g(f(x)) = e^{1+\tilde{f}(x)} = e \cdot e^{\tilde{f}(x)}$. The series

$$y = \tilde{f}(x) = \cos x - 1 = \sum_{n=1}^{\infty} (-1)^n \frac{x^{2n}}{(2n)!} = -\frac{x^2}{2!} + \frac{x^4}{4!} - \frac{x^6}{6!} + \cdots, \quad \forall x \in \mathbb{R}$$

and

$$g(y) = e^y = \sum_{n=0}^{\infty} \frac{y^n}{n!} = 1 + \frac{y}{1!} + \frac{y^2}{2!} + + \frac{y^3}{3!} + \cdots, \quad \forall y \in \mathbb{R}$$

satisfy the conditions of the Theorem of composite power series (both series have the radius of convergence equal to ∞). Therefore, we can obtain the series of $\tilde{h}(x) = g(\tilde{f}(x))$ by substituting the series of $\tilde{f}(x) = \cos x - 1$ for y in the series of e^y, grouping the terms with the same powers of x, and keeping the terms until the sixth power:

$$\tilde{h}(x) = e^{\cos x - 1} = 1 + \frac{1}{1!}\left(-\frac{x^2}{2!} + \frac{x^4}{4!} - \frac{x^6}{6!} + \cdots\right) + \frac{1}{2!}\left(-\frac{x^2}{2!} + \frac{x^4}{4!} + \cdots\right)^2$$

$$+ \frac{1}{3!}\left(-\frac{x^2}{2!} + \cdots\right)^3 + \cdots$$

$$= 1 - \frac{1}{1!}\frac{1}{2!}x^2 + \left(\frac{1}{1!}\frac{1}{4!} + \frac{1}{2!}\left(\frac{1}{2!}\right)^2\right)x^4$$

$$+ \left(-\frac{1}{1!}\frac{1}{6!} - 2\frac{1}{2!}\frac{1}{2!}\frac{1}{4!} - \frac{1}{3!}\left(\frac{1}{2!}\right)^3\right)x^6 + \cdots$$

$$= 1 - \frac{1}{2}x^2 + \frac{1}{6}x^4 - \frac{31}{720}x^6 + \cdots$$

Finally,

$$h(x) = e \cdot \tilde{h}(x) = e \cdot \left(1 - \frac{1}{2}x^2 + \frac{1}{6}x^4 - \frac{31}{720}x^6 + \cdots\right).$$

Example 8t Find the first terms of the expansion of the function $\frac{1}{\cos x}$ in the Taylor series at the point 0.

Since the Taylor series of $\cos x$ at the point $x_0 = 0$ contains the non-zero constant term, we use the following representation $h(x) = g(f(x)) = \frac{1}{\cos x} = \frac{1}{1-f(x)}$, where $y = f(x) = 1 - \cos x$ and $g(y) = \frac{1}{1-y}$. The series

$$y = f(x) = 1 - \cos x = \sum_{n=1}^{\infty}(-1)^{n+1}\frac{x^{2n}}{(2n)!} = \frac{x^2}{2!} - \frac{x^4}{4!} + \frac{x^6}{6!} - \frac{x^8}{8!} + \cdots, \ \forall x \in \mathbb{R}$$

and

$$g(y) = \frac{1}{1-y} = \sum_{n=0}^{\infty} y^n = 1 + y + y^2 + y^3 + \cdots, \ \forall y \in (-1, 1)$$

satisfy the conditions of the Theorem of composite power series (Sect. 3.2). The first series has the radius of convergence $R_x = \infty$ and the second—$R_y = 1$. Therefore, in a neighborhood of the point $x = 0$, we can obtain the series of $h(x) = g(f(x))$ by substituting the series of $f(x) = 1 - \cos x$ instead of y in the series of $\frac{1}{1-y}$ and joining the terms with the same powers of x:

$$h(x) = \frac{1}{\cos x} = 1 + \left(\frac{x^2}{2!} - \frac{x^4}{4!} + \frac{x^6}{6!} - \frac{x^8}{8!} + \dots\right) + \left(\frac{x^2}{2!} - \frac{x^4}{4!} + \frac{x^6}{6!} + \dots\right)^2$$

$$+ \left(\frac{x^2}{2!} - \frac{x^4}{4!} + \dots\right)^3 + \left(\frac{x^2}{2!} + \dots\right)^4 + \dots$$

$$= 1 + \frac{1}{2!}x^2 + \left(-\frac{1}{4!} + \left(\frac{1}{2!}\right)^2\right)x^4 + \left(\frac{1}{6!} - 2\frac{1}{2!}\frac{1}{4!} + \left(\frac{1}{2!}\right)^3\right)x^6 +$$

$$\left(-\frac{1}{8!} + 2\frac{1}{2!}\frac{1}{6!} + \left(\frac{1}{4!}\right)^2 - 3\left(\frac{1}{2!}\right)^2\frac{1}{4!} + \left(\frac{1}{2!}\right)^4\right)x^8 + \dots$$

$$= 1 + \frac{1}{2}x^2 + \frac{5}{24}x^4 + \frac{61}{720}x^6 + \frac{277}{8064}x^8 + \dots .$$

The convergence of the series for $g(y)$ is guaranteed when $|y| = |1 - \cos x| < 1$. In a neighborhood of $x = 0$ this is equivalent to the condition $|x| < \frac{\pi}{2}$ that gives the interval of convergence of the series for $h(x) = \frac{1}{\cos x}$.

Example 9t Find the first terms of the expansion of $\tan x$ in the power series at 0. This problem was solved in Sect. 5.3 using the expressions of the first derivatives of the function $\tan x$. Now we apply the alternative procedure which allows us to determine the interval of convergence of the obtained series. We make use of the result of the preceding example

$$\frac{1}{\cos x} = 1 + \frac{1}{2}x^2 + \frac{5}{24}x^4 + \frac{61}{720}x^6 + \frac{277}{8064}x^8 + \dots, \forall |x| < \frac{\pi}{2}$$

and the series of $\sin x$:

$$\sin x = \sum_{n=0}^{\infty}(-1)^n \frac{x^{2n+1}}{(2n+1)!} = x - \frac{x^3}{3!} + \frac{x^5}{5!} - \frac{x^7}{7!} + \dots, \forall x \in \mathbb{R}.$$

For both series the intervals of convergence were found. Multiplying these two series and joining the terms with the same powers of x, we obtain the required representation:

$$\tan x = \sin x \cdot \frac{1}{\cos x} = \left(x - \frac{x^3}{3!} + \frac{x^5}{5!} - \frac{x^7}{7!} + \dots\right)\left(1 + \frac{1}{2}x^2 + \frac{5}{24}x^4 + \frac{61}{720}x^6 + \frac{277}{8064}x^8 + \dots\right)$$

$$= x + \frac{x^3}{3} + \frac{2x^5}{15} + \frac{17x^7}{315} + \dots.$$

Naturally, we arrive at the same result as obtained before, but the new approach allows us to specify the interval of convergence of the resulting series: the obtained series converges on the intersection of the intervals of convergence of the two series that form the product, that is, on the interval $(-\frac{\pi}{2}, \frac{\pi}{2})$.

Example 10t Find the first terms in the power series expansion of $\cot x$ at the point 0.

First, notice that $\cot x$ is not defined at 0 and cannot be defined in a smooth way at the origin because $\lim_{x \to 0} \cot x = \infty$, that is, $x = 0$ is a singular point of $\cot x$ called the pole. Therefore, $\cot x$ is not an analytic function in a neighborhood of 0. However, we can try to separate this singularity and search for development in the Taylor series of a modified function.

Obviously, the singularity is caused by the part $\frac{1}{\sin x}$. If we multiply this term by x, then we can define a smooth function $h(x)$ in a neighborhood of 0: $h(x) = \frac{x}{\sin x}$, $\forall x \neq 0$ and $h(0) = 1$, where the value at the origin is chosen to guarantee the continuity at 0. Now we show that $h(x)$ is analytic in a neighborhood of 0 by finding its expansion with the help of the Theorem of composite power series (Sect. 3.2). To do this, we represent $h(x)$ in the form

$$h(x) = \frac{x}{\sin x} = \frac{1}{\frac{\sin x}{x}} = \frac{1}{1 - \left(1 - \frac{\sin x}{x}\right)} = g(f(x)), \; f(x) = 1 - \frac{\sin x}{x}, \; g(y) = \frac{1}{1-y}.$$

Then we consider the formal series of the function $f(x) = 1 - \frac{\sin x}{x}$, based on the series of $\sin x$ (recall that $f(0) = 0$ according to the definition $h(0) = 1$):

$$f(x) = 1 - \frac{1}{x} \sum_{n=0}^{\infty} (-1)^n \frac{x^{2n+1}}{(2n+1)!} = \sum_{n=1}^{\infty} (-1)^{n+1} \frac{x^{2n}}{(2n+1)!} = \frac{x^2}{3!} - \frac{x^4}{5!} + \frac{x^6}{7!} - \frac{x^8}{9!} + \dots.$$

The uniform convergence of this series on any interval $[-\tilde{R}, \tilde{R}]$, $\tilde{R} > 0$ follows from the Weierstrass test: $\left|\frac{x^{2n}}{(2n+1)!}\right| \leq \frac{\tilde{R}^{2n}}{(2n+1)!}$ and the majorant numerical series $\sum_{n=0}^{\infty} \frac{\tilde{R}^{2n}}{(2n+1)!}$ is convergent. This implies the convergence of the series for $f(x)$ on $\mathbb{R}$. Moreover, for every $x \neq 0$ the series converges to $f(x)$ and at the point $x = 0$ the sum of the series is 0, that corresponds to the definition of $f(x)$ at 0. Hence, the series converges to $f(x)$ on $\mathbb{R}$.

The second function $g(y)$ has the series expansion

$$g(y) = \frac{1}{1-y} = \sum_{n=0}^{\infty} y^n = 1 + y + y^2 + y^3 + y^4 + \dots$$

convergent on $|y| < 1$. Therefore, the conditions of the Theorem of composite series are satisfied, and consequently, in a neighborhood of $x = 0$, the function $h(x) = g(f(x))$ can be developed in the Taylor series by substituting the series of $f(x)$ for y in the series of $g(y)$. We can even specify the radius of this neighborhood: for the series $g(y)$, the condition $|y| = \left|1 - \frac{\sin x}{x}\right| < 1$ should be satisfied, which is equivalent to $0 < \frac{\sin x}{x} < 2$. The right-hand inequality always holds and the left-hand one leads to the restriction $x \in (-\pi, \pi)$ (we search for a solution only in a neighborhood of 0). Hence, on the interval $(-\pi, \pi)$ we can represent $h(x) = g(f(x))$ in a power series by substituting the series of $f(x) = 1 - \frac{\sin x}{x}$ instead of y in the series of $g(y) = \frac{1}{1-y}$ and joining the terms with the same powers of x (we keep here the powers until the eighth):

$$h(x) = \frac{x}{\sin x} = 1 + \left(\frac{x^2}{3!} - \frac{x^4}{5!} + \frac{x^6}{7!} - \frac{x^8}{9!} + \dots\right) + \left(\frac{x^2}{3!} - \frac{x^4}{5!} + \frac{x^6}{7!} + \dots\right)^2$$

$$+ \left(\frac{x^2}{3!} - \frac{x^4}{5!} + \dots\right)^3 + \left(\frac{x^2}{3!} + \dots\right)^4 + \dots$$

$$= 1 + \frac{1}{3!}x^2 + \left(-\frac{1}{5!} + \left(\frac{1}{3!}\right)^2\right)x^4 + \left(\frac{1}{7!} - 2\frac{1}{3!}\frac{1}{5!} + \left(\frac{1}{3!}\right)^3\right)x^6 +$$

$$\left(-\frac{1}{9!} + 2\frac{1}{3!}\frac{1}{7!} + \left(\frac{1}{5!}\right)^2 - 3\left(\frac{1}{3!}\right)^2\frac{1}{5!} + \left(\frac{1}{3!}\right)^4\right)x^8 + \dots$$

$$= 1 + \frac{1}{6}x^2 + \frac{7}{360}x^4 + \frac{31}{72 \cdot 210}x^6 + \frac{127}{8! \cdot 15}x^8 + \dots .$$

Next, we multiply the found series of $h(x)$ by the series of $\cos x$ and leave specified the terms until the eighth power:

$$\varphi(x) = h(x)\cos x$$

$$= \left(1 + \frac{1}{6}x^2 + \frac{7}{360}x^4 + \frac{31}{72 \cdot 210}x^6 + \frac{127}{8! \cdot 15}x^8 + \dots\right)\left(1 - \frac{x^2}{2!} + \frac{x^4}{4!} - \frac{x^6}{6!} + \frac{x^8}{8!} + \dots\right)$$

$$= 1 - \frac{1}{3}x^2 - \frac{1}{45}x^4 - \frac{2}{945}x^6 - \frac{1}{4725}x^8 + \dots .$$

Finally, to return to $\cot x$, we divide $\varphi(x)$ by x and obtain

$$\cot x = \frac{1}{x} - \frac{1}{3}x - \frac{1}{45}x^3 - \frac{2}{945}x^5 - \frac{1}{4725}x^7 + \dots ,$$

where the singular part of the original function is reflected in the term $\frac{1}{x}$ and the remaining part is the power series convergent on the interval $(-\pi, \pi)$.

5.5   The List of the Derived Formulas of Taylor Series

$$\frac{1}{1-x} = \sum_{n=0}^{\infty} x^n = 1 + x + x^2 + x^3 + \dots,$$

converges $\forall x \in (-1, 1)$, converges uniformly on $[-a, a]$, $0 < \forall a < 1$.

$$\frac{1}{(1-x)^2} = \sum_{n=1}^{\infty} nx^{n-1} = 1 + 2x + 3x^2 + 4x^3 + \dots,$$

converges $\forall x \in (-1, 1)$, converges uniformly on $[-a, a]$, $0 < \forall a < 1$.

$$\frac{1}{(1-x)^p} = \sum_{m=0}^{\infty} \frac{(m+p-1)(m+p-2)\cdot \dots \cdot (m+1)}{(p-1)!} x^m, \forall p \in \mathbb{N},$$

converges $\forall x \in (-1, 1)$, converges uniformly on $[-a, a]$, $0 < \forall a < 1$.

$$\ln(1-x) = -\sum_{n=1}^{\infty} \frac{x^n}{n} = -x - \frac{x^2}{2} - \frac{x^3}{3} - \frac{x^4}{4} - \dots,$$

converges $\forall x \in [-1, 1)$, converges uniformly on $[-1, a]$, $-1 < \forall a < 1$.

$$\frac{1}{1+x} = \sum_{n=0}^{\infty} (-1)^n x^n = 1 - x + x^2 - x^3 + \dots,$$

converges $\forall x \in (-1, 1)$, converges uniformly on $[-a, a]$, $0 < \forall a < 1$.

$$\frac{1}{(1+x)^2} = \sum_{n=1}^{\infty} (-1)^{n-1} nx^{n-1} = 1 - 2x + 3x^2 - 4x^3 + \dots,$$

converges $\forall x \in (-1, 1)$, converges uniformly on $[-a, a]$, $0 < \forall a < 1$.

$$\frac{1}{(1+x)^p} = \sum_{n=0}^{\infty} (-1)^n \frac{(n+p-1)(n+p-2)\cdot \dots \cdot (n+1)}{(p-1)!} x^n, \forall p \in \mathbb{N},$$

converges $\forall x \in (-1, 1)$, converges uniformly on $[-a, a]$, $0 < \forall a < 1$.

$$\ln(1+x) = \sum_{n=1}^{\infty}(-1)^{n-1}\frac{x^n}{n} = x - \frac{x^2}{2} + \frac{x^3}{3} - \frac{x^4}{4} + \cdots,$$

converges $\forall x \in (-1,1]$, *converges uniformly on* $[a,1]$, $-1 < \forall a < 1$.

$$\frac{1}{1+x^2} = \sum_{n=0}^{\infty}(-1)^n x^{2n} = 1 - x^2 + x^4 - x^6 + \cdots,$$

converges $\forall x \in (-1,1)$, *converges uniformly on* $[-a,a]$, $0 < \forall a < 1$.

$$\frac{1}{(1+x^2)^p} = \sum_{n=0}^{\infty}(-1)^n \frac{(n+p-1)(n+p-2)\cdot\ldots\cdot(n+1)}{(p-1)!} x^{2n}, \forall p \in \mathbb{N},$$

converges $\forall x \in (-1,1)$, *converges uniformly on* $[-a,a]$, $0 < \forall a < 1$.

$$\ln(1+x^2) = \sum_{n=1}^{\infty}(-1)^{n-1}\frac{x^{2n}}{n} = x^2 - \frac{x^4}{2} + \frac{x^6}{3} - \frac{x^8}{4} + \cdots,$$

converges $\forall x \in [-1,1]$, *converges uniformly on* $[-1,1]$.

$$\arctan x = \sum_{n=0}^{\infty}(-1)^n \frac{x^{2n+1}}{2n+1} = x - \frac{x^3}{3} + \frac{x^5}{5} - \frac{x^7}{7} + \cdots,$$

converges $\forall x \in [-1,1]$, *converges uniformly on* $[-1,1]$.

$$\mathrm{arccot}\, x = \frac{\pi}{2} - \sum_{n=0}^{\infty}(-1)^n \frac{x^{2n+1}}{2n+1} = \frac{\pi}{2} - x + \frac{x^3}{3} - \frac{x^5}{5} + \frac{x^7}{7} + \cdots,$$

converges $\forall x \in [-1,1]$, *converges uniformly on* $[-1,1]$.

$$(1+x)^p = \sum_{n=0}^{\infty}\frac{p(p-1)\cdot\ldots\cdot(p-n+1)}{n!} x^n$$

$$= 1 + \frac{p}{1!}x + \frac{p(p-1)}{2!}x^2 + \frac{p(p-1)(p-2)}{3!}x^3 + \cdots, \forall p \notin \mathbb{N}, p \neq 0,$$

converges $\forall x \in (-1,1)$, *converges uniformly on* $[-a,a]$, $0 < \forall a < 1$.

$$\frac{1}{\sqrt{1+x}} = (1+x)^{-\frac{1}{2}} = \sum_{n=0}^{\infty}(-1)^n \frac{1 \cdot 3 \cdot \ldots \cdot (2n-1)}{n!} \left(\frac{x}{2}\right)^n = \sum_{n=0}^{\infty}(-1)^n \frac{(2n-1)!!}{(2n)!!}x^n,$$

converges $\forall x \in (-1, 1]$, *converges uniformly on* $(a, 1]$, $-1 < \forall a < 1$.

$$\frac{1}{\sqrt{1-x}} = (1-x)^{-\frac{1}{2}} = \sum_{n=0}^{\infty} \frac{1 \cdot 3 \cdot \ldots \cdot (2n-1)}{n!} \left(\frac{x}{2}\right)^n = \sum_{n=0}^{\infty} \frac{(2n-1)!!}{(2n)!!}x^n,$$

converges $\forall x \in [-1, 1)$, *converges uniformly on* $[-1, a)$, $-1 < \forall a < 1$.

$$e^x = \sum_{n=0}^{\infty} \frac{x^n}{n!} = 1 + \frac{x}{1!} + \frac{x^2}{2!} + \frac{x^3}{3!} + \ldots,$$

converges $\forall x \in \mathbb{R}$, *converges uniformly on* $[-a, a]$, $\forall a > 0$.

$$b^x = \sum_{n=0}^{\infty} \frac{\ln^n b}{n!} x^n = 1 + \frac{\ln b}{1!}x + \frac{\ln^2 b}{2!}x^2 + \frac{\ln^3 b}{3!}x^3 + \ldots, \quad b > 0, b \neq 1,$$

converges $\forall x \in \mathbb{R}$, *converges uniformly on* $[-a, a]$, $\forall a > 0$.

$$e^{-x^2} = \sum_{n=0}^{\infty}(-1)^n \frac{x^{2n}}{n!} = 1 - \frac{x^2}{1!} + \frac{x^4}{2!} - \frac{x^6}{3!} + \ldots,$$

converges $\forall x \in \mathbb{R}$, *converges uniformly on* $[-a, a]$, $\forall a > 0$.

$$\cosh x = \sum_{n=0}^{\infty} \frac{x^{2n}}{(2n)!} = 1 + \frac{x^2}{2!} + \frac{x^4}{4!} + \ldots + \frac{x^{2n}}{(2n)!} + \ldots,$$

converges $\forall x \in \mathbb{R}$, *converges uniformly on* $[-a, a]$, $\forall a > 0$.

$$\sinh x = \sum_{n=0}^{\infty} \frac{x^{2n+1}}{(2n+1)!} = x + \frac{x^3}{3!} + \frac{x^5}{5!} + \ldots + \frac{x^{2n+1}}{(2n+1)!} + \ldots,$$

converges $\forall x \in \mathbb{R}$, *converges uniformly on* $[-a, a]$, $\forall a > 0$.

$$\arcsin x = \sum_{n=0}^{\infty} \frac{(2n-1)!!}{(2n)!!} \frac{x^{2n+1}}{2n+1} = x + \frac{1}{6}x^3 + \frac{3}{40}x^5 + \frac{5}{112}x^7 + \ldots,$$

$$converges\ \forall x \in [-1, 1], converges\ uniformly\ on\ [-1, 1].$$

$$\arccos x = \frac{\pi}{2} - \sum_{n=0}^{\infty} \frac{(2n-1)!!}{(2n)!!} \frac{x^{2n+1}}{2n+1} = \frac{\pi}{2} - x - \frac{1}{6}x^3 - \frac{3}{40}x^5 - \frac{5}{112}x^7 - \ldots,$$

$$converges\ \forall x \in [-1, 1], converges\ uniformly\ on\ [-1, 1].$$

$$\sin x = \sum_{n=0}^{\infty} (-1)^n \frac{x^{2n+1}}{(2n+1)!} = x - \frac{x^3}{3!} + \frac{x^5}{5!} - \frac{x^7}{7!} + \ldots,$$

$$converges\ \forall x \in \mathbb{R}, converges\ uniformly\ on\ [-a, a], \forall a > 0.$$

$$\cos x = \sum_{n=0}^{\infty} (-1)^n \frac{x^{2n}}{(2n)!} = 1 - \frac{x^2}{2!} + \frac{x^4}{4!} - \frac{x^6}{6!} + \ldots,$$

$$converges\ \forall x \in \mathbb{R}, converges\ uniformly\ on\ [-a, a], \forall a > 0.$$

$$\tan x = x + \frac{x^3}{3} + \frac{2x^5}{15} + \frac{17x^7}{315} + \ldots,$$

$$converges\ \forall x \in (-\frac{\pi}{2}, \frac{\pi}{2}), converges\ uniformly\ on\ [-a, a], 0 < \forall a < \frac{\pi}{2}.$$

$$\cot x - \frac{1}{x} = -\frac{1}{3}x - \frac{1}{45}x^3 - \frac{2}{945}x^5 - \frac{1}{4725}x^7 + \ldots,$$

$$converges\ \forall x \in (-\frac{\pi}{2}, \frac{\pi}{2}), converges\ uniformly\ on\ [-a, a], 0 < \forall a < \frac{\pi}{2}.$$

6   Applications of Taylor Series

> After the Fourier series, other series have entered the domain of
> analysis; they entered by the same door; they have been imagined
> in view of applications.
> Jules Henri Poincaré, 1905

6.1   Approximation of Functions

In this part, the application of power series is quite similar to the use of the Taylor formula. An additional advantage which the theory of series offers to us is the possibility to employ alternative methods of evaluation of the approximation errors: in addition to the estimate of the remainder in the Taylor formula, we can apply the very simple evaluation for the remainder of alternative series and use simple majorant series in general cases.

Example 1a Approximate the function e^{-x^2} in a neighborhood of the origin.
First, represent e^{-x^2} in the Taylor series using the series of e^x:

$$e^{-x^2} = \sum_{n=0}^{\infty} (-1)^n \frac{x^{2n}}{n!}, \ \forall x \in \mathbb{R}.$$

Like the series of e^x, the last series converges on $\mathbb{R}$ and converges uniformly on any closed interval. Since the series is alternating, its remainder can be estimated by the first disregarded term:

$$|r_n(x)| \leq \frac{x^{2(n+1)}}{(n+1)!}, \ \forall x \in \mathbb{R}.$$

If we are interested in finding the approximation on the interval $[-R, R]$ with a required accuracy, say 10^{-3}, then we have to calculate the partial sum of such index n that $\max_{[-R,R]} \frac{x^{2(n+1)}}{(n+1)!} < 10^{-3}$ (the uniform convergence of the series on $[-R, R]$ guarantees that such n exists for any required accuracy). Notice that the function $x^{2(n+1)}$ is even, positive and increasing on $[0, R]$, which ensures that it attains its maximum at the endpoint $x = R$. Hence, we need to find such n that $\frac{R^{2(n+1)}}{(n+1)!} < 10^{-3}$. This inequality does not have a general analytical solution with respect to n, but specifying any R, we can find the required value of n by trial and error. For example, if $R = 1$, then $(n + 1)! > 10^3$ and $n \geq 6$. In the case $R = 2$, we have $\frac{(n+1)!}{2^{2(n+1)}} > 10^3$ and $n \geq 14$. Notice that the choice of $n = 6$ and $n = 14$ in the series of e^{-x^2} corresponds to the Taylor polynomials T_{12} and T_{28}, of the 12-th and 28-th degrees, respectively.

The results of the computations, shown in Figs. 5.1 and 5.2, confirm the theoretical evaluations: for $n = 14$ the Taylor polynomial T_{28} differs from e^{-x^2} by less than $7 \cdot 10^{-4}$ on the entire interval $[-2, 2]$ (the dashed line), while for $n = 13$ the difference is greater than 10^{-3} near the endpoints (not shown in the picture); for $n = 6$ the difference between T_{12} and e^{-x^2} is less than $9 \cdot 10^{-4}$ on the entire interval $[-1, 1]$ (the dotted-dashed line), but for $n = 5$ the difference is still greater than 10^{-3} in a neighborhood of the endpoints (not shown in the picture).

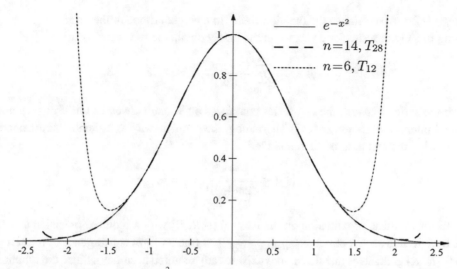

Fig. 5.1 Example 1a: function e^{-x^2} and its approximations

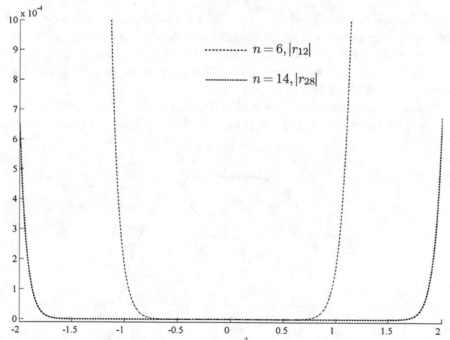

Fig. 5.2 Example 1a: errors of approximation of e^{-x^2}

Example 2a Approximate the function $\cos x^2$ in a neighborhood of the origin.
Using the Taylor series for $\cos x$, we immediately obtain the series for $\cos x^2$:

$$\cos x^2 = \sum_{n=0}^{\infty} \frac{(-1)^n x^{4n}}{(2n)!}, \ \forall x \in \mathbb{R}.$$

Like the series of $\cos x$, the last series converges on $\mathbb{R}$ and converges uniformly on any closed interval. Since the series is alternating, the simplest way to estimate its remainder is by using the first term of the remainder:

$$|r_n(x)| \le \frac{x^{4n+4}}{(2n+2)!}, \ \forall x \in \mathbb{R}.$$

To obtain the approximation on the interval $[-R, R]$ with a required precision $\varepsilon > 0$, we have to find such n that $\max_{[-R,R]} \frac{x^{4n+4}}{(2n+2)!} < \varepsilon$ (the uniform convergence of the series on $[-R, R]$ guarantees that this n does exist for any required accuracy). Since the function x^{4n+4} is even, positive and increasing on $[0, R]$, it achieves its maximum at $x = R$. Therefore, we need to find n such that $\frac{R^{4n+4}}{(2n+2)!} < \varepsilon$. Choosing, for instance, $R = \frac{\pi}{2}$ and $\varepsilon = 10^{-3}$, we have $(2n+2)! \cdot \left(\frac{2}{\pi}\right)^{4n+4} > 10^3$ and this inequality holds for $n \ge 6$. The accuracy of approximation improves fast with growth of n: for instance, to elevate the accuracy up to $\varepsilon = 10^{-5}$, it is sufficient to take $n \ge 7$. Notice that the choice of $n = 6$ and $n = 7$ for the series of $\cos x^2$ corresponds to the use of the Taylor polynomials T_{24} and T_{28} of the degrees 24 and 28, respectively.

The results of computations presented in Figs. 5.3 and 5.4 reveal that the obtained theoretical evaluations are slightly exaggerated: the Taylor polynomial of the 20-th degree

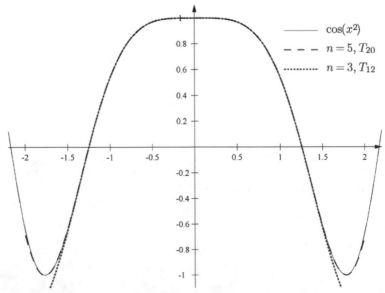

Fig. 5.3 Example 2a: function $\cos x^2$ and its approximation

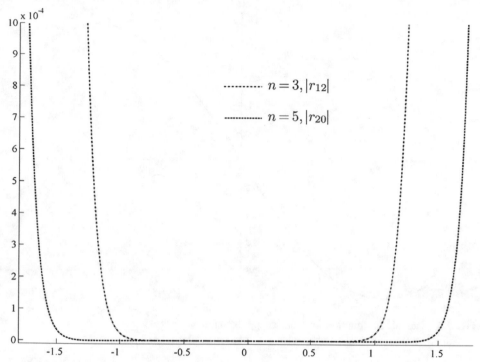

Fig. 5.4 Example 2a: errors of approximation of $\cos x^2$

that corresponds to $n = 5$ (the dashed line) already guarantees the accuracy of 10^{-3} on the interval $[-\frac{\pi}{2}, \frac{\pi}{2}]$, while the Taylor polynomial of 12-th degree ($n = 3$, the dotted-dashed line) has the error of approximation less than 10^{-3} on the interval $[-\frac{\pi}{3}, \frac{\pi}{3}]$.

Example 3a Approximate the function $\sin x$ on the interval $[-\frac{\pi}{4}, \frac{\pi}{4}]$.
The Taylor series of $\sin x$ has the form

$$\sin x = \sum_{n=0}^{\infty} (-1)^n \frac{x^{2n+1}}{(2n+1)!} = x - \frac{x^3}{3!} + \frac{x^5}{5!} - \frac{x^7}{7!} + \dots, \; \forall x \in \mathbb{R}.$$

This series converges on $\mathbb{R}$ and converges uniformly on any closed interval. Since the series is alternating, its remainder hasv a simple evaluation through the first disregarded

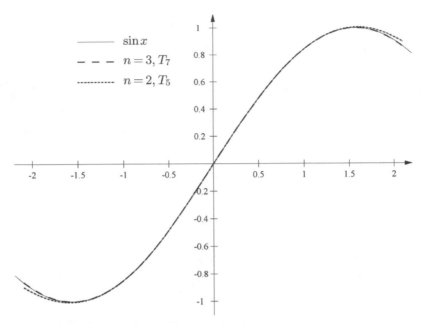

Fig. 5.5 Example 3a: function sin x and its approximations

term:

$$|r_n(x)| \le \frac{|x|^{2n+3}}{(2n+3)!}, \forall x \in \mathbb{R}.$$

To find the approximation on the interval $[-\frac{\pi}{4}, \frac{\pi}{4}]$ within a degree of accuracy $\varepsilon > 0$, we have to determine n that satisfies the inequality $\max_{[-\frac{\pi}{4}, \frac{\pi}{4}]} \frac{|x|^{2n+3}}{(2n+3)!} < \varepsilon$ (the uniform convergence of the series ensures that such n exists for any required degree of accuracy). Since $\frac{\pi}{4} < 0.8$, it is sufficient to solve the inequality $\frac{0.8^{2n+3}}{(2n+3)!} < \varepsilon$. If $\varepsilon = 10^{-6}$, then the inequality $(2n+3)! \left(\frac{5}{4}\right)^{2n+3} > 10^6$ holds for $n \ge 3$, that is, using the Taylor polynomial of the seventh degree

$$\sin x \approx T_7(x) = x - \frac{x^3}{3!} + \frac{x^5}{5!} - \frac{x^7}{7!}$$

we guarantee that the approximation error is less than 10^{-6} at every point $x \in [-\frac{\pi}{4}, \frac{\pi}{4}]$.

The obtained theoretical evaluation corresponds well to the numerical results shown in Figs. 5.5 and 5.6: the Taylor polynomial of the seventh degree T_7 (the dashed line) differs from the original function by less than $0.5 \cdot 10^{-6}$ over the interval $[-\frac{\pi}{4}, \frac{\pi}{4}]$, while the polynomial of the fifth degree T_5 (the dotted-dashed line) keeps the same accuracy only within the interval $[-\frac{\pi}{7}, \frac{\pi}{7}]$.

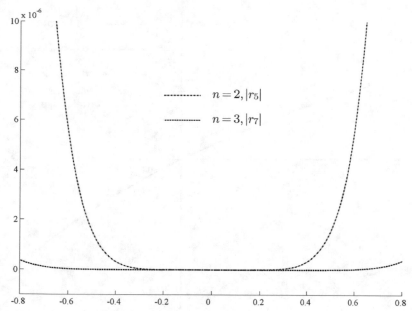

Fig. 5.6 Example 3a: errors of approximation of $\sin x$

Example 4a Approximate the function $\ln(1 - x)$ on the interval $[-0.5, 0.3]$.
The Taylor series for $\ln(1 - x)$ has the form

$$\ln(1 - x) = \sum_{n=1}^{\infty} \frac{x^n}{n} = x - \frac{x^2}{2} - \frac{x^3}{3} - \frac{x^4}{4} + \dots, \ \forall x \in [-1, 1).$$

This series converges on $[-1, 1)$ and converges uniformly on any interval $[-1, b)$, $-1 < b < 1$, in particular, on the given interval $[-0.5, 0.3]$. Since the series is negative, a better way to estimate its remainder is by using the Lagrange form in the Taylor formula:

$$r_n(x) = \frac{f^{(n+1)}(c)}{(n + 1)!} x^{n+1}, \ \forall x \in [-0.5, 0.3],$$

where $|c| < |x| < 0.5$. The general formula for derivatives of $\ln(1 - x)$ is $f^{(n+1)}(x) = -\frac{n!}{(1-x)^{n+1}}$, $\forall n \in \mathbb{N}$. Therefore,

$$r_n(x) = -\frac{1}{(1 - c)^{n+1}} \frac{1}{n + 1} x^{n+1}, \ \forall x \in [-0.5, 0.3].$$

If $x \geq 0$, then $c \geq 0$ and, in this case, $\max_{[0,0.3]} \frac{1}{(1-c)^{n+1}} = \frac{1}{(1-0.3)^{n+1}}$ and $\max_{[0,0.3]} x^{n+1} = 0.3^{n+1}$. Consequently,

$$|r_n(x)| \leq \frac{1}{(0.7)^{n+1}} \frac{1}{n + 1} 0.3^{n+1} = \frac{1}{n + 1} \left(\frac{3}{7}\right)^{n+1}, \ \forall x \in [0, 0.3].$$

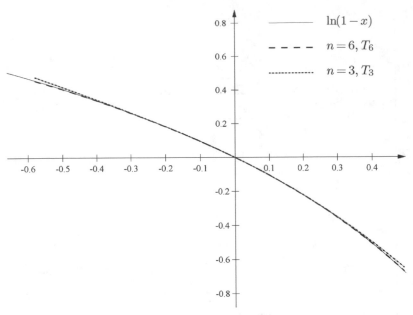

Fig. 5.7 Example 4a: function $\ln(1-x)$ and its approximations

On the other hand, if $x \leq 0$, then $c \leq 0$, and we get $\max_{[-0.5,0]} \frac{1}{(1-c)^{n+1}} = \frac{1}{(1-0)^{n+1}}$ and $\max_{[-0.5,0]} |x|^{n+1} = 0.5^{n+1}$. Therefore,

$$|r_n(x)| \leq \frac{1}{1^{n+1}} \frac{1}{n+1} 0.5^{n+1} = \frac{1}{n+1} \left(\frac{1}{2}\right)^{n+1}, \forall x \in [-0.5, 0].$$

Thus, on the entire interval $[-0.5, 0.3]$ we have the evaluation

$$|r_n(x)| \leq \frac{1}{n+1} \left(\frac{1}{2}\right)^{n+1}, \forall x \in [-0.5, 0.3].$$

It means that the condition $|r_n(x)| < \varepsilon, \forall x \in [-0.5, 0.3]$ is guaranteed by the inequality $\frac{1}{n+1} \left(\frac{1}{2}\right)^{n+1} < \varepsilon$.

For instance, choosing $n = 4$ for the approximation of $\ln(1-x)$:

$$\ln(1-x) \approx -x - \frac{x^2}{2} - \frac{x^3}{3} - \frac{x^4}{4}$$

we have the error less than $\frac{1}{4+1} \left(\frac{1}{2}\right)^{4+1} = \frac{1}{160}$ for $\forall x \in [-0.5, 0.3]$. If, to the contrary, a certain accuracy is required, for instance, $\varepsilon = 10^{-3}$, then solving the inequality $(n+1)2^{n+1} > 10^3$ we obtain the values $n \geq 7$ which guarantee the desired level of approximation.

The results of computations presented in Figs. 5.7 and 5.8 show that in this case the theoretical analysis overestimates the required degree of the polynomial: the Taylor

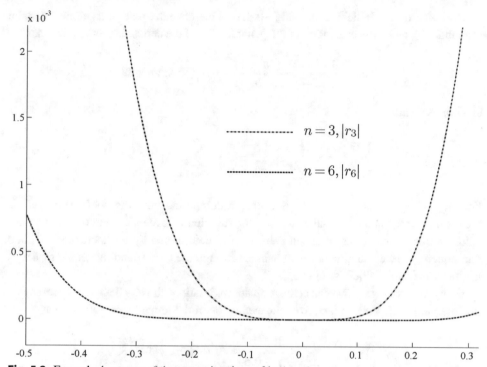

Fig. 5.8 Example 4a: errors of the approximations of $\ln(1 - x)$

polynomial T_6 (the dashed line) already has the error of approximation less than $0.6 \cdot 10^{-3}$ over the interval $[-0.5, 0.3]$.

Example 5a Approximate the function $\arcsin x$ on the interval $[-0.8, 0.8]$.
The Taylor series

$$\arcsin x = \sum_{n=0}^{\infty} \frac{1 \cdot 3 \cdot \ldots \cdot (2n - 1)}{2 \cdot 4 \cdot \ldots \cdot 2n} \frac{x^{2n+1}}{2n + 1}$$

converges uniformly on $[-1, 1]$—the entire domain of $\arcsin x$. This series is positive for $x > 0$ and negative for $x < 0$. Therefore, a simple formula of evaluation of the remainder of alternating series is not applicable in this case. Moreover, an evaluation of the remainder by any form of the Taylor theorem is unavailable for $\arcsin x$ since we do not know a general form of n-th derivative (recall that the power series for $\arcsin x$ was derived using the integration of a specific binomial series).

However, on the indicated interval $[-0.8, 0.8]$ we have the possibility to evaluate the remainder by using the evaluation of the general term of the series. Indeed, notice that

$$c_n \equiv \frac{1 \cdot 3 \cdot \ldots \cdot (2n-1)}{2 \cdot 4 \cdot \ldots \cdot 2n} \frac{1}{2n+1} < 1, \forall n \in \mathbb{N}$$

and consequently,

$$|r_n(x)| = \sum_{k=n+1}^{\infty} \frac{1 \cdot 3 \cdot \ldots \cdot (2k-1)}{2 \cdot 4 \cdot \ldots \cdot 2k} \frac{|x|^{2k+1}}{2k+1} < \sum_{k=n+1}^{\infty} |x|^{2k+1} = \frac{|x|^{2n+3}}{1-x^2}, \forall n \in \mathbb{N}.$$

For any $|x| < 1$, the term x^{2n+1} is principal comparing with the coefficient c_n (this means that x^{2n+1} is much smaller than c_n for sufficiently large n). Therefore, the values of the terms of the series of $\arcsin x$ are mainly determined by the factors x^{2n+1}, and consequently, the evaluation of the remainder through the corresponding geometric series is quite reasonable, albeit excessive.

Using this evaluation, we can achieve an approximation with any degree of accuracy on the interval $[-0.8, 0.8]$. In fact, from the evaluation of the remainder it follows that

$$|r_n(x)| < \frac{|x|^{2n+3}}{1-x^2} \le \frac{0.8^{2n+3}}{1-0.64}, \forall n \in \mathbb{N}, \forall x \in [-0.8, 0.8].$$

For example, choosing $\varepsilon = 10^{-3}$, we obtain the inequality $1.25^{2n+3} > \frac{1}{0.36} 10^3$, which is satisfied for $n \ge 17$.

Naturally, when we shrink the interval around the origin, the accuracy increases significantly. For instance, if we consider the interval $[-0.4, 0.4]$ instead of $[-0.8, 0.8]$, then we obtain the following evaluation of the remainder:

$$|r_n(x)| < \frac{|x|^{2n+3}}{1-x^2} \le \frac{0.4^{2n+3}}{1-0.16}, \forall n \in \mathbb{N}, \forall x \in [-0.4, 0.4].$$

Therefore, for the same admissible error $\varepsilon = 10^{-3}$, in this case we have the inequality $2.5^{2n+3} > \frac{1}{0.84} 10^3$, which is satisfied for $n \ge 3$.

With a more intensive effort, we can improve the obtained theoretical evaluation. First, we show by induction that

$$d_n \equiv \frac{1 \cdot 3 \cdot \ldots \cdot (2n-1)}{2 \cdot 4 \cdot \ldots \cdot 2n} < \frac{1}{\sqrt{2n+1}}, \forall n \in \mathbb{N}.$$

In fact, for $n = 1$ we have $d_1 = \frac{1}{2} < \frac{1}{\sqrt{3}}$. Suppose now that the inequality is true for some n and show its validity for $n + 1$:

$$d_{n+1} = \frac{1 \cdot 3 \cdot \ldots \cdot (2n-1) \cdot (2n+1)}{2 \cdot 4 \cdot \ldots \cdot 2n \cdot (2n+2)} < \frac{1}{\sqrt{2n+1}} \cdot \frac{2n+1}{2n+2}$$

$$= \frac{1}{\sqrt{2n+3}} \cdot \frac{\sqrt{2n+3}\sqrt{2n+1}}{2n+2} = \frac{1}{\sqrt{2n+3}} \cdot \frac{\sqrt{4n^2+8n+3}}{\sqrt{4n^2+8n+4}} < \frac{1}{\sqrt{2n+3}}.$$

Therefore, the inequality is proved. We can employ it in the evaluation of the remainder:

$$|r_n(x)| = \sum_{k=n+1}^{\infty} d_k \frac{|x|^{2k+1}}{2k+1} < \sum_{k=n+1}^{\infty} \frac{|x|^{2k+1}}{(2k+1)^{3/2}} < \frac{1}{(2n+3)^{3/2}} \sum_{k=n+1}^{\infty} |x|^{2k+1}$$

$$= \frac{1}{(2n+3)^{3/2}} \frac{|x|^{2n+3}}{1-x^2}, \forall n \in \mathbb{N}.$$

From this estimate, on the interval $[-0.8, 0.8]$ we get

$$|r_n(x)| < \frac{1}{(2n+3)^{3/2}} \frac{|x|^{2n+3}}{1-x^2} \leq \frac{1}{(2n+3)^{3/2}} \frac{0.8^{2n+3}}{0.36}, \forall n \in \mathbb{N}, \forall x \in [-0.8, 0.8]$$

and on the interval $[-0.4, 0.4]$ we have

$$|r_n(x)| < \frac{1}{(2n+3)^{3/2}} \frac{|x|^{2n+3}}{1-x^2} \leq \frac{1}{(2n+3)^{3/2}} \frac{0.4^{2n+3}}{0.84}, \forall n \in \mathbb{N}, \forall x \in [-0.4, 0.4].$$

Choosing again $\varepsilon = 10^{-3}$ we find $n \geq 7$ for $[-0.8, 0.8]$ and $n \geq 2$ for $[-0.4, 0.4]$. Notice that $n = 2$ and $n = 7$ in the series of $\arcsin x$ correspond to the Taylor polynomials T_5 and T_{15}, of degrees 5 and 15, respectively.

In this case, the second theoretical evaluation, which makes possible to diminish essentially the number of terms required to guarantee a desired accuracy, is very close to the computations shown in Figs. 5.9 and 5.10: the Taylor polynomial T_5 ($n = 2$, the dashed line) approximates $\arcsin x$ over the interval $[-0.4, 0.4]$ with the errors smaller than 10^{-4}, while T_3 has errors greater than 10^{-3} at the endpoints (not shown in the picture); the polynomial T_{15} ($n = 7$, the dotted-dashed line) approximates $\arcsin x$ over the interval $[-0.8, 0.8]$ with an accuracy of $6 \cdot 10^{-4}$, but T_{13} differs from $\arcsin x$ by an amount greater than 10^{-3} near the boundary points (not shown in the picture).

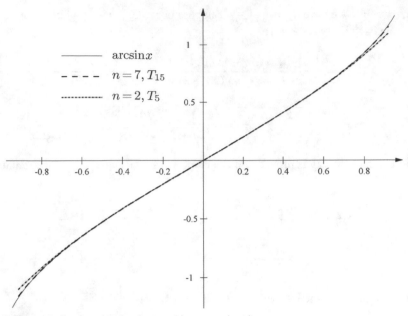

Fig. 5.9 Example 5a: function arcsin x and its approximations

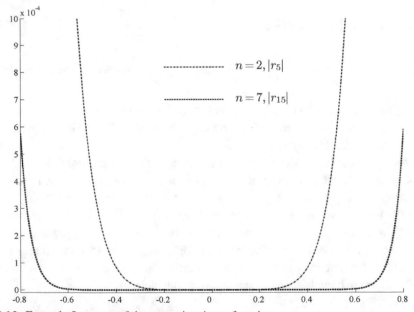

Fig. 5.10 Example 5a: errors of the approximations of arcsin x

Example 6a Find the error of approximation of the function $\tan x$ by the Taylor polynomial of the fifth degree on the interval $[-\frac{\pi}{6}, \frac{\pi}{6}]$.

Although theoretically $\tan x$ can be developed in the Taylor series convergent on $(-\frac{\pi}{2}, \frac{\pi}{2})$, but the expression for n-th derivative of $\tan x$ is unknown, and a general formula for the coefficients of its Taylor series is unavailable. (The formulas of these coefficients expressed in the terms of the Bernoulli numbers, which can be found in some sources, have only a theoretical significance, since the values of these coefficients cannot be found in a general form, although they are tabulated for many indices.) Under these conditions, the evaluation of the remainder in a general form, using the properties of the series or the form of the remainder in the Taylor formula, are not applicable. Even so, for a given order (not elevated) of the approximating Taylor polynomial, we can find a specific derivative of $\tan x$ and perform the evaluation in a neighborhood of a central point. This procedure is applied in the solutions of this example.

Recall that for $f(x) = \tan x$ the first seven derivatives were already calculated:

$$f'(x) = \frac{1}{\cos^2 x}, \; f''(x) = \frac{2 \sin x}{\cos^3 x},$$

$$f'''(x) = \frac{2 \cos^2 x + 6 \sin^2 x}{\cos^4 x}, \; f^{(4)}(x) = 8 \sin x \cdot \frac{3 - \cos^2 x}{\cos^5 x},$$

$$f^{(5)}(x) = \frac{16 \cos^4 x + 120 \sin^2 x}{\cos^6 x}, \; f^{(6)}(x) = 16 \sin x \frac{2 \cos^4 x - 30 \cos^2 x + 45}{\cos^7 x},$$

$$f^{(7)}(x) = \frac{16 \cos^2 x (2 \cos^4 x - 30 \cos^2 x + 45) + 16 \sin^2 x (6 \cos^4 x - 150 \cos^2 x + 315)}{\cos^8 x}.$$

Since $f^{(7)}(x) > 0$, it implies that $f^{(6)}(x)$ is a strictly decreasing function on $[-\frac{\pi}{6}, \frac{\pi}{6}]$. Taking into account that $f^{(6)}(x)$ is also an odd function, we conclude that $|f^{(6)}(x)|$ attains the maximum values at the endpoints of the interval $[-\frac{\pi}{6}, \frac{\pi}{6}]$. Therefore,

$$|f^{(6)}(x)| \le \max_{[-\frac{\pi}{6}, \frac{\pi}{6}]} |f^{(6)}(x)| = f^{(6)}(\frac{\pi}{6}) \approx \frac{189}{0.365} \approx 518.$$

Then, in the approximation

$$\tan x \approx T_5(x) = x + \frac{x^3}{3} + \frac{2x^5}{15}$$

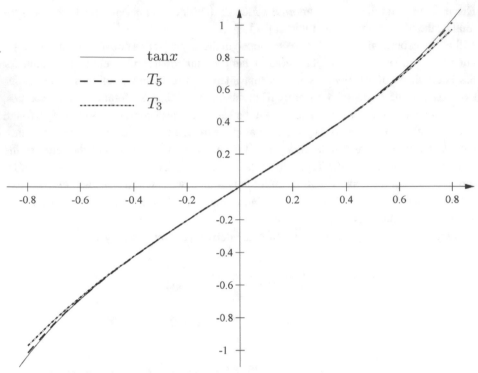

Fig. 5.11 Example 6a: function $\tan x$ and its approximations

the remainder $r_5(x)$ satisfies the evaluation

$$|r_5(x)| = \frac{|f^{(6)}(c)|}{6!}|x|^6 \leq \frac{518}{6!}\left(\frac{\pi}{6}\right)^6 \approx 0.72 \cdot 0.021 \approx 0.015, \forall x \in [-\frac{\pi}{6}, \frac{\pi}{6}].$$

In this example, the theoretical evaluation of the approximation error is visibly overestimated as it is illustrated by Figs. 5.11 and 5.12: in practice, the Taylor polynomial of the fifth degree T_5 (the dashed line) differs from the original function by an amount less than $7 \cdot 10^{-4}$ over the entire interval $[-\frac{\pi}{6}, \frac{\pi}{6}]$, and the polynomial T_3 (the dotted-dashed line) has the approximation error less than $6 \cdot 10^{-3}$ over the same interval.

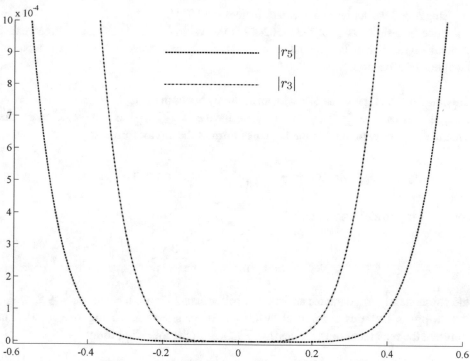

Fig. 5.12 Example 6a: errors of approximation of $\tan x$

6.2 Numerical Approximations

Example 1m Find the value of $\sin 3^o$ with an error less than 10^{-8}.
Using the series $\sin x = \sum_{n=0}^{\infty}(-1)^n \frac{x^{2n+1}}{(2n+1)!}$, $\forall x \in \mathbb{R}$, we can write

$$\sin 3^o = \sin \frac{\pi}{60} = \sum_{n=0}^{\infty} \frac{(-1)^n}{(2n+1)!} \left(\frac{\pi}{60}\right)^{2n+1}.$$

Since this series is alternating, its remainder satisfies a simple evaluation:

$$|r_n| \leq \frac{1}{(2n+3)!} \left(\frac{\pi}{60}\right)^{2n+3}.$$

To ensure the required accuracy, we have to solve the inequality $(2n+3)! \left(\frac{60}{\pi}\right)^{2n+3} > 10^8$, which is satisfied for $n \geq 1$. Then,

$$\sin 3^o \approx \frac{\pi}{60} - \frac{1}{3!}\left(\frac{\pi}{60}\right)^3 \approx 0.052335953$$

is the approximate value with eight exact decimal digits.

Using $n \geq 2$ we achieve an accuracy better than 10^{-12}.

Notice that $\sin 3^o \approx 0.0523359562429$ is the value of $\sin 3^o$ with the thirteen exact decimal digits (it can be computed on calculator or notebook or found in a table of trigonometric functions).

Example 2m Find the value of e with an accuracy better than 10^{-2}.

Using the series $e^x = \sum_{n=0}^{\infty} \frac{x^n}{n!}, \forall x \in \mathbb{R}$, we have $e = \sum_{n=0}^{\infty} \frac{1}{n!}$. Since $(e^x)^{(n)} = e^x$, the remainder can be expressed in the Lagrange form of the Taylor formula:

$$|r_n(x)| = \frac{e^c}{(n+1)!} x^{n+1}, 0 < c < x \leq 1.$$

For $x = 1$ it provides the evaluation

$$|r_n| < \frac{e^1}{(n+1)!} 1^{n+1} < \frac{3}{(n+1)!},$$

and consequently, to guarantee an accuracy better than 10^{-2} we have to use $n \geq 5$.

It happens that this theoretical evaluation is exaggerated. To improve it, we can represent the remainder of the series $e = \sum_{n=0}^{\infty} \frac{1}{n!}$ in the following form:

$$|r_n| = \sum_{k=n+1}^{\infty} \frac{1}{k!} = \frac{1}{(n+1)!} + \frac{1}{(n+2)!} + \dots = \frac{1}{(n+1)!} \left(1 + \frac{1}{n+2} + \frac{1}{(n+2)(n+3)} + \dots \right)$$

$$< \frac{1}{(n+1)!} \left(1 + \frac{1}{n+2} + \frac{1}{(n+2)^2} + \dots \right) = \frac{1}{(n+1)!} \frac{1}{1 - \frac{1}{n+2}} = \frac{n+2}{(n+1)!(n+1)}.$$

Here, the factor $\frac{n+2}{n+1}$ enters instead of 3 in the first evaluation, that is, for sufficiently large n the second evaluation is almost three times better than the first one. If $\varepsilon = 10^{-2}$, the second evaluation allows the use of the indices $n \geq 4$. Thus, we obtain that

$$e \approx 1 + \frac{1}{1!} + \frac{1}{2!} + \frac{1}{3!} + \frac{1}{4!} \approx 2.7083$$

is the value that approximates e with the desired accuracy.

Notice that $e \approx 2.71828$ with the six exact digits (calculated on a notebook or calculator or found in a table).

Example 3m Find the value of $\ln 2$ with an error less than 10^{-5}.

Using the series $\ln(1 + x) = \sum_{n=1}^{\infty}(-1)^{n-1}\frac{x^n}{n}$, $\forall x \in (-1, 1]$, we obtain at $x = 1$:

$$\ln 2 = \sum_{n=1}^{\infty}(-1)^{n-1}\frac{1}{n}.$$

The last series is alternating, and consequently, we obtain a simple estimate for its remainder

$$|r_n| \leq \frac{1}{n + 1}.$$

To achieve a required precision, we have to use $n \geq 10^5$ terms of the numerical series, which demands a large volume of computations.

To obtain a desired approximation with much faster calculations, we can use the following transformation. Recall that $\ln(1 - x) = -\sum_{n=1}^{\infty}\frac{x^n}{n}$, $\forall x \in [-1, 1)$ and combine the two formulas to obtain

$$\ln\frac{1 + x}{1 - x} = \ln(1 + x) - \ln(1 - x) = \sum_{n=1}^{\infty}(-1)^{n-1}\frac{x^n}{n} + \sum_{n=1}^{\infty}\frac{x^n}{n} = 2\sum_{n=0}^{\infty}\frac{x^{2n+1}}{2n + 1}, \forall x \in (-1, 1).$$

The value of $\ln 2$ corresponds to $x = \frac{1}{3}$ in the last formula, that is,

$$\ln 2 = 2\sum_{n=0}^{\infty}\frac{1}{(2n + 1)3^{2n+1}}.$$

Although this series is not alternating, the evaluation of its remainder still can be made without any complication:

$$|r_n| = 2\sum_{k=n+1}^{\infty}\frac{1}{(2k + 1)3^{2k+1}} = 2\left(\frac{1}{(2n + 3)3^{2n+3}} + \frac{1}{(2n + 5)3^{2n+5}} + \ldots\right)$$

$$< \frac{2}{(2n + 3)3^{2n+3}}\left(1 + \frac{1}{3^2} + \frac{1}{3^4} + \ldots\right) = \frac{2}{(2n + 3)3^{2n+3}}\frac{1}{1 - 1/9} = \frac{1}{4(2n + 3)3^{2n+1}}.$$

To guarantee an accuracy of 10^{-5}, we have to solve the inequality $4(2n+3)3^{2n+1} > 10^5$, which holds for $n \geq 4$. Thus,

$$\ln 2 \approx 2 \left(\frac{1}{3} + \frac{1}{3 \cdot 3^3} + \frac{1}{5 \cdot 3^5} + \frac{1}{7 \cdot 3^7} + \frac{1}{9 \cdot 3^9} \right) \approx 0.693144.$$

Notice that the value of $\ln 2$ with six exact leading digits is $\ln 2 \approx 0.6931472$.

Example 4m Find the value of $\sqrt{3}$ with an error less than 10^{-3}.
The binomial series related to the root operations

$$(1+x)^p = \sum_{n=0}^{\infty} \frac{p\,(p-1)\cdot \ldots \cdot (p-n+1)}{n!} x^n$$

$$= 1 + \frac{p}{1!}x + \frac{p\,(p-1)}{2!}x^2 + \frac{p(p-1)(p-2)}{3!}x^3 + \ldots, \forall p \notin \mathbb{N},$$

converges only on $(-1, 1)$. Besides, the form of the series shows that the remainder is significantly smaller when x is located near 0. For these reasons, we start with the following transformation of the given root:

$$\sqrt{3} = \sqrt{\frac{48}{16}\frac{49}{49}} = \frac{7}{4}\sqrt{\frac{48}{49}} = \frac{7}{4}\frac{1}{\sqrt{1+\frac{1}{48}}} = 1.75 \left(1 + \frac{1}{48}\right)^{-\frac{1}{2}}.$$

The corresponding binomial series is

$$\frac{1}{\sqrt{1+x}} = (1+x)^{-\frac{1}{2}} = \sum_{n=0}^{\infty}(-1)^n \frac{1 \cdot 3 \cdot \ldots \cdot (2n-1)}{n!} \left(\frac{x}{2}\right)^n$$

$$= 1 - \frac{x}{2} + \frac{1 \cdot 3}{2!}\left(\frac{x}{2}\right)^2 + \ldots, \forall x \in (-1, 1).$$

In the case $x = \frac{1}{48}$ it takes the form

$$\frac{1}{\sqrt{1+\frac{1}{48}}} = \sum_{n=0}^{\infty}(-1)^n \frac{1 \cdot 3 \cdot \ldots \cdot (2n-1)}{n!} \left(\frac{1}{96}\right)^n = 1 - \frac{1}{96} + \frac{3}{2}\left(\frac{1}{96}\right)^2 + \ldots.$$

Since the last series is alternating and its third summand is already less than $0.3 \cdot 10^{-3}$, it is sufficient to take the first two terms to guarantee the approximation of $\dfrac{1}{\sqrt{1+\frac{1}{48}}}$ with an

accuracy better than 10^{-3}. To return to $\sqrt{3}$ we have to multiply the result of the obtained approximation by 1.75, that still keep the approximation error below 10^{-3}. Therefore, we have

$$\sqrt{3} \approx 1.75 \cdot \left(1 - \frac{1}{96}\right) \approx 1.7318.$$

Notice that $\sqrt{3} \approx 1.73205$ with six exact digits.

Example 5m Find an approximate value of π accurate to within 10^{-4}.
To solve this task, it is natural to employ the series for one of the inverse trigonometric functions. For example, using the series $\arctan x = \sum_{n=0}^{\infty}(-1)^n \frac{x^{2n+1}}{2n+1}, \forall x \in [-1, 1]$, at the point $x = 1$, we find the following simple representation

$$\frac{\pi}{4} = \arctan 1 = \sum_{n=0}^{\infty}(-1)^n \frac{1}{2n+1}.$$

However, this series has a relatively slow convergence, and to obtain the desired accuracy of 10^{-4} for the approximation of π, it is necessary to satisfy the inequality

$$|r_n| \leq \frac{1}{2n+1} < \frac{1}{4} \cdot 10^{-4}$$

for the representation of $\frac{\pi}{4}$. This means that we have to use a very large number $n \geq 2 \cdot 10^4$ of the terms in the series for $\arctan 1$.

Evidently, the problem of a slow convergence is caused by the fact that the chosen point $x = 1$ is located at the endpoint of the interval of convergence. The situation is radically improved if we take $x = \frac{1}{\sqrt{3}}$ that generates the following representation

$$\frac{\pi}{6} = \arctan \frac{1}{\sqrt{3}} = \sum_{n=0}^{\infty}(-1)^n \frac{1}{(\sqrt{3})^{2n+1}} \frac{1}{2n+1} = \frac{1}{\sqrt{3}} \sum_{n=0}^{\infty}(-1)^n \frac{1}{3^n(2n+1)}.$$

The found series is again alternating, and consequently, its remainder satisfies the following simple evaluation:

$$|r_n| \leq \frac{1}{3^{n+1}(2n+3)}.$$

To ensure the required accuracy, we have to solve the inequality $3^{n+1}(2n+3) > \frac{6}{\sqrt{3}} \cdot 10^4$, which is satisfied for $n \geq 6$. Then

$$\pi \approx \frac{6}{\sqrt{3}}\left(1 - \frac{1}{3 \cdot 3} + \frac{1}{3^2 \cdot 5} - \frac{1}{3^3 \cdot 7} + \frac{1}{3^4 \cdot 9} - \frac{1}{3^5 \cdot 11} + \frac{1}{3^6 \cdot 13}\right) \approx 3.14167$$

is the value of π approximated within the error bound of 10^{-4}. Notice that the value of π with the six exact digits is $\pi \approx 3.14159$.

6.3   Finding Sums of Series of Functions

Example 1f Find the sum of the series $\sum_{n=1}^{\infty} n x^n$.
The sum of this series was already calculated in the introductory section of this chapter using a rudimentary technique. Let us derive the same result by applying the analytical properties of power series. The given series has an intimate connection with the geometric series $\sum_{n=0}^{\infty} x^n$: if we substitute x^{n-1} for x^n in the original series, the obtained series would be the result of differentiation of the geometric series. Then, bringing one power of x outside the infinite sum, we have

$$\sum_{n=1}^{\infty} n x^n = x \sum_{n=1}^{\infty} n x^{n-1} = x \sum_{n=1}^{\infty} (x^n)' = x \left(\sum_{n=0}^{\infty} x^n \right)' = x \left(\frac{1}{1-x} \right)' = \frac{x}{(1-x)^2}.$$

The uniform convergence of the geometric series $\sum_{n=0}^{\infty} x^n$ on any interval $[-a, a], 0 < a < 1$ justifies the obtained result for any $x \in (-1, 1)$. Of course, we have arrived at the same sum as in the introductory section.

Example 2f Find the sum of the series $\sum_{n=0}^{\infty} \frac{x^n}{n+1}$.
This series has a direct connection with the geometric series: if we multiply its general term by x, we obtain the integral of the general term of the geometric series: $\frac{x^{n+1}}{n+1} = \int_0^x t^n dt$. Therefore, we get

$$\sum_{n=0}^{\infty} \frac{x^n}{n+1} = \frac{1}{x} \sum_{n=0}^{\infty} \frac{x^{n+1}}{n+1} = \frac{1}{x} \sum_{n=0}^{\infty} \int_0^x t^n dt$$

$$= \frac{1}{x} \int_0^x \left(\sum_{n=0}^{\infty} t^n \right) dt = \frac{1}{x} \int_0^x \frac{1}{1-t} dt = -\frac{1}{x} \ln(1-x).$$

The uniform convergence of the geometric series $\sum_{n=0}^{\infty} x^n$ on any interval $[-a, a], 0 < a < 1$ justifies the obtained result for any $x \in (-1, 1), x \neq 0$.

Additionally, let us verify the obtained relation at the point $x = 0$: on the one hand we have $\left(\sum_{n=0}^{\infty} \frac{x^n}{n+1} \right) |_{x=0} = 1$, and on the other hand we get $\lim_{x \to 0} -\frac{\ln(1-x)}{x} = \lim_{x \to 0} \frac{1}{1-x} = 1$ (using L'Hospital's rule). Therefore, on the entire interval $(-1, 1)$ we obtain the continuous function

$$\sum_{n=0}^{\infty} \frac{x^n}{n+1} = \begin{cases} -\frac{1}{x} \ln(1-x), & x \neq 0 \\ 1, & x = 0 \end{cases}.$$

Example 3f Find the sum of the series $1 - x^3 + x^6 - x^9 + \dots$.
Notice that after the change of the variable $t = x^3$ the series takes the form of the geometric series

$$1 - t + t^2 - t^3 + \dots = \sum_{n=0}^{\infty}(-t)^n = \frac{1}{1+t}.$$

Therefore,

$$1 - x^3 + x^6 - x^9 + \dots = \frac{1}{1+x^3}.$$

Since the series $\sum_{n=0}^{\infty}(-t)^n$ converges on $(-1, 1)$, the series for x is also convergent on $(-1, 1)$.

Example 4f Find the sum of the series $\frac{x^2}{2} - \frac{x^3}{3 \cdot 2} + \frac{x^4}{4 \cdot 3} - \frac{x^5}{5 \cdot 4} + \dots$.
Notice that the series of derivatives represents the Taylor series for $\ln(1 + x)$:

$$\left(\frac{x^2}{2}\right)' - \left(\frac{x^3}{3 \cdot 2}\right)' + \left(\frac{x^4}{4 \cdot 3}\right)' - \left(\frac{x^5}{5 \cdot 4}\right)' + \dots = x - \frac{x^2}{2} + \frac{x^3}{3} - \frac{x^4}{4} + \dots$$

$$= \sum_{n=1}^{\infty}(-1)^{n-1}\frac{x^n}{n} = \ln(1 + x).$$

Therefore, for the original series we have

$$\frac{x^2}{2} - \frac{x^3}{3 \cdot 2} + \frac{x^4}{4 \cdot 3} - \frac{x^5}{5 \cdot 4} + \dots = \sum_{n=1}^{\infty}(-1)^{n-1}\int_0^x \frac{t^n}{n}dt$$

$$= \int_0^x \left(\sum_{n=1}^{\infty}(-1)^{n-1}\frac{t^n}{n}\right)dt = \int_0^x \ln(1 + t)dt = (1 + x)\ln(1 + x) - x.$$

The validity of the made transformations is guaranteed on $(-1, 1)$ by the uniform convergence of the series of $\ln(1 + x)$ on any closed interval inside the interval $(-1, 1)$.

Example 5f Find the sum of the series $\sum_{n=0}^{\infty}(-1)^n \frac{2n^2+1}{(2n)!}x^{2n}$.
Let us separate this series into two parts:

$$\sum_{n=0}^{\infty}(-1)^n\frac{2n^2+1}{(2n)!}x^{2n} = \sum_{n=1}^{\infty}(-1)^n\frac{n}{(2n-1)!}x^{2n} + \sum_{n=0}^{\infty}(-1)^n\frac{1}{(2n)!}x^{2n}.$$

Noting that the second series is the expansion of $\cos x$, we focus on the first series which also resembles the form of trigonometric series. To eliminate the factor n, we bring one x outside of the series and integrate the remaining series $f(x) = \sum_{n=1}^{\infty}(-1)^n \frac{n}{(2n-1)!}x^{2n-1}$ term by term:

$$\int_0^x f(t)dt = \int_0^x \left(\sum_{n=1}^{\infty}(-1)^n \frac{n}{(2n-1)!}t^{2n-1} \right) dt = \sum_{n=1}^{\infty}(-1)^n \frac{n}{(2n-1)!} \int_0^x t^{2n-1} dt$$

$$= \sum_{n=1}^{\infty}(-1)^n \frac{n}{(2n-1)!} \frac{x^{2n}}{2n} = \frac{x}{2} \sum_{n=1}^{\infty}(-1)^n \frac{x^{2n-1}}{(2n-1)!} = -\frac{x}{2} \sum_{n=0}^{\infty}(-1)^n \frac{x^{2n+1}}{(2n+1)!} = -\frac{x}{2} \sin x.$$

Therefore, $f(x) = \left(-\frac{x}{2} \sin x \right)' = -\frac{1}{2}\sin x - \frac{x}{2}\cos x$. Finally, the original series is equal to

$$xf(x) + \cos x = -\frac{x}{2} \sin x - \frac{x^2}{2} \cos x + \cos x.$$

The validity of all the performed transformations is guaranteed by the uniform convergence of the series for $\cos x$ and $\sin x$ on any closed interval.

6.4   Sums of Series of Numbers

Example 1n Find the sum of the series $\sum_{n=0}^{\infty} \frac{(-1)^n 2^{2n} \pi^{2n}}{(2n)!}$.

Comparing to the series for $\cos x$: $\cos x = \sum_{n=0}^{\infty} \frac{(-1)^n x^{2n}}{(2n)!}$, we conclude immediately that

$$\sum_{n=0}^{\infty} \frac{(-1)^n 2^{2n} \pi^{2n}}{(2n)!} = \cos 2\pi = 1.$$

Example 2n Find the sum of the series $\sum_{n=0}^{\infty} \frac{(-1)^n}{(2n)!}$.

Using the series $\cos x = \sum_{n=0}^{\infty} \frac{(-1)^n x^{2n}}{(2n)!}$, we choose $x = 1$ and obtain directly that

$$\sum_{n=0}^{\infty} \frac{(-1)^n}{(2n)!} = \cos 1.$$

Example 3n Find the sum of the series $\sum_{n=0}^{\infty} \frac{1}{(2n)!}$.

Comparing with the series for $\cosh x = \sum_{n=0}^{\infty} \frac{x^{2n}}{(2n)!}$, we choose $x = 1$ and conclude that

$$\sum_{n=0}^{\infty} \frac{1}{(2n)!} = \cosh 1.$$

Example 4n Find the sum of the series $\sum_{n=0}^{\infty} \frac{n}{2^n}$.

This series can be considered as the power series $\sum_{n=0}^{\infty} nx^n$ evaluated at $x = \frac{1}{2}$. The last series can be obtained by term-by-term differentiation of the geometric series and subsequent multiplication of the result by x:

$$\sum_{n=0}^{\infty} nx^n = \sum_{n=0}^{\infty} x \cdot (x^n)' = x \left(\sum_{n=0}^{\infty} x^n \right)' = x \left(\frac{1}{1-x} \right)' = \frac{x}{(1-x)^2}.$$

(One can also use the result of Example 1f in Sect. 6.3.) Then,

$$\sum_{n=0}^{\infty} \frac{n}{2^n} = \frac{\frac{1}{2}}{(1 - \frac{1}{2})^2} = 2.$$

The validity of all these transformations is ensured by the uniform convergence of the geometric series on any closed subinterval of the interval $(-1, 1)$ (this includes the point of evaluation $x = \frac{1}{2}$).

Example 5n Find the sum of the series $\sum_{n=0}^{\infty} \frac{2n+1}{3^n n!}$.

We connect this series with the power series $\sum_{n=0}^{\infty} \frac{(2n+1)x^n}{n!}$ considering that $\frac{1}{3^n}$ is the value of x^n at the point $x = \frac{1}{3}$. The introduced power series can be represented via a known series for the exponential function using the following transformations:

$$\sum_{n=0}^{\infty} \frac{(2n+1)x^n}{n!} = \sum_{n=0}^{\infty} \frac{2nx^n}{n!} + \sum_{n=0}^{\infty} \frac{x^n}{n!} = 2x \sum_{n=1}^{\infty} \frac{x^{n-1}}{(n-1)!} \mid e^x = 2x \sum_{n=0}^{\infty} \frac{x^n}{n!} \mid e^x = 2xe^x + e^x.$$

Then,

$$\sum_{n=0}^{\infty} \frac{2n+1}{3^n n!} = \left(2 \cdot \frac{1}{3} + 1 \right) e^{\frac{1}{3}} = \frac{5}{3} e^{\frac{1}{3}}.$$

The validity of all these transformations is guaranteed by the uniform convergence of the series for the exponential function on any closed interval.

6.5   Calculation of Limits

When we need to calculate the limits which represent indeterminate forms it may happen that the representation of the involved functions in power series is a more efficient (and sometimes the unique) way to find the result. The general procedure consists of expansion of the relevant functions in power series centered at the accumulation point of the limit (these series should converge uniformly in a neighborhood of the accumulation point).

Then, the obtained expressions involving series are simplified usually employing the arithmetic properties of series. Finally, the limit of the simplified expression is calculated by applying the analytical property of limit of power series. In the last step, the secondary power terms disappear, and consequently, there is no need to keep many terms of the series expansion in an explicit form. However, the number of terms of the Taylor series, which should be treated explicitly, depends strongly on the form of the original limit and this number may not be evident since the beginning. Therefore, if a first attempt shows that the chosen number of explicit terms is not sufficient, one has to return to the power series expansions and take in consideration an increased number of terms required to perform the calculation of a given limit.

Example 11 Calculate the limit $\lim\limits_{x \to 0} \frac{\sin x - x}{x^3}$.

This limit is the indeterminate form $\frac{0}{0}$, and consequently, the arithmetic rules of the limits are not applicable. One of the ways to solve this problem is by applying L'Hospital's rule (three times in a row):

$$\lim_{x \to 0} \frac{\sin x - x}{x^3} = \lim_{x \to 0} \frac{\cos x - 1}{3x^2} = \lim_{x \to 0} \frac{-\sin x}{6x} = \lim_{x \to 0} \frac{-\cos x}{6} = -\frac{1}{6}.$$

Another approach is to use the power series expansion of $\sin x$ (that converges on $\mathbb{R}$ and, in particular, converges uniformly in a neighborhood of the origin), simplify the obtained expression by canceling the same powers in the numerator and denominator (that can be done according to the arithmetic properties of power series), and finally calculate the limit of the simplified expression which already has no indetermination (using the analytical properties of power series):

$$\lim_{x \to 0} \frac{\sin x - x}{x^3} = \lim_{x \to 0} \frac{\left(x - \frac{1}{3!}x^3 + \frac{1}{5!}x^5 - \ldots\right) - x}{x^3} = \lim_{x \to 0} \left(-\frac{1}{3!} + \frac{1}{5!}x^2 - \ldots\right) = -\frac{1}{6}.$$

As is seen, in this case, the volume of calculations involved in the two methods is practically the same.

Example 21 Calculate the limit $\lim\limits_{x \to 0} \frac{\cos x^4 - 1 + \frac{1}{2}x^8}{x^{16}}$.

This limit represents the indeterminate form $\frac{0}{0}$, and consequently, the arithmetic rules of the limits are not applicable. Developing $\cos x^4$ in the Taylor series (based on the known series for $\cos x$), and simplifying the involved series, we obtain the following result:

$$\lim_{x \to 0} \frac{\cos x^4 - 1 + \frac{1}{2}x^8}{x^{16}} = \lim_{x \to 0} \frac{\left(1 - \frac{1}{2!}x^8 + \frac{1}{4!}x^{16} - \frac{1}{6!}x^{24} + \ldots\right) - 1 + \frac{1}{2}x^8}{x^{16}}$$

$$= \lim_{x \to 0} \left(\frac{1}{4!} - \frac{1}{6!}x^8 + \ldots\right) = \frac{1}{4!}.$$

Notice that the same value can be found by applying L'Hospital's rule four times in a row, which requires a larger volume of work, but does not generate any principal difficulty. We leave this approach to the reader.

Example 31 Calculate the limit $\lim\limits_{x \to 2} \dfrac{\cos \frac{3}{4}\pi x - \frac{3}{2}\pi \ln \frac{x}{2}}{(4 - x^2)^2}$.

To solve this indeterminate form $\frac{0}{0}$, we can use L'Hospital's rule (applied twice):

$$\lim_{x \to 2} \frac{\cos \frac{3}{4}\pi x - \frac{3}{2}\pi \ln \frac{x}{2}}{(4 - x^2)^2} = \lim_{x \to 2} \frac{-\frac{3}{4}\pi \sin \frac{3}{4}\pi x - \frac{3}{2}\pi \frac{1}{x}}{4x^3 - 16x} = \lim_{x \to 2} \frac{-\frac{9}{16}\pi^2 \cos \frac{3}{4}\pi x + \frac{3}{2}\pi \frac{1}{x^2}}{12x^2 - 16} = \frac{3\pi}{256}.$$

Another approach is the expansion of the functions $\cos \frac{3}{4}\pi x$ and $\ln \frac{x}{2}$ in the Taylor series centered at $a = 2$, that is, in the powers of $(x - 2)^n$ responsible for the indeterminate form of this limit. Simplifying these powers in the numerator with the factor $(x - 2)^2$ in the denominator, we can eliminate an indetermination and calculate the limit by the arithmetic rules.

To obtain the Taylor series at the point $a = 2$, we can construct these series directly using the method of the Taylor coefficients, or we can take advantage of the already known series for trigonometric and logarithmic functions. Let us use the second option. First, the original expansion for $\sin x$: $\sin x = x - \frac{1}{3!}x^3 + \ldots$ can be transformed via trigonometric relations to the following form

$$\cos \frac{3}{4}\pi x = \sin \frac{3}{4}\pi \, (x - 2) = \frac{3}{4}\pi (x - 2) - \frac{1}{3!}\left(\frac{3}{4}\pi(x - 2)\right)^3 + \ldots.$$

Second, from the basic development for $\ln(1 + x)$: $\ln(1 + x) = x - \frac{1}{2}x^2 + \frac{1}{3}x^3 - \ldots$ it follows that $\ln x = (x - 1) - \frac{1}{2}(x - 1)^2 + \frac{1}{3}(x - 1)^3 - \ldots$, and consequently,

$$\ln \frac{x}{2} = (\frac{x}{2} - 1) - \frac{1}{2}(\frac{x}{2} - 1)^2 + \frac{1}{3}(\frac{x}{2} - 1)^3 - \ldots = \frac{1}{2}(x - 2) - \frac{1}{2}\left(\frac{1}{2}(x - 2)\right)^2 + \frac{1}{3}\left(\frac{1}{2}(x - 2)\right)^3 - \ldots.$$

Notice that both series in the powers of $(x - 2)^n$ converge uniformly in a neighborhood of the point 2. Therefore, substituting these series in the original limit and applying the arithmetic rules of power series to simplify the expressions inside the limit, we obtain

$$\lim_{x \to 2} \frac{\cos \frac{3}{4}\pi x - \frac{3}{2}\pi \ln \frac{x}{2}}{(4 - x^2)^2}$$

$$= \lim_{x \to 2} \frac{\left(\left[\frac{3}{4}\pi(x - 2) - \frac{1}{3!}\left(\frac{3}{4}\pi(x - 2)\right)^3 + \ldots\right] - \frac{3}{2}\pi\left[\frac{1}{2}(x - 2) - \frac{1}{2}\left(\frac{1}{2}(x - 2)\right)^2 + \frac{1}{3}\left(\frac{1}{2}(x - 2)\right)^3 - \ldots\right]\right)}{(x - 2)^2(x + 2)^2}$$

$$= \lim_{x \to 2} \frac{\frac{3}{16}\pi(x-2)^2 - \frac{9}{128}\pi^3(x-2)^3 - \frac{1}{16}\pi(x-2)^3 + \dots}{(x-2)^2(x+2)^2}$$

$$= \lim_{x \to 2} \frac{\frac{3}{16}\pi - \frac{9}{128}\pi^3(x-2) - \frac{1}{16}\pi(x-2) + \dots}{(x+2)^2} = \frac{3\pi}{256}.$$

Example 4l Calculate the limit $\lim_{x \to 0} \frac{\cos x - 1 + \frac{1}{2}x \sin x}{\ln^4(1+x)}$.

This limit represents the indeterminate form $\frac{0}{0}$, which makes impossible an application of the arithmetic rules of limits. The solution can be found by expanding the involved functions $\cos x$, $\sin x$ and $\ln(1+x)$ in the Taylor series, then simplifying expressions inside the limit and eliminating the indeterminate form, and finally calculating the limits of the simplified series. Notice that the simplest way to discover the first (principal) powers in the expansion of $\ln^4(1+x)$ is by starting with the power series of $\ln(1+x)$ and applying subsequently the fourth power, instead of calculation of the product of the logarithmic series. Following this path, we obtain:

$$\lim_{x \to 0} \frac{\cos x - 1 + \frac{1}{2}x \sin x}{\ln^4(1+x)}$$

$$= \lim_{x \to 0} \frac{\left(1 - \frac{1}{2!}x^2 + \frac{1}{4!}x^4 - \frac{1}{6!}x^6 + \dots\right) - 1 + \frac{1}{2}x\left(x - \frac{1}{3!}x^3 + \frac{1}{5!}x^5 - \dots\right)}{\left(x - \frac{1}{2}x^2 + \frac{1}{3}x^3 - \dots\right)^4}$$

$$= \lim_{x \to 0} \frac{\left(\frac{1}{4!} - \frac{1}{2}\frac{1}{3!}\right)x^4 + \left(-\frac{1}{6!} + \frac{1}{2}\frac{1}{5!}\right)x^6 + \dots}{x^4(1 - 2x + \dots)}$$

$$= \lim_{x \to 0} \frac{\frac{1}{4!}(1-2) - \frac{1}{6!}(1-3)x^2 + \dots}{1 - 2x + \dots} = -\frac{1}{4!}.$$

To calculate this limit we can also apply L'Hospital's rule four times in a row (we leave this task to the reader).

Example 5l Calculate the limit $\lim_{x \to 0} \left(e^{x^7} + \sin^2 x - \sinh^2 x\right)^{\frac{1}{x^4}}$.

This limit is the indeterminate form 1^∞. The forms of this type are usually transformed to the indeterminate forms $\frac{0}{0}$ or $\frac{\infty}{\infty}$. In this case, we can make such a transformation applying the logarithmic function to the function $f(x)$ inside the limit:

$$g(x) = \ln f(x) = \frac{1}{x^4} \ln \left(e^{x^7} + \sin^2 x - \sinh^2 x\right).$$

If we find the limit of $g(x)$, then we can return to $f(x)$ using the properties of limits. The new limit $\lim_{x \to 0} g(x)$ represents the indeterminate form $\frac{0}{0}$, which can be solved by expanding the involved functions in the power series centered at 0. First, we find expansions for the three functions inside the logarithm and then we develop the logarithmic function. The formula for e^{x^7} is obtained directly by substituting x^7 for x in the basic power series expansion of e^x: $e^{x^7} = 1 + x^7 + \frac{1}{2}x^{14} + \dots$. The series for $\sin^2 x$ can be found by calculating the product of the basic series for $\sin x$ with itself, but it is easier to employ the trigonometric formula $\sin^2 x = \frac{1 - \cos 2x}{2}$ and the series for $\cos x$ with double argument $2x$: $\cos 2x = 1 - \frac{1}{2!}(2x)^2 + \frac{1}{4!}(2x)^4 - \frac{1}{6!}(2x)^6 + \dots$. To find the series for $\sinh^2 x$ we apply a similar technique: $\sinh^2 x = \frac{\cosh 2x - 1}{2}$ and $\cosh 2x = 1 + \frac{1}{2!}(2x)^2 + \frac{1}{4!}(2x)^4 + \frac{1}{6!}(2x)^6 + \dots$. Hence, the power series for the function $h(x) = e^{x^7} + \sin^2 x - \sinh^2 x$ can be written in the form

$$h(x) = \left[1 + x^7 + \frac{1}{2}x^{14} + \dots \right] + \frac{1}{2}\left[1 - 1 + \frac{1}{2!}(2x)^2 - \frac{1}{4!}(2x)^4 + \frac{1}{6!}(2x)^6 + \dots \right]$$

$$-\frac{1}{2}\left[-1 + 1 + \frac{1}{2!}(2x)^2 + \frac{1}{4!}(2x)^4 + \frac{1}{6!}(2x)^6 + \dots \right] = 1 - \frac{2}{3}x^4 + x^7 + \dots .$$

Using now the Taylor series for $\ln(1 + y)$ only with the first explicit term $\ln(1 + y) = y + \dots$ and changing its argument by the formula $y = -\frac{2}{3}x^4 + x^7 + \dots$, we obtain (by the Theorem of composite power series)

$$\ln h(x) = \ln\left(1 - \frac{2}{3}x^4 + x^7 + \dots \right) = -\frac{2}{3}x^4 + x^7 + \dots .$$

Substituting this representation in the limit of $g(x)$, we get

$$\lim_{x \to 0} g(x) = \lim_{x \to 0} \frac{1}{x^4} \ln\left(e^{x^7} + \sin^2 x - \sinh^2 x \right)$$

$$= \lim_{x \to 0} \frac{1}{x^4}\left(-\frac{2}{3}x^4 + x^7 + \dots \right) = \lim_{x \to 0}\left(-\frac{2}{3} + x^3 + \dots \right) = -\frac{2}{3}.$$

Since the limit $\lim_{x \to 0} g(x)$ is calculated, we can return to the function $f(x)$ and obtain the following final result:

$$\lim_{x \to 0}\left(e^{x^7} + \sin^2 x - \sinh^2 x \right)^{\frac{1}{x^4}} = \lim_{x \to 0} e^{g(x)} = e^{\lim_{x \to 0} g(x)} = e^{-\frac{2}{3}}.$$

Notice that an application of L'Hospital's rule also leads to the same result. We start with the same transformation of the original indeterminate form 1^∞ to the indeterminate form $\frac{0}{0}$ by introducing the function $g(x)$ instead of $f(x)$: $g(x) = \ln f(x) =$

$\frac{1}{x^4} \ln \left(e^{x^7} + \sin^2 x - \sinh^2 x \right)$. Applying L'Hospital's rule to $g(x)$ and combining the terms inside the limit in an appropriate way, we obtain

$$\lim_{x \to 0} g(x) = \lim_{x \to 0} \frac{\ln \left(e^{x^7} + \sin^2 x - \sinh^2 x \right)}{x^4}$$

$$= \lim_{x \to 0} \frac{7x^6 e^{x^7} + \sin 2x - \sinh 2x}{\left(e^{x^7} + \sin^2 x - \sinh^2 x \right) 4x^3}$$

$$= \lim_{x \to 0} \frac{1}{e^{x^7} + \sin^2 x - \sinh^2 x} \cdot \left(\lim_{x \to 0} \frac{7}{4} x^3 e^{x^7} + \lim_{x \to 0} \frac{\sin 2x - \sinh 2x}{4x^3} \right)$$

$$= 1 \cdot \left(0 + \lim_{x \to 0} \frac{2 \cos 2x - 2 \cosh 2x}{12x^2} \right)$$

$$= \lim_{x \to 0} \frac{-4 \sin 2x - 4 \sinh 2x}{24x} = \lim_{x \to 0} \frac{-8 \cos 2x - 8 \cosh 2x}{24} = -\frac{2}{3}.$$

Example 61 Calculate the limit $\lim_{x \to 0} \frac{\sin(\tan x) - \tan(\sin x)}{\arcsin(\arctan x) - \arctan(\arcsin x)}$.

This limit represents the indeterminate form $\frac{0}{0}$. Let us use the power series expansion of composite functions keeping up to the seventh power of x. First we recall the Taylor formulas centered at 0 for the four involved functions:

$$\sin x = x - \frac{x^3}{3!} + \frac{x^5}{5!} - \frac{x^7}{7!} + \ldots, \forall x \in \mathbb{R},$$

$$\tan x = x + \frac{x^3}{3} + \frac{2x^5}{15} + \frac{17x^7}{315} + \ldots, \forall x \in \left(-\frac{\pi}{2}, \frac{\pi}{2} \right),$$

$$\arcsin x = x + \frac{x^3}{6} + \frac{3x^5}{40} + \frac{5x^7}{112} + \ldots, \forall x \in (-1, 1),$$

$$\arctan x = x - \frac{x^3}{3} + \frac{x^5}{5} - \frac{x^7}{7} + \ldots, \forall x \in [-1, 1].$$

Then, by the Theorem of composite power series, we obtain the following resulting series for the numerator:

$$f(x) = \sin(\tan x) - \tan(\sin x) =$$

$$\left(x + \frac{x^3}{3} + \frac{2x^5}{15} + \frac{17x^7}{315} + \ldots \right) - \frac{1}{3!} \left(x + \frac{x^3}{3} + \frac{2x^5}{15} + \ldots \right)^3$$

$$+ \frac{1}{5!} \left(x + \frac{x^3}{3} + \ldots \right)^5 - \frac{1}{7!} (x + \ldots)^7$$

$$-\left[\left(x-\frac{x^3}{3!}+\frac{x^5}{5!}-\frac{x^7}{7!}+\ldots\right)+\frac{1}{3}\left(x-\frac{x^3}{3!}+\frac{x^5}{5!}+\ldots\right)^3\right.$$

$$\left.+\frac{2}{15}\left(x-\frac{x^3}{3!}+\ldots\right)^5+\frac{17}{315}(x+\ldots)^7\right]+\ldots$$

$$=\left(\frac{17}{315}-\frac{11}{3!\cdot 15}+\frac{1}{4!\cdot 3}-\frac{1}{7!}+\frac{1}{7!}-\frac{1}{3}\cdot\frac{13}{120}+\frac{2}{15}\cdot\frac{5}{6}-\frac{17}{315}\right)x^7$$

$$+\ldots=-\frac{1}{30}x^7+\ldots.$$

Proceeding in a similar way, we also derive the representation for the denominator:

$$g(x)=\arcsin(\arctan x)-\arctan(\arcsin x)$$

$$=\left(x-\frac{x^3}{3}+\frac{x^5}{5}-\frac{x^7}{7}+\ldots\right)+\frac{1}{6}\left(x-\frac{x^3}{3}+\frac{x^5}{5}+\ldots\right)^3$$

$$+\frac{3}{40}\left(x-\frac{x^3}{3}+\ldots\right)^5+\frac{5}{112}(x+\ldots)^7$$

$$-\left[\left(x+\frac{x^3}{6}+\frac{3x^5}{40}+\frac{5x^7}{112}+\ldots\right)-\frac{1}{3}\left(x+\frac{x^3}{6}+\frac{3x^5}{40}+\ldots\right)^3\right.$$

$$\left.+\frac{1}{5}\left(x+\frac{x^3}{6}+\right)^5-\frac{1}{7}(x+\ldots)^7\right]+\ldots$$

$$=\left(-\frac{1}{7}+\frac{1}{6}\cdot\frac{14}{15}-\frac{3}{40}\cdot\frac{5}{3}+\frac{5}{112}-\frac{5}{112}+\frac{1}{3}\cdot\frac{37}{120}-\frac{1}{5}\cdot\frac{5}{6}+\frac{1}{7}\right)x^7$$

$$+\ldots=-\frac{1}{30}x^7+\ldots.$$

Notice that in both series the remaining (non-explicit) terms have the powers of x greater than 7. Substituting these evaluations in the original limit, we arrive to the following result:

$$\lim_{x\to 0}\frac{f(x)}{g(x)}=\lim_{x\to 0}\frac{-\frac{1}{30}x^7+\ldots}{-\frac{1}{30}x^7+\ldots}=1.$$

This limit still can be calculated (at least theoretically) by L'Hospital's rule applying it seven times in a row, but the required technical work in this procedure is much more voluminous and cumbersome than in the method presented above.

6.6   Calculation of Integrals

Many indefinite integrals cannot be found in a closed form (that is, expressed as a finite term combination of elementary functions), even when the integrability conditions are satisfied. For many others, analytical calculations are too technically complicated and even practically impossible, although theoretically they can be performed. In these cases, a power series expansion can offer an alternative form of the presentation of result or simplification of the involved calculations.

Example 1i Calculate the integral $\int e^{-x^2} dx$.
This integral (frequently called the Gauss integral) cannot be expressed in finite terms of elementary functions, but can be found using the Taylor series. First, we represent e^{-x^2} in a power series using the series of e^x with $-x^2$ instead of x:

$$e^{-x^2} = \sum_{n=0}^{\infty} (-1)^n \frac{x^{2n}}{n!}, \; \forall x \in \mathbb{R}.$$

Like the series for e^x, this series converges on $\mathbb{R}$ and converges uniformly on any closed interval. Then, we can use term-by-term integration to obtain:

$$\int e^{-x^2} dx = \int \left(\sum_{n=0}^{\infty} (-1)^n \frac{x^{2n}}{n!} \right) dx$$

$$= \sum_{n=0}^{\infty} \int (-1)^n \frac{x^{2n}}{n!} dx = \sum_{n=0}^{\infty} (-1)^n \frac{x^{2n+1}}{n!(2n+1)} + C, \; \forall x \in \mathbb{R}.$$

According to the properties of power series, the last series converges on $\mathbb{R}$ and converges uniformly on any closed interval.

Example 2i Calculate the integral $\int \frac{\sin x}{x} dx$.
This is one more integral (called the Dirichlet integral) that cannot be found in a closed form, but it can be solved using the Taylor series. First, to eliminate the indetermination at the point $x = 0$, we define the integrand at this point in the continuous mode: $f(x) = \begin{cases} \frac{\sin x}{x}, x \neq 0 \\ 1, x = 0 \end{cases}$. Consider a formal series of this function based on the series of $\sin x$:

$$f(x) = \frac{1}{x} \sum_{n=0}^{\infty} (-1)^n \frac{x^{2n+1}}{(2n+1)!} = \sum_{n=0}^{\infty} (-1)^n \frac{x^{2n}}{(2n+1)!}.$$

Since the function $\frac{1}{x}$ is unbounded in any neighborhood of 0, the uniform convergence does not follow from arithmetic properties of series. However, we can show that the series converges uniformly on any interval $[-\tilde{R}, \tilde{R}]$, $\tilde{R} > 0$ by using the Weierstrass test: $\left| \frac{x^{2n}}{(2n+1)!} \right| \leq \frac{\tilde{R}^{2n}}{(2n+1)!}$ and the majorant numerical series $\sum_{n=0}^{\infty} \frac{\tilde{R}^{2n}}{(2n+1)!}$ converges. This implies that the series for $f(x)$ converges on $\mathbb{R}$. Besides, at every $x \neq 0$ the series converges to $f(x)$ (because the series without the denominator converges to $\sin x$) and at the point $x = 0$ the sum of the series is 1, which corresponds to the definition of $f(x)$. Thus, the series converges to $f(x)$ on $\mathbb{R}$.

Knowing the series of $f(x)$ and establishing its convergence, we can perform term-by-term integration:

$$\int f(x)dx = \int \left(\sum_{n=0}^{\infty} (-1)^n \frac{x^{2n}}{(2n+1)!} \right) dx = \sum_{n=0}^{\infty} \int (-1)^n \frac{x^{2n}}{(2n+1)!} dx$$

$$= \sum_{n=0}^{\infty} (-1)^n \frac{x^{2n+1}}{(2n+1)!(2n+1)} + C, \ \forall x \in \mathbb{R}.$$

Since the series of $f(x)$ converges on $\mathbb{R}$ and converges uniformly on any closed interval, the obtained series for the integral has the same convergence properties.

Example 3i Calculate the integral $\int_0^1 e^{-x^2} dx$.

Since the function e^{-x^2} has no antiderivative in elementary functions, we integrate the corresponding Taylor series term-by-term in a similar way it was done in Example 1i, but using the definite integral. The representation of e^{-x^2} in the Taylor series was already found (see Example 1i):

$$e^{-x^2} = \sum_{n=0}^{\infty} (-1)^n \frac{x^{2n}}{n!}, \ \forall x \in \mathbb{R}$$

and it was noted that this series converges on $\mathbb{R}$ and converges uniformly on any closed interval. Therefore, we can apply term-by-term integration:

$$\int_0^1 e^{-x^2} dx = \int_0^1 \left(\sum_{n=0}^{\infty} (-1)^n \frac{x^{2n}}{n!} \right) dx = \sum_{n=0}^{\infty} \int_0^1 (-1)^n \frac{x^{2n}}{n!} dx$$

$$= \sum_{n=0}^{\infty} \frac{(-1)^n}{n!(2n+1)} x^{2n+1} \Big|_0^1 = \sum_{n=0}^{\infty} \frac{(-1)^n}{n!(2n+1)}.$$

The obtained alternating series can be evaluated with any desired degree of accuracy. For example, if the error of approximation should be less than 10^{-3}, we have the following evaluation of the remainder: $|r_n| \leq \frac{1}{(n+1)!(2n+3)} < 10^{-3}$, whence $n \geq 4$, that is, the partial sum $s_4 = 1 - \frac{1}{3} + \frac{1}{10} - \frac{1}{42} + \frac{1}{216} = \frac{5651}{7560} \approx 0.7475$ ensures the required accuracy.

Example 4i Calculate the integral $\int_0^\pi \sin x^2 dx$.

The integral $\int \sin x^2 dx$ (called the Fresnel integral) cannot be calculated using a finite combination of elementary functions. Therefore, we will follow the steps of the preceding example: we represent $\sin x^2$ in a Taylor series, verify its convergence, apply term-by-term integration and obtain the answer in the form of a series of numbers.

First, the Taylor series of $\sin x$ can be used to find the series of $\sin x^2$:

$$\sin x^2 = \sum_{n=0}^\infty \frac{(-1)^n x^{4n+2}}{(2n+1)!}, \quad \forall x \in \mathbb{R}.$$

Knowing the convergence properties of the series of $\sin x$, we conclude that the series of $\sin x^2$ converges on $\mathbb{R}$ and converges uniformly on any closed interval, in particular, on $[0, \pi]$. Therefore, we can use term-by-term integration to get the following result:

$$\int_0^\pi \sin x^2 dx = \int_0^\pi \left(\sum_{n=0}^\infty \frac{(-1)^n x^{4n+2}}{(2n+1)!} \right) dx = \sum_{n=0}^\infty \int_0^\pi (-1)^n \frac{x^{4n+2}}{(2n+1)!} dx$$

$$= \sum_{n=0}^\infty \frac{(-1)^n}{(2n+1)!(4n+3)} x^{4n+3} \Big|_0^\pi = \sum_{n=0}^\infty \frac{(-1)^n}{(2n+1)!(4n+3)} \pi^{4n+3}.$$

The obtained alternating series can be approximated with any given accuracy. If we need the approximation with an error less than 10^{-5}, we have to solve the following estimate for the remainder: $|r_n| \leq \frac{\pi^{4n+7}}{(2n+3)!(4n+7)} < 10^{-5}$. Due to the relatively large factor π^{4n+7} it is necessary to take $n \geq 15$ to satisfy the inequality for the remainder. However, for sufficiently large n the accuracy increases fast: for instance, for $n \geq 20$ the accuracy is better than 10^{-11}.

Example 5i Calculate the integral $\int_0^{1/9} \sqrt{x} e^x dx$.

The given integral cannot be expressed in a closed form, and for this reason we will follow the procedure used in the two last examples. First we use the Taylor series of e^x to find a series of $\sqrt{x} e^x$ and determine its convergence properties. Then we integrate the series for the integrand term by term and obtain the answer in the form of a series of numbers.

Recall that $e^x = \sum_{n=0}^\infty \frac{x^n}{n!}$, $\forall x \in \mathbb{R}$, which implies that

$$\sqrt{x} e^x = \sum_{n=0}^\infty \sqrt{x} \frac{x^n}{n!}, \quad \forall x \geq 0.$$

Since the series for e^x converges on $\mathbb{R}$ and converges uniformly on any closed interval, multiplication of this series by the function $\sqrt{x}$ defined on $[0, +\infty)$ and bounded on any interval $[0, b]$, $b > 0$, results in a series convergent on $[0, +\infty)$ and convergent uniformly on any interval $[0, b]$, $b > 0$, in particular, on $[0, \frac{1}{9}]$. The obtained series is not a power series, but still we can use the integral property of series of functions to integrate term by term and obtain

$$\int_0^{1/9} \sqrt{x} e^x dx = \int_0^{1/9} \left(\sum_{n=0}^{\infty} \sqrt{x} \frac{x^n}{n!} \right) dx = \sum_{n=0}^{\infty} \frac{1}{n!} \int_0^{1/9} x^{n+1/2} dx$$

$$= \sum_{n=0}^{\infty} \frac{1}{n!} \left(\frac{x^{n+3/2}}{n+3/2} \right) \Big|_0^{1/9} = \sum_{n=0}^{\infty} \frac{2}{n!(2n+3)} \frac{1}{3^{2n+3}}.$$

Let us find the number of terms of the obtained series of numbers which ensures an approximation of the given integral with an accuracy better than 10^{-5}. We can evaluate the remainder in the following manner:

$$|r_n| = \sum_{k=n+1}^{\infty} \frac{2}{k!(2k+3)} \frac{1}{3^{2k+3}}$$

$$< \frac{2}{(n+1)!(2n+5)} \frac{1}{3^{2n+5}} \left(1 + \frac{1}{(n+1)3^2} + \frac{1}{(n+1)^2 3^4} + \cdots \right)$$

$$= \frac{2}{(n+1)!(2n+5)3^{2n+5}} \frac{1}{1 - \frac{1}{9(n+1)}} - \frac{2}{n!(2n+5)3^{2n+3}} \frac{1}{9n+8}.$$

To satisfy the restriction $|r_n| < 10^{-5}$, we have to solve the inequality $n!(2n+5)3^{2n+3}(9n+8) > 2 \cdot 10^5$, which is true for $n \geq 2$. Hence, the approximation

$$\int_0^{1/9} \sqrt{x} e^x dx \approx \frac{2}{3 \cdot 3^3} + \frac{2}{5 \cdot 3^5} + \frac{2}{2! \cdot 7 \cdot 3^7} \approx 0.026403$$

has an error less than 10^{-5}.

6.7 Solution of Ordinary Differential Equations

Recall that there is no universal method of solution of ordinary differential equations even for explicit equations of the first order. There are different particular methods designed for solving specific types of equations, but equations which do not belong to any of these types are ubiquitous, it happens even with linear equations, whose theory is relatively simple. In the case of linear equations, their solutions can be found in the form of power series even in the cases when the methods designed for specific types of equations do not work. Below we recall some known results of the theory of linear ordinary differential equations.

An n-th order linear homogeneous equation has the form

$$y^{(n)} + a_{n-1}(x)y^{(n-1)} + \ldots + a_1(x)y' + a_0(x)y = 0, \tag{5.30}$$

where $a_0(x), a_1(x), \ldots, a_{n-1}(x)$ are the coefficients depending only on x. We will consider this equation on an interval I where all the coefficients are continuous functions. Equation (5.30) has exactly n linearly independent particular solutions $y_1(x), y_2(x), \ldots, y_n(x)$ in I and its general solution $y_g(x)$ in I, which contains all the particular solutions, can be expressed through the linear combination of $y_1(x), y_2(x), \ldots, y_n(x)$:

$$y_g(x) = C_1 y_1(x) + C_2 y_2(x) + \ldots + C_n y_n(x),$$

where $C_1, C_2, \ldots, C_n$ are arbitrary constants.

An n-th order linear non-homogeneous equation

$$y^{(n)} + a_{n-1}(x)y^{(n-1)} + \ldots + a_1(x)y' + a_0(x)y = b(x) \tag{5.31}$$

with the coefficients $a_0(x), a_1(x), \ldots, a_{n-1}(x)$ and right-hand side $b(x)$ (all of them continuous functions in an interval I) has a general solution $y_n(x)$ in I, which contains all the particular solutions and can be represented in the form

$$y_n(x) = y_g(x) + y_p(x),$$

where $y_g(x)$ is a general solution of the homogeneous equation (5.30) and $y_p(x)$ is an arbitrary particular solution of the non-homogeneous equation (5.31).

The initial value problem for Eq. (5.31) on an interval I consists of Eq. (5.31) itself (with the coefficients $a_0(x), a_1(x), \ldots, a_{n-1}(x)$ and right-hand side $b(x)$ continuous in I) together with the initial conditions $y(x_0) = b_0, y'(x_0) = b_1, \ldots, y^{(n-1)}(x_0) = b_{n-1}$, where $x_0 \in I$. This problem has a unique solution in I for any set of initial values $b_0, b_1, \ldots, b_{n-1}$.

Any particular solution of a linear homogeneous equation (5.30) with analytic coefficients $a_0(x), a_1(x), \ldots, a_{n-1}(x)$ of the radii of convergence $R_0, R_1, \ldots, R_{n-1}$, respectively (all of them with respect to the central point x_0), can be represented in a power series centered at x_0 with the radius of convergence $R \geq \min\{R_0, R_1, \ldots, R_{n-1}\}$.

Any particular solution of a linear non-homogeneous equation (5.31) with analytic coefficients $a_0(x), a_1(x), \ldots, a_{n-1}(x)$ of the radii of convergence $R_0, R_1, \ldots, R_{n-1}$, respectively, and analytic right-hand side $b(x)$ of the radius of convergence R_b (all of them with respect to the central point x_0) can be represented in a power series centered at x_0 with the radius of convergence $R \geq \min\{R_0, R_1, \ldots, R_{n-1}, R_b\}$.

In general, the search for a power series solution of a linear differential equation consists of representation of particular solutions in the form of a formal power series with unknown coefficients and subsequent specification of these coefficients by substituting the series in the original equation. After the coefficients are found, we have to analyze the convergence of the obtained power series. Let us consider some examples.

Example 1e The first order equation $y' - y = 0$.
This is a trivial separable equation (that is, the equation of the type $y' = f(x)g(y)$), whose solution can be found immediately by separating y and x, and integrating independently both sides of the relation:

$$\int \frac{y'}{y} dx = \int \frac{dy}{y} = \int 1 dx \Rightarrow \ln|y| = x + A; \ y \equiv 0 \Rightarrow y = Ce^x, \forall C \in \mathbb{R}.$$

This simple equation is a good case for illustration of the method of power series. Let us search for a solution in the form $y(x) = \sum_{n=0}^{\infty} c_n x^n$, where the coefficients of the series should be determined substituting this series in the original equation:

$$\left(\sum_{n=0}^{\infty} c_n x^n\right)' - \sum_{n=0}^{\infty} c_n x^n = 0.$$

Assuming that the series has a non-zero radius of convergence, we use term-by-term differentiation inside an interval of convergence and obtain

$$\sum_{n=1}^{\infty} n c_n x^{n-1} = \sum_{n=0}^{\infty} c_n x^n.$$

Then, using the uniqueness of the power series, we equal the coefficients with the same powers on the left-hand and right-hand sides, that gives the following recurrence relation: $nc_n = c_{n-1}, \forall n \in \mathbb{N}$. This relation can be solved for c_n in terms of c_0: $c_n = \frac{1}{n} c_{n-1} = \frac{1}{n(n-1)} c_{n-2} = \ldots = \frac{1}{n!} c_0$, where the coefficient c_0 is an arbitrary parameter. Therefore, we obtain the general solution in the form

$$y(x) = C \sum_{n=0}^{\infty} \frac{1}{n!} x^n = Ce^x, \forall C = c_0 \in \mathbb{R},$$

that is, we arrive at the same solution that was derived above in a simpler way. Since the generated series is already recognized as the Taylor series of e^x, there is no need to check its convergence: we know that this series converges on $\mathbb{R}$ and converges uniformly on any closed interval. This justifies all the operations performed with this series on $\mathbb{R}$. (Actually, the presented earlier general statement regarding the existence of solutions in the form of power series guarantees that the found power series has the radius of convergence $R = \infty$, since the equation coefficient $a_0 \equiv -1$ is an analytic function on $\mathbb{R}$.)

Example 2e The second order equation $y'' + y = 0$.
This is an elementary linear homogeneous equation with constant coefficients, whose standard method of solution consists of finding the two roots $\lambda_{1,2} = \pm i$ of the characteristic equation $\lambda^2 + 1 = 0$ and using these roots to compose the corresponding (linearly independent) particular solutions $y_1(x) = \cos x$ and $y_2(x) = \sin x$. After this, the general solution is found in the standard form $y_g(x) = C_1 \cos x + C_2 \sin x$.

Now we apply the power series method. We search for a solution in the form $y(x) = \sum_{n=0}^{\infty} c_n x^n$, where the coefficients c_n should be determined by substituting the series in the original equation:

$$\left(\sum_{n=0}^{\infty} c_n x^n\right)'' + \sum_{n=0}^{\infty} c_n x^n = 0.$$

Assuming that the series has a non-zero radius of convergence, we use term-by-term differentiation inside an interval of convergence and obtain

$$\sum_{n=2}^{\infty} n(n-1)c_n x^{n-2} = -\sum_{n=0}^{\infty} c_n x^n.$$

The uniqueness Theorem of the power series implies that the coefficients on left-hand and right-hand sizes are equal, that leads to the following recurrence relation: $n(n-1)c_n = -c_{n-2}$, $\forall n = 2, 3, \ldots$, where c_0 and c_1 are arbitrary parameters. This recurrence relation can be split into two independent groups: the first starts with c_0 and involves all the even-numbered coefficients, while the second starts with c_1 and involves all the odd-numbered coefficients. Solving the relations of the first set, we have for $\forall k \in \mathbb{N}$

$$c_n = c_{2k} = -\frac{1}{2k(2k-1)}c_{2k-2} = (-1)^2 \frac{1}{2k(2k-1)(2k-2)(2k-3)}c_{2k-4} = \ldots = (-1)^k \frac{1}{(2k)!}c_0,$$

and, in the same manner, for the second group we get

$$c_n = c_{2k+1} = -\frac{1}{(2k+1)(2k)}c_{2k-1} = (-1)^2 \frac{1}{(2k+1)2k(2k-1)(2k-2)}c_{2k-3} = \ldots$$

$$= (-1)^k \frac{1}{(2k+1)!}c_1.$$

Hence, we obtain two linearly independent solutions in the form of power series (we choose here $c_0 = 1$ and $c_1 = 1$):

$$y_1(x) = \sum_{n=0}^{\infty} (-1)^n \frac{1}{(2n)!} x^{2n}$$

and

$$y_2(x) = \sum_{n=0}^{\infty} (-1)^n \frac{1}{(2n+1)!} x^{2n+1}.$$

It is easy to recognize that these two series are the expansions of the functions $y_1(x) = \cos x$ and $y_2(x) = \sin x$, respectively (recall that both series converge uniformly on any closed interval). Thus, we arrive at the same general solution derived earlier by the traditional method.

Example 3e The second order equation $y'' - xy = 0$.
This is a linear homogeneous equation with variable coefficients of type (5.30) (called Airy's equation). The solution of this equation requires the use of power series: $y(x) = \sum_{n=0}^{\infty} c_n x^n$. To find the values of c_n we substitute the series in the original equation

$$\sum_{n=2}^{\infty} n(n-1)c_n x^{n-2} = \sum_{n=0}^{\infty} c_n x^{n+1}$$

and equal the coefficients of the same powers on the left- and right-hand sides: $n(n-1)c_n = c_{n-3}, \forall n = 3, 4, \ldots$. Evidently, these recurrence relations can be split into the three groups: the first one contains the indices $3k$, $k \in \mathbb{N}$ and can be uniquely defined by the coefficient c_0, the second contains the indices $3k + 1$ and is determined by c_1, and the third, with the indices $3k+2$, is defined by c_2. Notice that $c_2 = 0$, since the coefficient with x^0 on the left-hand side is $2c_2$ and on the right-hand side is 0. The other two parameters—c_0 and c_1—can be chosen arbitrarily. In this way, only the two sets of the coefficients remain:

$$c_{3k} = \frac{1}{3k(3k-1)} c_{3k-3} = \frac{1}{3k(3k-1)(3k-3)(3k-4)} c_{3k-6} = \cdots$$

$$= \frac{1}{2 \cdot 3 \cdot \ldots \cdot (3k-4)(3k-3)(3k-1)3k} c_0$$

and

$$c_{3k+1} = \frac{1}{(3k+1)3k}c_{3k-2} = \frac{1}{(3k+1)3k(3k-2)(3k-3)}c_{3k-5} = \ldots$$

$$= \frac{1}{3 \cdot 4 \cdot \ldots \cdot (3k-3)(3k-2)3k(3k+1)}c_1.$$

Consequently, the two linearly independent solutions has the following form (we set here $c_0 = 1$ and $c_1 = 1$)

$$y_1(x) = \sum_{n=0}^{\infty} \frac{x^{3n}}{2 \cdot 3 \cdot \ldots \cdot (3n-4)(3n-3)(3n-1)3n}$$

and

$$y_2(x) = \sum_{n=0}^{\infty} \frac{x^{3n+1}}{3 \cdot 4 \cdot \ldots \cdot (3n-3)(3n-2)3n(3n+1)}.$$

The general solution is found in the usual form of the linear combination: $y_g(x) = C_1 y_1(x) + C_2 y_2(x)$.

To justify the used operations, we have to determine the interval of uniform convergence of both involved series. It can be done in two ways. First, we can employ the theory of analytic solutions of differential equations. Indeed, notice that the coefficients $a_0 = -x$ and $a_1 = 0$ of the original equation are analytic functions on $\mathbb{R}$. Then, we can use the general statement that ensures that a power series solution has the same radius of convergence, and consequently, both series of y_1 and y_2 converge uniformly on (at least) any closed interval. Another way is to verify directly convergence of the series, applying one of the tests. For example, using D'Alembert's test for the first series, we have

$$\frac{|x^{3n+3}|}{2 \cdot 3 \cdot \ldots \cdot (3n-1)3n(3n+2)(3n+3)} \cdot \frac{2 \cdot 3 \cdot \ldots \cdot (3n-1)3n}{|x^{3n}|}$$

$$= |x^3| \frac{1}{(3n+2)(3n+3)} \underset{n \to \infty}{\to} 0, \forall x \in \mathbb{R},$$

which means that the first series converges on $\mathbb{R}$ and converges uniformly on (at least) any closed interval. Using the same test, the reader can check that the same is true for the second series. This justifies all the steps of the applied method of power series solution.

Example 4e The second order equation $(x^2 + 1)y'' + xy' - y = 0$.
This is a linear homogeneous equation with variable coefficients, that can be solved using the power series method. Searching for a solution in the form $y(x) = \sum_{n=0}^{\infty} c_n x^n$ with the

undetermined coefficients c_n, we substitute this series in the original equation:

$$(x^2 + 1) \sum_{n=2}^{\infty} n(n-1)c_n x^{n-2} + x \sum_{n=1}^{\infty} n c_n x^{n-1} - \sum_{n=0}^{\infty} c_n x^n = 0.$$

Separating the terms with different powers

$$\sum_{n=2}^{\infty} n(n-1)c_n x^n + \left[2c_2 x^0 + 3 \cdot 2c_3 x^1 + \sum_{n=2}^{\infty} (n+2)(n+1)c_{n+2} x^n \right]$$

$$+ \left[c_1 x^1 + \sum_{n=2}^{\infty} n c_n x^n \right] - \left[c_0 x^0 + c_1 x^1 + \sum_{n=2}^{\infty} c_n x^n \right] = 0$$

and grouping the terms with the same powers

$$(2c_2 - c_0) + 6c_3 x + \sum_{n=2}^{\infty} [n(n-1)c_n + (n+2)(n+1)c_{n+2} + n c_n - c_n] x^n = 0,$$

we obtain the following relations for the coefficients c_n:

$$2c_2 - c_0 = 0, c_3 = 0; (n+1)(n-1)c_n + (n+2)(n+1)c_{n+2} = 0, \forall n = 2, 3, \ldots.$$

The last set of equations are the recurrence relations that can be simplified to the form $c_{n+2} = -\frac{n-1}{n+2}c_n, \forall n = 2, 3, \ldots$. Then, we can divide these relations into two groups— with the even-numbered coefficients $n = 2k$ and with the odd-numbered ones $n = 2k + 1$. In the first group we have

$$c_{2k} = -\frac{2k-3}{2k}c_{2k-2} = (-1)^2 \frac{2k-3}{2k} \frac{2k-5}{2k-2} c_{2k-4} = \cdots$$

$$= (-1)^{k-1} \frac{(2k-3)(2k-5) \cdot \ldots \cdot 3 \cdot 1}{2k(2k-2) \cdot \ldots \cdot 6 \cdot 4} c_2, \forall k = 2, 3, \ldots; \quad c_2 = \frac{1}{2}c_0.$$

In the second group, all the coefficients starting from the index 3 vanish because $c_3 = 0$, and the unique remaining (arbitrary) coefficient is c_1:

$$c_{2k+1} = 0, \forall k = 1, 2, \ldots; \quad \forall c_1.$$

Hence, we have the two linearly independent solutions: $y_1 = x$ and the series

$$y_2(x) = 2 + x^2 + \sum_{n=2}^{\infty} (-1)^{n-1} \frac{1 \cdot 3 \cdot \ldots \cdot (2n-3)}{4 \cdot 6 \cdot \ldots \cdot 2n} x^{2n}$$

(in the last series we have chosen $c_2 = 1$). To justify the used procedures we have to find the interval of uniform convergence of the obtained series. There are two ways to do this. To apply the theory of analytic solutions, we note that the coefficients of the normalized differential equation are $a_1 = \frac{x}{x^2+1}$ and $a_0 = -\frac{1}{x^2+1}$, and these functions are analytic on $(-1, 1)$. Consequently, we can use the general statement that guarantees the uniform convergence of the series solution on any closed subinterval of $(-1, 1)$. Another option is to check the convergence directly using, for instance, D'Alembert's test:

$$\frac{|x^{2n+2}| \cdot 1 \cdot 3 \cdot \ldots \cdot (2n-3)(2n-1)}{4 \cdot 6 \cdot \ldots \cdot 2n(2n+2)} \cdot \frac{4 \cdot 6 \cdot \ldots \cdot 2n}{|x^{2n}| \cdot 1 \cdot 3 \cdot \ldots \cdot (2n-3)} = x^2 \frac{2n-1}{2n+2} \underset{n \to \infty}{\to} x^2, \forall x \in \mathbb{R}.$$

Hence, the series converges for $|x| < 1$ and diverges for $|x| > 1$. This implies that the series solution converges uniformly on any closed interval inside $(-1, 1)$.

Example 5e   The initial value problem: the second order equation $y'' + \cos x \cdot y = 0$ subject to the initial conditions $y(0) = 1$, $y'(0) = 0$.
The given equation is linear homogeneous with the coefficients $a_0(x) = \cos x$, $a_1(x) = 0$, which are analytic functions on $\mathbb{R}$. Since the coefficient $\cos x$ is not a polynomial, we will need to use the Cauchy product of series to find the undetermined coefficients c_n in the series solution $y(x) = \sum_{n=0}^{\infty} c_n x^n$. Substituting this series and the expansion of the coefficient

$$a_0(x) = \cos x = \sum_{n=0}^{\infty} (-1)^n \frac{x^{2n}}{(2n)!} = 1 - \frac{1}{2!}x^2 + \frac{1}{4!}x^4 - \frac{1}{6!}x^6 + \ldots$$

in the differential equation, we have

$$\sum_{n=2}^{\infty} n(n-1)c_n x^{n-2} + \sum_{n=0}^{\infty} (-1)^n \frac{x^{2n}}{(2n)!} \cdot \sum_{n=0}^{\infty} c_n x^n = 0$$

or

$$\sum_{n=0}^{\infty} (n+2)(n+1)c_{n+2}x^n + \sum_{n=0}^{\infty} e_n x^n = 0,$$

where e_n are the coefficients of the Cauchy product. The relations for c_n take the form

$$(n+2)(n+1)c_{n+2} + e_n = 0, \ n = 0, 1, \ldots .$$

To find the coefficients c_n we need to know the expressions for e_n. Since the odd-numbered coefficients in the series of $\cos x$ are zero, the expressions of e_n is convenient to separate in even- and odd-numbered. If $n = 2k$, $k \in \mathbb{N}$, then

$$e_{2k} = c_{2k} - \frac{1}{2!}c_{2k-2} + \frac{1}{4!}c_{2k-4} - \ldots + \frac{(-1)^{k-1}}{(2k-2)!}c_2 + \frac{(-1)^k}{(2k)!}c_0.$$

If $n = 2k + 1$, $k \in \mathbb{N}$, then

$$e_{2k+1} = c_{2k+1} - \frac{1}{2!}c_{2k-1} + \frac{1}{4!}c_{2k-3} - \ldots + \frac{(-1)^{k-1}}{(2k-2)!}c_3 + \frac{(-1)^k}{(2k)!}c_1.$$

Respectively, we separate the relations for c_n in even- and odd-numbered:

$$n = 2k : \ (2k+2)(2k+1)c_{2k+2} + e_{2k} = 0$$

$$\Rightarrow (2k+2)(2k+1)c_{2k+2} = - \left(c_{2k} - \frac{1}{2!}c_{2k-2} + \frac{1}{4!}c_{2k-4} - \ldots + \frac{(-1)^{k-1}}{(2k-2)!}c_2 + \frac{(-1)^k}{(2k)!}c_0 \right);$$

$$n = 2k + 1 : \ (2k+3)(2k+2)c_{2k+3} + e_{2k+1} = 0$$

$$\Rightarrow (2k+3)(2k+2)c_{2k+3} = - \left(c_{2k+1} - \frac{1}{2!}c_{2k-1} + \frac{1}{4!}c_{2k-3} - \ldots + \frac{(-1)^{k-1}}{(2k-2)!}c_3 + \frac{(-1)^k}{(2k)!}c_1 \right).$$

Hence, we have two decoupled sets of the recurrence relations—for the even and odd indices. To see better what kind of the relations is derived, we specify some first formulas for the even and odd indices:

$$n = 0 : 2c_2 = -c_0; \ n = 2 : 12c_4 = - \left(c_2 - \frac{1}{2!}c_0 \right);$$

$$n = 4 : 30c_6 = - \left(c_4 - \frac{1}{2!}c_2 + \frac{1}{4!}c_0 \right); \ \ldots;$$

$$n = 1 : \ 6c_3 = -c_1; \ n = 3 : 20c_5 = - \left(c_3 - \frac{1}{2!}c_1 \right);$$

$$n = 5 : 42c_7 = - \left(c_5 - \frac{1}{2!}c_3 + \frac{1}{4!}c_1 \right); \ \ldots.$$

Starting from the coefficient c_0 and using the first set of relations, we find one by one all the even-numbered coefficients; in the same way, specifying c_1 and using the second set, we determine all the odd-numbered coefficients. The arbitrariness of the choice of two coefficients corresponds to definition of the two initial conditions.

Now we apply the initial conditions. From $y(0) = 1$ it follows that $c_0 = 1$ and from $y'(0) = 0$ we get $c_1 = 0$. Consequently, all the odd coefficients vanish and the solution of the initial value problem can be expressed in the form $y(x) = \sum_{k=0}^{\infty} c_{2k}x^{2k}$, where $c_0 = 1$

and the remaining coefficients are found successively by the recurrence relations

$$c_{2k+2} = -\frac{1}{(2k+2)(2k+1)}\left(c_{2k} - \frac{1}{2!}c_{2k-2} + \frac{1}{4!}c_{2k-4} - \cdots\right.$$

$$\left. +\frac{(-1)^{k-1}}{(2k-2)!}c_2 + \frac{(-1)^k}{(2k)!}c_0\right), \quad k = 0, 1, \ldots.$$

To finalize, notice that the coefficient $a_0 = \cos x$ of the original equation is an analytic function on $\mathbb{R}$. Then, by the general statement, the power series solution has the same radius of convergence $R = \infty$, and consequently, the obtained series converges uniformly on (at least) any closed interval.

Example 6e The initial value problem: the second order equation $y'' - xy' + y = 1$ together with the initial conditions $y(0) = 0$, $y'(0) = 0$.

The given equation is linear non-homogeneous of type (5.31). The coefficients of the equations and its right-hand part are analytic functions on $\mathbb{R}$. Let us find the solution in the form of a power series $y(x) = \sum_{n=0}^{\infty} c_n x^n$. Substituting this series in the original equation, we have

$$\sum_{n=2}^{\infty} n(n-1)c_n x^{n-2} - x\sum_{n=1}^{\infty} nc_n x^{n-1} + \sum_{n=0}^{\infty} c_n x^n = 1$$

or

$$2c_2 + c_0 + \sum_{n=1}^{\infty} ((n+2)(n+1)c_{n+2} - nc_n + c_n)x^n = 1.$$

Equaling the coefficients with the same powers of x, we get

$$2c_2 + c_0 = 1; \quad (n+2)(n+1)c_{n+2} + (1-n)c_n = 0, \forall n \in \mathbb{N}.$$

Employing the initial conditions, we find $c_0 = 0$ and $c_1 = 0$. Then, the first equation of the system of coefficients gives $c_2 = \frac{1}{2}$ and the remaining equations form the recurrence relations

$$c_{n+2} = \frac{n-1}{(n+2)(n+1)}c_n, \forall n \in \mathbb{N},$$

that can be decoupled into two groups—with the odd indices starting from c_1, and with the even indices starting from c_2. Since c_1 and c_2 were already found, all the coefficients c_n are specified: for the odd-numbered coefficients we have $c_{2k-1} = 0, \forall k \in \mathbb{N}$ (since

$c_1 = 0$), and for the even-numbered coefficients we get

$$c_{2k} = \frac{2k-3}{2k(2k-1)}c_{2k-2} = \frac{(2k-3)(2k-5)}{2k(2k-1)(2k-2)(2k-3)}c_{2k-4}$$

$$= \dots = \frac{(2k-3)(2k-5)\dots 1}{2k(2k-1)(2k-2)(2k-3)\cdot\dots\cdot 4\cdot 3}c_2, \forall k \in \mathbb{N}.$$

Since $c_2 = \frac{1}{2}$, it follows that

$$c_{2k} = \frac{1\cdot 3\cdot\dots\cdot(2k-5)(2k-3)}{3\cdot 4\cdot\dots\cdot(2k-1)2k}\cdot\frac{1}{2}, \forall k \in \mathbb{N},$$

and consequently,

$$y(x) = \frac{x^2}{2} + \sum_{n=2}^{\infty}\frac{1\cdot 3\cdot\dots\cdot(2n-5)(2n-3)}{(2n)!}x^{2n}.$$

To finalize, notice that the coefficients $a_1 = -x$ and $a_0 = 1$ as well as the right-hand side $b(x) = 1$ of the original equation are analytic functions on $\mathbb{R}$. Then, by the general statement, the power series solution has the same radius of convergence $R = \infty$, and consequently, the obtained series converges uniformly on (at least) any closed interval. The reader can check that the same result follows from the application of a convergence test.

6.8   Complement: The Number e Is Irrational

Using the representation of e^x in the power series

$$e^x = \sum_{n=0}^{\infty}\frac{x^n}{n!}, \quad \forall x \in \mathbb{R}$$

we can show that the number e is irrational.

Suppose, by contradiction, that e is rational and can be written in the form $e = \frac{p}{q}$, $p, q \in \mathbb{N}$. Then,

$$e^{-1} = \frac{q}{p} = \sum_{n=0}^{\infty}\frac{(-1)^n}{n!}.$$

Separating the last series into two parts—the finite sum up to $n = p$ and p-th remainder—and multiplying both sides of the last equality by $p!$, we obtain

$$p!\left(\frac{q}{p} - \sum_{n=0}^{p}\frac{(-1)^n}{n!}\right) = p!\sum_{n=p+1}^{\infty}\frac{(-1)^n}{n!}$$

or

$$q(p-1)! - \sum_{n=0}^{p}(-1)^n\frac{p!}{n!} = \sum_{n=p+1}^{\infty}(-1)^n\frac{p!}{n!}.$$

On the left-hand side all the numbers are integers, and consequently, their sum is also an integer. At the same time, on the right-hand side we have an alternating series whose sum s is located between 0 and its first term (see evaluations of alternating series in Sect. 4.1 of Chap. 2), that is, $0 < |s| \le \frac{p!}{(p+1)!} = \frac{1}{p+1}$. Since $p \in \mathbb{N}$, it follows that $\frac{1}{p+1}$ is not an integer. Therefore, we arrive at the contradiction.

6.9   Complement: The Number π Is Irrational

The demonstration of this result does not use power series, instead it employs a specific result of sequences of numbers. However, we find it appropriate to place this statement in this part of the text, since it does not interfere with other results and its meaning is similar to the preceding statement on the irrationality of e.

Let us suppose, by contradiction, that π is a rational number, that is, $\pi = \frac{p}{q}$, $p, q \in \mathbb{N}$. We introduce the two types of polynomials

$$f(x) = \frac{x^n(p - qx)^n}{n!}$$

and

$$F(x) = \sum_{j=0}^{n}(-1)^j f^{(2j)}(x) = f(x) - f^{(2)}(x) + f^{(4)}(x) - \ldots + (-1)^n f^{(2n)}(x)$$

and investigate their properties. First, notice that $n!f(x)$ is a polynomial of degree $2n$, containing the terms whose degrees vary from n to $2n$, and with the integer coefficients (because $p, q \in \mathbb{N}$). Therefore, $f(x)$ can be written in the form $f(x) = \frac{x^n(p-qx)^n}{n!} =$

$\frac{1}{n!} \sum_{i=n}^{2n} c_i x^i$. Then, derivatives of $f(x)$ can be expressed in the following manner:

$$f^{(k)}(x) = \frac{1}{n!} \left(c_n \cdot n \cdot \ldots \cdot (n-k+1)x^{n-k} + terms\ of\ x\ of\ degree \right.$$

$$\left. greater\ than\ n-k \right), \ if\ k < n;$$

$$f^{(n)}(x) = \frac{1}{n!} \left(c_n \cdot n! + terms\ of\ x \right) ;$$

$$f^{(n+1)}(x) = \frac{1}{n!} \left(c_{n+1} \cdot (n+1)! + terms\ of\ x \right) ;$$

$$\cdots$$

$$f^{(2n)}(x) = \frac{1}{n!} c_{2n} \cdot (2n)! ;$$

$$f^{(k)}(x) = 0, \ if\ k > 2n.$$

Evaluating these derivatives at the origin, we have

$$f^{(k)}(0) = 0, \ if\ k < n; \ f^{(k)}(0) = 0, \ if\ k > 2n;$$

$$f^{(n)}(0) = c_n; \ f^{(n+1)}(0) = (n+1)c_{n+1}; \ \ldots f^{(2n)}(0) = (n+1) \cdot \ldots \cdot 2n \cdot c_{2n}.$$

This shows that all derivatives $f^{(k)}(0)$ have integer values. The same can be said about derivatives at the point π. Indeed, $f(x)$ satisfies the following relation

$$f(\pi - x) = f(\frac{p}{q} - x) = \frac{\left(\frac{p}{q} - x\right)^n \left(p - q\left(\frac{p}{q} - x\right)\right)^n}{n!} = \frac{\left(\frac{p}{q} - x\right)^n (qx)^n}{n!}$$

$$= \frac{(p - qx)^n x^n}{n!} = f(x),$$

which implies the corresponding relations for derivatives: $f^{(k)}(\pi - x) = (-1)^k f^{(k)}(x)$. The same conclusion is valid for derivatives of any order of $F(x)$ at the points 0 and π, because they are the sums and differences of derivatives of $f(x)$.

Now, by performing elementary calculations, we reveal one more property of $F(x)$ and $f(x)$:

$$\left(F'(x)\sin x - F(x)\cos x\right)' = F''(x)\sin x + F(x)\sin x$$

$$= \left(f^{(2)}(x) - f^{(4)}(x) - \ldots + (-1)^{n-1}f^{(2n)}(x)\right)\sin x$$

$$+ \left(f(x) - f^{(2)}(x) + f^{(4)}(x) - \ldots + (-1)^n f^{(2n)}(x)\right)\sin x = f(x)\sin x .$$

Then, integrating the last relation on $[0, \pi]$, we get

$$\int_0^\pi f(x)\sin x\,dx = \left(F'(x)\sin x - F(x)\cos x\right)\big|_0^\pi = F(\pi) + F(0).$$

On the right-hand side we have an integer and this integer is positive, because $f(x)\sin x > 0$, $\forall x \in (0, \pi)$. On the other hand, $x^n < \pi^n$ and $(p - qx)^n = p^n\left(1 - \frac{q}{p}x\right)^n = p^n\left(1 - \frac{x}{\pi}\right)^n < p^n$, $\forall x \in (0, \pi)$, which implies the following evaluation of the integrand:

$$0 < f(x)\sin x < \frac{\pi^n p^n}{n!} , \quad \forall x \in (0, \pi).$$

Since $\frac{(\pi p)^n}{n!} \xrightarrow[n\to\infty]{} 0$, the integral on the left-hand side can be made as small as we want by choosing sufficiently large values of n, and consequently, this integral cannot be a positive integer. Thus, we arrive at the contradiction.

7   Complement: Borel's Theorem

. . . the idea acquired in the study of functions of a complex variable – that Taylor's series is valuable for every function which has some derivatives well defined in a domain – is imprinted on the mind to the point of becoming identical with it and appears as a notion as simple as that of continuity for the functions of a real variable that are encountered in physics.
Félix Édouard Justin Émile Borel, 1895

7.1   Smooth Non-analytic Function

Consider the function

$$f(x) = \begin{cases} 0, \ x \le 0 \\ e^{-\frac{1}{x}}, \ x > 0 \end{cases},$$

whose domain is $\mathbb{R}$ and show that this function is infinitely differentiable on $\mathbb{R}$ (let us call such functions smooth), but is not analytic on $\mathbb{R}$. Evidently $f(x)$ is infinitely differentiable at every point $x \ne 0$, since it is constant on the interval $(-\infty, 0)$ and it is a composition of infinitely differentiable functions $t = \frac{1}{x}$ on $x \in (0, +\infty)$ (with the image $t \in (0, +\infty)$) and e^{-t} on $t \in (0, +\infty)$. Hence, the unique problematic point is 0. Since the function is zero on the left of the origin, all its left-sided derivatives at 0 are equal to 0. Therefore, to prove the infinite differentiability at the origin, we can show that $f(x)$ is continuous at 0 and all the right-sided limits of the derivatives at 0 are equal to 0.

Notice first that $f(x)$ is continuous at the origin:

$$\lim_{x \to 0^+} f(x) = \lim_{x \to 0^+} e^{-\frac{1}{x}} = \lim_{t \to +\infty} e^{-t} = 0 = \lim_{x \to 0^-} 0 = \lim_{x \to 0^-} f(x).$$

Now we find a general form of the derivatives when $x > 0$. For the first derivatives we have:

$$f'(x) = \frac{1}{x^2} e^{-\frac{1}{x}}, \ f''(x) = \frac{1 - 2x}{x^4} e^{-\frac{1}{x}}, \ f'''(x) = \frac{1 - 6x + 6x^2}{x^6} e^{-\frac{1}{x}},$$

that make possible to guess that a general form of k-th derivative is

$$f^{(k)}(x) = \frac{P_{k-1}(x)}{x^{2k}} e^{-\frac{1}{x}},$$

where $P_{k-1}(x)$ is a polynomial of degree $k - 1$. Let us demonstrate this by induction. For the first value of k the formula was already derived. Assuming now that the expression for $f^{(k)}(x)$ is true, we demonstrate the validity of the formula for $(k + 1)$-th derivative:

$$f^{(k+1)}(x) = \left(\frac{P_{k-1}(x)}{x^{2k}} e^{-\frac{1}{x}} \right)' = \frac{P_{k-1}(x)}{x^{2k}} \frac{1}{x^2} e^{-\frac{1}{x}}$$

$$+ \frac{P'_{k-1}(x) \cdot x^{2k} - P_{k-1}(x) \cdot 2kx^{2k-1}}{x^{4k}} e^{-\frac{1}{x}}$$

$$= \frac{P_{k-1}(x) \cdot (1 - 2kx) + P'_{k-1}(x) \cdot x^2}{x^{2(k+1)}} e^{-\frac{1}{x}} = \frac{P_k(x)}{x^{2(k+1)}} e^{-\frac{1}{x}}.$$

Here $P_k(x)$ is a polynomial of degree k, because its principal coefficient is different from zero:

$$P_k(x) = a_k x^k + \ldots = P_{k-1}(x) \cdot (1 - 2kx) + P'_{k-1}(x) \cdot x^2$$

$$= -a_{k-1} x^{k-1} \cdot 2kx + a_{k-1}(k-1)x^{k-2} \cdot x^2 + \ldots = -a_{k-1}(1+k)x^k + \ldots,$$

that is, $a_k = -(1+k)a_{k-1} \neq 0$. Using the found formula for $f^{(k)}(x)$, $x > 0$, we can prove that the right-sided limits of these derivatives at 0 are equal to 0:

$$\lim_{x \to 0^+} f^{(k)}(x) = \lim_{x \to 0^+} \frac{P_{k-1}(x)}{x^{2k}} e^{-\frac{1}{x}} = \lim_{t \to +\infty} P_{k-1}\left(\frac{1}{t}\right) \cdot t^{2k} e^{-t}$$

$$= \lim_{t \to +\infty} P_{k-1}\left(\frac{1}{t}\right) \cdot \lim_{t \to +\infty} t^{2k} e^{-t} = 0.$$

(Here, we have used the two auxiliary results: $\lim_{t \to +\infty} P_{k-1}\left(\frac{1}{t}\right) = P_{k-1}(0) = 1$ due to the continuity of a polynomial, and $\lim_{t \to +\infty} t^{2k} e^{-t} = 0$ according to L'Hospital's rule applied $2k$ times in a row). Thus, $f(x)$ is continuous at the origin and the one-sided limits of the derivatives coincide:

$$\lim_{x \to 0^+} f^{(k)}(x) = 0 = \lim_{x \to 0^-} f^{(k)}(x), \ \forall k = 0, 1, \ldots .$$

Therefore, $f(x)$ is infinitely differentiable at 0 and all its derivatives (including the function value) are zeros. Hence, all the Taylor coefficients of $f(x)$ are zeros at the central point 0. However, the series of zeros converges to zero, that is, the sum of the Taylor series of $f(x)$ differs from the values of $f(x)$ at every point $x > 0$. For this reason, the function $f(x)$ cannot be represented in the Taylor series centered at the origin, that is, $f(x)$ is not analytic at the origin.

To complete the study of this function, we investigate its monotonicity, concavity and horizontal asymptotes when $x > 0$. Notice that $f'(x) = \frac{1}{x^2} e^{-\frac{1}{x}} > 0$, $x > 0$ implies the strict increasing of $f(x)$ on $[0, +\infty)$. The second derivative $f''(x) = \frac{1-2x}{x^4} e^{-\frac{1}{x}}$, $x > 0$ vanishes only at the point $x = \frac{1}{2}$, keeping the positive sign for $0 < x < \frac{1}{2}$ and the negative for $x > \frac{1}{2}$. This means that on the interval $(0, \frac{1}{2})$ the function has the upward concavity, on the interval $(\frac{1}{2}, +\infty)$—the downward concavity, and $x = \frac{1}{2}$ is the inflection point. Finally, the limit at infinity

$$\lim_{x \to +\infty} f(x) = \lim_{x \to +\infty} e^{-\frac{1}{x}} = \lim_{t \to 0^+} e^{-t} = 1$$

shows that $y = 1$ is the horizontal asymptote of $f(x)$. The graph of $f(x)$ is sketched in Fig. 5.13.

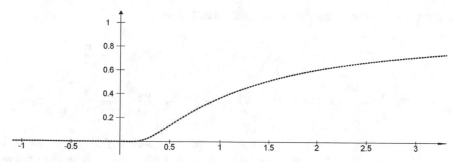

Fig. 5.13 Function $f(x)$

7.2 Transition Function

Now we take the function

$$f(x) = \begin{cases} 0, & x \le 0 \\ e^{-\frac{1}{x}}, & x > 0 \end{cases},$$

whose essential properties were analyzed, and form the new function

$$h(x) = \frac{f(1-x)}{f(x) + f(1-x)}.$$

Let us clarify some properties of this function. First, $h(x)$ is defined on $\mathbb{R}$, since its denominator is always positive: for $x \le 0$ $(1 - x \ge 1)$ we have $f(x) + f(1 - x) = f(1-x) = e^{-\frac{1}{1-x}} > 0$; for $x \ge 1$ $(1-x \le 0)$ we get $f(x)+f(1-x) = f(x) = e^{-\frac{1}{x}} > 0$; and for $x \in (0, 1)$ $(1 - x \in (0, 1))$ both summands are positive. The image of $h(x)$ is the interval $[0, 1]$. More specifically, when $x \le 0$ we have $h(x) = \frac{f(1-x)}{f(1-x)} = 1$; when $x \ge 1$ we obtain $h(x) = \frac{0}{f(x)} = 0$; and for $x \in (0, 1)$ all the terms of the numerator and denominator are positive with the denominator greater than the numerator, and consequently $h(x) = \frac{f(1-x)}{f(x)+f(1-x)} \in (0, 1)$. Since $f(x)$ is a smooth function (infinitely differentiable on $\mathbb{R}$), the function $h(x)$ is also smooth. Let us show that $h(x)$ is strictly decreasing on $(0, 1)$. Indeed, the derivative of $h(x)$ can be expressed in the following manner (we denote $t = 1 - x$):

$$h'(x) = \left(\frac{f(t)}{f(x) + f(t)} \right)' = \frac{f_t(t) \cdot (-1) \cdot (f(x) + f(t)) - f(t) \cdot (f_x(x) + f_t(t) \cdot (-1))}{(f(x) + f(t))^2}$$

$$= -\frac{f_t(t)f(x) + f(t)f_x(x)}{(f(x) + f(t))^2}.$$

For $x \in (0, 1)$ the numerator can be represented as follows:

$$f_t(t)f(x) + f(t)f_x(x) = e^{-\frac{1}{t}}\frac{1}{t^2} \cdot e^{-\frac{1}{x}} + e^{-\frac{1}{t}} \cdot e^{-\frac{1}{x}}\frac{1}{x^2}$$

$$= e^{-\frac{1}{1-x}}\frac{1}{(1-x)^2} \cdot e^{-\frac{1}{x}} + e^{-\frac{1}{1-x}} \cdot e^{-\frac{1}{x}}\frac{1}{x^2} = e^{-\frac{1}{1-x}}e^{-\frac{1}{x}}\left(\frac{1}{(1-x)^2} + \frac{1}{x^2}\right) > 0.$$

Consequently, $h'(x) < 0$ on $(0, 1)$, which means that $h(x)$ is strictly decreasing on $(0, 1)$.

Thus, $h(x)$ is a smooth function on $\mathbb{R}$, equal to 1 on $(-\infty, 0]$, to 0 on $[1, +\infty)$ and strictly decreasing on $(0, 1)$:

$$h(x) = \begin{cases} 1, x \leq 0 \\ 0, x \geq 1 \\ strictly\ decreasing,\ x \in (0, 1) \end{cases}.$$

Such functions are called the transition functions from 1 to 0 on the interval $(0, 1)$. They are also called "bridging" or "smoothing" functions. Naturally, $\bar{h}(x) = -h(x)$ is a transition function from -1 to 0 on the interval $(0, 1)$: $-h(x)$ is a smooth function on $\mathbb{R}$, equal to -1 on $(-\infty, 0]$, to 0 on $[1, +\infty)$ and strictly increasing on $(0, 1)$:

$$-h(x) = \begin{cases} -1, x \leq 0 \\ 0, x \geq 1 \\ strictly\ increasing,\ x \in (0, 1) \end{cases}.$$

The graph of $h(x)$ is shown in Fig. 5.14.

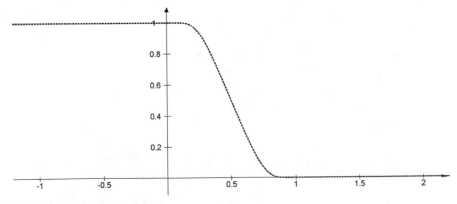

Fig. 5.14 Transition function $h(x)$

Using a transition function from 1 to 0 on the interval $(0, 1)$, it is easy to find a transition function from $a \neq 0$ to 0 on the interval (α, β) $(\alpha < \beta)$. To modify the magnitude of the transition, we just multiply $h(x)$ by the new constant a, and to change the interval of the transition we make the change of the independent variable: $x = \frac{t-\alpha}{\beta-\alpha}$. Then, the function $\tilde{h}(t) = ah\left(\frac{t-\alpha}{\beta-\alpha}\right)$ is smooth on $\mathbb{R}$ and satisfies the properties of the transition:

$$\tilde{h}(t) = \begin{cases} a, t \leq \alpha \\ 0, t \geq \beta \\ strictly\ monotone,\ t \in (\alpha, \beta) \end{cases}.$$

Finally, we introduce a symmetric transition function. If the interval (α, β) lies on the right of the origin $(0 < \alpha < \beta)$, then we can compose the function that makes transition from a to 0 on (α, β) and, symmetrically with respect to the ordinate axis, the transition from 0 to a on $(-\beta, -\alpha)$. To obtain such a function, we can take the part of $\tilde{h}(x)$ that corresponds to $x \geq 0$ and reflect it with respect to the ordinate axis. Translating this geometric construction to the analytic form, we obtain $\varphi(x) = \tilde{h}(|x|)$. According to the properties of $\tilde{h}(x)$, the function $\varphi(x)$ is smooth on $\mathbb{R}$, equal to 0 on $(-\infty, -\beta]$, equal to a on $(-\alpha, \alpha)$, and again equal to 0 on $[\beta, +\infty)$. Besides, it is strictly monotone on $(-\beta, -\alpha)$ and (α, β): if $a > 0$, $\varphi(x)$ is strictly increasing on $(-\beta, -\alpha)$ and strictly decreasing on (α, β); if $a < 0$, the monotonicity is of inverse direction. The analytic description of the symmetric transition function from a to 0 on the interval (α, β) is as follows:

$$\varphi(x) = \begin{cases} a, |x| \leq \alpha \\ 0, |x| \geq \beta \\ strictly\ monotone,\ \alpha < |x| < \beta \end{cases}$$

and $\varphi(x)$ is smooth on $\mathbb{R}$. The graph of $\varphi(x)$ for $a = 1$, $\alpha = 1$ and $\beta = 2$ is shown in Fig. 5.15.

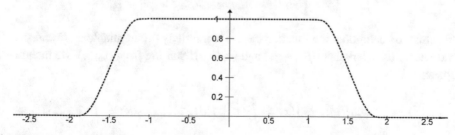

Fig. 5.15 Symmetric transition function $\varphi(x)$

7.3   Borel's Theorem

Émile Borel have proved the following statement: for any sequence of real numbers c_n, there exists a smooth (infinitely differentiable on $\mathbb{R}$) function such that $f^{(n)}(0) = c_n$, $\forall n$. In the context of power series, we reword Borel's statement in the following (equivalent) form:

Borel's Theorem *Any power series is a Taylor series.*

Proof As usual, to simplify considerations we perform the proof for a series centered at the origin (following the original Borel's formulation).

Given a power series $\sum_{n=0}^{\infty} c_n x^n$, the Borel's theorem states that there exists a function $f(x)$ such that this series can be written in the form $\sum_{n=0}^{\infty} \frac{f^n(0)}{n!} x^n$, that is, there exists $f(x)$ such that $c_n = \frac{f^n(0)}{n!}$, $\forall n = 0, 1, \ldots$. It is evident that $f(x)$ should be a smooth (infinitely differentiable) function in a neighborhood of 0. We will construct the required function in the form of a series $f(x) = \sum_{n=0}^{\infty} f_n(x)$ whose general term $f_n(x)$ is related to the general coefficient c_n of the given power series. As the elements of this construction we will use the symmetric transition functions $\varphi(x)$ defined above, that is, the infinitely differentiable on $\mathbb{R}$ functions which are constant outside the intervals $(-\beta, -\alpha)$ and (α, β) ($\beta > \alpha > 0$) and strictly monotone on these intervals.

Pick one of the coefficients c_n and associate with it the function $f_n(x)$ formed in the following way. We start with the auxiliary function $\varphi_{n0}(x)$, which is the symmetric transition function from $c_n \cdot n!$ to 0 on the interval (α_n, β_n), where $\alpha_n = \frac{1}{2|c_n| \cdot n! + 1}$ and $\beta_n = 2\alpha_n$. The description of this function is as follows (see the preceding subsection):

$$\varphi_{n0}(x) = \begin{cases} c_n \cdot n!, \ |x| \leq \alpha_n \\ 0, \ |x| \geq \beta_n \\ strictly \ monotone, \ \alpha_n < |x| < \beta_n \end{cases}.$$

Notice that, by definition, the function $\varphi_{n0}(x)$ is infinitely differentiable on $\mathbb{R}$ and strictly monotone on the intervals $(-\beta_n, -\alpha_n)$ and (α_n, β_n). Starting from $\varphi_{n0}(x)$, we integrate it n times:

$$\varphi_{n1}(x) = \int_0^x \varphi_{n0}(t)dt, \ \varphi_{n2}(x) = \int_0^x \varphi_{n1}(t)dt, \ \ldots, \ \varphi_{nn}(x) = \int_0^x \varphi_{n,n-1}(t)dt$$

and then define $f_n(x) = \varphi_{nn}(x)$, $\forall x \in \mathbb{R}$. If $n = 0$, then we simply set $f_0(x) = \varphi_{00}(x)$, $\forall x \in \mathbb{R}$. Evidently, $f_n(x)$ is infinitely differentiable on $\mathbb{R}$ and its derivatives satisfy the following relations:

$$f_n'(x) = \varphi_{n,n-1}(x), \ f_n''(x) = \varphi_{n,n-2}(x), \ \ldots, \ f_n^{(n-1)}(x) = \varphi_{n1}(x), \ f_n^{(n)}(x) = \varphi_{n0}(x).$$

We can show by induction that these derivatives can be evaluated in the form

$$|f_n^{(k)}(x)| \leq \frac{|x|^{n-k-1}}{(n-k-1)!}, \ \forall k = 0, 1, \ldots, n-1, \ \forall x \in \mathbb{R}.$$

In fact, for $f_n^{(n-1)}(x)$ we have

$$|f_n^{(n-1)}(x)| = \left| \int_0^x \varphi_{n0}(t)dt \right| \leq \int_0^{\beta_n} |c_n| \cdot n! dt = |c_n| \cdot n! \cdot \beta_n < 1.$$

Therefore,

$$|f_n^{(n-2)}(x)| = \left| \int_0^x f_n^{(n-1)}(t)dt \right| \leq \int_0^{|x|} 1 dt \leq |x|.$$

Assume now that $|f_n^{(k)}(x)| \leq \frac{|x|^{n-k-1}}{(n-k-1)!}$ for some $0 < k < n-1$. Then

$$|f_n^{(k-1)}(x)| = \left| \int_0^x f_n^{(k)}(t)dt \right| \leq \int_0^{|x|} \frac{|t|^{n-k-1}}{(n-k-1)!} dt = \frac{|x|^{n-k}}{(n-k)!},$$

which demonstrates the validity of the inequality for $\forall k = 0, 1, \ldots, n-1, \forall n = 0, 1, \ldots$. In particular, $|f_n(x)| \leq \frac{|x|^{n-1}}{(n-1)!}$.

Now we define the desired function $f(x)$ as the series $f(x) = \sum_{n=0}^{\infty} f_n(x)$. First we show that this series converges uniformly on any closed interval. Indeed, if $|x| \leq K$, then $|f_n(x)| \leq \frac{|x|^{n-1}}{(n-1)!} \leq \frac{K^{n-1}}{(n-1)!}$, and consequently, the convergence of the majorant numerical series $\sum_{n=1}^{\infty} \frac{K^{n-1}}{(n-1)!}$ implies (by the Weierstrass test) the uniform convergence of the series $\sum_{n=0}^{\infty} f_n(x)$ on $[-K, K]$. In a similar manner we can prove that the series of derivatives $\sum_{n=0}^{\infty} f_n^{(k)}(x)$ also converges uniformly on $[-K, K]$ for any positive K. Indeed, take an arbitrary k and fix it for a moment. For all the elements $f_n^{(k)}(x)$ with $n \geq k+1$ the inequality $|f_n^{(k)}(x)| \leq \frac{|x|^{n-k-1}}{(n-k-1)!}$ is satisfied (this means that there is only a finite number of the terms from $n = 0$ to $n = k$ that, possibly, do not satisfy this evaluation). Then $|f_n^{(k)}(x)| \leq \frac{|x|^{n-k-1}}{(n-k-1)!} \leq \frac{K^{n-k-1}}{(n-k-1)!}$, and consequently, the convergence of the majorant numerical series $\sum_{n=k+1}^{\infty} \frac{K^{n-k-1}}{(n-k-1)!}$ implies (by the Weierstrass test) the uniform convergence of the series $\sum_{n=0}^{\infty} f_n^{(k)}(x)$ on $[-K, K]$. Thus, the series $\sum_{n=0}^{\infty} f_n(x)$ and the series of its derivatives of any order $\sum_{n=0}^{\infty} f_n^{(k)}(x)$ converge uniformly on any closed interval. This property allows the term-by-term differentiation of $f(x)$:

$$f^{(k)}(x) = \sum_{n=0}^{\infty} f_n^{(k)}(x), \ \forall k = 0, 1, \ldots .$$

Using the last formula for the derivatives, we evaluate the values of $f^{(k)}(x)$ at the origin. Choosing an arbitrary n and fixing it, we have the following relations for the derivatives of $f_n(x)$ up to order $n - 1$ at 0:

$$f_n(0) = \int_0^0 \varphi_{n,n-1}(t)dt = 0, \ \ldots, \ f_n^{(n-1)}(0) = \varphi_{n1}(0) = \int_0^0 \varphi_{n0}(t)dt = 0.$$

For the n-th derivative we use the definition of the function $\varphi_{n0}(x)$:

$$f_n^{(n)}(0) = \varphi_{n0}(0) = c_n \cdot n!.$$

Finally, any derivative of order $m > n$ is the derivative of order $m - n$ of $\varphi_{n0}(x)$, which is a constant function in a neighborhood of 0, and consequently,

$$f_n^{(m)}(0) = \varphi_{n0}^{(m-n)}(0) = 0, \forall m > n.$$

Summarizing,

$$f_n^{(k)}(0) = \begin{cases} c_n \cdot n!, k = n \\ 0, k \neq n \end{cases}, \forall n.$$

Therefore,

$$f(0) = f_0(0) = c_0, \ f'(0) = f_1'(0) = c_1 \cdot 1!, \ \ldots, \ f^{(n)}(0) = f_n^{(n)}(0) = c_n \cdot n!, \ \ldots,$$

that is, $f(x)$ is the required function. $\qquad\qquad\qquad\qquad\qquad\qquad\qquad\qquad\qquad\quad \square$

Remark The uniform convergence of the series for $f(x)$: $f(x) = \sum_{n=0}^{\infty} f_n(x)$ has no connection with the convergence of the original series $\sum_{n=0}^{\infty} c_n x^n$. It may happen that the latter converges only at its central point, like, for instance, the series $\sum_{n=0}^{\infty} n! x^n$.

Corollary 1 *There is an infinite number of functions whose Taylor series is a given power series (or rewording, there is an infinite number of functions with the same Taylor series).*

Indeed, the functions $h(x) = f(x) + \lambda e^{-\frac{1}{x^2}}$, where $f(x)$ is the function constructed in the proof of Borel's Theorem and λ is an arbitrary parameter, have the same Taylor series since the Taylor series of the function $e^{-\frac{1}{x^2}}$ is zero series.

Corollary 2 *In Sect. 4.3 we have considered an example of the Taylor series that diverges at all the points except at the central point. Using the result of Borel's Theorem, we can create many other functions with the same property: it is sufficient to take any divergent power series (for example, $\sum_{n=0}^{\infty} n! x^n$) and, following the proof of Borel's Theorem, construct a function $f(x)$, whose Taylor series coincides with the chosen power series.*

Historical Remarks Borel's theorem was first known from the Émile Borel dissertation published in 1895. It was discovered very recently that Peano has demonstrated practically the same theorem using completely different approach based on a generalization of the example of infinitely differentiable everywhere and analytic nowhere function given by du Bois-Reymond in 1876. Peano's formulation and prove were hidden in the book on differential and integral calculus published in 1884 and remained unnoticed until 2014. However, by tradition, this result is attributed to Émile Borel.

Exercises

1. Find the interval of convergence of the power series and investigate the behavior at the endpoints:

   (1) $\sum_{n=1}^{\infty} \frac{5^n + (-3)^n}{n+1} x^n$;

   (2*) $\sum_{n=1}^{\infty} \frac{(2n)!!}{(2n+1)!!} x^n$;

   (3) $\sum_{n=0}^{\infty} \frac{(-3)^n}{\sqrt[3]{3n+1}} (x-1)^n$;

   (4) $\sum_{n=0}^{\infty} \frac{(x+1)^n}{3^n (n+3) \ln^3(n+3)}$;

   (5) $\sum_{n=0}^{\infty} \frac{(2n+1)(x+2)^n}{2^n (3n+2)(5n-1)}$;

   (6) $\sum_{n=1}^{\infty} \frac{(x-1)^n}{4^n n^p}$, $\forall p \in \mathbb{R}$;

   (7) $\sum_{n=1}^{\infty} \frac{n!}{2^n n^n} (1-3x)^n$;

   (8) $\sum_{n=1}^{\infty} 2^n \sin \frac{1}{3^n} x^n$;

   (9) $\sum_{n=2}^{\infty} \frac{(2x-3)^n}{5^n n \ln^p n}$, $\forall p \in \mathbb{R}$;

   (10*) $\sum_{n=1}^{\infty} \left(1 + \frac{1}{n}\right)^{n^2} x^n$;

   (11) $\sum_{n=0}^{\infty} \frac{2^n}{(n+1)!} (2-5x)^n$;

   (12) $\sum_{n=0}^{\infty} \frac{2^{n^2}}{(n+1)!} (x-1)^n$;

   (13) $\sum_{n=1}^{\infty} (-1)^n \frac{(n!)^2}{(2n)!} (x-4)^n$;

(14*) $\sum_{n=1}^{\infty}(-1)^n \frac{(2n)!}{(n!)^2}(x-4)^n$;

(15) $\sum_{n=0}^{\infty}(-1)^n \frac{(x-2)^{2n}}{2^n(n+2)}$;

(16) $\sum_{n=1}^{\infty} \frac{5^n}{n^n} x^{n^2}$;

(17) $\sum_{n=2}^{\infty} \frac{(2+(-1)^n)^n}{n \ln n}\left(\frac{x}{2}+1\right)^n$;

(18*) $\sum_{n=0}^{\infty} \frac{(3+4\cos \frac{n\pi}{4})^n}{\sqrt[3]{n+2}}\left(\frac{2x-6}{5}\right)^n$;

(19) $\sum_{n=1}^{\infty} \frac{(2+3(-1)^{\frac{n(n+1)}{2}})^n}{n^{3/2}}(x-1)^n$;

(20*) $\sum_{n=1}^{\infty} \frac{(x+1)^n}{4\sqrt{n}\sqrt[3]{n^2+2}}$.

2. Suppose the series $\sum_{n=0}^{\infty} a_n x^n$ and $\sum_{n=0}^{\infty} b_n x^n$ have the radii of convergence R_a and R_b, respectively. Show that:

   (1) the radius of convergence of the series $\sum_{n=0}^{\infty}(a_n + b_n)x^n$ satisfies the condition
   $R_s \geq \min(R_a, R_b)$;

   (2) the radius of convergence of the series $\sum_{n=0}^{\infty}(a_n b_n)x^n$ satisfies the condition
   $R_p \geq R_a \cdot R_b$.

   Provide examples of series when the equality is attained and when the inequality is strict.

3. Represent the function as a power series centered at 0 using the power series expansions of elementary function. Find the set of convergence of the obtained series:

   (1) $f(x) = \sin^3 x$;

   (2) $f(x) = \cos^3 x$;

   (3) $f(x) = \cos^4 x$;

   (4) $f(x) = \frac{x^5}{1-3x}$;

   (5) $f(x) = \frac{x}{x^2-5x+6}$;

   (6) $f(x) = \frac{1}{x^2+x+1}$;

   (7*) $f(x) = \frac{2x^2}{\sqrt[3]{1-2x^2}}$;

   (8) $f(x) = \frac{3x-5}{(x-1)^3}$;

   (9) $f(x) = \ln(1 - 2x + x^2 - 2x^3)$;

   (10) $f(x) = \ln\sqrt{\frac{1+x}{1-x}}$;

   (11) $f(x) = \ln(x^2 + 5x + 6)$;

   (12*) $f(x) = x\sqrt[4]{16+x^2}$;

   (13) $f(x) = \sin 2x + 2x \cos 2x$;

   (14) $f(x) = (1+x)e^{-2x}$.

4. Represent the function as a power series centered at 0 using the power series expansion of its derivative. Find the set of convergence of the obtained series:

   (1) $f(x) = \ln\left(x + \sqrt{1+x^2}\right)$;

   (2) $f(x) = \frac{1}{4}\ln\frac{1+x}{1-x} + \frac{1}{2}\arctan x$;

   (3) $f(x) = \arctan\frac{2-2x}{1+4x}$;

   (4) $f(x) = x \arctan x - \ln\sqrt{1+x^2}$;

(5) $f(x) = 2x \arccos 2x - \sqrt{1 - 4x^2}$;

(6) $f(x) = 2x \arcsin(1 - 8x^4)$;

(7) $f(x) = x \ln\left(x + \sqrt{1 + x^2}\right) - \sqrt{1 + x^2}$;

(8*) $f(x) = \tan x$ (find the first five nonzero terms).

5. Express the function in the form of a power series of a given quantity. Find the set of convergence of the obtained series:

(1) $f(x) = \frac{x}{x^2 - 5x + 6}$, powers of $(x - 5)$;

(2) $f(x) = \frac{1}{x^2 + 3x + 2}$, powers of $(x + 4)$;

(3) $f(x) = x^3 - 3x^2 + 2x - 5$, powers of $(x - 5)$;

(4*) $f(x) = \ln x$, powers of $\frac{1-x}{1+x}$;

(5*) $f(x) = \frac{x}{\sqrt{1+x}}$, powers of $\frac{x}{1+x}$;

(6) $f(x) = \frac{1}{1-x}$, powers of $\frac{1}{x}$;

(7*) $f(x) = \arctan x$, powers of $\frac{1}{x}$.

6. Find a power series expansion about 0 of the given function and determine its interval of convergence:

(1) $f(x) = (1 + x^2) \arctan x$;

(2) $f(x) = (1 - x)^2 \cosh \sqrt{x}$;

(3) $f(x) = \frac{\ln(1+x)}{1+x}$;

(4) $f(x) = \frac{\ln(1-x)}{1-x}$;

(5*) $f(x) = \ln^2(1 - x)$;

(6) $f(x) = \frac{\arctan x}{1+x^2}$;

(7*) $f(x) = \arctan^2 x$;

(8) $f(x) = \ln \cos x$ (find the first four nonzero terms);

(9) $f(x) = e^x \cos x$ (find the first six nonzero terms);

(10) $f(x) = e^x \sin x$ (find the first six nonzero terms).

7. Represent the given function as a power series centered at the origin and find its interval of convergence:

(1) $f(x) = \int_0^x \frac{dt}{\sqrt{1-t^4}}$;

(2) $f(x) = \int_0^x \frac{\arctan t}{t} dt$;

(3) $f(x) = \int_0^x \sin t^2 dt$;

(4) $f(x) = \int_0^x \cos t^2 dt$;

(5) $f(x) = \int_0^x \frac{\ln(1+t)}{t} dt$;

(6) $f(x) = \int_0^x \frac{\arcsin t}{t} dt$.

8. Determine the set of convergence of the given series and find its sum:

(1) $\sum_{n=1}^{\infty} \frac{x^{2n-1}}{2n-1}$;

(2) $\sum_{n=1}^{\infty} \left(x^n + \frac{1}{2^n x^n}\right)$;

(3) $\sum_{n=0}^{\infty} (-1)^n e^{-n \sin x}$;

(4) $\sum_{n=1}^{\infty} \frac{n}{n+1} \frac{x^n}{2^n}$;

(5) $\sum_{n=1}^{\infty} \frac{x^{4n}}{(4n)!}$;

(6) $\sum_{n=0}^{\infty} \frac{(n+1)^2}{2n+1} x^{2n}$;

(7) $\sum_{n=1}^{\infty} \frac{x^{n+1}}{(1-x^n)(1-x^{n+1})}$;

(8) $\sum_{n=1}^{\infty} \frac{x^n}{n(n+1)}$;

(9) $\sum_{n=0}^{\infty} \frac{x^{2n+1}}{2n+1}$;

(10) $\sum_{n=0}^{\infty} (-1)^n \frac{x^{2n+1}}{2n+1}$;

(11) $\sum_{n=0}^{\infty} \frac{x^{2n}}{(2n)!}$;

(12) $\sum_{n=0}^{\infty} \frac{(2n-1)!!}{(2n)!!} x^n$;

(13) $\sum_{n=1}^{\infty} nx^n$;

(14) $\sum_{n=1}^{\infty} (-1)^{n-1} n^2 x^n$;

(15) $\sum_{n=1}^{\infty} n(n+1)x^n$;

(16) $\sum_{n=0}^{\infty} \frac{n^2+1}{2^n n!} x^n$;

(17) $\sum_{n=0}^{\infty} \frac{(-1)^n n^2}{(n+1)!} x^n$;

(18) $\sum_{n=1}^{\infty} (-1)^{n-1} \frac{x^{2n}}{n(2n-1)}$;

(19) $\sum_{n=0}^{\infty} \frac{x^{4n+1}}{4n+1}$;

(20) $\sum_{n=1}^{\infty} \frac{1 \cdot 4 \cdot 7 \cdot \ldots \cdot (3n-2)}{3 \cdot 6 \cdot 9 \cdot \ldots \cdot 3n} \frac{x^n}{2^n}$;

(21) $\sum_{n=0}^{\infty} \frac{2n+1}{n!} x^{2n}$;

(22) $\sum_{n=0}^{\infty} n(n+2)x^n$;

(23) $\sum_{n=0}^{\infty} \frac{n^2}{(2n+1)!} x^{2n}$;

(24) $\sum_{n=0}^{\infty} (-1)^n \frac{2n^2+1}{(2n)!} x^{2n}$.

9. Find the sum of the series of numbers:

(1) $\sum_{n=1}^{\infty} \frac{1}{(2n-1)(2n+1)}$;

(2) $\sum_{n=1}^{\infty} (-1)^{n-1} \frac{1}{(2n-1)3^{n-1}}$;

(3) $\sum_{n=1}^{\infty} \frac{2n-1}{2^n}$;

(4) $\sum_{n=1}^{\infty} (-1)^{n-1} \frac{1}{n(n+1)}$;

(5) $\sum_{n=0}^{\infty} \frac{2^n(n+1)}{n!}$;

(6) $\sum_{n=0}^{\infty} (-1)^n \frac{n}{(2n+1)!}$;

(7) $\sum_{n=1}^{\infty} \frac{n^2}{n!}$;

(8) $\sum_{n=0}^{\infty} \frac{(-1)^n}{2n+1}$;

(9) $\sum_{n=0}^{\infty} (-1)^n \frac{(2n-1)!!}{(2n)!!}$;

(10) $\sum_{n=0}^{\infty} \frac{(2n-1)!!}{(2n)!!} \cdot \frac{1}{2n+1}$;

(11) $\sum_{n=1}^{\infty} (-1)^{n-1} \frac{1}{3n-2}$;

(12) $\sum_{n=1}^{\infty} \frac{1}{n(2n+1)}$;

(13) $\sum_{n=0}^{\infty} \frac{2n+1}{n!}$;

(14) $\sum_{n=2}^{\infty} \frac{1}{n^2-1}$;

(15) $\sum_{n=1}^{\infty} \frac{1}{n(n+1)(n+2)}$;

(16) $\sum_{n=1}^{\infty} \frac{1}{n(n+m)}$, $\forall m \in \mathbb{N}$;

(17) $\sum_{n=2}^{\infty} \frac{(-1)^n}{n^2+n-2}$;

(18) $\sum_{n=0}^{\infty} (-1)^n \frac{3n^2+5}{(2n)!}$.

10. Find the limit using the power series expansions of elementary functions:

(1) $\lim_{x \to 0} \frac{\cos x - e^{-x^2/2}}{x^4}$;

(2) $\lim_{x \to 0} \frac{e^x \sin x - x(1+x)}{x^3}$;

(3) $\lim_{x \to +\infty} x^{3/2}(\sqrt{x+1} + \sqrt{x-1} - 2\sqrt{x})$;

(4) $\lim_{x \to +\infty} (\sqrt[6]{x^6 + x^5} - \sqrt[6]{x^6 - x^5})$;

(5) $\lim_{x \to +\infty} [(x^3 - x^2 + \frac{x}{2})e^{1/x} - \sqrt{x^6 + 1}]$;

(6) $\lim_{x \to 0} \frac{a^x + a^{-x} - 2}{x^2}$, $a > 0$;

(7) $\lim_{x \to \infty} [x - x^2 \ln(1 + \frac{1}{x})]$;

(8) $\lim_{x \to 0} (\frac{1}{x} - \frac{1}{\sin x})$;

(9) $\lim_{x \to 0} \frac{1}{x}(\frac{1}{x} - \cot x)$;

10) $\lim_{x \to 0} \frac{\sin(\sin x) - x\sqrt[3]{1-x^2}}{x^5}$;

(11*) $\lim_{x \to 0} \frac{1 - (\cos x)^{\sin x}}{x^3}$;

(12) $\lim_{x \to 0} \frac{\sinh \tan x - x}{x^3}$;

(13) $\lim_{x \to 0} \frac{1 - \cos x^2}{x^2 \sin x^2}$;

(14) $\lim_{x \to 0} \frac{\arcsin 2x - 2 \arcsin x}{x^3}$;

(15) $\lim_{x \to 0} \frac{1}{x\sqrt{x}} \left(\sqrt{a} \arctan \sqrt{\frac{x}{a}} - \sqrt{b} \arctan \sqrt{\frac{x}{b}} \right)$, $a, b \in \mathbb{R}$;

(16) $\lim_{x \to 0} \frac{\cos \sin x - \cos x}{x^4}$;

(17) $\lim_{x \to 0} \frac{1}{x} \left(\frac{1}{\tanh x} - \frac{1}{\tan x} \right)$;

(18) $\lim_{x \to 0} \left(\frac{1}{\ln(x+\sqrt{1+x^2})} - \frac{1}{\ln(1+x)} \right)$;

(19) $\lim_{x \to +\infty} \left(\sqrt[3]{x^3 + x^2 + x + 1} - \sqrt{x^2 + x + 1} \frac{\ln(e^x + x)}{x} \right)$;

(20) $\lim_{x \to 0} \left(\frac{\cos x}{\cosh x} \right)^{\frac{1}{x^2}}$;

(21) $\lim_{x \to 0} \left(\frac{\arctan x}{x} \right)^{\frac{1}{x^2}}$;

(22) $\lim_{x \to 0} \frac{\sqrt[4]{1+\frac{x}{3}} - \sqrt[3]{1+\frac{x}{4}}}{1 - \sqrt{1 - \frac{x^2}{2}}}$;

(23) $\lim_{x \to 0} \frac{\sqrt[m]{1+\alpha x^3} - \sqrt[k]{1+\beta x^3}}{\sqrt[m]{1+\alpha x^3}\sqrt[k]{1+\beta x^3} - 1}$, $k, m \in \mathbb{N}$, $\alpha, \beta \in \mathbb{R}$, $\alpha^2 + \beta^2 \neq 0$;

(24) $\lim_{x \to 0} \frac{\sqrt{1+\frac{1}{2}x^3 \sin x} + \sqrt{\cos x^2} - 2}{e^{x^3} + e^{-x^3} - 2}$;

(25) $\lim\limits_{x \to 0} \dfrac{\ln(1+xe^x)-x-\frac{1}{2}x^2}{\ln(x+\sqrt{1+x^2})-x}$;

(26) $\lim\limits_{x \to 0} \dfrac{1-\cos^{\alpha+\beta} x}{\sqrt{(1-\cos^\alpha x)(1-\cos^\beta x)}}$, $\alpha, \beta > 0$;

(27) $\lim\limits_{x \to 0} \dfrac{\cos(x^2 e^{x^2})-\cos(x^2 e^{-x^2})}{x^6}$.

11. Find the definite integral by expanding the integrand in the power series. Calculate its value with a given accuracy δ:

(1) $\int_0^1 \frac{\ln(1+x)}{x}dx$, $\delta = 10^{-2}$;

(2) $\int_0^{1/2} e^{-x^2}dx$, $\delta = 10^{-4}$;

(3) $\int_0^1 \sqrt{x}\sin x\, dx$, $\delta = 10^{-4}$;

(4) $\int_0^1 \cos x^2 dx$, $\delta = 10^{-5}$;

(5*) $\int_2^4 e^{1/x}dx$, $\delta = 10^{-3}$;

(6) $\int_0^{1/2} \sqrt{1-x^3}dx$, $\delta = 10^{-5}$;

(7) $\int_0^{1/2} \frac{dx}{\sqrt{1+x^4}}$, $\delta = 10^{-5}$;

(8) $\int_3^{+\infty} \frac{dx}{\sqrt[4]{1+x^8}}$, $\delta = 10^{-5}$;

(9) $\int_0^{1/2} \frac{\arctan x}{x}dx$, $\delta = 10^{-4}$;

(10) $\int_0^{1/2} \frac{\arcsin x}{x}dx$, $\delta = 10^{-4}$;

(11) $\int_0^1 \frac{\sinh x}{x}dx$, $\delta = 10^{-5}$;

(12) $\int_0^1 \frac{\sin x}{x}dx$, $\delta = 10^{-4}$;

(13) $\int_0^1 e^{-x^4}dx$, $\delta = 10^{-4}$;

(14) $\int_0^1 \sqrt{x}\cos x\, dx$, $\delta = 10^{-5}$;

(15) $\int_0^1 \sqrt[3]{8+x^2}dx$, $\delta = 10^{-4}$;

(16) $\int_0^1 \sqrt[3]{x}\cos 2x\, dx$, $\delta = 10^{-3}$;

(17) $\int_0^1 \sqrt{x}e^{-x^2}dx$, $\delta = 10^{-3}$;

(18) $\int_0^1 \frac{\ln(x+\sqrt{1+x^2})}{x}dx$, $\delta = 10^{-2}$;

(19*) $\int_0^1 \frac{dx}{x^x}$, $\delta = 10^{-4}$;

(20*) $\int_0^1 \ln x \cdot \ln(1-x)dx$, $\delta = 10^{-2}$;

(21*) $\int_0^{+\infty} e^{-x^2}\sin \frac{x}{2}dx$, $\delta = 10^{-5}$;

(22) $\int_0^{2/5} \sqrt[3]{1+x^4}dx$, $\delta = 10^{-5}$;

(23) $\int_0^{1/2} \frac{dx}{\sqrt{1-x^3}}dx$, $\delta = 10^{-4}$.

12. Using the power series expansions of elementary functions, calculate an approximate value of the given expression with an error less than δ:

(1) $e^{1/2}$, $\delta = 10^{-3}$;

(2) e^{-1}, $\delta = 10^{-3}$;

(3) $e^{-1/3}$, $\delta = 10^{-5}$;

(4) $\ln \frac{3}{2}$, $\delta = 10^{-3}$;

(5) $\ln 1.2$, $\delta = 10^{-4}$;

(6) $\ln 0.8$, $\delta = 10^{-4}$;

(7) $\ln 3$, $\delta = 10^{-2}$;

(8) $\sqrt{24}$, $\delta = 10^{-6}$;

(9) $\sqrt[4]{630}$, $\delta = 10^{-6}$;

(10) $\frac{1}{\sqrt[3]{128}}$, $\delta = 10^{-5}$;

(11) $\sqrt[5]{30}$, $\delta = 10^{-5}$;

(12) $\sqrt[10]{1000}$, $\delta = 10^{-6}$;

(13) $\frac{1}{\sqrt{51}}$, $\delta = 10^{-5}$;

(14) $\sqrt[3]{130}$, $\delta = 10^{-6}$;

(15) $\arctan \frac{1}{3}$, $\delta = 10^{-5}$;

(16*) $\arctan 2.5$, $\delta = 10^{-4}$;

(17) $\arcsin \frac{1}{5}$, $\delta = 10^{-5}$;

(18) $\sinh 1$, $\delta = 10^{-3}$;

(19) $\cosh \frac{1}{2}$, $\delta = 10^{-4}$;

(20) $\cos \frac{1}{10}$, $\delta = 10^{-6}$;

(21) $\sin \frac{1}{4}$, $\delta = 10^{-5}$;

(22) $\sin 18^o$, $\delta = 10^{-5}$;

(23) $\cos 9^o$, $\delta = 10^{-5}$.

13*. Using the formula $\arctan a + \arctan b = \arctan \frac{a+b}{1-ab}$, which holds under the condition $\arctan a + \arctan b \in (-\frac{\pi}{2}, \frac{\pi}{2})$, find the number π with accuracy 10^{-4} for the specified values of a and b:

(1) $a = \frac{1}{2}, b = \frac{1}{3}$;

(2) $a = \frac{2}{5}, b = \frac{3}{7}$;

(3) $a = \frac{\sqrt{3}}{5}, b = \frac{1}{3\sqrt{3}}$;

(4) $a = \frac{1}{2\sqrt{3}}, b = \frac{\sqrt{3}}{7}$.

14. Find power series solutions of the ordinary differential equations and initial value problems:

(1) $y' + 4y = 0$;

(2) $(x - 2)y' + 3y = 0$;

(3) $(1 + x^2)y' = 1$, $y(0) = 0$;

(4) $y'' + p^2 y = 0$, $y(0) = 0$, $y'(0) = p$, $p \in \mathbb{R}$;

(5) $y'' + xy = 0$, $y(0) = 1$, $y'(0) = 0$;

(6) $(x^2 - 4)y'' + 3xy' + y = 0$;

(7) $(1 - x^2)y'' - xy' = 0$, $y(0) = 0$, $y'(0) = 1$;

(8) $y'' + \frac{1}{x}y' + y = 0$, $y(0) = 1$, $y'(0) = 0$;

(9) $(1 - x^2)y'' - 5xy' - 4y = 0$, $y(0) = 1$, $y'(0) = 0$;

(10) $(1 - x^2)y'' - 2xy' + p(p + 1)y = 0$, $p \in \mathbb{R}$ (the Legendre equation).

15. Using a power series solution of the initial value problem $(1-x^2)y'-xy = 1$, $y(0) = 0$, find a power series expansion about 0 of the function $f(x) = \frac{\arcsin x}{\sqrt{1-x^2}}$. Make the same for the function $g(x) = \arcsin^2 x$.

16*. Verify whether the following statements are true or false:
  (1) if two power series $\sum_{n=0}^{+\infty} a_n x^n$ and $\sum_{n=0}^{+\infty} b_n x^n$ take the same values at infinitely many points, then $a_n = b_n$, $\forall n$;
  (2) if two power series, convergent on the same interval, have different coefficients, then they assume different values at least at one point;
  (3) if two functions have the same Taylor series convergent on an interval, then they have the same values on this interval;
  (4) if a power series diverges at the endpoint $a + R$ of its interval of convergence $(a - R, a + R)$, then it converges non-uniformly on $[a, a + R)$.

Bibliography

Textbooks on Calculus, Real Analysis and Infinite Series

1. Abbott, S. (2015). *Understanding Analysis*. New York: Springer.
2. Boas, R. P. (1996). *A Primer of Real Functions*. Washington, DC: The Mathematical Association of America.
3. Bonar, D. D., Khoury, M. J. (2006). *Real Infinite Series*. Washington, DC: The Mathematical Association of America.
4. Bromwich, T. J. I. (2005). *An Introduction to the Theory of Infinite Series*. Providence, RI: The American Mathematical Society (The first edition: Bromwich, T. J. I. (1908). *An Introduction to the Theory of Infinite Series*. London: MacMillan)
5. Fikhtengol'ts, G. M. (1965). *The Fundamentals of Mathematical Analysis* (Vols. 1,2). Oxford: Pergamon Press.
6. Gaughan, E. D. (2009). *Introduction to Analysis*. Providence, RI: The American Mathematical Society.
7. Grigorieva, E. (2016). *Methods of Solving Sequence and Series Problems*. Basel: Birkhäuser.
8. Hyslop, J. M. (2006). *Infinite Series*. Mineola, NY: Dover (The first edition: Hyslop, J. M. (1942). *Infinite Series*. Edinburg: Oliver and Boyd).
9. Ilyin, V. A., Poznyak, E. G. (1982). *Fundamentals of Mathematical Analysis*, Parts 1,2. Moscow: Mir.
10. Knopp, K. (1990). *Theory and Applications of Infinite Series*. Mineola: Dover (The first edition: Knopp, K. (1921). *Theorie und Anwendung der Unendlichen Reihen*, in German. Berlin: Springer).
11. Little, C. H. C., Teo, K. L., Brunt, B. (2015). *Real Analysis via Sequences and Series*. New York: Springer.
12. Rudin, W. (1976). *Principles of Mathematical Analysis*. New York: McGraw-Hill.
13. Spivak, M. (2008). *Calculus*. Houston: Publish or Perish.
14. Stewart, J. (2015). *Calculus: Early Transcendentals*. Boston: Cengage Learning.
15. Wade, W. (2017). *An Introduction to Analysis*. Upper Saddle River, NJ: Pearson Prentice Hall.
16. Zorich, V. A. (2016). *Mathematical Analysis I,II*. Berlin: Springer.

© The Author(s), under exclusive license to Springer Nature Switzerland AG 2022
L. Bourchtein, A. Bourchtein, *Theory of Infinite Sequences and Series*,
https://doi.org/10.1007/978-3-030-79431-6

History of Analysis (Including Theory of Infinite Sequences and Series)

1. Bottazzini, U. (1986). *The Higher Calculus: A History of Real and Complex Analysis from Euler to Weierstrass*. New York: Springer.
2. Edwards, C. H. (1979). *The Historical Development of the Calculus*. New York: Springer.
3. Grabiner, J. V. (1981). *The Origins of Cauchy's Rigorous Calculus*. Cambridge, MA: Massachusetts Institute of Technology Press.
4. Grattan-Guiness, I. (1970). *The Development of the Foundations of Mathematical Analysis from Euler to Riemann*. Cambridge, MA: Massachusetts Institute of Technology Press.
5. Gray, J. (2015). *The Real and the Complex: A History of Analysis in the 19th Century*. Berlin: Springer.
6. Hairer, E., & Wanner, G. (2008). *Analysis by Its History*. New York: Springer.
7. Kline, M. (1990). *Mathematical Thought from Ancient to Modern Times* (Vols. 2,3). New York: Oxford University Press.
8. Korle, H.-H. (2015). *Infinite Series in a History of Analysis*. Berlin: De Gruyter.
9. Roy, R. (2011). *Sources in the Development of Mathematics. Infinite Series and Products from Fifteenth to the Twenty-first Century*. New York: Cambridge University Press.

Original Classical Sources on Real Analysis (Mentioned in the Text)

1. Cauchy, A. L. (1821). *Cours d'Analyse de L'École Royale Polytechnique* (in French). Paris: L'Imprimerie Royale (Bradley, R. E., Sandifer, C. E. (2009). *Cauchy's Cours d'Analyse: An Annotated Translation*. New York: Springer).
2. de Lagrange, J. L. (1797). *Théory de Fonctions Analytiques* (in French). Paris: L'Imprimerie de la république.
3. de L'Hospital, G. F. A. (1696). *Analyse des Infiniment Petits* (in French). Paris: L'Imprimerie Royale (Bradley, R. E., Petrilli, S. J., & Sandifer, C. E. (2015). *L'Hôpital's Analyse des Infiniments Petits: An Annotated Translation with Source Material by Johann Bernoulli*. Basel: Birkhäuser).
4. Euler, L. (1748). *Introductio in Analysin Infinitorum* (Vols. 1 and 2, in Latin). Lausanne: Bousquet (Euler, L. (1990). *Introduction to Analysis of the Infinite*, Book 1, 2. New York: Springer).
5. Newton, I. (1687). *Philosophiæ Naturalis Principia Mathematica* (in Latin). London: Halley (Newton, I. (1999). *The Principia: Mathematical Principles of Natural Philosophy*. Berkeley, CA: University of California Press).

Further Reading (Fourier Series, Divergent Series, Summability Methods, Multiple Series, Complex Series, Laurent Series, and More)

1. Alexits, G. (1961). *Convergence Problems of Orthogonal Series*. New York: Pergamon Press.
2. Bhatia, R. (2005). *Fourier Series*. Washington, DC: The Mathematical Association of America.
3. Boos, J., & Cass, P. (2000). *Classical and Modern Methods in Summability*. New York: Oxford University Press.
4. Bourchtein, A., & Bourchtein, L. (2017). *Counterexamples on Uniform Convergence*. Hoboken, NJ: Wiley.

5. Bourchtein, A., & Bourchtein, L. (2021). *Complex Analysis*. Singapure: Springer/Hindustan Publishing.
6. Budak, S. V., & Fomin, B. M. (1973). *Multiple Integrals, Field Theory and Series*. Moscow: Mir.
7. Evgrafov, M. A. (2019) *Analytic Functions*. Mineola, NY: Dover.
8. Gelbaum, B. R., Olmsted, J. M. H. (2003). *Counterexamples in Analysis*. Mineola, NY: Dover.
9. Hardy, G. H. (2000). *Divergent Series*. Providence, RI: The American Mathematical Society.
10. Hille, E. (2012). *Analytic Function Theory* (Vols. I, II). Providence, RI: The American Mathematical Society.
11. Kaczor, W. J., & Novak, M. T. (2000). *Problems in Mathematical Analysis I, II*. Providence, RI: The American Mathematical Society.
12. Lanczos, C. (2016). *Discourse on Fourier Series*. Philadelphia: The Society for Industrial and Applied Mathematics.
13. Markushevich, A. I. (2005). *Theory of Functions of a Complex Variable* (Vols. 1–3). Providence, RI: The American Mathematical Society.
14. Remmert, R. (1998). *Theory of Complex Functions*. New York: Springer.
15. Shawyer, B., & Watson B. (1994). *Borel's Methods of Summability. Theory and Applications*. New York: Oxford University Press.
16. Tolstov, G. P. (1976). *Fourier Series*. Mineola, NY: Dover.
17. Zygmund, A. (2002). *Trigonometric Series* (Vols. I, II). Cambridge: Cambridge University Press.

Index

A

Abel
  Lemma, 242
  Lemma, extension for derivatives, 248
  summation by parts formula, 108
  summation by parts formula, series of
      functions, 208
  test for series of numbers, 110
  test for uniform convergence, 210
  theorem of convergence at endpoints, 250
Alternating series
  definition, 45, 105
  Leibniz's test, 105
  remainder estimate, 106
Analytic function, 269

C

Cauchy
  chain of tests, 97
  criterion for convergence of a series, 54
  criterion for sequence of numbers, 18
  criterion for uniform convergence of
      sequences, 158
  criterion for uniform convergence of series,
      202
  integral test for positive series, 58
  product of series of numbers, 54, 116
  remainder, Taylor formula, 272
  sequence of numbers, 18
  test for positive series with limit, 76
  test for positive series with lim sup,
      84
  test for positive series without limit, 76
  uniform sequence, 158

D

D'Alembert
  characteristic, 87
  test for positive series, 87
  test for positive series with limit, 73
  test for positive series with lim sup and
      lim inf, 83, 93
  test for positive series without limit, 72
Dirichlet
  example, commutative property of series,
      124
  test for series of numbers, 109
  test for uniform convergence, 208
  theorem 1 of commutative property of series,
      123
  theorem 2 of commutative property of series,
      123
Double series
  absolute convergence, 134
  definition, 132

I

Irrationality of e, 349
Irrationality of π, 350

N

Negative series
  definition, 45

P

Positive series
  associative property, 122

© The Author(s), under exclusive license to Springer Nature Switzerland AG 2022
L. Bourchtein, A. Bourchtein, *Theory of Infinite Sequences and Series*,
https://doi.org/10.1007/978-3-030-79431-6

Bertrand characteristic, 88
Bertrand test, 88
Bertrand test with lim sup and lim inf, 94
Cauchy chain, 97
Cauchy condensation test, 69
Cauchy's test, 100
Cauchy's test with limit, 76
Cauchy's test with lim sup, 84
Cauchy's test without limit, 76
commutative property, 123
comparison test with limit, 64
comparison test without limit, 63
D'Alembert characteristic, 87
D'Alembert's (ratio test), 72
D'Alembert's and Cauchy's tests,
      comparison, 79
D'Alembert's test, 87
D'Alembert's test with limit, 73
D'Alembert's test with lim sup and lim inf,
      83, 93
D'Alembert's test without limit, 72
definition, 45
Gauss test, 89
general criterion for convergence, 58
integral test (Cauchy-Maclauren test), 58
integral test, remainder evaluation, 62
Jamet test, 100
Kummer chain, 85
Kummer theorem, 86
Kummer theorem with lim sup and lim inf,
      93
Raabe characteristic, 88
Raabe test, 88
Raabe test with lim sup and lim inf, 93
Schlömilch's test, 70
Power series, 240
Abel's Theorem, 250
analytic function, 269
analytic properties, 260
arithmetic properties, 251
centered at the origin, 240
convergence at endpoints, 250
criterion of expansion, 277
derivatives
      extension of Abel's Lemma, 248
      convergence, 249
derivatives of higher order, convergence, 250
functional properties, 254
interval of convergence, 245

properties
   boundedness of sum, 260
   change of central point, 259
   change of variable, 254
   composition of series, 255
   continuity at endpoints, 262
   continuity of sum, 261
   differentiability, 263
   even and odd functions, 268
   indefinite integral, 263
   limit of sum, 261
   multiplication by function, 251
   product, 252
   Riemann integral, 262
   sum, 251
   term-by-term differentiation, 263
   term-by-term integration, 262, 263
radius of convergence, 245
set of convergence, 243, 245
Taylor coefficients, 270
test for expansion, 277
uniqueness
   theorem 1, 266
   theorem 2, 267

R
Radius of convergence
   Cauchy formula, 246
   Cauchy-Hadamard formula, 247
   D'Alembert formula, 246
Repeated series
   absolute convergence, 133, 135
   definition, 132

S
Sequence of functions
   convergence at a point, 142
   definition, 141
   divergence at a point, 142
   domain of sequence, 141
   limit function, 142
   non-uniform convergence, 150, 151
   notations, 141
   pointwise convergence, 142, 149
   pointwise divergence, 142
   uniform convergence, 150, 151
      arithmetic properties, 155

Weierstrass approximation theorem, 182
Sequence of numbers
    alternating, 2
    arithmetic means, 19
    arithmetic properties, 11
    Bolzano-Weierstrass theorem, 17
    bounded, 2
    bounded above, 2
    bounded below, 2
    boundedness, 9
    boundedness and convergence, 13
    Cauchy criterion, 18
    Cauchy sequence, 18
    common properties, 8
    comparison properties, 8
    convergence, 4
    convergent, 4
    convergent subsequence, 17
    decreasing, 2
    definition, 1
    divergence, 4
    divergent, 4, 5
    finite limit, 4
    general limit, 15
    geometric, 7
    geometric means, 19
    increasing, 2
    indeterminate form $\frac{0}{0}$, 23
    indeterminate form $\frac{\infty}{\infty}$, 23
    indeterminate form $\infty - \infty$, 24
    indeterminate forms, 22
        division by the highest power, 25
        L'Hospital's rules, 27
        Stolz-Cesàro theorems, 30
    infinite limit, 5
    limit, 4
    monotone, 2
    monotone subsequence, 17
    monotonicity and convergence, 13
    negative, 2
    notation, 2
    partial limit, 15
    positive, 2
    properties of product and quotient, 11
    properties of sum and difference, 11
    property of absolute value, 12
    property of composite function, 12
    property of exponent, 12
    property of generating function, 13

    property of power, 12
    special properties, 13
    squeeze theorem, 10
    subsequence, 14
    techniques of solution of indeterminate
        forms, 25
    unbounded, 2
    uniqueness of the limit, 8
Series of functions
    Abel's summation by parts formula, 208
    absolute convergence, 194
    convergence at point, 193
    definition, 192
    divergence, 193
    divergence at point, 193
    domain of series, 192
    general term, 192
    non-uniform convergence, 198
    partial sum, 193
    pointwise convergence, 193, 197
    remainder, 193
    sum of series, 193
    test for non-uniform convergence, 198
    uniform convergence, 197
        arithmetic properties, 202
Series of numbers
    Abel's summation by parts formula, 108
    Abel's test, 110
    absolute convergence, 114
    alternating, 45, 105
    alternating harmonic, 49
    arithmetic properties, 53
    associative property, 120
    Cauchy criterion, 54
    Cauchy product, 54, 116
    commutative property, 123
        Dirichlet example, 124
        Dirichlet theorem 1, 123
        Dirichlet theorem 2, 123
        Riemann theorem, 125
    conditional convergence, 115
    convergence, 45
    convergence and remainder, 55
    convergence through remainders, 56
    convergent, 45
    definition, 44
    Dirichlet's test, 109
    divergence, 45
    divergence test, 55

divergent, 45
double series, 132
geometric, 46
harmonic, 47
harmonic series, 60
necessary condition of convergence, 55
negative, 45
negative part, 118
partial sum, 44
positive, 45
positive part, 118
product, convergence, 117
property of linear combination, 53
property of product, 54
p-series, 48, 61
remainder, 44
repeated series, 132
tests for absolute convergence, 115
Set of convergence
  Cauchy-Hadamard theorem, 247
  theorem for power series, 244

T
Taylor
  coefficients, 271
  formula, 272
  polynomial, 271
  series, 271
  theorem, 272
Taylor formula, 272
  remainder
    Cauchy form, 272
    Lagrange form, 272
Taylor polynomial, 271
Taylor series, 271
  approximation of functions, 308
  Borel's theorem, 352
  calculation of limits, 329
  convergent to different function, 275
  criterion of expansion, 277
  divergent at different points, 275
  divergent everywhere, 275
  expansion of elementary functions, 280
  function $(1 + x)^p$, 285, 287
  function $\frac{1}{1+x^2}$, 283
  function $\frac{1}{1+x}$, 283
  function $\frac{1}{1-x}$, 281, 284

function $\frac{1}{\sqrt{1-x}}$, 291
function a^x, 297
function e^x, 287, 292
function e^{-x^2}, 296
function $\cos x$, 294
function $\cos^2 x$, 297
function $\cot x$, 296, 302
function $\ln(1 - x)$, 282
function $\sin x$, 294
functions, list of formulas, 304
function $\tan x$, 295, 301
function $x \cos x^2$, 297
integral calculation, 336
Maclaurin series, 271
numerical approximation, 321
remainders, 274
solution of differential equations, 339
sums of series of functions, 326
sums of series of numbers, 328
Taylor coefficients, 271
Taylor formula, 272
Taylor polynomials, 274
Taylor's theorem, 272
test for expansion, 277

U
Uniform convergence
  Abel's test, 210
  arithmetic properties
    sequences, 155
    series, 202
  boundedness of limit function, 161
  boundedness of sum, 215
  Cauchy criterion for sequences, 158
  Cauchy criterion for series, 202
  continuity of limit function, 166
  continuity of sum, 220
  definition for sequence, 150
  definition for series, 197
  differentiability by parameter, 176
  differentiability of limit function, 176
  differentiability of sum, 224
  differentiation term by term, 224
  Dini's theorem for sequences, 160
  Dini's theorem for series, 214
  Dirichlet's test, 208
  general comparison test for series, 205
  generalized Leibniz's test, 211

improper integral, 173
integrability of limit function, 169
integrability of sum, 221
integration by parameter, 169
integration term by term, 221
limit of limit function, 163
limit of sum, 218
majorant series, 205
product of sequences, 157
product of sequence with bounded function,
    156
product of series with bounded functions,
    202
properties of limit functions, 161
properties of sum of series, 215

relation with absolute convergence, 203
sum of sequences, 155
sum of series, 202
uniform Cauchy sequence, 158
Weierstrass function, 227
Weierstrass M-test, 205
Weierstrass test, 205

W
Weierstrass
    approximation theorem, 182
    function everywhere continuous and
        nowhere differentiable, 227
    test for uniform convergence, 205

Printed in the United States
by Baker & Taylor Publisher Services